Methods in Molecular Biology

Series Editor
John M. Walker
School of Life and Medical Sciences
University of Hertfordshire
Hatfield, Hertfordshire, AL10 9AB, UK

For further volumes:
http://www.springer.com/series/7651

Heterologous Gene Expression in *E. coli*

Methods and Protocols

Edited by

Nicola A. Burgess-Brown

SGC, Nuffield Department of Clinical Medicine, University of Oxford, Oxford, UK

Humana Press

Editor
Nicola A. Burgess-Brown
SGC
Nuffield Department of Clinical Medicine
University of Oxford
Oxford, UK

ISSN 1064-3745 ISSN 1940-6029 (electronic)
Methods in Molecular Biology
ISBN 978-1-4939-6885-5 ISBN 978-1-4939-6887-9 (eBook)
DOI 10.1007/978-1-4939-6887-9

Library of Congress Control Number: 2017934051

Printed on acid-free paper

This Humana Press imprint is published by Springer Nature
The registered company is Springer Science+Business Media LLC
The registered company address is: 233 Spring Street, New York, NY 10013, U.S.A.

Preface

Heterologous gene expression in *E. coli* has been one of the most widely used methods for generating recombinant proteins for many scientific analyses and still remains the first choice for most laboratories around the world. The ease of use and low cost of production often lead researchers to initially attempt to express their proteins of interest in *E. coli* rather than opting for a eukaryotic host. Decades of development have seen the variety of methods for expressing genes in *E. coli* broaden, with improved media and optimized conditions for growth, a choice of promoter systems to regulate expression, fusion tags to aid solubility and purification, and *E. coli* host strains to accommodate more challenging or toxic proteins.

Having worked in the area of protein production for structural genomics for the past 12 years, and also having a requirement to generate human proteins, I have seen a shift from expression of many genes in *E. coli* to use of the baculovirus expression system using insect cells and more recently to mammalian cells. This revolution from prokaryotic to eukaryotic expression has been visible throughout the protein production field and is largely due to the requirement to obtain specific proteins linked to disease, for functional assays as well as structures, which may be larger, or require machinery to enable specific post-translational modifications. It is perhaps important to note, however, that the structural output from the SGC in Oxford today is still ~80% derived from *E. coli*.

This book is aimed at molecular biologists, biochemists, and structural biologists, both from the beginning of their research careers to those in their prime, to give both an historical and modern overview of the methods available to express their genes of interest in this exceptional organism. The topics are largely grouped under four parts: (I) high-throughput cloning, expression screening, and optimization of expression conditions, (II) protein production and solubility enhancement, (III) case studies to produce challenging proteins and specific protein families, and (IV) applications of *E. coli* expression. This volume provides scientists with a toolbox for designing constructs, tackling expression and solubility issues, handling membrane proteins and protein complexes, and innovative engineering of *E. coli*. It will hopefully prove valuable both in small laboratories and in higher throughput facilities. I would like to thank all the authors for their contributions and for making this a global effort.

Oxford, UK — ***Nicola A. Burgess-Brown***

Contents

Contributors

THOMAS B. ACTON • *Evotec (US), Princeton, NJ, USA*
RAVEENDRA ANANGI • *Institute for Molecular Bioscience, The University of Queensland, Brisbane, QLD, Australia*
EMILY P. BALSKUS • *Department of Chemistry and Chemical Biology, Harvard University, Cambridge, MA, USA*
JOHN J. BARKER • *Evotec Ltd, Abingdon, Oxfordshire, UK*
DOMINIKA BEDNARCZYK • *Department of Bimolecular Sciences, Weizmann Institute of Science, Rehovot, Israel*
IMRE BERGER • *The School of Biochemistry, University Walk, University of Bristol, Clifton, UK; The European Molecular Biology Laboratory (EMBL), BP 181, Unit of Virus Host Cell Interactions (UVHCI), Horowitz, Grenoble Cedex, France*
FRANK BERNHARD • *Centre for Biomolecular Magnetic Resonance, Institute for Biophysical Chemistry, Goethe-University of Frankfurt/Main, Frankfurt/Main, Germany*
HÜSEYIN BESIR • *Protein Expression and Purification Core Facility, EMBL Heidelberg, Heidelberg, Germany*
LOUISE E. BIRD • *Oxford Protein Production Facility-UK, Research Complex at Harwell, Rutherford Appleton Laboratory, Oxfordshire, UK; Division of Structural Biology, Henry Wellcome Building for Genomic Medicine, University of Oxford, Oxford, UK*
CHARIKLEIA BLACK • *Evotec Ltd, Abingdon, Oxfordshire, UK*
MAREN BLECKMANN • *Helmholtz Zentrum für Infektionsforschung GmbH, Braunschweig, Germany*
JOSHUA BRITTON • *Department of Chemistry, University of California, Irvine, CA, USA; Centre for NanoScale Science and Technology, School of Chemical and Physical Sciences, Flinders University, Adelaide, SA, Australia*
DAVID CASAS-MAO • *Research Complex at Harwell, Rutherford Appleton Laboratory, Oxfordshire, UK; School of Biosciences, University of Nottingham, Loughborough, Leicestershire, UK*
ROD CHALK • *Structural Genomics Consortium (SGC), Nuffield Department of Medicine, University of Oxford, Oxford, UK*
SCOTT CHERRY • *Macromolecular Crystallography Laboratory, Center for Cancer Research, National Cancer Institute at Frederick, Frederick, MD, USA*
IAN COLLINSON • *The School of Biochemistry, University Walk, University of Bristol, Clifton, UK*
CHRISTOPHER D.O. COOPER • *Department of Biological Sciences, School of Applied Sciences, University of Huddersfield, Huddersfield, West Yorkshire, UK*
BEN CRISTOFORI-ARMSTRONG • *Institute for Molecular Bioscience, The University of Queensland, QLD, Australia*
JOHN-PAUL DENSON • *Protein Expression Laboratory, Cancer Research Technology Program, Frederick National Laboratory for Cancer Research, Frederick, MD, USA*
NEIL DIXON • *Manchester Institute of Biotechnology, University of Manchester, Manchester, UK*
CHARLOTTE A. DODSON • *Molecular Medicine, National Heart & Lung Institute, Imperial College London, London, UK*

EKAITZ ERRASTI-MURUGARREN • *Institute for Research in Biomedicine (IRB Barcelona), Barcelona Institute of Science and Technology, Barcelona, Spain*
DOMINIC ESPOSITO • *Protein Expression Laboratory, Cancer Research Technology Program, Frederick National Laboratory for Cancer Research, Frederick, MD, USA*
SAM FESTENSTEIN • *Evotec Ltd, Abingdon, Oxfordshire, UK*
TOSHIMICHI FUJIWARA • *Institute for Protein Research, Osaka University, Osaka, Japan*
JAN-WILLEM DE GIER • *Department of Biochemistry and Biophysics, Center for Biomembrane Research, Stockholm University, Stockholm, Sweden; Xbrane Biopharma AB, Solna, Sweden*
OPHER GILEADI • *Structural Genomics Consortium, University of Oxford, Headington, Oxford, UK*
SUSANNE GRÄSLUND • *Structural Genomics Consortium, Department of Biochemistry and Biophysics, Karolinska Institutet, Solna, Sweden*
ALEXANDER GRÄWE • *Centre for Biomolecular Magnetic Resonance, Institute for Biophysical Chemistry, Goethe-University of Frankfurt/Main, Frankfurt/Main, Germany*
DARREN J. HART • *Institut de Biologie Structurale (IBS), CNRS, CEA, Université Grenoble Alpes, Grenoble, France*
ERIK HENRICH • *Centre for Biomolecular Magnetic Resonance, Institute for Biophysical Chemistry, Goethe-University of Frankfurt/Main, Frankfurt/Main, Germany*
CHRISTOPH HERWIG • *Research Division Biochemical Engineering, Institute of Chemical Engineering, TU Wien, Vienna, Austria; Christian Doppler Laboratory for Mechanistic and Physiological Methods for Improved Bioprocesses, Institute of Chemical Engineering, TU Wien, Vienna, Austria*
JOOP VAN DEN HEUVEL • *Helmholtz Zentrum für Infektionsforschung GmbH, Braunschweig, Germany*
RICHARD B. HITCHMAN • *Evotec Ltd, Abingdon, Oxfordshire, UK*
ANNA HJELM • *Department of Biochemistry and Biophysics, Center for Biomembrane Research, Stockholm University, Stockholm, Sweden*
NURZIAN BERNSEL ISMAIL • *Xbrane Biopharma AB, Solna, Sweden*
STEPHEN R.P. JAFFÉ • *Department of Chemical and Biological Engineering, ChELSI Institute, University of Sheffield, Sheffield, UK*
QUIYANG JIANG • *The European Molecular Biology Laboratory (EMBL), BP 181, and Unit of Virus Host Cell Interactions (UVHCI), Horowitz, Grenoble Cedex, France*
KATHARINA KARSTE • *Helmholtz Zentrum für Infektionsforschung GmbH, Braunschweig, Germany*
GLENN F. KING • *Institute for Molecular Bioscience, The University of Queensland, QLD, Australia*
CHOJIRO KOJIMA • *Institute for Protein Research, Osaka University, Osaka, Japan*
GRIETJE KUIPERS • *Department of Biochemistry and Biophysics, Center for Biomembrane Research, Stockholm University, Stockholm, Sweden; Xbrane Biopharma AB, Solna, Sweden*
HOK SAU KWONG • *Evotec Ltd, Abingdon, Oxfordshire, UK*
JAMES D. LOVE • *Department of Biochemistry, Albert Einstein College of Medicine at Yeshiva University, Bronx, NY, USA; ATUM, Newark, CA, USA*
JOEN LUIRINK • *The Amsterdam Institute of Molecules, Medicines and Systems, VU University Amsterdam, Amsterdam, The Netherlands*
ARIO DE MARCO • *Department of Biomedical Sciences and Engineering, University of Nova Gorica, Vipava, Slovenia*

BRIAN D. MARSDEN • *Structural Genomics Consortium, Nuffield Department of Medicine, University of Oxford, Oxford, Oxfordshire, UK; Nuffield Department of Orthopaedics, Rheumatology and Musculoskeletal Sciences, Kennedy Institute of Rheumatology, University of Oxford, Oxford, Oxfordshire, UK*
PHILIPPE J. MAS • *Integrated Structural Biology Grenoble (ISBG), CNRS, CEA, Université Grenoble Alpes, EMBL, Grenoble, France*
ROSA MORRA • *Manchester Institute of Biotechnology, University of Manchester, Manchester, UK*
SUSANNE MÜLLER-KNAPP • *Target Discovery Institute and Structural Genomics Consortium, Oxford University, Oxford, UK; Goethe-University Frankfurt, Buchmann Institute for Life Sciences, Frankfurt am Main, Germany*
ANTJE NEUBAUER • *Enpresso GmbH, Berlin, Germany*
HOANG NGUYEN • *VNUHCM-University of Science, Hochiminh City, Vietnam*
JAGROOP PANDHAL • *Department of Chemical and Biological Engineering, ChELSI Institute, University of Sheffield, Sheffield, UK*
YOAV PELEG • *The Israel Structural Proteomics Center (ISPC), Weizmann Institute of Science, Rehovot, Israel*
VINIT J. PEREIRA • *Abcam plc, Cambridge Bioscience, Cambridge, UK*
MARKUS PESCHKE • *The Amsterdam Institute of Molecules, Medicines and Systems, VU University Amsterdam, Amsterdam, The Netherlands*
TRANG PHAN • *VNUHCM-University of Science, Hochiminh City, Vietnam*
PHUONG HUYNH • *VNUHCM-University of Science, Hochiminh City, Vietnam*
VADIVEL PRABAHAR • *Migal-Galilee Research Institute, Kiryat Shmona, Israel*
SREEJITH RARAN-KURUSSI • *Macromolecular Crystallography Laboratory, Center for Cancer Research, National Cancer Institute at Frederick, Frederick, MD, USA*
COLIN L. RASTON • *Centre for NanoScale Science and Technology, School of Chemical and Physical Sciences, Flinders University, Adelaide, SA, Australia*
ARTURO RODRÍGUEZ-BANQUERI • *Institute for Research in Biomedicine (IRB Barcelona), Barcelona Institute of Science and Technology, Barcelona, Spain; Unitat de Proteòmica Aplicada i Enginyeria de Proteïnes, Institut de Biotecnologia i Biomedicina (IBB), Universitat Autònoma de Barcelona (UAB), Barcelona, Spain*
RALF-BERNHARDT RUES • *Centre for Biomolecular Magnetic Resonance, Institute for Biophysical Chemistry, Goethe-University of Frankfurt/Main, Frankfurt/Main, Germany*
NATALIE J. SAEZ • *Institute for Molecular Bioscience, The University of Queensland, QLD, Australia*
PAVEL SAVITSKY • *Target Discovery Institute and Structural Genomics Consortium, Oxford University, Oxford, UK*
CHRISTIANE SCHAFFITZEL • *The School of Biochemistry, University Walk, University of Bristol, Clifton, UK; The European Molecular Biology Laboratory (EMBL), BP 181, and Unit of Virus Host Cell Interactions (UVHCI), Grenoble Cedex, France*
SUSAN SCHLEGEL • *Molecular Microbial Ecology, Institute of Biogeochemistry and Pollutant Dynamics, ETH Zurich, Dübendorf, Switzerland*
RYAN J. SCHULZE • *Department of Biochemistry and Molecular Biology, Mayo Clinic, Rochester, MN, USA*
JOSHUA N. SMITH • *Department of Molecular Biology and Biochemistry, University of California, Irvine, CA, USA*

OLIVER SPADIUT • *Research Division Biochemical Engineering, Institute of Chemical Engineering, TU Wien, Vienna, Austria; Christian Doppler Laboratory for Mechanistic and Physiological Methods for Improved Bioprocesses, Institute of Chemical Engineering, TU Wien, Vienna, Austria*

BENJAMIN STRUTTON • *Department of Chemical and Biological Engineering, ChELSI Institute, University of Sheffield, Sheffield, UK*

TOSHIHIKO SUGIKI • *Institute for Protein Research, Osaka University, Osaka, Japan*

TROY TAYLOR • *Protein Expression Laboratory, Cancer Research Technology Program, Frederick National Laboratory for Cancer Research, Frederick, MD, USA*

TUOM TRUONG • *VNUHCM-University of Science, Hochiminh City, Vietnam*

KAISA UKKONEN • *BioSilta Oy, Oulu, Finland*

TAMAR UNGER • *The Israel Structural Proteomics Center (ISPC), Weizmann Institute of Science, Rehovot, Israel*

ANTTI VASALA • *BioSilta Oy, Oulu, Finland*

JOSÉ LUIS VÁZQUEZ-IBAR • *Institute for Research in Biomedicine (IRB Barcelona), Barcelona Institute of Science and Technology, Barcelona, Spain; Institute for Integrative Biology of the Cell (I2BC), iBiTec-S/SB2SM, CEA Saclay CNRS UMR 9198, University Paris-Sud, University Paris-Saclay, Cedex, France*

DAVID VIKSTRÖM • *Xbrane Biopharma AB, Solna, Sweden*

STEPHEN WALLACE • *Department of Chemistry and Chemical Biology, Harvard University, MA, USA; Institute of Quantitative Biology, Biochemistry and Biotechnology, School of Biological Sciences, University of Edinburgh, Edinburgh, UK*

DAVID S. WAUGH • *Macromolecular Crystallography Laboratory, Center for Cancer Research, National Cancer Institute at Frederick, Frederick, MD, USA*

GREGORY A. WEISS • *Department of Chemistry, University of California, Irvine, CA, USA; Department of Molecular Biology and Biochemistry, University of California, Irvine, CA, USA*

PHILLIP C. WRIGHT • *Faculty of Science, Agriculture and Engineering, Newcastle University, Newcastle, Upon Tyne, UK*

DAVID J. WURM • *Research Division Biochemical Engineering, Institute of Chemical Engineering, TU Wien, Vienna, Austria*

KATE YOUNG • *Manchester Institute of Biotechnology, University of Manchester, Manchester, UK*

DI ZHANG • *Macromolecular Crystallography Laboratory, Center for Cancer Research, National Cancer Institute at Frederick, Frederick, MD, USA*

Part I

High-Throughput Cloning, Expression Screening, and Optimization

Chapter 1

Recombinant Protein Expression in *E. coli*: A Historical Perspective

Opher Gileadi

Abstract

This introductory chapter provides a brief historical survey of the key elements incorporated into commonly used *E. coli*-based expression systems. The highest impact in expression technology is associated with innovations that were based on extensively studied biological systems, and where the tools were widely distributed in the academic community.

Key words *E. coli*, Promoter, Recombinant protein, Protein engineering, Expression vectors

1 Introduction

Early studies on purified proteins depended on proteins found in relatively high abundance, or with distinct solubility and stability profiles, such as hemoglobin, albumin, and casein. Even with the expansion of interest into a wider universe of enzymes, hormones, and structural proteins, researchers have sought to purify proteins from sources (organisms, tissues, and organelles) containing the highest abundance of the desired protein. It was recognized, even before the era of genetic engineering, that microorganisms and cultured cells could be ideal sources for protein production. A remarkable example, just before the development of recombinant DNA technologies, was the overproduction of the lactose repressor (product of the lacI gene). This protein is normally produced in *E. coli* at ~10 copies/cell. Muller-Hill and colleagues [1] used clever selection techniques to isolate promoter mutations that led to a tenfold increase in protein expression; this allele ($lacI^q$) was then transferred to a lysis-deficient bacteriophage, allowing achieving very high copy numbers of the phage (and the $lacI^q$ gene), leading to the target protein being ~0.5 % of total cellular protein [1]; all this—without restriction enzymes and in vitro DNA recombination! The emergence of precision recombinant DNA techniques

Nicola A. Burgess-Brown (ed.), *Heterologous Gene Expression in E. coli: Methods and Protocols*, Methods in Molecular Biology, vol. 1586, DOI 10.1007/978-1-4939-6887-9_1,

led to the production of the first biotechnology-derived drugs, insulin, growth hormone, and interferons, subsequently expanding to 23 FDA-approved biologic drugs produced in *E. coli* [2]. Concurrently, thousands of other proteins were produced in bacteria for research purposes. In this chapter, I will briefly review the major innovations that created the toolkit for recombinant protein expression in *E. coli*.

2 Expression from *E. coli* RNAP Promoters

We have already seen the first principles driving high-efficiency recombinant gene expression: strong promoters, and high gene copy numbers. A third principle that became important early on is inducible gene expression; typically, an expression process will involve growth of cells in the absence of expression, then induction of gene expression through transcriptional regulatory elements or by infection or activation of viruses. Expression vectors were developed based on a small number of well-studied gene promoter systems, which remain popular to this day (reviewed in ref. 3). The Lac promoter/operator and its derivatives (UV5, tac) are inducible by galactose or Isopropyl β-D-1-thiogalactopyranoside (IPTG), and repressed by glucose. The phage lambda P_L promoter is one of the strongest promoters known for *E. coli* RNA polymerase (RNAP). When combined with a temperature-sensitive repressor (cI847), the P_L promoter can be induced by a temperature shift, avoiding the use of chemical inducers [4]. The araBCD promoter, tightly regulated by the araC repressor/activator, avoids leaky expression in the absence of the inducer arabinose [5]. Interestingly, synthetic *E. coli* RNAP promoters based on a consensus derived from multiple sequence alignment perform rather poorly [6, 7]; rather, it is a combination of the canonical −35 and −10 elements with less defined downstream sequences, as well as an optimal environment for protein synthesis initiation and elongation that drives the highest levels of expression.

3 Maximizing Expression Levels

For most applications, *E. coli* RNAP promoters have been superseded by expression systems using bacteriophage promoters and RNA polymerases. The bacteriophage T7 polymerase is highly selective for cognate phage promoters, and achieves very high levels of expression [8]. The commonly used T7 expression systems are regulated by a double-lock: lac operators (repressor-binding sites) are placed at the promoter driving the target gene as well as the promoter driving the expression of the T7 RNA polymerase [9]. Expression is repressed in the absence of inducer, and is rapidly turned on when IPTG is added. There is some expression in the absence of inducer, which can be further reduced by including glucose in the growth medium (catabolite repression) [10] and by expressing T7 lysozyme, an inhibitor of T7 RNA polymerase, from plasmids pLysS or pLysL [9]. With the successful implementation of these principles, other issues become rate-limiting. High-level expression of foreign genes may be hampered by codon usage that is nonoptimal for the host cell. This makes a real difference [11], and has been addressed using either synthetic, codon-optimized genes, or by co-expressing a set of tRNA molecules that recognize some of the codons that are rare in *E. coli* (available as commercial strains, such as Rosetta™ and CodonPlus). Sequence optimization may also affect other impediments to gene expression, such as mRNA secondary structure or mRNA degradation, as well as secondary advantages such as eliminating or introducing restriction sites.

4 Fusion Tags

The next major development has been the introduction of generic purification tags. The general principle is to genetically fuse the protein of interest to another protein or peptide, for which affinity purification reagents are available. The tags introduced in the late 1980s are still very widely used. The earliest were epitope tags [12]: short peptides that are recognized by monoclonal antibodies, allowing affinity purification and elution with free peptides (e.g., FLAG [13], HA [14], and myc [15] tags). These were followed by the hexahistidine tag [16] which allows purification by immobilized-metal affinity chromatography (IMAC), and the full-length protein glutathione S-transferase (GST) [17] which binds to glutathione-sepharose. Short peptide tags are sometimes concatenated to provide better avidity of binding to the affinity columns, allowing more stringent washes and better purity, but these are mostly used for expression in eukaryotic cells. Tags can be removed using sequence-specific proteases (enterokinase, the blood-clotting factors X and thrombin, viral proteases such as TEV and the rhinovirus 3C protease, SUMO protease, engineered subtilisin, or inteins). Fusion tags seem to perform at least two functions: first, providing a handle for affinity purification; and second, promoting the solubility of the target protein by changing the overall hydrophobicity and charge and by providing chaperone-like functions. Because the selectivity and the solubilizing effect are context-dependent, there has been a continuing development of new fusion tags to address specific goals in different cell types.

It is frequently observed that the highest expression levels of a recombinant protein do not necessarily correlate with the highest yields of soluble, properly folded protein. In fact, rapid production of heterologous proteins more often leads to aggregation and precipitation, with no recovery of active protein. This problem has been addressed using three approaches: modulating growth and induction conditions; modifying the host strain; and engineering the target protein. Many eukaryotic proteins expressed in *E. coli* are only soluble when induced at low temperatures, typically 15–25 °C. Other changes in induction conditions, such as the use of carefully calibrated autoinduction media [10] and the use of moderately active promoters, have on occasion led to higher yields. Host strains have been engineered to over-express chaperone proteins [18–20], to encourage disulfide bond formation [21], or to remove autophosphorylated sites from active protein kinases [22]. Finally, proteins can be recovered from denatured precipitates using refolding techniques following solubilization in guanidine or urea; however, refolding methods seem to be mostly effective only for a subset of proteins, predominantly extracellular domains or proteins. The recent application of high-throughput and design of experiment methods to optimize refolding conditions may help to rescue more proteins that cannot be properly folded during expression in bacteria.

5 The Protein Is the Most Important Variable

The most dramatic improvements in recovery of soluble proteins have come from optimizing the sequence of the expressed protein. The degree of flexibility in the engineering of the target protein depends on the purpose of the project. In many cases, a truncated protein that contains a well-folded globular domain will be solubly expressed, while the full-length protein may contain intrinsically disordered and hydrophobic regions that drive aggregation. This is particularly relevant when expressing proteins for crystallization, and it has been noted that constructs truncated to include the structured domains tend to express and crystallize well [23]. In addition to truncations, internal mutations that stabilize the protein can dramatically affect the yields of soluble proteins [24] as well as membrane proteins [25, 26]; identifying these mutants most often requires molecular evolution techniques, as there is rarely any solid basis for rational design, especially if the structure of the protein is unknown. A more natural version relies on natural diversity: very often, systematic cloning and test-expression of multiple orthologues of the target protein can lead to the identification of a related protein that does express well in *E. coli*. Alternatively, synthetic versions of the target proteins based on multiple sequence alignments have been used in some instances to generate better yields.

6 High-Throughput Methods

With the advent of genomic-scale studies, there was a need to streamline and parallelize the cloning process. New methods were developed to enable cloning of PCR-generated DNA fragments into vectors without prior cleavage by restriction enzymes, and cloning of each fragment into multiple vectors. These methods include variants of ligation-independent cloning (LIC) [23, 27–29] and site-specific recombination methods [30]. The choice of method depends on the details of the experimental goals: LIC methods require only minimal (or no) additions to the cloned sequence, while recombinase-based methods (e.g., the Gateway® method) [30] add obligatory sequences within the encoded protein. On the other hand, when there is a need to repeatedly clone the same fragment into multiple vectors, recombinase-based methods allow a sequence-verified DNA insert to be transferred in a virtually non-mutagenic manner. An additional development to enable efficient cloning with low background has been the introduction of toxic genes in cloning vectors that are inactivated by the insertion of the cloned fragments [31, 32].

7 Heteromeric Complexes

It has been realized for a long time that attempts to express individual polypeptides in heterologous cells may fail because the native structure of the protein requires hetero-oligomerization. Techniques for co-expression of several components of a protein complex were applied sporadically, combining more than one protein/transcription unit on a single plasmid, or by combining separate compatible plasmids in a bacterial cell (or a combination of both). Recently developed systems for recombining multiple coding sequences into one plasmid [33] will allow generating protein complexes efficiently and systematically in *E. coli*.

8 One Method Fits All?

A search of GenBank for organism/vector yields >8000 hits; it would be safe to estimate the number of *E. coli* expression vectors is at least 1000. There are probably $>10^4$ publications describing the expression and purification of individual proteins, all differing at least slightly in the experimental details; the information is very difficult to collate. The structural genomics projects in the US, Europe, and Japan have systematically expressed and purified proteins from a variety of organisms, with extensive documentation and several benchmarking studies to evaluate the success of different approaches. A paper published jointly in 2008 by most of the big players [34] shows that a fairly narrow range of techniques accounts for the vast majority of successfully produced proteins. Some more detailed comparative studies (e.g., [29, 35]) have shown that by far the most common combination is BL21(DE3)-derived host strains supplemented with rare-codon tRNAs; growth in rich medium, with either IPTG-driven or autoinduction at 20–25 °C. The biggest impact on the yield of soluble protein is linked to (1) construct selection (truncation/mutation); (2) fusion tags, and (3) lowering the temperature during induction. Do these statistics mean that more than 35 years of method development is almost redundant, beyond a handful of core methods that cover all our needs? Probably not; the aggregate statistics hide the fact that the parameters of the structural genomics projects allowed for a considerable failure rate; in practice, the core methods (and the variants used) could recover soluble proteins for less than 50 % of eukaryotic target proteins that were attempted. Individual proteins may be rescued by more sophisticated solutions developed over the years, as documented in this volume. However, it is likely that these methods will have a marginal effect on the overall success rates of expressing eukaryotic proteins in *E. coli*, leaving us with a sizeable fraction of proteins that cannot be productively expressed.

9 Future Prospects

What are the future prospects? On one hand, it is sensible to transfer proteins that consistently fail to be produced in *E. coli* to other expression systems, which are becoming more efficient and cost-effective. However, it is likely that bacteria will continue to be a major workhorse for recombinant protein expression. One point that emerges from this historical survey is that most significant developments were based on thorough knowledge of particular biological systems. Indeed, the choice of *E. coli* and Coliphage-derived elements was a consequence of decades of fundamental research on these organisms, starting from the 1940s [36]. A recent splendid example of the use of in-depth fundamental research is the development of CRISPR-Cas9 systems for gene editing [37, 38]. So, true innovation in expanding the universe of proteins that can be produced in bacterial cells is likely to come from unexpected areas, based on in-depth knowledge. I would hazard a guess that big developments will come from synthetic biology. The engineering of *E. coli* host strains has proceeded piecemeal, typically adding or modifying individual proteins or pathways [39, 40]. Yet, a variety of other bacteria are used as host strains, including *Pseudomonas* and *Bacillus subtilis*, which provide specific advantages. With the advent of fully engineered bacterial cells [41] and the reconstitution of complex metabolic pathways [42, 43], it is plausible that novel "protein factories" will be designed to incorporate features from a variety of expression systems, to provide features that are missing or suboptimal in current *E. coli* hosts. These may include posttranslational modifications, chaperone functions, incorporation into membranes with controllable lipid composition, and secretion to the culture media. Parallel efforts will include extensive protein evolution to derive well-behaved and highly expressed versions of the proteins of interest.

As a final note, it is maybe obvious that the most widely adapted techniques and expression systems are those that were widely available to the academic community (at least), either through open distribution (by organizations such as Addgene [44]) or through reasonably priced vendors. It is imperative that future core technologies are not protected to an extent that makes them practically inaccessible to the majority of researchers. A sensible mix of commercial licensing and academic freedom-to-operate can benefit both the inventors and the society at large.

References

1. Muller-Hill B, Crapo L, Gilbert W (1968) Mutants that make more lac repressor. Proc Natl Acad Sci U S A 59:1259–1264
2. Baeshen MN, Al-Hejin AM, Bora RS et al (2015) Production of biopharmaceuticals in *E. coli*: current scenario and future perspectives. J Microbiol Biotechnol 25:953–962
3. Baneyx F (1999) Recombinant protein expression in *Escherichia coli*. Curr Opin Biotechnol 10:411–421
4. Remaut E, Stanssens P, Fiers W (1983) Inducible high level synthesis of mature human fibroblast interferon in *Escherichia coli*. Nucleic Acids Res 11:4677–4688
5. Guzman LM, Belin D, Carson MJ et al (1995) Tight regulation, modulation, and high-level expression by vectors containing the arabinose PBAD promoter. J Bacteriol 177:4121–4130
6. Brunner M, Bujard H (1987) Promoter recognition and promoter strength in the *Escherichia coli* system. EMBO J 6:3139–3144
7. Deuschle U, Kammerer W, Gentz R et al (1986) Promoters of *Escherichia coli*: a hierarchy of in vivo strength indicates alternate structures. EMBO J 5:2987–2994
8. Rosenberg AH, Lade BN, Chui DS et al (1987) Vectors for selective expression of cloned DNAs by T7 RNA polymerase. Gene 56: 125–135
9. Dubendorff JW, Studier FW (1991) Controlling basal expression in an inducible T7 expression system by blocking the target T7 promoter with lac repressor. J Mol Biol 219:45–59
10. Studier FW (2014) Stable expression clones and auto-induction for protein production in *E. coli*. Methods Mol Biol 1091:17–32
11. Burgess-Brown NA, Sharma S, Sobott F et al (2008) Codon optimization can improve expression of human genes in *Escherichia coli*: a multi-gene study. Protein Expr Purif 59: 94–102
12. Munro S, Pelham HR (1984) Use of peptide tagging to detect proteins expressed from cloned genes: deletion mapping functional domains of *Drosophila* hsp 70. EMBO J 3: 3087–3093
13. Hopp TP, Prickett KS, Price VL et al (1988) A short polypeptide marker sequence useful for recombinant protein identification and purification. Nat Biotechnol 6:1204–1210
14. Field J, Nikawa J, Broek D et al (1988) Purification of a RAS-responsive adenylyl cyclase complex from *Saccharomyces cerevisiae* by use of an epitope addition method. Mol Cell Biol 8:2159–2165
15. Robertson D, Paterson HF, Adamson P et al (1995) Ultrastructural localization of ras-related proteins using epitope-tagged plasmids. J Histochem Cytochem 43:471–480
16. Hochuli E, Dobeli H, Schacher A (1987) New metal chelate adsorbent selective for proteins and peptides containing neighbouring histidine residues. J Chromatogr 411:177–184

17. Smith DB, Johnson KS (1988) Single-step purification of polypeptides expressed in *Escherichia coli* as fusions with glutathione S-transferase. Gene 67:31–40
18. Lee SC, Olins PO (1992) Effect of overproduction of heat shock chaperones GroESL and DnaK on human procollagenase production in *Escherichia coli*. J Biol Chem 267:2849–2852
19. Nishihara K, Kanemori M, Kitagawa M et al (1998) Chaperone coexpression plasmids: differential and synergistic roles of DnaK-DnaJ-GrpE and GroEL-GroES in assisting folding of an allergen of Japanese cedar pollen, Cryj2, in *Escherichia coli*. Appl Environ Microbiol 64: 1694–1699
20. Ferrer M, Chernikova TN, Timmis KN et al (2004) Expression of a temperature-sensitive esterase in a novel chaperone-based *Escherichia coli* strain. Appl Environ Microbiol 70: 4499–4504
21. Bessette PH, Aslund F, Beckwith J et al (1999) Efficient folding of proteins with multiple disulfide bonds in the *Escherichia coli* cytoplasm. Proc Natl Acad Sci U S A 96:13703–13708
22. Shrestha A, Hamilton G, O'Neill E et al (2012) Analysis of conditions affecting autophosphorylation of human kinases during expression in bacteria. Protein Expr Purif 81:136–143
23. Savitsky P, Bray J, Cooper CD et al (2010) High-throughput production of human proteins for crystallization: the SGC experience. J Struct Biol 172:3–13
24. Tsai J, Lee JT, Wang W et al (2008) Discovery of a selective inhibitor of oncogenic B-Raf kinase with potent antimelanoma activity. Proc Natl Acad Sci U S A 105:3041–3046
25. Schlinkmann KM, Hillenbrand M, Rittner A et al (2012) Maximizing detergent stability and functional expression of a GPCR by exhaustive recombination and evolution. J Mol Biol 422:414–428
26. Serrano-Vega MJ, Magnani F, Shibata Y et al (2008) Conformational thermostabilization of the beta1-adrenergic receptor in a detergent-resistant form. Proc Natl Acad Sci U S A 105: 877–882
27. Aslanidis C, de Jong PJ (1990) Ligation-independent cloning of PCR products (LIC-PCR). Nucleic Acids Res 18:6069–6074
28. Klock HE, Lesley SA (2009) The polymerase incomplete primer extension (PIPE) method applied to high-throughput cloning and site-directed mutagenesis. Methods Mol Biol 498:91–103
29. Unger T, Jacobovitch Y, Dantes A et al (2010) Applications of the restriction free (RF) cloning procedure for molecular manipulations and protein expression. J Struct Biol 172:34–44
30. Hartley JL, Temple GF, Brasch MA (2000) DNA cloning using in vitro site-specific recombination. Genome Res 10:1788–1795
31. Bernard P, Gabant P, Bahassi EM et al (1994) Positive-selection vectors using the F plasmid ccdB killer gene. Gene 148:71–74
32. Gay P, Le Coq D, Steinmetz M et al (1985) Positive selection procedure for entrapment of insertion sequence elements in gram-negative bacteria. J Bacteriol 164:918–921
33. Haffke M, Marek M, Pelosse M et al (2015) Characterization and production of protein complexes by co-expression in *Escherichia coli*. Methods Mol Biol 1261:63–89
34. Structural Genomics C, China Structural Genomics C, Northeast Structural Genomics C et al (2008) Protein production and purification. Nat Methods 5:135–146
35. Vincentelli R, Cimino A, Geerlof A et al (2011) High-throughput protein expression screening and purification in *Escherichia coli*. Methods 55:65–72
36. Cairns J, Stent GS, Watson JD (2007) In: Centennial (ed) Phage and the origins of molecular biology. Cold Spring Harbor Laboratory Press, Cold Spring Harbor, NY
37. Cong L, Ran FA, Cox D et al (2013) Multiplex genome engineering using CRISPR/Cas systems. Science 339:819–823
38. Mali P, Yang L, Esvelt KM et al (2013) RNA-guided human genome engineering via Cas9. Science 339:823–826
39. Chen R (2012) Bacterial expression systems for recombinant protein production: *E. coli* and beyond. Biotechnol Adv 30:1102–1107
40. Makino T, Skretas G, Georgiou G (2011) Strain engineering for improved expression of recombinant proteins in bacteria. Microb Cell Fact 10:32
41. Hutchison CA 3rd, Chuang RY, Noskov VN et al (2016) Design and synthesis of a minimal bacterial genome. Science 351: aad6253
42. Galanie S, Thodey K, Trenchard IJ et al (2015) Complete biosynthesis of opioids in yeast. Science 349:1095–1100
43. Paddon CJ, Westfall PJ, Pitera DJ et al (2013) High-level semi-synthetic production of the potent antimalarial artemisinin. Nature 496:528–532
44. Kamens J (2015) The Addgene repository: an international nonprofit plasmid and data resource. Nucleic Acids Res 43:D1152–D1157

Chapter 2

N- and C-Terminal Truncations to Enhance Protein Solubility and Crystallization: Predicting Protein Domain Boundaries with Bioinformatics Tools

Christopher D.O. Cooper and Brian D. Marsden

Abstract

Soluble protein expression is a key requirement for biochemical and structural biology approaches to study biological systems in vitro. Production of sufficient quantities may not always be achievable if proteins are poorly soluble which is frequently determined by physico-chemical parameters such as intrinsic disorder. It is well known that discrete protein domains often have a greater likelihood of high-level soluble expression and crystallizability. Determination of such protein domain boundaries can be challenging for novel proteins. Here, we outline the application of bioinformatics tools to facilitate the prediction of potential protein domain boundaries, which can then be used in designing expression construct boundaries for parallelized screening in a range of heterologous expression systems.

Key words Bioinformatics, Protein expression, Protein solubility, Protein structure, Domain, BLAST, PSIPRED, Hidden Markov Model (HMM), Alignment, Secondary structure

1 Introduction

In order to study proteins by structural, biochemical, or biophysical approaches, a key requirement is the ability to produce sufficient levels of purified protein, ranging from the microgram to milligram levels depending on the technique in question [1]. It is costly, inefficient, and often impossible to obtain sufficiently pure and adequate quantities from native sources [2]. Modern approaches frequently utilize heterologous protein expression systems such as *Escherichia coli*, optimized to produce large quantities of protein from plasmid expression vectors containing a cloned and defined sequence [3, 4]. It is well known, however, that sequence of the protein is one of the most important determinants of successful protein expression, solubility, or crystallization potential [1, 5]. Results vary greatly between the expression constructs used (encoding fragments of defined protein sequence length and

Nicola A. Burgess-Brown (ed.), *Heterologous Gene Expression in E. coli: Methods and Protocols*, Methods in Molecular Biology, vol. 1586, DOI 10.1007/978-1-4939-6887-9_2, © Springer Science+Business Media LLC 2017

context) [6] due to differing protein physicochemical properties and biological factors such as protein folding, export, or toxicity in the host cell. Indeed, studies on heterologous expression in *E. coli* show that less than half of proteins from prokaryotes and one fifth from eukaryotes can be expressed in a soluble form as full-length proteins [7].

In such circumstances researchers often turn to alternative expression hosts, often closer to the original organism of the protein of interest [8], such as other bacterial systems (e.g., *Bacillus* [9] and *Lactococcus* [10]), or eukaryotic systems (e.g., baculovirus/insect cells [11] and protozoa [12]). Furthermore, a wide range of solubility-enhancing and affinity fusion tags have also been successfully applied to heterologous expression systems, such as GST, MBP, and thioredoxin [13]. Different levels of expression between fusion tags and target proteins in comparative screens, however, suggest the necessity of screening multiple tags [14].

Eukaryotic proteins are often comprised of modular structures of defined, folded domains, linked by flexible or unstructured stretches of sequence. Protein domains are thought to fold independently, exhibit globularity (e.g., contain a hydrophobic core and hydrophilic exterior), and perform a specific function (e.g., binding), such that the combination and juxtapositioning of domains determines overall protein function [15]. There is a long-held premise that well-ordered or compact domains or fragments will yield better-behaving proteins than full-length proteins for protein expression and structural studies, in relation to solubility and crystallization potential [16]. For instance, rigid proteins have a greater propensity to crystallize than flexible or highly disordered proteins [5], resulting from increased flexibility either between domains in multi-domain proteins, or from within domains (e.g., unstructured N- or C-termini or internal loops) entropically hampering crystallization [17]. Furthermore, many proteins exist in complexes with other partners, exhibiting poor expression or solubility when expressed alone and/or in alternative hosts due to, for example, the exposing of hydrophobic patches that the interacting partner normally protects [16]. This may occur even if such regions are localized to a single domain.

Therefore, delineation of independent, folded, and compact protein domains for expression as individual units is a key tool in protein and structural biochemistry. Significant attempts have been undertaken to predict optimal protein constructs for expression, many of which involve multiple truncations of full-length proteins from either, or both, the N- and C-termini to express individual domains [7]. Parallel analysis of multiple domains and domain fragments has been simplified with the advent of high-throughput cloning and expression/purification methods [18]. Iterative but random trial and error approaches toward constructing N- or C-terminal truncation, however, can be costly and time-consuming.

A more informed approach, which we call "domain boundary analysis" or DBA, involves the interrogation of multiple bioinformatics methods to predict protein structural features. This targeted approach to delimit protein domain boundaries and their subsequent combinatorial arrangement is more likely to result in ordered, defined, and globular protein fragments [6, 19]. DBA has been very successful in our hands, with nearly half of human proteins attempted being successfully expressed and purified, and around 20% of those attempted resulting in a solved high-resolution X-ray structure [1]. Here, we take the reader through practical usage of a range of common bioinformatics approaches used in DBA, toward defining well-behaving protein domains for biochemical and structural analysis.

2 Materials

All analyses described here can be performed on any standard PC, Mac OS X, or Linux-based operating system on a standard desktop or laptop computer with an internet connection. Most common web browsers (Explorer, Safari, Chrome, etc.) work with the bioinformatics servers described. Many of the platforms described can be downloaded and installed locally on Linux-based systems or incorporated into bespoke web services, but we are restricting our descriptions to individual web-based analyses for ease of use. The sole requirement from the user is the protein sequence of interest, with residues represented in the IUPAC single-letter code format [20]. In a minority of cases, it may be necessary to provide the sequence in FASTA format [21] which can be facilitated by the simple addition of an identifier (name) preceded with the character ">," required as the first and separate line in the sequence:

```
>sequence_name
MTGHYTHHAYGRETYIPSDFGNMKILPSSWQ
```

Protein three-dimensional structure visualization can be performed also using web-based software or via software that is either provided specifically for an operating system (e.g., Windows, OS/X, Linux) or in an independent form using a platform such as Java.

3 Methods

Our approach to defining construct boundaries by DBA utilizes a range of common bioinformatics approaches, all freely available online. A hierarchical approach is taken to define boundaries (Fig. 1), initially identifying domains using a combination of homology-based and Hidden Markov Model (HMM) approaches,

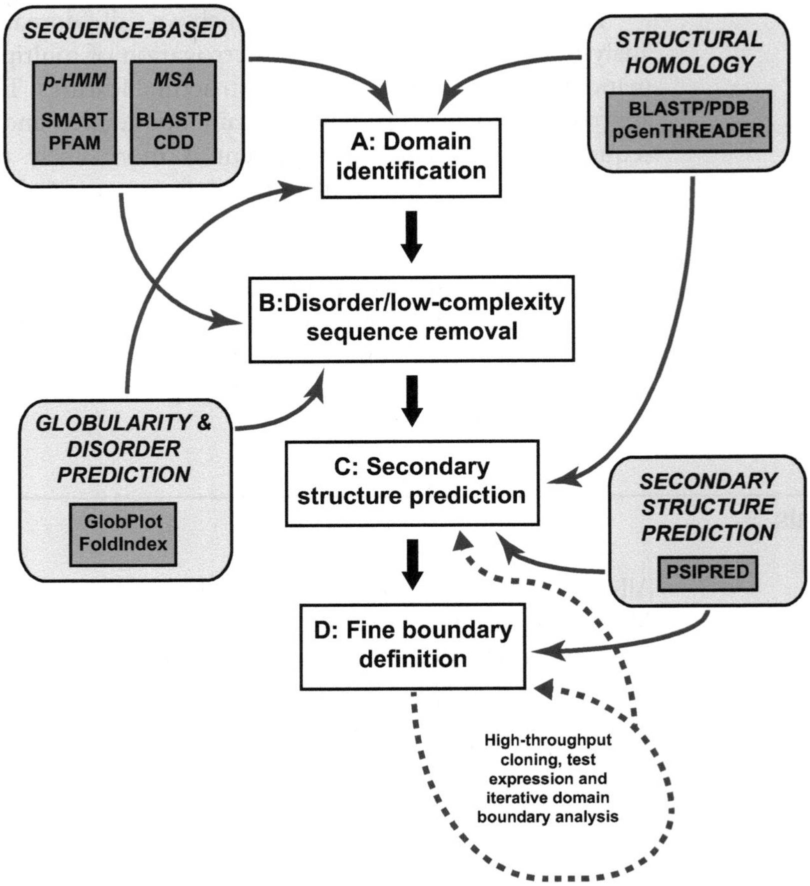

Fig. 1 Representation of the hierarchical approach to domain boundary analysis. The workflow is shown by *boxed rectangles* (A to D) connected by *solid black arrows*. The involvement of bioinformatics tools at various pipeline stages (*dark gray boxes*, grouped by type of method (*rounded light gray boxes*)) is represented by *gray arrows*. *Dashed gray arrows* represent iteration of secondary element/fine boundary redesign following cloning and protein test expression, where necessary. *p-HMM* profile-Hidden Markov Model; *MSA* Multiple Sequence Alignment; *PDB* Protein Data Bank

supplemented by disorder prediction to suggest protein globularity, a reliable indicator of folded domains. Once potential domains are identified, multiple finer-grained boundaries are defined using predicted secondary structural elements as termini, again supplemented with disorder propensity information. Sequence and structural homology information can further supplement to help guide the determination of likely soluble or crystallizable protein boundaries.

Parallel testing of multiple constructs with different domain boundaries can increase experimental success (Fig. 2) [1]. Our DBA approach is designed to be used in conjunction with Ligation-Independent Cloning (LIC) or other high-throughput cloning methods to construct N- and C-terminal tagged fusions, combined with small-scale parallel expression in multiple systems (*E. coli*,

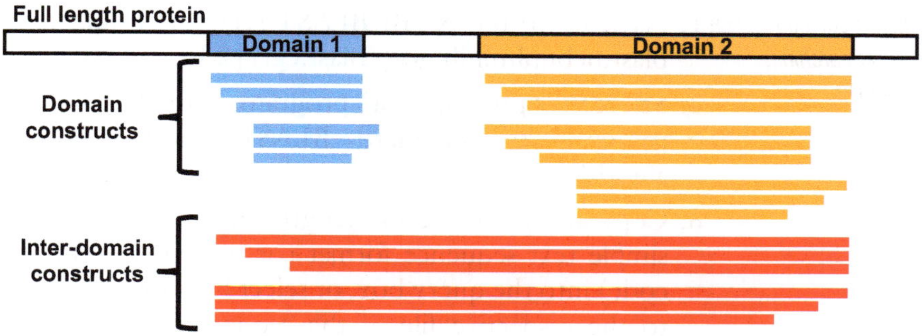

Fig. 2 Representation of domain boundary analysis. Individual domains in a full-length protein sequence are identified (*blue/orange*), then combinatorial sets of N- and C-terminal truncations are made. Constructs containing tandem domains (*red*) may also be used

baculovirus-infected insect cells) [1, 7, 18, 22]. The number of domain boundaries attempted is determined by the researcher in relation to resources and time available but, from our experience, 12–40 constructs per domain is typical, normally matched to multiple domain-defining secondary structural elements [1]. If multiple tandem domains are present, the respective N- and C-terminal boundaries can also be combined for multiple-domain constructs (Fig. 2). In addition, it is also worth attempting the full-length protein itself in expression trials, perhaps with multiple small N- and C-terminal DBA-defined truncations.

3.1 Prediction of Protein Secondary Structure and Domains Using Sequence and Structural Homology

Since the concept of the "domain hypothesis," a number of experimental and de novo computational/statistical methods have been used to attempt to predict protein domain boundaries [15]. The simplest approach to assign boundaries, however, is often by similarity to previously defined domains. Hence, the approach we take for DBA uses a number of complementary approaches, either based on direct sequence-based homology (BLAST [23], Conserved Domain Database (CDD) [24]), or profile HMM-based approaches (SMART [25], PFAM [26]). The CDD is a database of annotated multiple sequence alignments, allowing alignment of query sequences to previously detected or characterized domains. The HMM-based SMART and PFAM databases provide a complementary, but often more sensitive, detection of domains including many not found in the CDD, alongside a number of predicted but uncharacterized "Domains of Unknown Function" (DUFs). These approaches are particularly useful to identify "core" domain regions, the precise boundaries of which can be subsequently explored with disorder/secondary element prediction tools described later.

Where strong sequence homology to existing characterized domains may not exist, predicted secondary structure (PSIPRED [27]) and homologies both to close (BLAST/Protein Data Bank (PDB) [28]) and remote structural templates (pGenTHREADER [29]) can potentially be identified, to guide construct termini design.

3.1.1 Domain Prediction Using Homology Searching: BLAST and the CDD

1. Navigate to the NCBI BLAST server web interface (http://blast.ncbi.nlm.nih.gov/Blast.cgi) [23].
2. Select the "protein blast" program, in the Basic BLAST section to open the standard BLAST interface to the blastp algorithm.
3. Copy and paste the full-length query sequence in FASTA or simple text sequence format (or the NCBI protein accession code) into the query box, or select "Choose File" and navigate to the respective file, if the sequence is saved as a text file (*see* **Note 1**).
4. Select the database to be searched from the dropdown menu of the Database option of the Choose Search Set section. Choose "Protein Data Bank proteins(pdb)" to search within potential homologous structures (*see* **Note 2**).
5. The BLAST search can be optionally taxonomically limited should the user require, by starting to type either the common or Latin species/taxon name into the Organism field (e.g., *Homo sapiens*). On typing, taxon options pop up, and select the most relevant one (*see* **Note 3**).
6. Leave the algorithm and general parameters as default for blastp (protein-protein BLAST), with BLOSUM62 matrix and gap parameters as 11/1 (*see* **Note 4**).
7. Press the blue "BLAST" button to run the search.
8. Once the search is complete, the results are graphically displayed as an overview distribution of BLAST hits mapped onto the query sequence (Fig. 3a). The color represents the homology between query sequence and identified sequence, with red matches as closest and the longest significant match at the top of the matched sequences (color key is above at the top of the distribution image). Multiple matched regions represent the presence of multiple domains in the query sequence.
9. Select a match on the distribution image to automatically scroll down the page to respective alignment HSP report (Fig. 3b), representing a homologous sequence for which a protein structure is present in the PDB database (*see* **Note 5**). The corresponding aligned residue positions of the query and match ("Sbjct") are displayed flanking the alignment.
10. Click on the link beginning "pdb" next to "Sequence ID " in the HSP report to access the corresponding protein structure information, linking to the PDB structure file.
11. The query sequence is also searched against the CDD [24] with the graphical output arranged above the distribution report (top frame, Fig. 3c). This displays CDD matches and also strong matches from the SMART and PFAM databases (*see* Subheading 3.1.2). Click on the CDD output image to

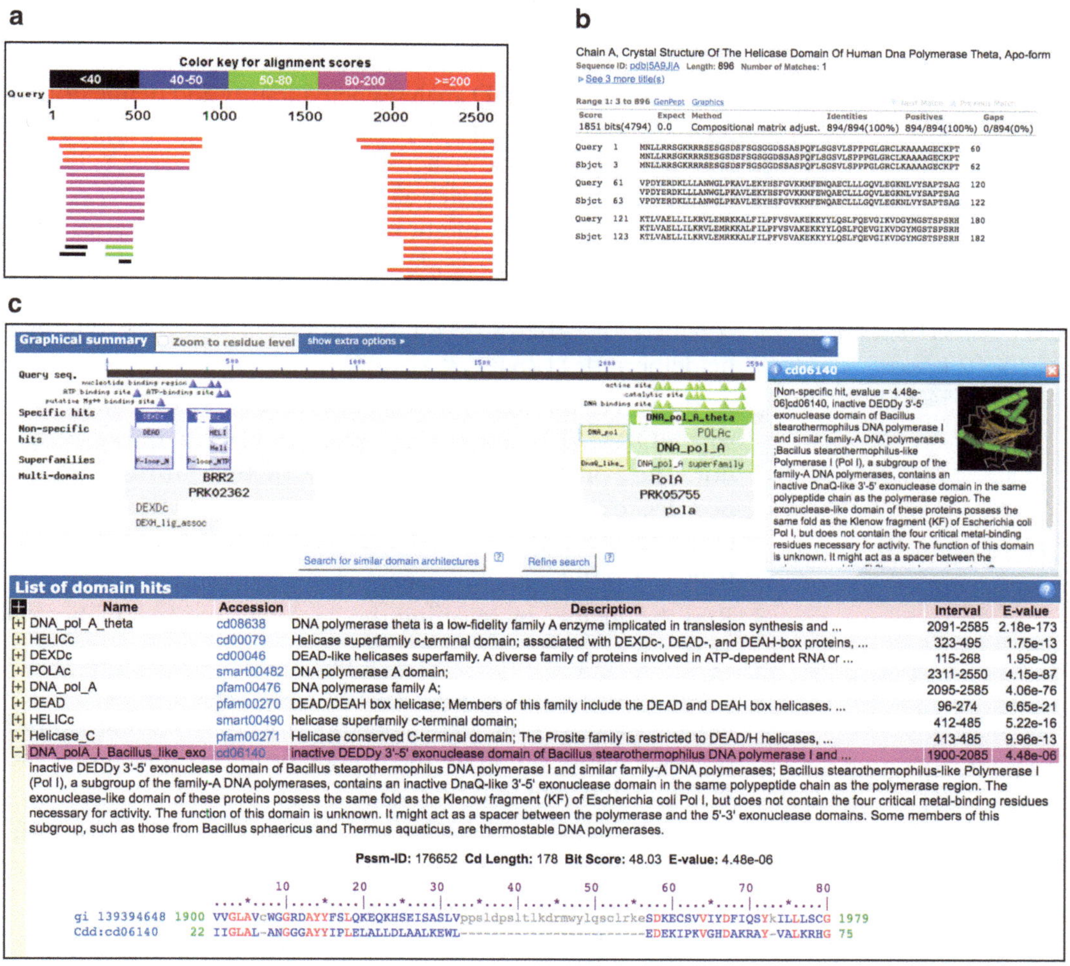

Fig. 3 Screenshot from NCBI BLAST output using the human POLQ protein as input to search against the PDB database. (**a**) Distribution of BLAST hits mapped onto the input sequence, color coded for strength of alignment. (**b**) Detailed BLAST HSP alignment. (**c**) CDD output (*top frames*, domain annotations with example pop up window for cd06140 CDD entry; *lower frames*, domain lists with example expansion showing input sequence alignment against CDD consensus)

open a new browser window with the same graphical display and an additional detailed list of matched domains (lower panel, Fig. 3c), detailing the boundary regions of the query that matches the domain ("interval") and E-value match significance (*see* **Note 6**).

12. Position the mouse pointer over the domain image in the CDD graphical output, whereby a popup window appears with available biological information (right side window in top frame, Fig. 3c). Alternatively, click on the "+" of a domain in the list to expand the list to provide biological descriptions, with an alignment of the query sequence against the consensus for this domain (lower panel, Fig. 3c), with the boundaries

shown flanking the alignment. Minimize the expansion by clicking "–."

The results from CDD analyses help identify and define domain boundaries (contributing to step A of DBA, Fig. 1), with BLASTP searches identifying close structural homologues (step A, Fig. 1). CDD and HSP local sequence alignments help to identify consensus residue positions that might indicate domain boundaries (steps A and C, Fig. 1).

3.1.2 Domain Prediction with HMM Databases: SMART and PFAM

1. Navigate to the SMART webserver (http://smart.embl-heidelberg.de) [25].
2. At the top of the web interface, ensure the SMART mode is set to "NORMAL" and the webpage displays a query box. If not, click on the "NORMAL" link in the "SMART mode" box. Paste the full-length protein sequence into the query box, ensuring all search options are selected in the Sequence Analysis pane (*see* **Note 7**).
3. Run the analysis by selecting the "Sequence SMART" button.
4. SMART output displays a graphical representation of recognized domains from the SMART database, with an approximate residue scale bar (Fig. 4a). Mouse over the domain representation to pop up the residue positions and significance of the match (Fig. 4a).
5. If search options were selected (this section, **step 2**) domains not present in SMART may be recognized, e.g., PFAM and transmembrane (TM) regions (Fig. 4b, *see* **Note 8**).
6. Click the domain in the graphical output to link to detailed domain information (Fig. 4c).
7. Click on the "Align your sequence against the SMART alignment" button, to generate a similar alignment to the consensus sequence as performed with the CDD software (Subheading 3.1.1, **step 12**).

Results from SMART/PFAM searches may identify both characterized and predicted (DUF) domains, with consensus alignments helping delineate domain boundaries (steps A and C, Fig. 1), similar but often more sensitive than CDD (*see* Subheading 3.1.1). In addition, SMART/PFAM also predict low-complexity sequences (often disordered, *see* Subheading 3.2), used in step B (Fig. 1) (*see* **Note 9**).

3.1.3 The PSIPRED Workbench for Protein Domain and Secondary Structure Prediction

PSIPRED [27] and pGenTHREADER [29, 30] are part of the UCL PSIPRED suite of tools [31], for protein fold and secondary structure prediction (http://bioinf.cs.ucl.ac.uk/psipred/) (*see* **Note 10**). The advantage of this server is that multiple algorithms may be run simultaneously from a single-query sequence submission. PSIPRED is among the most accurate predictors of

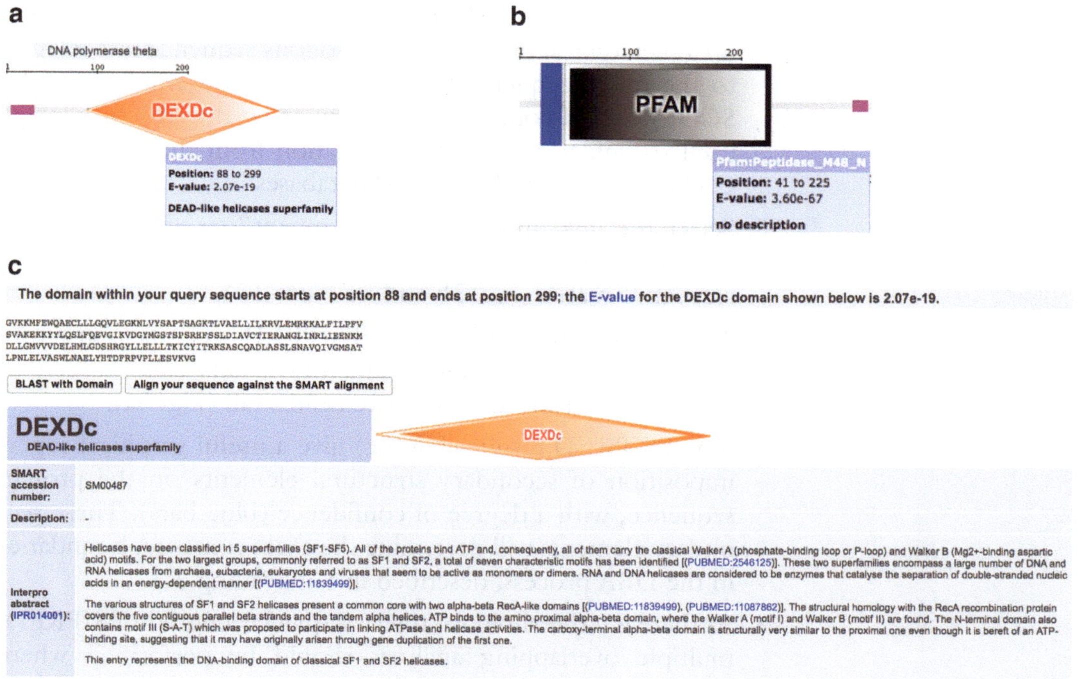

Fig. 4 Screenshot from SMART output, using human POLQ protein as input. (**a**) Graphical output showing recognized SMART domain, with popup window on mouse over. (**b**) Graphical output showing recognized transmembrane region (*blue*) and PFAM domain, with popup window on mouse-over. (**c**) Expansion on clicking SMART domain from Fig. 4a

protein secondary structural elements, critical for the DBA procedure described here, and in more detail in Subheading 3.3. Like BLAST searches of the PDB database (Subheading 3.1.1.), pGenTHREADER is particularly useful to find PDB templates for structural considerations in DBA (Subheading 3.3), but has the advantage of using PSI-BLAST and threading methods to help determine remote structural homologies (*see* **Note 11**) [32], increasing sensitivity compared with BLAST in our hands.

1. In the web interface , select PSIPRED and pGenTHREADER and paste the protein sequence into the "Input Sequence" window as FASTA or raw sequence format (*see* **Note 12**). Multiple sequences may also be posted.
2. Enter a valid email address in "Submission Details" pane (recommended, *see* **Note 13**) and click "Predict" to run the analysis.
3. Once the submission is complete, the results page (Fig. 5a) displays results from different algorithms in different tabs, with the option to download the results (see respective tab) as text or printable PostScipt/PDF files.

4. For pGenTHREADER, click on the respective tab, bringing up a hierarchical display of homologous sequence hits relating to the query sequence (*see* **Note 14**). Click the links under SCOP/CATH codes, CATH entry or on the structure image itself to link to structural information from the SCOP [33], CATH [34], or PDBsum [35] databases.
5. Select the link under "View Alignment" to open a window displaying a structural alignment of the query sequence to the respective match (Fig. 5b and *see* **Note 15**).
6. The pGenTHREADER uses a PSIPRED secondary structure prediction in its operation, and full results can be seen or downloaded from the respective results tab (Fig. 5a).
7. Raw PSIPRED results (Fig. 5c) give a useful graphical superimposition of secondary structural elements on the protein sequence, with a degree of confidence (blue bars). These secondary elements will determine the exact construct boundaries in the DBA process, described in Subheading 3.3.
8. As there is a threshold for query sequence length in PSIPRED, multiple overlapping analyses should be performed where appropriate (*see* **Note 12**).

pGenTHREADER matches thus help identify homologous domains (step A, Fig. 1) and along with resulting PSIPRED predictions, help identify secondary structural elements and fine domain boundaries (steps C and D respectively, Fig. 1).

3.2 Protein Domain Identification Using Globularity and Disorder Prediction

The methods described for domain identification have so far been based on prior experimental data, often as a consequence of advances in genome sequencing and structural genomics. That is, identifying protein domains using previously identified related or homologous domains using HMMs or alignments, or from structural homology to previously solved structures of proteins. However, to delineate domains that lack well-defined annotation in the literature, unbiased techniques are required. It is well known that protein domains are usually made up of globular well-ordered cores of secondary structure, with inter-domain linkers often disordered [36]. Here, we describe the use of the FoldIndex [37] and GlobPlot 2 [38] webservers that provide complementary approaches to predict order (globularity) to define domain boundaries and regions of proteins that may negatively influence protein crystallization.

3.2.1 Disorder Analysis with FoldIndex

1. Paste the protein sequence directly into the "Sequence area" window of the FoldIndex webserver (http://bioportal.weizmann.ac.il/fldbin/findex) [37].
2. Default parameters are advised for the sequence window and step, but enable the "graph Phobic values" and "graph charge values" options (*see* **Note 16**).

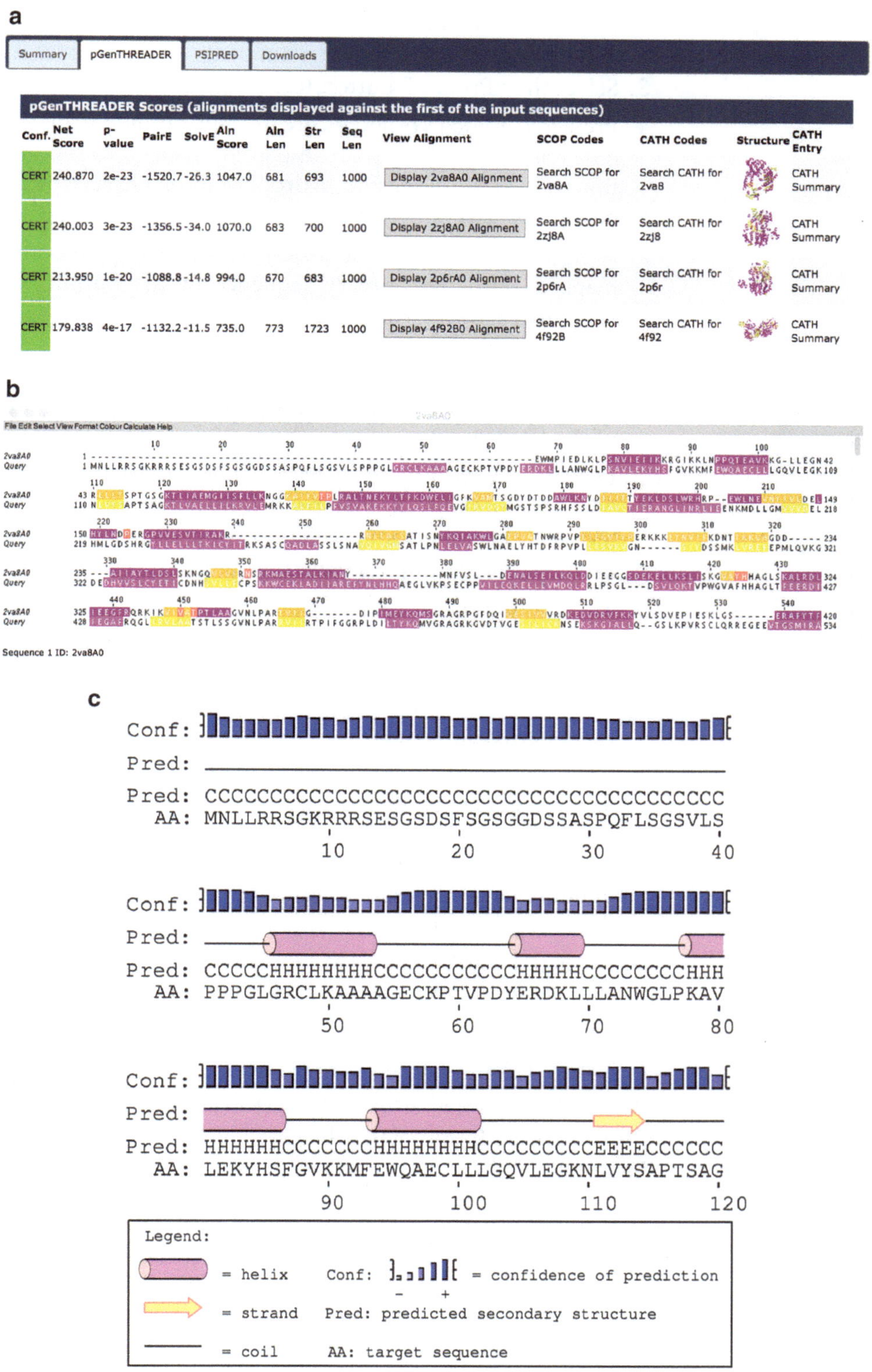

Fig. 5 Screenshots of graphical outputs from the PSIPRED suite of programs. (**a**) pGenTHREADER table output, with most identical/homologous sequence ranked highest (lowest p-value is most significant), with high confidence hits in *green* (medium in *orange* and weak in *red*, not shown). (**b**) Structural alignment output following selection of “View Alignment” in (**a**). Predicted or structurally determined α-helices (*purple*) and β-strands (*yellow*) are mapped onto query and matched sequences, respectively. (**c**) Detailed PSIPRED output for query sequence with same color scheme as for (**b**), with secondary elements definitions: C, coil, H, α-helix, E, β-strands, and “Conf” representing prediction confidence

3. Select the "Process" button to run the analysis.
4. Predicted folded (ordered, green) and unfolded (disordered, red) regions are graphically displayed, mapped to residue position (Fig. 6a), alongside hydrophobic or charged regions if previously selected. This image may be saved as a PNG file.
5. Alongside prediction statistics, (dis)order predictions are mapped onto the primary sequence in the output window (Fig. 6b), allowing (dis)order to be mapped onto the sequence (*see* **Note 16**).

3.2.2 Disorder Analysis with GlobPlot

1. Paste the protein sequence directly into the "Sequence" window of the GlobPlot 2 webserver (http://globplot.embl.de) [38].
2. Default parameters are advised, but otherwise enable the "Russell/Linding" disorder propensity option and the "Perform SMART/Pfam domain prediction" options (*see* **Note 17**).
3. Select the "GlobPlot NOW!" button to run the analysis.
4. As with FoldIndex (Subheading 3.2.1), ordered/disordered regions are mapped onto the protein primary sequence (Fig. 6c), in this case green/black respectively (*see* **Note 18**). In addition, predicted ordered sequences ("GlobDoms") are listed above the sequence.
5. Graphical results (which can also be downloaded in PostScript format) display predicted globularity/disorder as green/blue blocks respectively, alongside residue number (Fig. 6d). Disorder propensity is plotted as a white line, with downhill and uphill regions corresponding to predicted globular regions or disorder, respectively.
6. Predicted SMART/PFAM domains are superimposed onto this plot according to the included key, allowing simple combination of de novo globularity and HMM approaches.

FoldIndex and GlobPlot approaches thus help identify globular regions, toward identification of (sub)-domains (step A, Fig. 1) and disordered termini (step B, Fig. 1), in the domain boundary analysis hierarchy.

3.3 Combining Bioinformatics Approaches for Domain Boundary Prediction

Once bioinformatics analyses have been completed, results should be combined cohesively as part of the DBA process. Figure 1 demonstrates the overall DBA workflow, and the contribution of each bioinformatics tool to the process. Most aspects of the procedure have been duplicated with multiple algorithms, increasing the accuracy of domain boundary prediction. Important considerations are illustrated using human POLQ (DNA polymerase θ, UniProt ID: O75417) as an example (Fig. 7) [39].

1. Alignment and HMM-based approaches identify predicted domains by homology (A, Fig. 1), with improved confidence conferred if multiple servers predict domains in the same

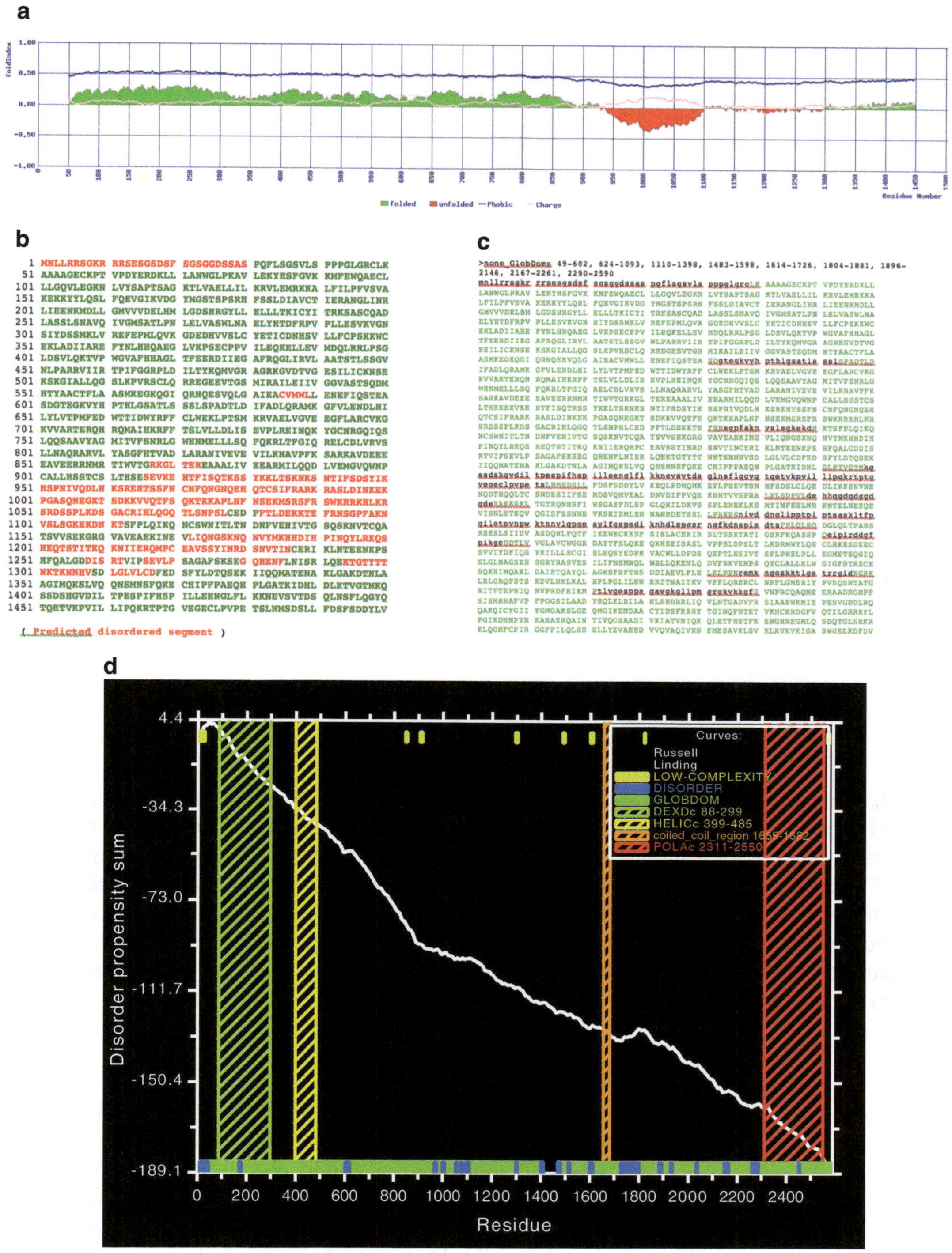

Fig. 6 Output from FoldIndex and GlotPlot servers, using residues 1–1500 or full-length human POLQ as a query sequence, respectively. (**a**) FoldIndex PNG file graphical output, with *green* and *red* regions as folded/unfolded respectively. Hydrophobic and charge propensity are plotted as *blue* and *pink* traces respectively. (**b**) FoldIndex output screenshot with predicted ordered/disordered regions plotted onto the query sequence as *green*/*red* text respectively. (**c**) GlobPlot output screenshot with predicted globular/disordered regions plotted onto the query sequence as *green capitalized*/*black small case text* respectively. (**d**) GlobPlot graphical output for full-length POLQ as query sequence. Globular domains are *green blocks*, disordered regions as *blue blocks* and recognized SMART domains according to the key. Disorder propensity is plotted as the *white line*, described in the main text

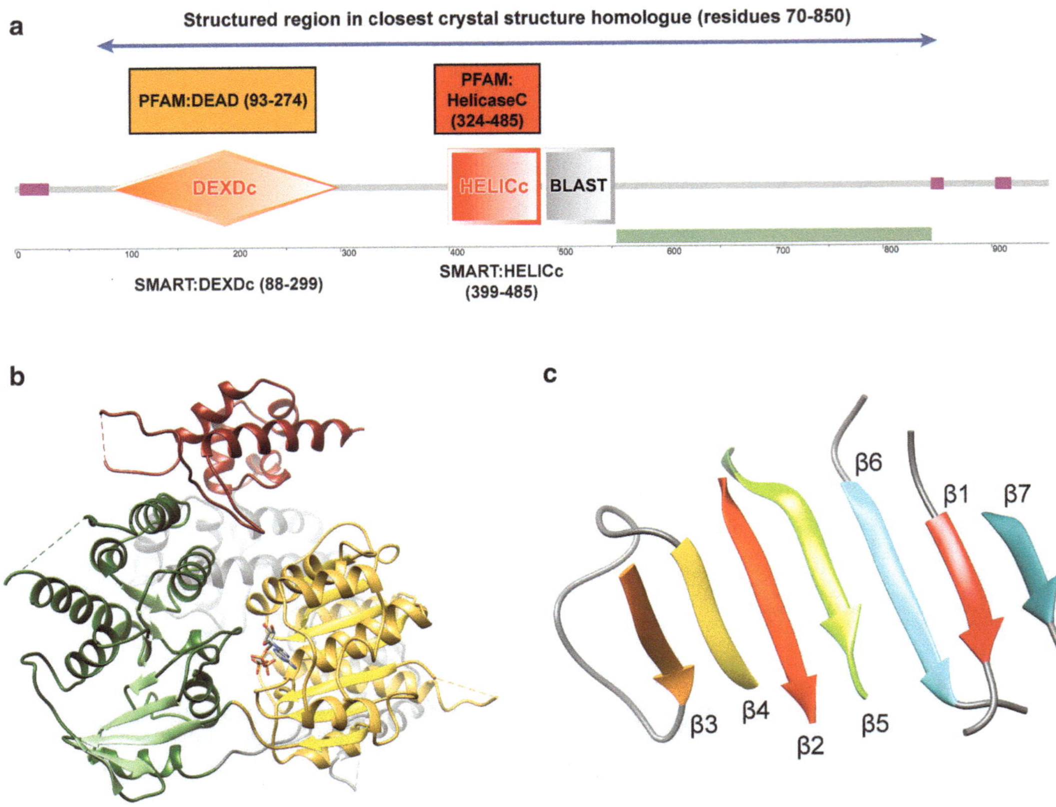

Fig. 7 Considerations in domain boundary analysis. (**a**) Representation of PFAM and SMART detected domains mapped to the first 1000 residues of human POLQ (base image generated by SMART server [25]). Numbers in parentheses denote predicted domain boundaries from respective analyses, with low-complexity regions in purple. The closest structure homologue is PDB:2P6R *A. fulgidus* HEL308. (**b**) (Sub) domain crystallized structure of human POLQ (~residues 70–900, PDB:5AGA [39]), showing RecA and helix-hairpin-helix subdomains rendered in *green/yellow* and *red*, respectively. (**c**) Parallel β-sheet from human POLQ structure showing non-contiguous β-strand arrangement, with strands numbered from N- to C-terminus (β1–β7). Images in (**b**) and (**c**) were rendered with Chimera [40]

sequence neighborhood (e.g., PFAM:DEAD and SMART: DEXDc domains, Fig. 7a). Additional non-HMM domains (e.g., "BLAST," Fig. 7a) should also be taken into account, even if only found by a single algorithm. Low-complexity sequences are found at the extreme ends of the 1–900aa region and are recommended not to be included in designed constructs (B, Fig. 1). In this example, the analysis suggests two to three domains in POLQ from ~80 to 550 residues.

2. Disorder prediction with both GlobPlot2 and FoldIndex suggests the protein is predominantly globular up to 900aa (step A, Fig. 1 and Fig. 6). Biologically inferred data from the most homologous structure (*Archaeoglobus fulgidus* HEL308, found from both BLAST searches to the PDB database and pGen-

THREADER) suggests that the entire region from ~70 to 850aa is globular from its expression and structural determination; hence, the HMM-derived domains such as SMART:DEXDc are likely to be sub-domains (step A, Fig. 1) (*see* **Note 19**).

3. Domain boundaries can in principle focus on the sub-domains, but examination of homologous structures (Fig. 7b) suggests that if this was the case, significant biological information would be lost (*see* **Note 20**). Here, the expected substrate (an ATP analogue) is bound between the RecA sub-domains (green/yellow) corresponding to the two predicted PFAM/SMART sub-domains in Fig. 7a. Hence, the more biologically relevant domain boundaries should span these two sub-domains. Furthermore, a cryptic domain not detected in HMM-based searches can only be noted by comparison to the homologous HEL308 structure, seen here in the final POLQ structure (helix-hairpin-helix, red in Fig. 7b). Hence, analysis of sequence similarity in homologous protein structures can yield important information in addition to sequence-based HMM searches (step A, Fig. 1).
4. Co-localization of domains to the same region of sequence may have different local boundaries (e.g., PFAM:DEAD 93–274aa and SMART:DEXDc 88–299aa). In such cases, we recommend using the longer of the two regions if within 10–20 residues as the boundary (*see* **Note 9**).
5. Once approximate domain boundaries are predicted, use PSIPRED secondary structure predictions to delineate secondary elements as the next level of construct boundary, serially expanding the boundaries in both directions one element at a time (step C, Fig. 1). It is important to compare PSIPRED predictions to the actual elements in homologous determined structures, e.g., with the structural alignment output of pGenTHREADER (*see* **Note 21**), to avoid bisecting secondary structural elements.
6. If homologous structures are found from BLAST or pGenTHREADER searches, PSIPRED secondary element predictions should be compared to those in the known structure in case removing a specific element destabilizes the protein (*see* **Notes 22** and **23**).
7. The final stage of DBA is to choose the residue positions to determine the precise construct boundaries (step D, Fig. 1). It is critical that full secondary elements are considered when determining the termini of boundaries, e.g., in this example the first α-helix as a boundary should begin at GRCLK (Fig. 5c). If resources allow, a further boundary should be designed by the addition of a small amount of coil/non-element structure, e.g., GLGRCLK (Fig. 5c). Close additional

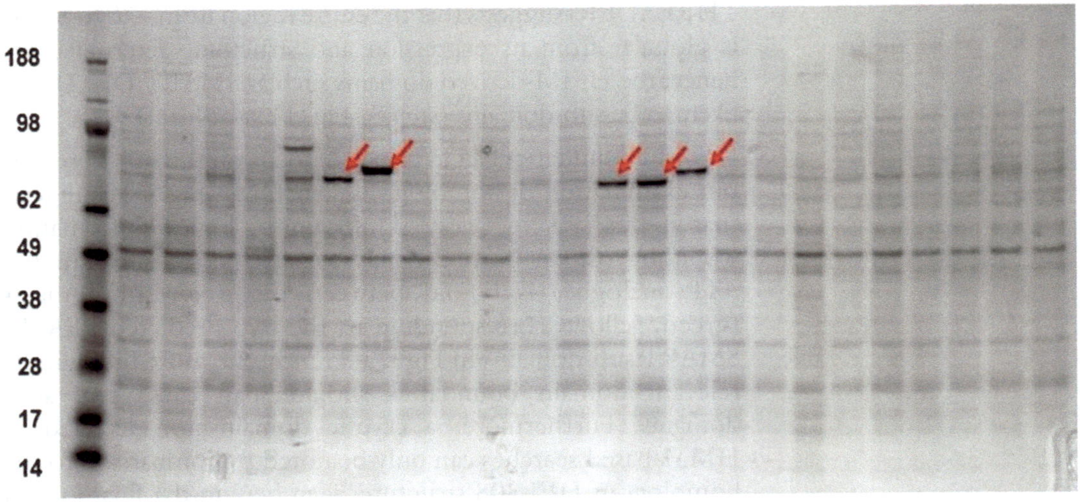

Fig. 8 Typical small-scale protein expression screening. SDS-PAGE analysis of 3 ml test expression from Sf9 insect cell of various N- and C-terminal construct truncations of human POLQ, following no soluble expression in *E. coli*. *Red arrows* denote successful and correctly sized proteins

boundaries may be useful, as such regions are often not structured in crystals and the true secondary element may in fact comprise this additional sequence, among other factors (*see* **Note 24**).

3.4 Further Methods for Domain Boundary Analysis: Beyond Bioinformatics

The DBA approach we have outlined here to delineate protein domains is designed to be used in conjunction with high-throughput parallel cloning and expression methods, as described earlier [1]. *E. coli* systems are predominantly used in initial expression screening, moving to baculovirus-mediated insect cell expression if not successful. Although such approaches frequently lead to respectable success rates in small-scale tests (Fig. 8) [1], reiteration of the DBA procedure may be required for protein expression optimization for difficult targets. Analogous approaches have been attempted by others, often bringing together similar bioinformatics approaches but in automated pipelines, such as ProteinCCD [19], or by our colleagues at the Structural Genomics Consortium [6]. However for small-scale domain prediction, the use of individual bioinformatics tools allows the user a great deal of analytical flexibility, depending on the protein in question.

A range of experimental data may also be applied to protein domain delineation. If full-length protein is available, limited proteolysis combined with mass spectrometric (MS) approaches can determine core folded domains, as connecting unfolded sequence or disordered termini may be trimmed away by proteases, with core domains identified by MS [41]. In addition, the advent of powerful high-throughput screening of random or combinatorial protein truncation or mutation libraries allows an unbiased approach with no prior knowledge required [42]. Rather than replacing bioinformatics approaches to domain boundary analysis, these experimental techniques may facilitate the accuracy of domain prediction for difficult proteins, especially if used in combination with in-silico approaches described here.

4 Notes

1. Single or lists of multiple sequences can also be entered in this manner. Sub-sequences may be selected in the "Query subrange" box.
2. The full NCBI protein sequence database can be searched instead if homologous structures are not required or found, by selecting the "Nonredundant protein sequences (nr)" drop-down option.
3. We normally leave the "Organism" option blank, to give the greatest chance of finding a close homologue.
4. Blastp algorithm parameters can be changed if using protein sequences with few close homologues, but we find default parameters are adequate for most sequences, especially for mammalian proteins.
5. HSP (High-scoring Segment Pair) is the alignment of the query to database sequence, generally representing a single domain. However, multiple HSPs may be present within a domain if variable intervening sequences are present (e.g., loop regions or low-complexity sequences). Significance of matches ("Expect" or "E-value") is greater the smaller the number, with zero being most significant. The length of the match (both for identity and similarity ("positives")) is also displayed.
6. Expect (E)-values are an estimate of the significance of a BLAST match, i.e., the number of hits expected by chance in a particular database. Hence, the lower the number and closer to zero the E-value, the more significant the match, e.g., $1e^{-6}$ is a good starting point for a significant hit.
7. Optional tick boxes engage additional database searching, including PFAM [26], membrane protein signal sequences [43], repeats, and outlier homologues.
8. Identification of TM regions is beneficial, as following their high hydrophobicity, their removal increases the likelihood of soluble protein domain expression.
9. IMPORTANT: CDD/SMART/PFAM methods and domain definitions are very conservative, often defining domains as core regions and hence removing surrounding regions that

may in fact be true domain boundaries. Hence, if multiple methods coincide with approximate boundaries, the longest prediction should be used. Furthermore, predicted secondary structural elements (Subheading 3.1.3) around these predicted domain boundaries should extend away from, rather than into these regions, in order to prevent shortened and therefore erroneous domain boundary predictions.

10. Additional software, useful for construct design and run simultaneously, is available in the PSIPRED workbench package [31], particularly for transmembrane helix and topology prediction (e.g., MEMSAT3/MEMSATSVM) and additional orthogonal disorder prediction (DISOPRED3) , but out of the scope of these protocols.
11. Although pGenTHREADER is useful for detecting remote structural homologies in the case of low sequence similarity, care should be taken in the interpretation of, or using such remote homologies, as false-positive hits may be prevalent with some hits bearing no real functional similarity.
12. An upper sequence length limit of 1500 residues exists for PSIPRED workbench servers. Hence, longer proteins should be broken down into shorter fragments for submission, ideally not comprising multiple domains. These should be arranged as tiles of fragments with 200–500 residue overlaps, to ensure that positioning at fragment ends does not influence prediction accuracy.
13. The PSIPRED workbench algorithms are computationally intensive and may take up to 2 h to run; hence, it is recommended to supply an email address for delivery of a weblink to results.
14. The color code on the left panel for pGenTHREADER results (Fig. 5a) gives a rapid idea of match confidence, with green being firm hits, followed by orange then yellow (weak). Orange/weak hits should only be used if green and confident matches are not found, suggesting that only remote structural homology has been found.
15. pGenTHREADER structural alignments are especially useful when only remote homologies are matched to query sequences, guiding alignment on the basis of (predicted) structure, rather than potentially biased or misguided poor sequence similarity. In such circumstances, the use of multiple weak/average matches should be used to reduce bias in PDB template choice.
16. Graphing the hydrophobic and charged regions in FoldIndex gives further information to solubility propensity, i.e., hydrophobic/charged regions are likely to negatively/positively influence protein solubility respectively.

17. The SMART/PFAM search is useful in GlobPlot, superimposing HMM-based domain searches (Subheading 3.1.2) onto globularity/disorder predictions and the query sequence.
18. Copying the colored alignment from FoldIndex and GlobPlot and pasting into word processing or text editing software with the "Courier" font preserves text formatting and spacing for useful documentation.
19. It should be noted that although a stretch of protein may be predicted to be (globally) globular, it could in fact comprise a string of local globular domains with very small linkers that do not show up in disorder prediction.
20. Many protein structure visualization platforms may be freely downloaded, and although this is out of the scope of this chapter, the authors recommend Chimera (cgl.ucsf.edu/chimera/) [40] or PyMOL (pymol.org).
21. If only remote homologues exist, such structural alignments in pGenTHREADER will considerably increase the accuracy of secondary element prediction.
22. Removing specific secondary structural elements could expose significant regions of hydrophobicity (or remove favorable charged regions), both of which could diminish protein solubility.
23. In parallel β-sheets in particular, the strand arrangement from one side to another does not necessarily follow the N- to C-terminal order. Hence, removal of the most N-terminal strand could destabilize a whole β-sheet if juxtaposed centrally in the β-sheet, with increased likelihood of protein insolubility (e.g., removal of N-terminal β1 or β2 from POLQ would split the β-sheet, Fig. 7c).
24. Terminal residue composition may influence protein expression [44], hence a range of alternative but close boundaries may be beneficial. Even if soluble protein is produced, some terminal residues may negatively influence crystal packing , e.g., PPPGLGRCLK (Fig. 5c) may cause a sharp N-terminal kink increasing disorder or decrease potential packing, due to the high proline content.

Acknowledgments

The SGC is a registered charity (number 1097737) that receives funds from AbbVie, Bayer Pharma AG, Boehringer Ingelheim, Canada Foundation for Innovation, Eshelman Institute for Innovation, Genome Canada, Innovative Medicines Initiative (EU/EFPIA) [ULTRA-DD grant no. 115766], Janssen, Merck &

Co., Novartis Pharma AG, Ontario Ministry of Economic Development and Innovation, Pfizer, São Paulo Research Foundation-FAPESP, Takeda, and Wellcome Trust [106169/ZZ14/Z]. C.D.O.C. thanks the University of Huddersfield for support.

References

1. Savitsky P, Bray J, Cooper CD et al (2010) High-throughput production of human proteins for crystallization: the SGC experience. J Struct Biol 172:3–13
2. Mesa P, Deniaud A, Montoya G et al (2013) Directly from the source: endogenous preparations of molecular machines. Curr Opin Struct Biol 23:319–325
3. Makrides SC (1996) Strategies for achieving high-level expression of genes in *Escherichia coli*. Microbiol Rev 60:512–538
4. Terpe K (2006) Overview of bacterial expression systems for heterologous protein production: from molecular and biochemical fundamentals to commercial systems. Appl Microbiol Biotechnol 72:211–222
5. Dale GE, Oefner C, D'Arcy A (2003) The protein as a variable in protein crystallization. J Struct Biol 142:88–97
6. Sagemark J, Kraulis P, Weigelt J (2010) A software tool to accelerate design of protein constructs for recombinant expression. Protein Expr Purif 72:175–178
7. Graslund S, Sagemark J, Berglund H et al (2008) The use of systematic N- and C-terminal deletions to promote production and structural studies of recombinant proteins. Protein Expr Purif 58:210–221
8. Fernandez FJ, Vega MC (2013) Technologies to keep an eye on: alternative hosts for protein production in structural biology. Curr Opin Struct Biol 23:365–373
9. Zweers JC, Barak I, Becher D et al (2008) Towards the development of Bacillus subtilis as a cell factory for membrane proteins and protein complexes. Microb Cell Fact 7:10
10. Morello E, Bermudez-Humaran LG, Llull D et al (2008) *Lactococcus lactis*, an efficient cell factory for recombinant protein production and secretion. J Mol Microbiol Biotechnol 14:48–58
11. Mahajan P, Strain-Damerell C, Gileadi O et al (2014) Medium-throughput production of recombinant human proteins: protein production in insect cells. Methods Mol Biol 1091: 95–121
12. Fernandez-Robledo JA, Vasta GR (2010) Production of recombinant proteins from protozoan parasites. Trends Parasitol 26: 244–254
13. Esposito D, Chatterjee DK (2006) Enhancement of soluble protein expression through the use of fusion tags. Curr Opin Biotechnol 17:353–358
14. Hammarstrom M, Hellgren N, van Den Berg S et al (2002) Rapid screening for improved solubility of small human proteins produced as fusion proteins in *Escherichia coli*. Protein Sci 11:313–321
15. Ingolfsson H, Yona G (2008) Protein domain prediction. Methods Mol Biol 426:117–143
16. Gopal GJ, Kumar A (2013) Strategies for the production of recombinant protein in *Escherichia coli*. Protein J 32:419–425
17. Derewenda ZS (2010) Application of protein engineering to enhance crystallizability and improve crystal properties. Acta Crystallogr D Biol Crystallogr 66:604–615
18. Gileadi O, Burgess-Brown NA, Colebrook SM et al (2008) High throughput production of recombinant human proteins for crystallography. Methods Mol Biol 426:221–246
19. Mooij WT, Mitsiki E, Perrakis A (2009) ProteinCCD: enabling the design of protein truncation constructs for expression and crystallization experiments. Nucleic Acids Res 37:W402–W405
20. IUPAC-IUB Commission on Biochemical Nomenclature (1969) A one-letter notation for amino acid sequences. Tentative rules. Biochem J 113:1–4
21. Lipman DJ, Pearson WR (1985) Rapid and sensitive protein similarity searches. Science 227:1435–1441
22. Keates T, Cooper CD, Savitsky P et al (2012) Expressing the human proteome for affinity proteomics: optimising expression of soluble protein domains and in vivo biotinylation. N Biotechnol 29:515–525
23. Altschul SF, Gish W, Miller W et al (1990) Basic local alignment search tool. J Mol Biol 215:403–410

24. Marchler-Bauer A, Derbyshire MK, Gonzales NR et al (2015) CDD: NCBI's conserved domain database. Nucleic Acids Res 43: D222–D226
25. Schultz J, Milpetz F, Bork P et al (1998) SMART, a simple modular architecture research tool: identification of signaling domains. Proc Natl Acad Sci U S A 95: 5857–5864
26. Finn RD, Coggill P, Eberhardt RY et al (2016) The Pfam protein families database: towards a more sustainable future. Nucleic Acids Res 44:D279–D285
27. Jones DT (1999) Protein secondary structure prediction based on position-specific scoring matrices. J Mol Biol 292:195–202
28. Rose PW, Prlic A, Bi C et al (2015) The RCSB Protein Data Bank: views of structural biology for basic and applied research and education. Nucleic Acids Res 43:D345–D356
29. Lobley A, Sadowski MI, Jones DT (2009) pGenTHREADER and pDomTHREADER: new methods for improved protein fold recognition and superfamily discrimination. Bioinformatics 25:1761–1767
30. Jones DT (1999) GenTHREADER: an efficient and reliable protein fold recognition method for genomic sequences. J Mol Biol 287:797–815
31. Buchan DW, Minneci F, Nugent TC et al (2013) Scalable web services for the PSIPRED protein analysis workbench. Nucleic Acids Res 41:W349–W357
32. McGuffin LJ, Jones DT (2003) Improvement of the GenTHREADER method for genomic fold recognition. Bioinformatics 19:874–881
33. Murzin AG, Brenner SE, Hubbard T et al (1995) SCOP: a structural classification of proteins database for the investigation of sequences and structures. J Mol Biol 247: 536–540
34. Sillitoe I, Lewis TE, Cuff A et al (2015) CATH: comprehensive structural and functional annotations for genome sequences. Nucleic Acids Res 43:D376–D381
35. Laskowski RA, Hutchinson EG, Michie AD et al (1997) PDBsum: a Web-based database of summaries and analyses of all PDB structures. Trends Biochem Sci 22:488–490
36. Dosztanyi Z, Tompa P (2008) Prediction of protein disorder. Methods Mol Biol 426: 103–115
37. Prilusky J, Felder CE, Zeev-Ben-Mordehai T et al (2005) FoldIndex: a simple tool to predict whether a given protein sequence is intrinsically unfolded. Bioinformatics 21: 3435–3438
38. Linding R, Russell RB, Neduva V et al (2003) GlobPlot: exploring protein sequences for globularity and disorder. Nucleic Acids Res 31:3701–3708
39. Newman JA, Cooper CD, Aitkenhead H et al (2015) Structure of the helicase domain of DNA Polymerase theta reveals a possible role in the microhomology-mediated end-joining pathway. Structure 23:2319–2330
40. Pettersen EF, Goddard TD, Huang CC et al (2004) UCSF Chimera—a visualization system for exploratory research and analysis. J Comput Chem 25:1605–1612
41. Gao X, Bain K, Bonanno JB et al (2005) High-throughput limited proteolysis/mass spectrometry for protein domain elucidation. J Struct Funct Genomics 6:129–134
42. Hart DJ, Tarendeau F (2006) Combinatorial library approaches for improving soluble protein expression in *Escherichia coli*. Acta Crystallogr D Biol Crystallogr 62:19–26
43. Petersen TN, Brunak S, von Heijne G et al (2011) SignalP 4.0: discriminating signal peptides from transmembrane regions. Nat Methods 8:785–786
44. Bivona L, Zou Z, Stutzman N et al (2010) Influence of the second amino acid on recombinant protein expression. Protein Expr Purif 74:248–256

Chapter 3

Harnessing the Profinity eXact™ System for Expression and Purification of Heterologous Proteins in *E. coli*

Yoav Peleg, Vadivel Prabahar, Dominika Bednarczyk, and Tamar Unger

Abstract

Highly purified recombinant proteins in large quantities are valuable material for biochemical and structural studies. To achieve this goal, versatile tools were developed to increase the expression of the recombinant proteins and to facilitate the purification process. Fusion tags are commonly used for enhancing expression and solubility and some can be used in the purification process. However, these tags may need to be removed by treatment with specific proteases in order to obtain the tag-free protein. The Profinity eXact™ system provides an alternative system for a fusion tag, enhancing expression and purification in one-step. Here we describe a set of new vectors in which the Profinity eXact™ tag, in addition to a 6× His-tag, with or without additional expression-enhancing sequences, could be used in the Profinity eXact™ system. We show that the solubility enhancing tags (Trx, GST, GB1) increase the yield of the purified tested protein compared to the vector containing only a His-tag upstream of the Profinity eXact™ fusion tag.

Key words Profinity eXact, Protein expression, Protein purification, Profinity vectors

1 Introduction

Affinity tags are used on a routine basis to enhance protein expression and facilitate purification of recombinant proteins. The polyhistidine (His) tag is the most commonly used, primarily due to its simplicity of handling and low costs. Other popular affinity tags include glutathione S-transferase (GST) and the maltose binding protein (MBP), which unlike the short His-tag, add a large (~24 and 40 KDa, respectively) and unrelated protein moiety to the fusion protein. Regardless of the tag used for purification, for many applications, the tag must be removed in order to obtain the native protein. Removal of the tag is done primarily by digestion with proteases, such as tobacco etch virus (TEV), Factor Xa, enterokinase, thrombin, or Sumo [1–3]. An additional purification step is needed to separate the protease and non-cleaved protein from the mature cleaved product.

Nicola A. Burgess-Brown (ed.), *Heterologous Gene Expression in E. coli: Methods and Protocols*, Methods in Molecular Biology, vol. 1586, DOI 10.1007/978-1-4939-6887-9_3, © Springer Science+Business Media LLC 2017

The Profinity eXact™ protein expression and purification system (Bio-Rad) offers an alternative approach, which combines both affinity purification and subsequent on-column cleavage of the tag, thus releasing a tag-free protein. The system is based on the immobilized Subtilisin engineered mutant (S189) protease, which recognizes and binds the engineered affinity tag prodomain of subtilisin fused to the N-terminus of the target protein [4]. Thus the Subtilisin is used as both the affinity ligand and the processing protease. The process includes: I. Cloning the engineered 8.2 kDa subtilisin prodomain at the N-terminus of the target of interest, and expression in bacteria. II. Binding of the fusion protein to the immobilized subtilisin column. III. Washing to remove free contaminants from the column. IV. Cleavage on the column, precisely at the C-terminus of a nine amino acid sequence (EEDKLFKAL) corresponding to the Subtilisin prodomain, in the presence of fluoride anions to release the target protein. The resulting product, produced in a single step, is a purified native protein that lacks any amino acid residues from the Profinity eXact™ tag.

Two commercial vectors, pPAL7 and pPAL8 (Bio-Rad), are available for expression of this tag in *E. coli*. These two vectors differ only in the codon usage of the tag, whereby the codon sequence in pPAL8 is optimized for *E. coli* expression. Both vectors utilize the T7 lac promoter for expression in *E. coli* cells. Any other bacterial or eukaryotic expression vector can be engineered to contain the Profinity eXact™ tag to obtain high expression levels and a simple purification procedure. The tag sequence can be modified, and the codon usage can be further optimized for a given expression system. In certain cases, a small spacer (e.g., Thr–Ser) at the N-terminus of the target protein may enhance the efficiency of tag cleavage. The Profinity eXact™ tag system was used successfully for expression and purification of target proteins in *E. coli*, resulting in production of homogenous proteins for functional and structural studies [5–9]. This one-step chromatographic purification system provides an easy and rapid process to obtain recombinant native target proteins, avoiding the multiple consecutive chromatographic purification steps usually required with other tags.

Here, we describe in detail the construction of novel expression vectors compatible with the Profinity eXact™ system, containing additional fusion-tags and providing enhanced solubility and expression, and describe the expression and purification process. The Profinity eXact™ system enables efficient binding of the fusion protein to the resin followed by on-column cleavage in one step. However, depending on the target protein sequence and the expression levels, optimization of the binding and cleavage conditions are sometimes necessary. Partial cleavage could occur when non-ideal N-terminal amino acids of the target protein exist adjacent to the cleavage recognition sequence. Following the on-column cleavage, protein contaminants in addition to the native protein may appear. In such a case, additional purification steps would be required.

2 Materials

2.1 Bacterial Strains and DNA Cloning

For all procedures involving DNA cloning and plasmid preparation, the DH5α strain of *E. coli* (Agilent Technology) is used. For protein expression, *E. coli* BL21(DE3) (Novagen/EMD Millipore Chemicals) is employed. All materials required for the Restriction Free (RF) and Transfer-PCR (TPCR) cloning procedures, including preparation of competent cells, have been described in detail previously [10–13] (*see* **Note 1**).

2.2 Profinity eXact™ Expression Vectors

In initial experiments conducted at the Israel Structural Proteomics Center (ISPC) the vectors provided by Bio-Rad, pPAL7 and the codon-optimized pPAL8 were used. These T7-based vectors occasionally resulted in low protein expression for some of the proteins tested. We therefore constructed a series of expression vectors containing the Profinity eXact™ tag in addition to a 6× His-tag, with or without additional expression-enhancing sequences. The Profinity eXact™ tag was introduced into the previously described vectors [14] harboring glutathione S-transferase (GST), the β1-domain of the streptococcal protein G (GB1), and thioredoxin (Trx). The newly established vectors, pET28-Profinity, pETTrx-Profinity, pETGST-Profinity, and pETGB1-Profinity are schematically represented in Fig. 1. Kanamycin was used as a selection marker for all the new expression vectors constructed. Maltose binding protein (MBP) fusion was also established, but testing of the MBP vector is not described in the current study.

2.3 Protein Expression

1. Expression vectors (listed in Fig. 1).
2. High efficiency *E. coli* BL21(DE3) competent cells for heat shock or electroporation-mediated transformation (*see* **Note 2**).

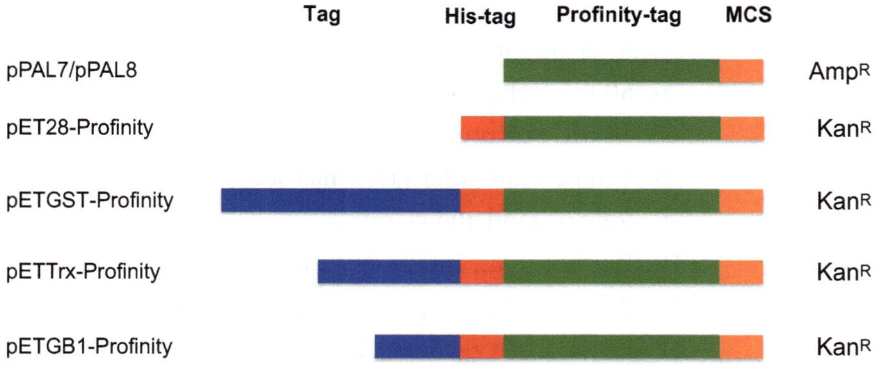

Fig. 1 Schematic representation of the Profinity eXact™ vectors. The Bio-Rad vectors pPAL7 and pPAL8 harbor only an N-terminal Profinity eXact**™** tag (*green*) followed by a multiple cloning site (MCS, *orange*). The other vectors were constructed in-house and are based on the pET vector backbone (Novagen) to contain 6× His-tag (marked in *red*) with or without an additional expression enhancing tag (*blue*). The components of the expression cassette are not drawn to scale. The MCS and the antibiotic resistance in the Profinity-pET based vectors differ from the ones in pPAL7 and pPAL8 (*see* **Note 4**)

3. Luria–Bertani (LB) liquid medium: Dissolve 20 g of LB broth (Lennox; contains 10 g of tryptone, 5 g of sodium chloride, and 5 g of yeast extract) in 1 L of ultrapure water. Adjust pH to 7.0 using 1 M NaOH and sterilize by autoclaving.
4. 30 mg/mL kanamycin stock solution (1000×): Dissolve 300 mg of kanamycin monosulfate powder in 10 mL of ultrapure water. Filter through a 0.2 μm syringe filter and store aliquots of 500 μL–1 mL at −20 °C.
5. 0.2 M IPTG (isopropyl-thio-β-D-galactopyranoside) stock solution.
6. 250 mL flasks.
7. Temperature-controlled (15–37 °C) shaker.
8. 50 mL polypropylene centrifuge tubes.
9. 14 mL polypropylene tubes.
10. Benchtop centrifuge.
11. Spectrophotometer.
12. Spectrophotometer cuvettes.

2.4 Cell Extraction, Protein Purification and Analysis

1. Cell-lysis buffer for sonication. Bacteria are lysed in 0.1 M sodium phosphate buffer pH 7.2 with 1 μL/mL protease inhibitor cocktail (Set IV, EMD Chemicals Inc.) (*see* **Note 3**).
2. Phosphate buffer: Prepared by mixing 360 mL of 0.2 M sodium phosphate dibasic stock solution (Na_2HPO_4) and 140 mL of 0.2 M sodium phosphate, monobasic, monohydrate stock solution ($NaH_2PO_4.H_2O$) in a total volume of 1 L of ultrapure water. The buffer was passed through 0.22 μm filter and stored at 4 °C.
3. Sonicator equipped with a micro-tip for 1.5 mL tubes.
4. 50 mL centrifuge tubes.
5. RC6 Sorvall floor centrifuge.
6. 50 mL polypropylene centrifuge tubes.
7. 14 mL polypropylene tubes.
8. Temperature-adjustable microcentrifuge and 1.5 mL tubes.
9. Retort stand and clamps.
10. Rotator (e.g., Intelli-mixer or equivalent).
11. Profinity eXact™ purification resin (Bio-Rad).
12. Econo-Pac® chromatography columns, 1.5 × 12 cm polypropylene column.
13. Column wash and storage buffer: 0.1 M sodium phosphate buffer, pH 7.2.
14. Elution buffer: 0.1 M sodium phosphate buffer pH 7.2, containing 0.1 M Sodium fluoride. To 100 mL of the 0.1 M sodium

phosphate buffer (pH 7.2) described above, add 420 mg of sodium fluoride, mix well, and store at 4 °C.

15. Column stripping buffer: 0.1 M phosphoric acid. The buffer was prepared by diluting 6.8 mL of concentrated phosphoric acid (H_3PO_4, 14.6 M) to a total volume of 1 L with ultrapure water. The buffer is passed 0.22 μm filter and stored at 4 °C.
16. Protein gel electrophoresis system (e.g., Bolt Mini Gel Tank, Novex—Life Technologies or similar).
17. Protein gels (Bolt 4–12% Bis-Tris Plus, Novex—Life Technologies, or similar).
18. Protein running buffer—MES SDS running buffer (× 20 concentrated, Novex—Life Technologies, or similar).
19. InstantBlue™, Coomassie based staining solution for protein gel (Expedeon or equivalent).
20. Gel-doc 2000 visualization system (Bio-Rad Laboratories or similar).
21. Ultrapure water for buffer preparation.
22. Protein sample buffer (SB), 4× concentrated (SB × 4); To make 10 mL of a SB × 4 stock solution: Mix 4.8 mL of 0.5 M Tris–HCl, pH 6.8, 0.8 g of SDS, 4.0 mL of glycerol, 0.4 mL of 14.7 M β-mercaptoethanol and 8 mg of bromophenol blue. Store in aliquots at −20 °C.
23. PageRuler™ Prestained Protein Ladder (Thermo Fisher Scientific).

2.5 Column Regeneration

1. Stripping buffer: 0.1 M phosphoric acid buffer.
2. Storage and wash buffer: 0.1 M sodium phosphate buffer pH 7.2 (0.02% (w/v) sodium azide is added only for long-term storage).

3 Methods

3.1 Establishment of Expression Vectors

Cloning of any gene of interest into the newly established Profinity eXact™ expression vectors can be performed by the Restriction Free (RF) or Transfer-PCR (TPCR) cloning techniques (*see* **Note 1**). Alternatively, cloning can be performed using restriction enzymes at the multiple cloning sites (MCS) (*see* **Note 4**), downstream to the Profinity eXact™ cassette. Cloning of the water soluble chlorophyll protein (WSCP) gene into the different Profinity eXact™ expression vectors, pET28-Profinity, pETTrx-Profinity, pETGST-Profinity, and pETGB1-Profinity, resulted in a precise and seamless integration without introduction of unnecessary sequences (*see* **Note 5**). Nevertheless, integrity of the expression cassettes must be verified by DNA sequencing before proceeding to protein expression.

3.2 Protein Expression

Optimization of protein expression is achieved by changing multiple parameters, including codon optimization, alteration of the induction temperature, expression strain, and the type of solubility tag. Therefore, when a new protein with the Profinity eXact™ tag is tested for expression, an initial screen should be performed using the vectors established in this study. Fig. 2 illustrates a comparative analysis of expression and purification of the WSCP protein using several of the vectors listed in Fig. 1. The solubility enhancing tags (Trx, GST, GB1) increase the yield of the WSCP compared to the pET28-Profinity vector containing only a His-tag (Fig. 2).

1. Transform expression vector into competent BL21(DE3) *E. coli* cells (*see* **Note 2**). Following the recovery stage, cells are transferred into a 14 mL tube containing 4 mL of LB plus 30 μg/mL kanamycin for all newly established vectors listed in Fig. 1. Cells are grown overnight for 14–16 h at 37 °C, and are used directly for the expression experiment.

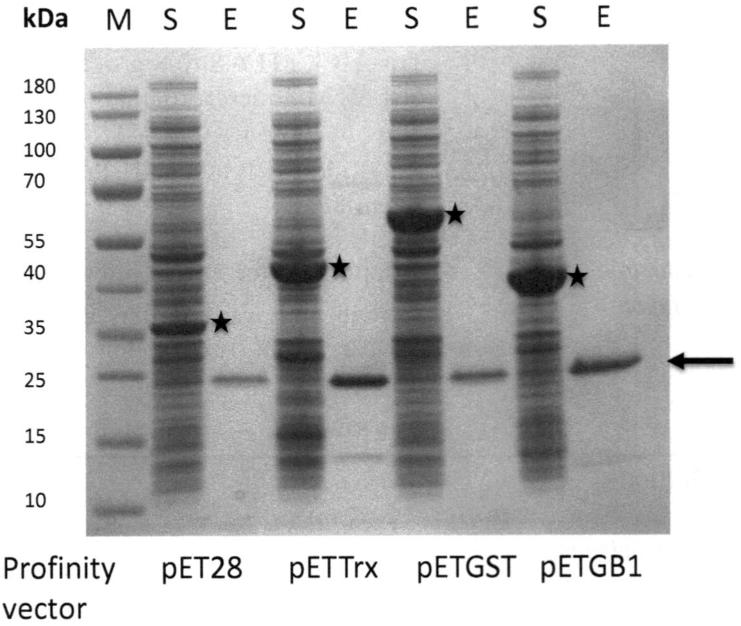

Fig. 2 Comparison of protein expression using different solubility tags. The Water Soluble Chlorophyll Protein gene (WSCP, amino acids 12–190, NCBI XP_013613804.1) was cloned into each of the vectors listed in Fig. 1 (*see* **Note 5**). Expression of proteins was performed at 37 °C for 3 h. Cell pellets were processed in parallel, as described in the text. Analysis was performed using Bolt 4–12% Bis-Tris plus gel (Invitrogen). S- Soluble fraction, following cell lysis. E- Elution fraction following cleavage from the Profinity eXact™ tag. *Arrow* indicates position of the WSCP following cleavage. Asterisk indicates position of the full-length fusion protein in each of the soluble fraction

2. The following morning, dilute the culture 1:100, into 250 mL flasks containing 100 mL of fresh LB medium plus 30 μg/mL kanamycin (*see* **Note 6**).
3. Incubate cultures at 37 °C with shaking until OD_{600} reaches 0.6–0.8.
4. Induce protein expression with 200 μM IPTG (1:1000 dilution of 0.2 M IPTG stock solution). For each clone, incubate one flask at 37 °C for 3–4 h (*see* **Note 7**).
5. Harvest cells by centrifugation at 4 °C for 15 min at 12,000 × *g*.
6. Store cell pellet at −20 °C, or proceed immediately to protein extraction and purification.

3.3 Protein Extraction, Purification, and Analysis

The buffer constituents used for protein purification using the Profinity eXact™ system should be carefully selected. While certain buffers are recommended for binding and cleavage (*see* Subheading 2.4), other buffers, and commonly used salts such as NaCl or KCl, must be avoided. Chloride ions act as slower cleavage-triggering ions. In addition, the time and temperature of the reaction should be monitored to ensure complete cleavage. Addition of additives such as 0.5 M L-Arginine was shown to enhance the binding of the protein to the resin and to support efficient cleavage [8]. For some proteins, a small spacer (e.g., Thr–Ser) at the N-terminus of the target protein (+1 and +2 positions) may enhance the cleavage efficiency of the tag. Proline must be avoided at the +1 position, since its presence will inhibit cleavage. Detailed information on reagents that are compatible with the system, and suitable amino acids at the +1 and +2 positions are described in the Profinity eXact™ manual (*see* **Note 8**). The amount of Profinity eXact™ resin used should be adjusted based on the predicted expression levels, determined in prior small-scale experiments. The amount of the fusion protein expression is expected to vary, depending on properties of the fusion partner (*see* **Note 9**). The Profinity eXact™ resin may be regenerated multiple times for repeated use. We have used the same resin, following regeneration, for more than 50 different proteins or protein variants, without apparent loss of binding capacity.

3.3.1 Preparation of Chromatography Resin

1. Mix the Profinity eXact™ affinity resin and transfer 2 mL of resin suspension (equivalent to 1 mL of settled beads) to the Bio-Rad Econo-Pac column. When packing the column, inclusion of any air bubbles should be avoided.
2. Equilibrate the column with 15 column volumes (CV) of wash buffer.

3.3.2 Cell Extraction by Sonication

1. Resuspend each cell pellet in 5 mL of sonication lysis buffer (*see* Subheading 2.4, **item 1**) and transfer to a 50 mL polypropylene tube.

2. Disrupt the cells on ice, by sonication using a micro-tip. Use 20% amplitude with four intervals of 30 sec ON and 15 s OFF. If the bacterial suspension is not clear, repeat the process (*see* **Note 10**).
3. Remove cell debris by centrifugation at 4 °C for 40 min at 12,000 × *g*. Transfer the clear supernatant into a new 15 mL tube.

3.3.3 Protein Purification and Analysis

1. Load the sample slowly into the equilibrated column; hold by clamp on a retort stand. Load without disturbing the resin. Allow the lysate to drain by gravity.
2. Add 10–15 column volumes (CV) of cold wash buffer and allow the buffer to flow through.
3. After washing, leave a small volume (~200–400 μL) of buffer on top of the resin, and tightly close the bottom-tip of the column with a cap. Add 2–3 CV of cold elution buffer and tightly close the top of the column with end-cap.
4. Gently shake the column so that the resin completely mixes with the buffer.
5. Keep the column at 4 °C, in a cold cabinet. Column can be slightly tilted during the incubation.
6. Collect the eluate containing the target protein after incubation for 3–24 h.
7. The eluted samples are analyzed in Bolt protein gels using 1× MES SDS running buffer alongside the pre-stained protein ladder. The voltage is set at 165 and electrophoresis allowed to run for 35 min.
8. The gel is stained using InstantBlue solution and de-stained using tap water.

3.3.4 Resin Regeneration

1. Following the protein cleavage step, wash the resin with 10 CV of wash buffer.
2. Optional step: Wash the column with 3–5 CV of 0.1 M NaOH (*see* **Note 11**) and then wash the column with 10 CV wash buffer.
3. Strip the column with 10 CV of stripping buffer.
4. Wash the column with 10 CV of wash buffer and store the column at 4 °C with wash buffer including 0.02% (w/v) sodium azide.

4 Notes

1. Restriction Free (RF) and Transfer-PCR (TPCR) cloning techniques are used on a routine basis at the ISPC. RF and TPCR cloning are highly efficient and robust. In addition, using these

approaches, cloning can be performed into any vector of choice and at any position, avoiding the need for restriction enzymes. Primer design and reaction conditions are described in detail in our previous publications [10–13]. In brief, primers (forward and reverse) for the RF or TPCR reactions include at the 5′-end a vector specific sequence (25–30 bases) complementary to the site of integration into the recipient vector, and at the 3′-end a sequence complementary to the gene of interest used for amplification of the gene (Tm 60–70 °C for the gene-specific sequence). Primers up to 60 bases are ordered with only basic desalting purification. Longer primers are purified either by HPLC or SDS-PAGE. An example of primer design for the RF or TPCR reactions is discussed in **Note 5.** The RF cloning is a two-stage procedure, in which, in the first stage, a set of mega-primers is generated and purified. In the second stage, the mega-primers are integrated into the destination vector. The generation of the mega-primer is performed using 20 ng of the donor vector and 0.5 μM of the forward and the reverse primers. Following purification of the mega-primer, a typical RF reaction includes 100 ng of mega-primer and 20 ng of destination vector. The TPCR is a single tube reaction in which all reaction components are included in the same tube. On a routine basis for the TPCR reaction, 20 μM of the forward and reverse primers are used in addition to 10 ng of the donor and destination vectors. The same amplification conditions are used for both RF and TPCR reactions: 95 °C for 30 s followed by 30 cycles of 95 °C for 30 s, 60 °C for 1 min, and 72 °C for 5 min. The reaction is continued with a single cycle at 72 °C for 7 min.

2. We found that preparation of competent cells according to a procedure described elsewhere [15] results in highly efficiency transformation. In this procedure, cells are grown at 18 °C prior to harvesting and preparation of the competent cells. Alternatively, BL21(DE3) cells can be transformed by electroporation. For preparation of cells for electroporation, consult the MicroPulser™ Applications Guide from Bio-Rad (http://www.biorad.com/webroot/web/pdf/lsr/literature/4006174B.pdf).

3. For details on buffers and reagents compatible with the Profinity eXact™ resin, consult the Profinity eXact™ instruction manual (http://www.bio-rad.com/webroot/web/pdf/lsr/literature/10011260.pdf). The protease inhibitor cocktail can be omitted from the lysis buffer if the protein is stable.

4. When the new Profinity eXact™ expression vectors (listed in Fig. 1) were established, we did not take into account the use of the unique MCS for cloning. The RF or TPCR methods, which we routinely use for cloning (*see* **Note 1**) do not rely on

the presence of restriction endonuclease sites. However, if one wishes to use restriction enzymes for cloning, combinations of *Hind*III, which is part of the Profinity eXact tag in pPAL8 (http://www.biorad.com/webroot/web/pdf/lsr/literature/Bulletin_6045.pdf), and *Not*I or *Xho*I can be used. These combinations can be used for all the newly established vectors, except for pETGB1-Profinity.

5. For cloning of the WSCP gene into the expression vectors listed in Fig. 1, we designed a set of primers: Forward, TPWSCP ProfinityF (5′-GTCGAAGAGGACAAGCTCTTCAAAGCTTTG*AGAGAACAGGTGAAGGACTCC*) and reverse, TPWSCPProfinityR (5′-GTGGTGGTGGTGGTGCTCG AGT GCGGCCGC**TTA***AGTAGCATCATCATCAACCTTC*). The underlined letters represent the vector-specific sequences, which determine the site of integration and can be added to any specific gene sequence for cloning into the expression vectors listed in Fig. 1. The stop codon is marked in bold in the reverse primer. Italic letters represent WSCP specific sequences. Note that for all the Profinity eXact™ vectors described in this study, a single set of primers is used for establishment of the mega-primer [10–13].
6. A glycerol stock should be prepared from the selected culture (final glycerol concentration 20–25% (v/v)). The stock should be stored at −80 °C until used.
7. It is highly recommended to optimize protein expression and to perform a small-scale expression screen to determine the optimal expression conditions [14].
8. Based on Bio-Rad data presented in the Profinity eXact™ manual, we cloned a different protein (LvWSCP) into the pETTrx-Profinity expression vector, using two tandem threonine (Thr–Thr) sequences following the Profinity cleavage site [7]. The expression and subsequent purification of the LvWSCP was highly successful, with a yield of about 10 mg/L culture.
9. The binding capacity of the resin was reported to be >3 mg of tag-free protein (for maltose binding protein) per mL resin (http://www.biorad.com/webroot/web/pdf/lsr/literature/Bulletin_5655.pdf).
10. If the proteins need to remain active following extraction, make sure not to overheat the extracts; perform extraction on ice, and when using a micro-tip, do not exceed 40% of the maximal amplitude. When multiple samples are handled simultaneously, a multiple-tip probe is a good option (available from Sonics).
11. The use of 0.1 M NaOH during column regeneration ensures complete removal of contaminants from the resin. This step can be performed, as well, following the stripping stage. The resin should be washed immediately after the NaOH treatment and the stripping step.

Acknowledgments

This research was supported by the Dana and Yossie Holander Center for Structural Proteomics.

References

1. Terpe K (2003) Overview of tag protein fusion: from molecular and biochemical fundamentals to commercial systems. Appl Microbiol Biotechnol 60:523–522
2. Waugh DS (2011) An overview of enzymatic reagents for the removal of affinity tags. Protein Expr Purif 80:283–293
3. Butt TR, Edavettal SC, Hall JP et al (2005) SUMO fusion technology for difficult-to-express proteins. Protein Expr Purif 43:1–9
4. Ruan B, Fisher KE, Alexander PA et al (2004) Engineering subtilisin into fluoride-triggered processing protease useful for one-step protein purification. Biochemistry 43:14539–14546
5. Huang L, Mao X, Abdulaev NG et al (2012) One-step purification of a functional, constitutively activated form of visual arrestin. Protein Expr Purif 82:55–60
6. Imaizumi K, Nishikawa S, Tarui H et al (2013) High level expression and efficient one-step chromatographic purification of soluble human leukemia inhibitor factor (LIF) in *Escherichia coli*. Protein Expr Purif 90:20–26
7. Bednarczyk D, Takahashi S, Satoh H et al (2015) Assembly of water-soluble chlorophyll-binding proteins with native hydrophobic chlorophylls in water-in-oil emulsions. Biochim Biophys Acta 1847:307–313
8. Ueda T, Akuta T, Kikuchi-Ueda T et al (2016) Improving the soluble expression and purification of recombinant human stem cell factor (SCF) in endotoxin-free *Escherichia coli* by disulfides shuffling with persulfide. Protein Expr Purif 120:99–105
9. Bednarczyk D, Dym O, Prabahar V et al (2016) Fine tuning of chlorophyll spectra by protein-induced ring deformation. Angew Chem Int Ed Engl 55:6901–6905
10. Unger T, Jacobovitch Y, Dantes A et al (2010) Applications of the restriction Free (RF) cloning procedure for molecular manipulations and protein expression. J Struct Biol 172:34–44
11. Peleg Y, Unger T (2014) Application of the restriction-free (RF) cloning for multicomponents assembly. Methods Mol Biol 1116: 73–87
12. Erijman A, Dantes A, Bernheim R et al (2011) Transfer-PCR (TPCR): a highway for DNA cloning and protein engineering. J Struct Biol 175:171–177
13. Erijman A, Shifman JM, Peleg Y (2014) A single-tube assembly of DNA using the Transfer-PCR (TPCR) platform. Methods Mol Biol 1116:89–101
14. Peleg Y, Unger T (2012) Resolving bottlenecks for recombinant protein expression in *E. coli*. Methods Mol Biol 800:173–186
15. Inoue H, Nojima H, Okayama H (1990) High efficiency transformation of *Escherichia coli* with plasmids. Gene 196:23–28

Chapter 4

ESPRIT: A Method for Defining Soluble Expression Constructs in Poorly Understood Gene Sequences

Philippe J. Mas and Darren J. Hart

Abstract

Production of soluble, purifiable domains or multi-domain fragments of proteins is a prerequisite for structural biology and other applications. When target sequences are poorly annotated, or when there are few similar sequences available for alignments, identification of domains can be problematic. A method called expression of soluble proteins by random incremental truncation (ESPRIT) addresses this problem by high-throughput automated screening of tens of thousands of enzymatically truncated gene fragments. Rare soluble constructs are identified by experimental screening, and the boundaries revealed by DNA sequencing.

Key words Protein expression, Random library, Directed evolution

1 Introduction

Disciplines including structural biology, inhibitor screening, biophysical analyses, and vaccinology require tens of milligrams of pure mono-disperse protein. Many methods exist, but most commonly *Escherichia coli* is used as the expression host due to ease of genetic manipulation, cell growth, and processing of biomass. Labeling protocols required in structural biology (seleno-methionine, isotopes for NMR) are also greatly simplified in *E. coli*. In order to achieve stable expression and subsequent crystallization, multi-domain proteins are often truncated into smaller sub-constructs comprising single or multiple domains that may be better expressed, more stable, and less flexible. Probably, the most common approach to define well-behaving sub-constructs of a protein is to identify domains by alignment of similar sequences which reveals evolutionarily conserved regions; these may correspond to structural domains. In the absence of similar sequences, this approach is not possible and tools such as computational order and disorder predictors may provide clues to domain location [1]. Where full-length protein is available, perhaps via lower yielding systems (e.g., expression in

Nicola A. Burgess-Brown (ed.), *Heterologous Gene Expression in E. coli: Methods and Protocols*, Methods in Molecular Biology, vol. 1586, DOI 10.1007/978-1-4939-6887-9_4, © Springer Science+Business Media LLC 2017

mammalian cells or yeast strains), limited proteolysis and mass spectrometry can identify stable domains. In both approaches, hypotheses on domain locations are followed by PCR cloning of sub-constructs and expression testing. While these underpin the production of soluble proteins, they are frequently insufficient for success. Notably, some of the most interesting proteins may be those for which little information is known, few similar sequences exist, or for proteins with undefined partners or conditions necessary for stable expression of full-length material. Even when a domain location can be predicted, constructs may require substantial optimization of their boundaries before expression is acceptable; a long trial-and-error process.

We developed a method called expression of soluble proteins by random incremental truncation (ESPRIT) that, in many cases, is able to identify well-behaving, high-yielding, soluble, and purifiable constructs without the need to hypothesize the locations of domain boundaries [2]. It uses principles of directed evolution and synthetic biology whereby large random truncation libraries are synthesized that contain rare constructs with the desired properties (*see* Fig. 1) [3, 4]. These hits are identified in a second step of automated high-throughput soluble expression screening using instruments developed during the early phase of genome sequencing and analysis (colony pickers and arrayers). During the ESPRIT process, the target gene is pre-cloned into a pET9a-derived mini-vector (*see* Fig. 2) and a library of nested gene truncations generated using exonuclease III and mung bean nuclease of the linearized vector arm bearing the gene [5]. Vectors containing gene fragments of an interesting size range are isolated by agarose gel electrophoresis, circular plasmids recovered by re-ligation with T4 DNA ligase, and libraries of tens of thousands of potential expression clones obtained by *E. coli* transformation (*see* Fig. 3). The expression screening step of ESPRIT uses an arabinose-inducible BL21 strain of *E. coli* into which the plasmid library is transformed and titrated onto agar trays. In a first automated process, colony picking robots transfer clones into 384-well micro-titer plates (typically ~28,000 individual clones). Inocula from freshly grown cultures are then printed onto nitrocellulose membranes and protein expression induced with arabinose during colony growth. Colonies are lysed in situ on the nitrocellulose membranes that are then hybridized with fluorescent probes to identify those expressing putatively soluble constructs (*see* Fig. 4). Detection of in vivo biotinylation of a C-terminally fused biotin acceptor peptide [6], together with the presence of the N-terminal hexahistidine peptide, suggests the construct being expressed by that truncation mutant is both soluble and intact. Typically, 95 clones are isolated from the plates containing the main library and tested, along with a positive control, for soluble protein production resulting in the identification of

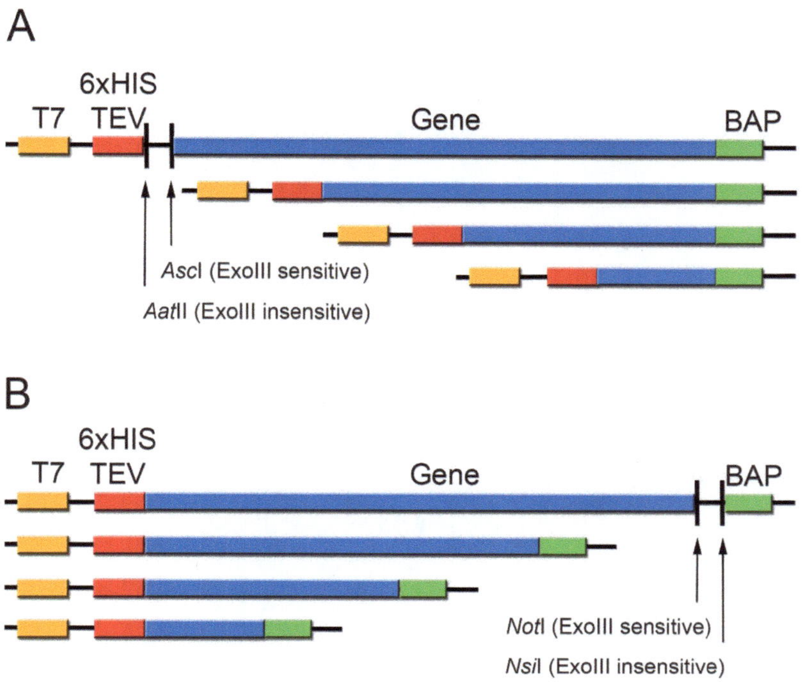

Fig. 1 Unidirectional ESPRIT truncation library strategies. (**a**) N-terminally truncated protein constructs are encoded by a 5′ DNA truncation library. The double digested plasmid leaves an exonuclease III resistant end (*Aat*II) and hydrolysable end (*Asc*I). After truncation steps and vector recircularization, the hexahistidine tag (*red*) is fused to each residue of the protein. (**b**) C-terminal truncations encoded by a 3′ DNA truncation library. The exonuclease III resistant end (*Nsi*I) and hydrolysable end (*Not*I) enable the unidirectional truncations, and vector recircularization results in fusion of the biotin acceptor peptide (*green*) to the protein. Cassettes are expressed under the control of the strong T7 promoter

clones that are characterized by DNA sequencing to identify their expression-compatible domain boundaries.

Further developments of the basic method that will not be detailed here include a co-expression format for binary complexes (CoESPRIT) [7], and a genetic selection to remove the out-of-frame constructs during library construction (ORF selector ESPRIT) [8]. Examples of successful structural determination following construct definition by ESPRIT include domains from influenza polymerase [9–11], HCMV terminase [12], *Helicobacter pylori* CagA [13], and *Bacillus subtilis* SpoIIE [14].

2 Materials

2.1 Library Construction

1. Enzymes: *Aat*II (20,000 units/mL), *Asc*I (10,000 units/mL), *Nsi*I (10,000 units/mL), *Not*I (10,000 units/mL), exonuclease III (100,000 units/mL), mung bean nuclease (10,000 units/

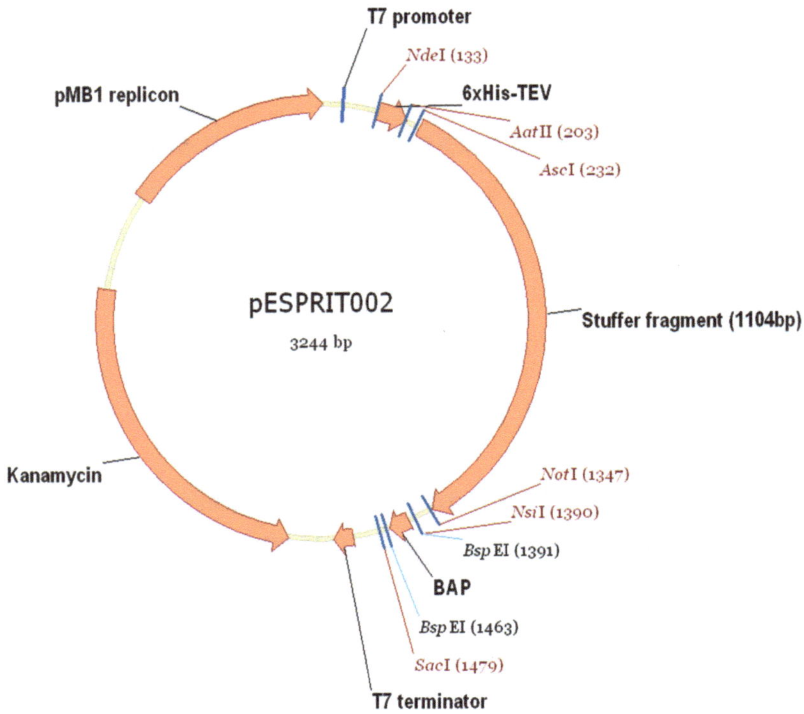

Fig. 2 The pESPRIT002 vector is a 2.1 kilobase pET9a derivative with unnecessary DNA removed to reduce its size. For 5′ deletion libraries, inserts are subcloned in-frame with the sequence encoding the biotin acceptor peptide using *AscI* and *NsiI*. For 3′ deletion libraries, inserts are in-frame with the hexahistidine TEV encoding sequence and subcloned using *AatII* and *NotI*

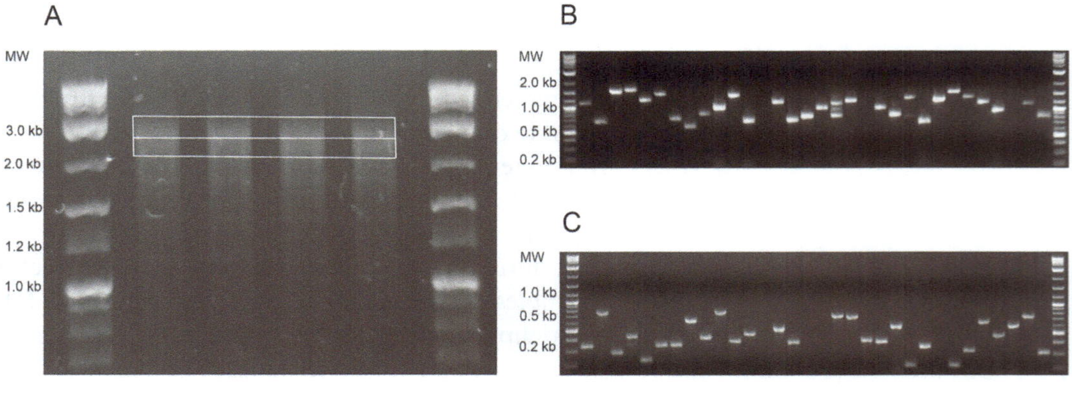

Fig. 3 (**a**) Exonuclease III and mung bean nuclease truncation products electrophoresed on a 0.5% (w/v) agarose TBE gel stained with ethidium bromide. *White lines* reveal the sub-libraries excised from the gel. Colony PCR measurements of insert sizes from plasmids recovered from (**b**) large insert and (**c**) small insert sub-libraries analyzed by 1% (w/v) agarose TBE gel

mL), Taq DNA polymerase (5000 units/mL) (New England Biolabs); T4 DNA ligase rapid ligation kit (5000 units/mL) (Roche Applied Science); *Pfu* DNA Polymerase (2500 units/mL) (Stratagene).

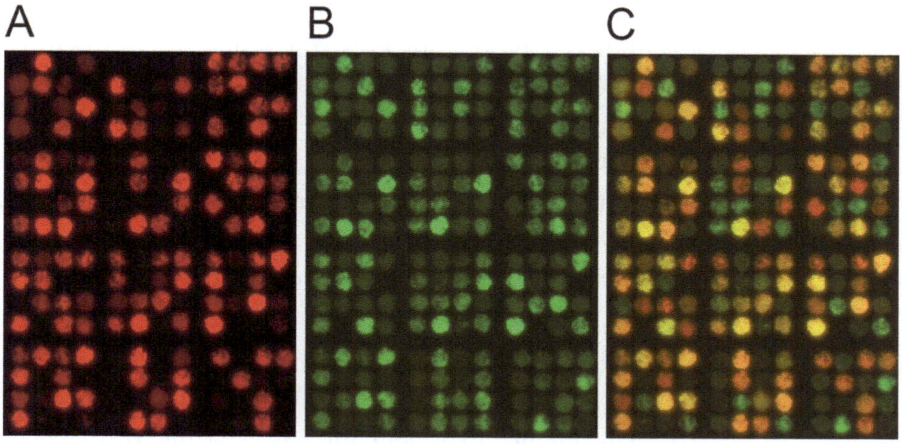

Fig. 4 Magnified image of colony array probed with fluorescent detection reagents against the N-terminal hexahistidine tag (**a**; *red*), C-terminal biotin acceptor peptide (**b**; *green*). Panel C is the merge of both individual detection channels where *yellow* corresponds to strong double tag signals

2. *E. coli* strains: One shot MACH1 chemically competent cells (Thermo Fisher Scientific), BL21 AI (Thermo Fisher Scientific) transformed with RIL plasmid (Agilent) and made electro-competent.
3. Oligos: T7For: 5′-GCGAA ATTAA TACGA CTCAC TATAG G-3′, T7Rev: 5′-GCTAG TTATT GCTCA GCGGT GGC-3′; NterFor1: 5′-GGTCG ATCGG CGCGC CTGA-[gene specific]-3′; NterRev1: 5′-GGTCG ATCAT GCATT-[gene specific]-3′; CterFor1: 5′-GGTCG ATCGA CGTCG A-[gene specific]-3′; CterRev15'-GGTCG ATCGC GGCCG CTCA-[gene specific]-3′. Restriction sites for cloning into pESPRIT002 are underlined. The gene-specific sequence should have a melting temperature of approximately 65 °C and respect the sense (for) and antisense (rev) orientation.
4. DNA purification: NucleoSpin gel and PCR clean-up kit (Macherey-Nagel), QIAprep mini-prep kit (Qiagen).
5. Buffers for high-quality, nick-free DNA purification: resuspension buffer (50 mM glucose, 25 mM Tris–HCl, and 10 mM EDTA, pH 8); lysis buffer (0.2 N NaOH, 1% (w/v) SDS); neutralization solution (prepared from 4 mL of 5 M K acetate, 766 μL of 100% acetic acid adjusted to a final volume of 6 mL).
6. Microbiological growth media can be bought as premixes from numerous suppliers. Examples are LB broth, terrific broth (TB), SOC medium.
7. 50 mg/mL Kanamycin stock solution: Prepared in H_2O and sterilized through a 0.2 μm filter.
8. 30 mg/mL Chloramphenicol stock solution: Prepared in ethanol.

9. 5 mM biotin: Prepared in H_2O. Add single drops of NaOH until the biotin dissolves, then sterilize through a 0.2 μm filter).
10. 20 μM arabinose: Prepared in H_2O. Sterilize through a 0.2 μm filter.
11. Plastic consumables for screening: Qtrays and 384-well plates (Molecular Devices).
12. Horizontal DNA electrophoresis system.
13. Glass beads (5 mm diameter) for spreading agar plates.
14. Colony picking robot with both picking and arraying functionality (e.g., Kbiosystems K3, Molecular Devices QPix).
15. Plate filling instrument for 384-well micro-titer plates (e.g., Genetix QFill2).
16. 10× HMFM: Prepare 31.5 g of K_2HPO_4, 9 g of KH_2PO_4, 4.5 g of Na_2Citrate, and 220 mL of glycerol in a final volume of 450 mL in distilled water and autoclaved. Add 50 mL of $MgSO_4$ (0.45 g $MgSO_4 \cdot 7H_2O$ in 50 mL H_2O). TB-HMFM medium is 9 parts TB plus 1 part 10× HMFM.
17. Fluorescence imager (e.g., GE Typhoon Trio imager).
18. Streptavidin Alexa Fluor 488 and rabbit anti-mouse Alexa Fluor 532 conjugates (Thermo Fisher Scientific). Anti-His antibody (GE Healthcare).
19. 10× PBS: Prepare 80 g of NaCl, 2 g of KCl, 14.4 g of Na_2HPO_4, and 2.4 g of KH_2PO_4 in 800 mL of sterile H_2O, adjust to pH 7.4 and increase volume to 1 L. PBS-T comprises 1 x PBS with 0.1% (v/v) Tween20.
20. Plate incubator compatible with a 384-well format (e.g., Digilab HiGro).
21. QReps 384-pin plastic plate replicators (Molecular Devices).
22. Supported nitrocellulose membranes (Amersham Protran 0.45 NC).
23. Solutions for preparation of colony blots: membrane denaturation solution (0.5 N NaOH and 1.5 M NaCl); membrane neutralization buffer (1 M Tris–HCl, 1.5 M NaCl, pH 7.5); 20 × SSC: Prepare 175.3 g of NaCl and 88.2 g of Na_3Citrate in 800 mL of water, adjust to pH 7.0 with HCl, then bring final volume to 1 L. Superblock (Thermo Fisher Scientific).
24. Software to fit grids over array images and extract pixel intensity data (e.g., Visual Grid, GPC Biotech).

2.2 Small-Scale Expression Screening

1. Microbiological growth media and antibiotics are described in the library screening section above.

2. 24-well blocks with 10 mL square wells and AirPore tape sheets (Qiagen).
3. Sphaeroplast preparation buffer: 20 mM Tris–HCl, 300 mM NaCl, 20% (w/v) sucrose, 1 mg/mL lysozyme, pH 8.0.
4. Sphaeroplast lysis buffer: 10 mM Tris–HCl and 0.5% (v/v) Brij58, pH 7.0.
5. Benzonase (Roche Applied Science).
6. Protease inhibitor cocktail EDTA-free (Roche Applied Science).
7. Ni^{2+} NTA agarose resin (Qiagen).
8. 96-well receiver plate (Macherey-Nagel).
9. Ni^{2+} NTA wash buffer: 50 mM $NaPO_4$, 300 mM NaCl, 5 mM imidazole, pH 7.0.
10. Ni^{2+} NTA elution buffer: 50 mM $NaPO_4$, 300 mM NaCl, 300 mM imidazole, pH 7.0.

3 Methods

3.1 Cloning of Target Insert into pESPRIT002

The plasmid pESPRIT002 plasmid is a reduced-size derivative of pET9a encoding an N-terminal hexahistidine tag and a C-terminal biotin acceptor peptide (*see* Fig. 2). For 3′ deletion libraries, the insert is subcloned using standard PCR, restriction digest and ligation procedures not detailed here. A key aspect is that the insert is subcloned with the end to be truncated adjacent to a pair of restriction sites that, because of their differing overhangs, permit unidirectional truncation with exonuclease [5] (*see* Fig. 1). Primer pairs for 5′ deletion are NterFor1 and NterRev1; for 3′ deletion primers CterFor1 and CterRev1 are used. The plasmid insert should be DNA sequenced and a cloning strain of *E. coli* transformed for subsequent midi-prep steps (e.g., One shot MACH1 cells).

3.2 Preparation of High-Quality, Nick-Free Plasmid DNA

Alkaline lysis, phenol chloroform extraction and isopropanol precipitation result in a glassy pellet of nucleic acids that is then processed further through a DNA mini-prep kit polish to remove residual salts and RNA (*see* **Note 1**).

1. Pellet cells from 100 mL of saturated overnight culture 13,000 × *g* for 10 min in 50 mL Falcon tubes. Resuspend in 4 mL of resuspension buffer.
2. Add 8 mL of lysis buffer and gently invert the tube for 5 min at 4 °C.
3. Add 6 mL of neutralization solution. Mix gently by inverting the tube and centrifuge lysates at 13,000 × *g* for 10 min.

Transfer the supernatant (about 17.5 mL) to a clean Falcon tube on ice.

4. Add 20 mL of phenol:chloroform:isoamyl alcohol (25:24:1). Invert the tube repeatedly to generate an emulsion for 5 min on a rotating wheel (set at 30). Centrifuge 3000 × *g* for 10 min at 4 °C. Transfer supernatant and repeat extraction a second time. Recover 10 mL of upper aqueous phase into a new Falcon tube.
5. Add 1 mL of 3 M Na acetate pH 5.2 and 2.5 volumes of isopropanol. Place tubes at −80 °C for 1–12 h. Centrifuge for 30 min at 13,000 × *g* at 4 °C, pour off supernatant gently, and add 50 mL of ice cold 70% (v/v) ethanol to the pellet without disturbing it. Centrifuge for 15 min at 8000 × *g* and 4 °C. Pour off ethanol and remove droplets by aspiration or pipetting. Air-dry pellet for 5–10 min.
6. To the pellet, add 1 mL of P1 buffer from the QIAprep kit and resuspend. All other buffers in this section are also present in the kit. Split into four Eppendorf tubes (250 μL each) and add 250 μL of P2 buffer and 350 μL of N3 buffer. Load into six mini-prep columns and proceed with both 500 μL PB and 750 μL PE washes as detailed in the kit instructions. Recentrifuge to dry column, then elute DNA from each column in 50 μL EB buffer into clean Eppendorf tubes. Pool the DNA in a single tube and determine concentration using standard methods. Only un-nicked super-coiled DNA should be visible by ethidium bromide stained agarose gel, with no RNA. Store at 4 °C and use within 1–2 days for the exonuclease steps (*see* **Note 2**).

3.3 Vector Linearization, Truncation, and End-Polishing

The plasmid pESPRIT002 (*see* Fig. 2) containing the gene is cut at the gene terminus to be truncated. The restriction enzyme used (*Asc*I or *Not*I) leaves a 5′ overhang (sticky end) which is an exonuclease III substrate. A second cut is made on the vector side of the insert (*Aat*II or *Nsi*I) adjacent to the hexahistidine or biotin acceptor peptide sequences respectively; the resulting 3′ overhangs are exonuclease III resistant. Consequently, when the exonuclease III time-course treatment is performed, only the gene side of the linearized plasmid is digested. Mung bean nuclease is then used to remove the single stranded DNA remaining and the ends of the plasmids are blunt-ended with *Pfu* polymerase. The protocol below assumes a 5′ truncation library is being generated; conditions for the 3′ library are in parentheses.

1. 10 μg of plasmid DNA is digested in a volume of 500 μL with 80 units of *Aat*II (*Nsi*I) for 2 h at 37 °C in NEB buffer 4. Digestion is confirmed by analyzing 50 ng of DNA on a 0.7% agarose gel alongside 50 ng of uncut vector; full loss of super-

coiling should be evident. Add 80 units of *Asc*I (*Nsi*I) to reaction and incubate for a further 2 h.

2. Clean up the digest using a NucleoSpin gel and PCR clean-up kit. Twice the reaction volume of NTI buffer is added to the DNA solution. The protocol is followed according to the kit instructions. Finally, DNA is eluted in 50 μL of NE buffer. Determine plasmid DNA concentration by standard methods (e.g., intensity on gel relative to standards, spectrophotometry). If necessary, DNA can be kept at 4 °C until next step, but do not freeze since this can lead to DNA nicking.
3. The exonuclease III truncation reaction is prepared in a PCR tube on ice, comprising 4 μg of linearized DNA (final concentration of 33.3 ng/μL), 1× NEB buffer 1. The salt concentration is adjusted according to the deletion rate (bp/min) required (*see* Table 1). Four hundred units of exonuclease III are added, resulting in a final reaction volume of 120 μL (*see* **Note 3**). Prepare the quenching tube comprising 250 μL of NT buffer from the NucleoSpin Gel and PCR clean-up kit and place on ice.

Table 1
Exonuclease III truncation rate is varied by adjusting the NaCl concentration supplementing the reaction mix. For a gene less than 2.4 kb, truncations are performed in 60 min; for 2.4–4.8 kb truncation reactions take 120 min

Additional NaCl/mM	Deletion rate (bp/min)
10	41
20	36
30	31
40	27
50	23
60	20
70	17
80	15
90	13
100	11
110	9
120	8
130	7
140	6
150	5

4. Transfer reaction to 22 °C (e.g., PCR machine heater block) to start truncation reaction. For 1 h, every 60 s, transfer 1/60th reaction volume (i.e., 2 μL) to the quenching tube on ice. For a 2 h truncation reaction on a longer gene, the aliquot is taken every 120 s.
5. Truncated DNA is purified by treating the quenched reaction with NucleoSpin gel and PCR cleanup according to the protocol of the kit with elution in 40 μL of NE buffer after 1 min of incubation (*see* **Note 4**).
6. The mung bean nuclease digest comprises the 40 μL of eluted DNA with 1× mung bean nuclease buffer, 5000 units of mung bean nuclease in a final volume of 50 μL; this is incubated at 30 °C for 30 min, then cleaned up with the NucleoSpin gel and PCR clean-up kit as in the step above.
7. The proofreading *Pfu* polymerase is used to blunt-end the linearized, truncated vector. A 50 μL reaction is prepared from the 40 μL of DNA eluted in the previous step, 1× *Pfu* buffer, dNTPs (final concentration 100 μM) and 2.5 units of *Pfu* polymerase; incubate at 72 °C for 20 min, then place on ice.

3.4 Gel Extraction, Ligation, and Transformation

The linearized pESPRIT002 plasmids containing truncated gene inserts can be size-selected by agarose gel electrophoresis, yielding libraries (or multiple sub-libraries) with inserts of defined size ranges. For example, a domain of 200 amino acids might be found in a sub-library with DNA inserts between 400 and 800 nucleotides in length. This procedure also removes small but highly soluble fragments (peptides) that may give unwanted signals later during screening.

1. Wash the gel tank and components with distilled water to remove possible contaminating plasmids. Prepare a 0.5% TBE agarose gel with ethidium bromide and precool tank and buffer to 4 °C in a cold room.
2. For accurate sizing, load the DNA across three lanes (approximately 12 mm) and load 20% of the normal quantity of DNA ladder so as to give sharper bands. Also load 20 ng of the pESPRIT002 vector minus insert in the lane on the far outside of the ladder (to indicate the size of the parent vector). Electrophorese at 80 V (8 V/cm) for 1 h in the cold room (*see* **Note 5**).
3. Under long wavelength UV trans-illumination (365 nm), excise the DNA from the agarose gel with a clean blade. Consider size fractionating this into two or three sub-libraries (e.g., plasmid plus insert size 0–1 kb, 1–2 kb, 2–3 kb). Extract the DNA from the gel using a NucleoSpin gel and PCR clean-up kit according to protocol of the kit and elute from the spin column in 20 μL of NE buffer.

4. Each preparation of linear plasmid DNA (a sub-library) is religated and *E. coli* transformed as follows: 8 μL of the eluted DNA from the above step is ligated with T4 DNA ligase using a Roche Rapid Ligation Kit according to the instructions provided. 2 μL of the ligation mix are used to transform highly competent *E. coli* cells by standard methods with recovery of transformed cells in 1 mL of LB or SOC medium (*see* **Note 6**). Spread 10 and 100 μL of the transformation recovery mix on LB agar Petri dishes with 50 μg/mL kanamycin and incubate plates overnight at 37 °C. Place remainder of transformation recovery mix at 4 °C for plating later.

3.5 Assessment and Harvesting Library

1. Using standard methods, either colony PCR using flanking T7For and T7Rev primers, or plasmid purification and digestion with *Nde*I and *Sac*I, insert sizes are determined from 24 to 48 clones chosen at random (*see* Fig. 3). Sizes are determined by agarose gel electrophoresis and can be plotted as a simple ordered histogram (insert sizes sorted small to large) to verify linearity of the truncation process (*see* **Note** 7). For a more detailed analysis, a scatter plot can be drawn of length in base pairs against sorted clone number—small to large. Box plots can be drawn to check if the data are normally distributed, a requirement to perform a parametric test. The theoretical mean size of fragments is compared with the experimentally determined lengths. A one-sample *t*-test set at 5% level of significance quantifies if these means are similar. The libraries are considered validated once box plots and *t*-test are satisfied.
2. The plasmid library is recovered by plating the remainder of the transformation mix stored at 4 °C on 22 × 22 cm square trays filled with 300 mL of LB agar (50 μg/mL kanamycin). Pre-wet 24 glass beads in sterile SOC medium to block their surfaces and add to the agar tray, followed by the transformation recovery mix at about 10,000 clones per plate (the dilutions plated on small Petri dishes previously will suggest the volume of mix to plate to obtain this number). Agitate and rock plates to distribute cell mix evenly across surface of agar, then leave plates partially open in a sterile environment to dry for 30 min. Close plates and incubate overnight at 37 °C or until small uniform colonies are visible.
3. Add 5 mL of sterile LB medium to plates and harvest the colonies with a spreader. Pipette into a Falcon tube and repeat a second time. Wash cells with 20 mL of PBS three times by resuspension and centrifugation. No lumps should be visible when the tube is held to the light. Make several 1.5 mL glycerol freezes of these cells, then process the remainder through a commercial midi-prep kit (or several mini-prep columns and pool) to extract plasmid. Aliquots of plasmid library can be frozen at −20 °C.

3.6 Robotic Steps: Colony Picking, Plate Incubation, Plate Replication, and Colony Arraying

Colony picking is necessary to isolate individual clones from the library and place them in wells of microtiter plates, thereby giving each clone a precise positional address (plate number, well number) that can be tracked through subsequent steps. We use a Kbiosystems K3 colony picker, but there are a number of alternative models available with the same functionalities. The exact programming of the instruments will be omitted in favor of a description of the processes necessary to screen libraries of colonies.

1. Transform 50 μL of electro-competent *E. coli* BL21 AI RIL with 5–15 ng of library plasmid. Recover for 1 h in 1 mL of SOC at 37 °C in a 2 mL Eppendorf tube at 250 rpm. Titrate fivefold dilutions on Petri dishes with both 50 μg/mL kanamycin and 30 μg/mL chloramphenicol to determine number of colony forming units (cfu) per mL. Place remaining transformation mix in a fridge.
2. Pour eight 22 × 22 cm Genetix plates using 300 mL of LB agar with 50 μg/mL kanamycin and 30 μg/mL chloramphenicol on a sterile flat surface (e.g., a microbiological hood). Allow plates to dry open in a microbiological hood for 20 min. Prepare 10 mL dilutions of transformation mix in SOC from 10^{-1} to 10^{-4} dilution. Add 24 glass beads, pre-wetted in SOC to block their surface and 0.8 mL of the transformation mix, aiming for 4–8000 colonies per plate. Distribute mix evenly on the agar, then dry open plates in a microbiological hood for 10 min. Close plates and incubate overnight at 37 °C.
3. Fill sufficient 384-well plates with 80 μL of TB-HMFM using a plate filler (e.g., Genetix QFill2) supplemented with 50 μg/mL kanamycin and 30 μg/mL chloramphenicol. Label plates and load into the colony picking robot. Pick colonies from LB agar colony trays according to the size of the experiment, for example 72 × 384-well plates are 27,648 colonies, requiring 5–10 agar trays. For the Kbiosystems K3 picker, this corresponds to approximately one full day of robot operation at a picking rate of 4000 colonies per hour (including manual plate handling steps).
4. Once all plates are picked, incubate them overnight at 37 °C in an incubator that is able to shake efficiently 384-well plates. We use a Digilab HiGro with a 9 mm rotational pitch for 96-well plates, but which functions efficiently for 384-well format also at 300 rpm. Once cultures have grown to saturation, plates can be frozen with lids at −80 °C, or kept for short durations (1–4 days) in a cold room.
5. Fresh replicates of the library generate better colony arrays (below) and also allow for the master copy to be kept in a frozen state, maintaining viability and limiting contamination. Therefore, the main library is replicated into fresh plates con-

taining LB with 50 μg/mL kanamycin and 30 μg/mL chloramphenicol using disposable 384-pin plastic replicators (*see* **Note 8**). These copies are grown overnight at 37 °C in the HiGro incubator resulting in fresh saturated cultures.

6. Prepare two 22 × 22 cm QTrays per membrane to be arrayed, typically four since a second backup membrane is preferred. Half of the plates contain 300 mL of LB agar with 50 μg/mL kanamycin and 30 μg/mL chloramphenicol, the other half with the same antibiotics but further supplemented with 50 μM biotin and 0.2% (w/v) arabinose (*see* **Note 9**). Cut a nitrocellulose membrane (e.g., Protran 0.45 NC) to the dimension of the QTray and lay it over the plate with antibiotics only, avoiding air bubbles.

7. To array the library onto the membranes, a gridding pattern should be selected compatible with the size of library. We typically aim for a threefold oversample of the truncation length sampled (*see* **Note 10**) and use a 4 × 4, or 5 × 5 arraying pattern with each clone present in duplicate (*see* Fig. 5). Array

A

Field 6

42	46	42	43
41	44	47	45
47	48	43	46
41	45	44	48

Field 4

26	30	26	27
25	28	31	29
31	32	27	30
25	29	28	32

Field 2

10	14	10	11
9	12	15	13
15	16	11	14
9	13	12	16

Field 5

34	38	34	35
33	36	39	37
39	40	35	38
33	37	36	40

Field 3

18	22	18	19
17	20	23	21
23	24	19	22
17	21	20	24

Field 1

2	6	2	3
1	4	7	5
7	8	3	6
1	5	4	8

B

Field 6

62	66	72	62	63
61	70	64	71	72
69	66	0	67	70
67	68	71	63	65
61	65	69	64	68

Field 4

38	42	48	38	39
37	46	40	47	48
45	42	0	43	46
43	44	47	39	41
37	41	45	40	44

Field 2

14	18	24	14	15
13	22	16	23	24
21	18	0	19	22
19	20	23	15	17
13	17	21	16	20

Field 5

50	54	60	50	51
49	58	52	59	60
57	54	0	55	58
55	56	59	51	53
49	53	57	52	56

Field 3

26	30	36	26	27
25	34	28	35	36
33	30	0	31	34
31	32	35	27	29
25	29	33	28	32

Field 1

2	6	12	2	3
1	10	4	11	12
9	6	0	7	10
7	8	11	3	5
1	5	9	4	8

Fig. 5 Patterns for a six field colony array arrayed onto a 22 × 22 cm nitrocellulose membrane. (**a**) The 4 × 4 pattern corresponds to forty-eight 384-well plates (18,432 clones); (**b**) the 5 × 5 pattern seventy-two plates 384-well plates (27,648 clones)

inocula from freshly replicated library plates onto nitrocellulose membranes on agar according to the protocol of the instrument used. This should be done in a semi-sterile environment (e.g., after pretreatment of robot deck with ultraviolet light) and using ethanol and heat treatment of the arraying pins between impressions.

8. Incubate the robot-inoculated membranes overnight at room temperature (20–22 °C) until colonies are just visible (approximately 0.5 mm in diameter), then carefully lift the membrane from the plate and lay it onto agar trays with antibiotics, biotin, and arabinose, pre-warmed to 30 °C, avoiding air bubbles. Incubate at 30 °C for a further 4 h during which time the colonies will grow to approximately 1.5 mm in diameter. If a pair of identical membranes is prepared, one can be put in an empty QTray at −80 °C as a backup, while the other is processed through subsequent steps.
9. Colonies are lysed according to a protocol described [15]. The nitrocellulose membrane bearing arrayed induced colonies is placed on filter paper soaked in membrane denaturation solution for 10 min. It is then transferred sequentially to two filter papers soaked in membrane neutralization buffer for 2 × 5 min each. The membrane is then immersed in 2× SSC solution for 15 min and, using a glass spreader, the colonies are scraped gently from the membranes until no cellular debris are visible. At this point, place in 50 mL of Superblock buffer overnight (*see* **Note 11**).

3.7 Identification of Potentially Soluble Clones from the Random Truncation Library

In this step, the membrane is hybridized with fluorescent probes against the C-terminal biotin acceptor peptide (Streptavidin Alexa488) and the N-terminal hexahistidine tag (anti-hexahistidine mouse monoclonal followed by Alexa532 labeled rabbit anti-mouse secondary antibody (*see* **Note 12**)). Detection is by fluorescence scanning and analysis of the array image is performed using software able to extract data from images (e.g., Visual Grid; *see* **Note 13**) and subsequent processing of data in a spreadsheet program. The output is a list of clones, defined by their plate and well number in the library, ranked for potential soluble yield. A positive control, pESPRIT002 with an in-frame *malE* insert encoding maltose-binding protein fused to the hexahistidine and biotin acceptor peptide tags, is also included in the array.

1. The Superblock reagent is poured away and replaced by 50 mL of PBS-T with 16 μL (1:3000 dilution) of anti-hexahistidine monoclonal antibody and incubated in a cold roller bottle for 1 h at 4 °C. The antibody solution is then removed and the membrane washed for 3 × 5 min in PBS-T. Then 50 mL of PBS-T supplemented with 10 μL of streptavidin Alexa Fluor 488 conjugate and 50 μL of rabbit anti-mouse Alexa Fluor 532

conjugate is added to the roller bottle and incubated for 1 h at 4 °C (*see* **Note 12**). The membrane is then washed for 3 × 5 min in PBS-T.

2. Fluorescence signals from the membrane are visualized using a flat-bed fluorescence imager (e.g., Typhoon Trio). Pre-scans at low resolution permit adjustment of the laser and photomultiplier tube settings, then higher detail scans are performed at 50 μm resolution to obtain quantitative data. For the Alexa Fluor conjugates above, the Typhoon Trio imager settings are λ_{ex} 488 nm, λ_{em} 520 nm, 40 nm band pass for streptavidin, and λ_{ex} 532 nm, λ_{em} 555 nm, 20 nm single pass for the Alexa 532 secondary antibody.
3. Fluorescent colony array membranes are quantified using software capable of extracting pixel data from gray-scale images (*see* **Note 13**). The spot intensity data from both streptavidin and hexahistidine channels is imported into a spreadsheet program for subsequent treatment. In a first step, duplicate signals are averaged, the hexahistidine data is represented with a frequency histogram. The bin range is 30 when library is ca. 30,000 clones and 20 for ca. 20,000 clones. The frequency histogram reveals the mode comprising noise and background signal. Data corresponding to these unwanted clones are deleted, and the remaining data is divided into four equal parts. Then, the third quartile, representing the most intense 25% of clones with respect to their hexahistidine tag intensities, is selected. This subset of the original data is then sorted according to streptavidin signal intensities, resulting in a ranked list of constructs that possess both biotin acceptor peptides and hexahistidine tags.

3.8 Purification Screening and Characterization of Soluble Clones

In this last step, positive clones from the array-based screen are grown in liquid culture, proteins expressed and purification trials performed with SDS-PAGE analysis by western blot and Coomassie blue-staining. In principle, this can be done using normal laboratory methods in small shake flasks (50–100 mL cultures), but here is described a simple high-throughput screen capable of analyzing 96 constructs that uses an efficient sphaeroplast lysis protocol.

1. Ninety-five positive clones from the array analysis and a positive control in the same cell strain (pESPRIT002 with an in-frame *malE* insert encoding maltose-binding protein fused to both hexahistidine and biotin acceptor peptide tags) are used to inoculate 1 mL of LB with antibiotics (50 μg/mL kanamycin and 30 μg/mL chloramphenicol) in a deep well 96-well plate that is sealed with an air-permeable adhesive membrane. This is incubated overnight at 37 °C in the HiGro incubator at 300 rpm.
2. Prepare 4 × 24-well blocks with 4 mL of TB broth with antibiotics (50 μg/mL kanamycin and 30 μg/mL chloramphenicol)

per well. Transfer 40 μL of the saturated overnight preculture to each well of the 24-well plate and incubate at 37 °C in the HiGro incubator at 300 rpm until an optical density (OD_{600nm}) of 0.6 is obtained for several wells tested (3–4 h). Induce protein expression by adding 50 μM biotin and 0.2% (w/v) arabinose (final concentrations) to each well, reducing the HiGro temperature to 25 °C and shaking at 300 rpm overnight.

3. Cells are harvested in a centrifuge with plate rotor at 2500 × *g*, 10 min at 4 °C. The supernatant is discarded and each pellet resuspended in 4 mL of sphaeroplast preparation buffer for 30 min. Sphaeroplasts (osmotically buffered intact cells lacking walls) are pelleted by centrifugation at 2500 × *g*, 10 min at 4 °C. The supernatant is carefully aspirated from the pellets and discarded; the plates are put in the −80 °C freezer for 30 min for a single cycle of freeze-thaw.
4. The plates are removed from the freezer and each pellet is resuspended in 800 μL of sphaeroplast lysis buffer supplemented with 1:1000 dilutions of Benzonase and EDTA-free inhibitor cocktail. Keep lysates on ice to limit unwanted proteolysis.
5. Dispense 75 μL of Ni^{2+} NTA agarose resin into each well of a receiver plate. Wash wells with 700 μL of water by placing filter plate over a deep 96-well block and centrifuging at 500 × *g* in a rotor adapted for plates. Wash with 700 μL of Ni^{2+} NTA wash buffer per well. Remove excess buffer by centrifuging again at 500 × *g*. Place plastic adhesive tape over the bottom of the plate to seal the base.
6. Centrifuge the lysates from above at 2500 × *g* for 30 min. Transfer the supernatants to individual wells of the receiver plate containing resin, carefully noting the identity of the clone in each well (due to the 24- to 96-well format change). Apply plastic adhesive tape to the top of the plate and incubate with agitation for 30 min in the cold room to allow hexahistidine-tagged proteins to bind the resin.
7. Remove the plastic tapes and place the receiver plate over a deep well block. Centrifuge at 500 × *g* for 5 min at 4 °C, then add 700 μL of Ni^{2+} NTA wash buffer per well, and centrifuge again. Place the receiver plate over a clean 96-well PCR plate, add 50 μL of Ni^{2+} NTA elution buffer, and elute proteins by centrifugation at 500 × *g* for 5 min at 4 °C.
8. Eluted protein samples are mixed with Laemmli loading buffer, heated and analyzed by SDS-PAGE with Coomassie blue-staining. Additional streptavidin or western blots may be performed using standard protocols. Constructs yielding visible purified proteins are sequenced with T7For and T7Rev primers to identify construct boundaries.

4 Notes

1. The combination of classical plasmid preparation by alkaline lysis coupled to a downstream mini-prep column step results in nick-free DNA that is not damaged by internal digestion during the exonuclease truncation step.
2. Plasmid DNA for exonuclease truncation should be kept at 4 °C and not frozen to reduce chance of nicking.
3. The deletion rate has been calibrated on DNA at a final concentration of 33.3 ng/μL and exonuclease III at 100 unit per μg of DNA.
4. Mung bean nuclease is sensitive to NaCl so it is necessary to remove the salt added to calibrate the exonuclease III truncation.
5. The smear of exonuclease III/mung bean nuclease truncated DNA observed by agarose gel electrophoresis should start at the size of the linearized vector and fade out at around 2–3 kb.
6. Any other T4 DNA ligase protocol should also suffice. The cell strain used for the transformation of the ligation is also not important other than the cell competency should be as high as possible. The commercial preparations of chemically competent MACH1 cells we use have an advertised competency of $>10^9$ colony forming units per μg of DNA.
7. Plasmid insert sizes can be calculated quite accurately from band sizes on agarose gels, or exactly by DNA sequencing. Subtraction of flanking regions present due to primer or restriction site positions results in the size of the target gene insert.
8. Disposable 384-pin replicators can be washed, wrapped in aluminum foil, and autoclaved many times.
9. Agar should be cooled to approximately 50 °C prior to the addition of antibiotics, biotin, and arabinose. This can be achieved by placing bottles of molten agar in a heated water bath. Plates should be poured on a flat surface to ensure accurate robotic arraying.
10. The size of the library to be screened is calculated as follows: For a gene of 1000 bp, where fragments in a size range of 200–1000 bp are being tested, there are 800 potential gene fragments. A threefold oversample requires 2400 clones to be tested. If three different fixed ends are pooled in the same library, 7200 clones would be tested. If both 5′ and 3′ deletion libraries were to be tested in parallel, each with three fixed ends against which the truncations were performed, 14,400 clones would be tested.

11. For blocking and hybridization of fluorescent probes, we recommend a roller incubator with glass tubes in which the 22 × 22 cm membranes can be inserted. This greatly economizes the volume of Superblock and probe solutions.
12. Alexa dyes are relatively resistant to photo bleaching, but is advisable to minimize exposure of the labeled membranes to strong light. Membranes should be kept in dark containers.
13. For quantification of fluorescence signals on colony arrays, we use Visual Grid from GPC Biotech, but this has been discontinued for several years. ImageQuant and similar programs may have the same function; however, software specifically designed for extracting data from DNA micro- or macro-arrays is better since it permits easy fitting of a mask over the data.

Acknowledgments

The ESPRIT platform of the Grenoble Instruct Centre (ISBG: UMS 3518 CNRS-CEA-UJF-EMBL) receives current financial support from FRISBI (ANR-10-INSB-05-02) and GRAL (ANR-10-LABX-49-01) within the Grenoble Partnership for Structural Biology (PSB). Recent developments were funded from the European Community Seventh Framework Programme (FP7/2007-2013) under BioStruct-X (grant agreement 283570).

References

1. Dosztányi Z, Sandor M, Tompa P et al (2007) Prediction of protein disorder at the domain level. Curr Protein Pept Sci 8:161–171
2. Yumerefendi H, Tarendeau F, Mas PJ et al (2010) ESPRIT: an automated, library-based method for mapping and soluble expression of protein domains from challenging targets. J Struct Biol 172:66–74
3. Hart DJ, Waldo GS (2013) Library methods for structural biology of challenging proteins and their complexes. Curr Opin Struct Biol 23:403–408
4. Yumerefendi H, Desravines DC, Hart DJ (2011) Library-based methods for identification of soluble expression constructs. Methods 55:38–43
5. Ostermeier M, Lutz S (2003) The creation of ITCHY hybrid protein libraries. Methods Mol Biol 231:129–141
6. Beckett D, Kovaleva E, Schatz PJ (1999) A minimal peptide substrate in biotin holoenzyme synthetase-catalyzed biotinylation. Protein Sci 8:921–929
7. An Y, Meresse P, Mas PJ et al (2011) CoESPRIT: a library-based construct screening method for identification and expression of soluble protein complexes. PLoS One 6:e16261
8. An Y, Yumerefendi H, Mas PJ et al (2011) ORF-selector ESPRIT: a second generation library screen for soluble protein expression employing precise open reading frame selection. J Struct Biol 175:189–197
9. Tarendeau F, Boudet J, Guilligay D et al (2007) Structure and nuclear import function of the C-terminal domain of influenza virus polymerase PB2 subunit. Nat Struct Mol Biol 14:229–233
10. Tarendeau F, Crepin T, Guilligay D et al (2008) Host determinant residue lysine 627 lies on the surface of a discrete, folded domain of influenza virus polymerase PB2 subunit. PLoS Pathog 4:e1000136
11. Guilligay D, Tarendeau F, Resa-Infante P et al (2008) The structural basis for cap binding by influenza virus polymerase subunit PB2. Nat Struct Mol Biol 15:500–506

12. Nadal M, Mas PJ, Blanco AG et al (2010) Structure and inhibition of herpesvirus DNA packaging terminase nuclease domain. Proc Natl Acad Sci U S A 107:16078–16083
13. Angelini A, Tosi T, Mas P et al (2009) Expression of *Helicobacter pylori* CagA domains by library-based construct screening. FEBS J 276:816–824
14. Rawlings AE, Levdikov VM, Blagova E et al (2010) Expression of soluble, active fragments of the morphogenetic protein SpoIIE from *Bacillus subtilis* using a library-based construct screen. Protein Eng Des Sel 23:817–825
15. Büssow K, Cahill D, Nietfeld W et al (1998) A method for global protein expression and antibody screening on high-density filters of an arrayed cDNA library. Nucleic Acids Res 26:5007–5008

Chapter 5

Optimizing Expression and Solubility of Proteins in *E. coli* Using Modified Media and Induction Parameters

Troy Taylor, John-Paul Denson, and Dominic Esposito

Abstract

The major goal of any protein expression experiment is to combine the maximum production per cell of soluble protein with the highest possible cell density to most efficiently obtain high yields of protein. A large number of parameters can be optimized in these experiments, but one of the most interesting parameters that have a strong effect on both per cell productivity and cell density is the cellular growth media coupled to the expression induction process. Using specialized media and testing multiple induction conditions, it is possible to significantly enhance the production of heterologous proteins from *E. coli*.

Key words Protein expression, Growth media, Autoinduction, IPTG, Solubility

1 Introduction

While numerous heterologous systems exist for the expression of proteins of biological and pharmaceutical relevance, expression in the bacterium *E. coli* remains one of the most popular methods for producing proteins due to its ease of use and minimal cost. Over the past several decades, many improvements have been made to the *E. coli* expression system, focusing on advances in generating high levels of expression with induction systems like the T7 polymerase [1], improvements in protein quality and yield via the addition of solubility fusion tags [2], and optimization of very large-scale fermentation processes for industrial-scale production of proteins [3]. Many of these advances were made in high-throughput operations that focused on structural biology [4], and which had the luxury of taking the successful proteins and ignoring those which failed. However, for many researchers focused on a particular essential target, it became clear that additional and targeted optimization of many parameters of *E. coli* expression were needed for each protein—in effect, a panel of different options had a much higher likelihood of leading to a successful and high-yield production process. For this reason, our laboratory

Nicola A. Burgess-Brown (ed.), *Heterologous Gene Expression in E. coli: Methods and Protocols*, Methods in Molecular Biology, vol. 1586, DOI 10.1007/978-1-4939-6887-9_5,

and many others began to examine many of the steps in the *E. coli* expression process to identify areas which could still benefit from optimization on a case-by-case basis, and to determine whether some of these tricks could be applied more universally across many protein targets.

While yield per unit cell is an important factor in *E. coli* expression, sometimes a lower yield can still be tolerated if one can increase the overall productivity of the system. Higher productivity requires higher cell densities in the same unit volume, and often this parameter can be limited by the amount of nutrients in the growth medium, or the accumulation of toxic by-products of metabolism. In the early days, simple minimal media were often used for protein expression, but ultimately, cell productivity was increased by the appearance of the so-called rich media which contained additional factors that stimulated growth of the cells to higher densities or stabilized growth of the cultures for longer periods of time [5]. Although the higher cell densities often produced more protein, it became quickly apparent that simply producing more cells was not necessarily related to the production of more high-quality protein. Instead, in a somewhat paradoxical manner, many high density processes led to the appearance of higher yields of protein, but much reduced quality, as the protein was either insoluble or aggregated due to either the large amount of protein produced, or the rapid production of protein overwhelming the cell's ability to fold it properly [6]. To solve these problems, a number of systems were developed which modified the way in which protein production was induced in the cell, allowing it to be timed better with the growth rate of the cells or permitting the reduction in temperature of the culture prior to protein expression. One of the most successful induction processes in the last decade has been the autoinduction system designed by Studier [7]. This system utilizes a cascade of carbon sources in the medium to ensure that protein induction begins only when the cells have reached an optimal point in their growth cycle. Using this system, investigators no longer need to assess the growth of cells continuously to decide when to add inducer. However, this process has some drawbacks in terms of expression flexibility. A common process used to enable heterologous expression is to induce cells at very low temperature (<20 °C) to improve solubility of proteins [8]. Temperature shifting with autoinduction eliminates the benefit of not having to constantly monitor cells, and actually proves to be quite difficult since timing of autoinduction is not always the same from culture to culture.

For these reasons, our laboratory has explored alternative growth media formulations and optimization of induction systems to attempt to identify conditions which enhance soluble protein production from *E. coli*. While a single system does not always work for a given protein of interest, we have identified media and conditions which allow for a thorough investigation of expression variables and which can result in significant increases in both protein yield and quality.

2 Materials

2.1 Construction of Expression Clones

1. LR Clonase II kit (Thermo Fisher; comes with LR Clonase® II enzyme mix, 2 μg/mL Proteinase K solution).
2. Qiagen Miniprep Spin Kit.
3. DH10B chemically competent cells (Thermo Fisher).
4. pDest-566 (Addgene) or other T7 promoter-based *E. coli* Gateway Destination vector.
5. Circlegrow-Amp medium: 40 g/L of Circlegrow medium (MP Biomedicals), autoclave for 20 min, cool to 55 °C, and add 100 μg/mL ampicillin.
6. LB-Amp agar plates: LB-agar petri plates with 100 μg/mL ampicillin.
7. 50 mL Falcon culture tubes.
8. *BsrG*I restriction enzyme.

2.2 Generation of E. coli Expression Strains

1. Microcentrifuge tubes.
2. 1 mL Nalgene cryogenic vials (for competent cells).
3. LB medium (per liter): 10 g of tryptone, 5 g of yeast extract, and 10 g of NaCl.
4. ThermoMixer F1.5 (Eppendorf).
5. 100 mg/mL ampicillin, sodium salt: 1000× stock solution prepared in water.
6. 50 mg/mL kanamycin sulfate: 1000× stock solution prepared in water.
7. 15 mg/mL chloramphenicol: 1000× stock solution prepared in 50% ethanol.
8. LB agar plates: 15 g of agar per liter of LB medium.
9. 3 mm glass beads.
10. Lazy-L Spreaders (Sigma-Aldrich).

2.3 E. coli Protein Expression

1. 1000× trace metals mix: Dissolve 1.35 g of $FeCl_3 \cdot 6H_2O$ in 50 mL of 0.1 M HCl, then add 36 mL of H_2O and dissolve 294 mg of $CaCl_2$, 198 mg of $MnCl_2 \cdot 4H_2O$, 288 mg of $ZnSO_4 \cdot 7H_2O$, 48 mg of $CoCl_2 \cdot 6H_2O$, 34 mg of $CuCl_2 \cdot 6H_2O$, and 48 mg of $NiCl_2 \cdot 6H_2O$. Add water to 100 mL total and sterilize through a 0.22 micron syringe filter. Store at room temperature for up to a year.

2. MDAG-135 (7): 2 mM $MgSO_4$, 25 mM Na_2HPO_4, 50 mM NH_4Cl, 5 mM Na_2SO_4, 0.1% aspartate, 25 mM KH_2PO_4, 0.35% (w/v) glucose, 0.2× trace metals mix and 200 μg/mL of each of 18 standard amino acids (no cysteine or tyrosine).
3. Dynamite Medium: Dissolve 12 g of tryptone, 24 g of yeast extract, 6.3 mL of glycerol, 3.8 g of KH_2PO_4, 12.5 g of K_2HPO_4, 5 g of glucose, and 0.195 g of $MgSO_4$ (anhydrous) in a total volume of 1 L of water, autoclave to sterilize.
4. Superior Broth (AthenaES): Add 35 g of Superior Broth powdered medium to 1 L of water, autoclave to sterilize.
5. Terrific Broth (Teknova, Hollister, CA): Add 40 g of Terrific Broth powdered medium to 1 L of water, autoclave to sterilize.
6. Circlegrow Broth (MP Biomedicals): Add 40 g of Circlegrow Broth powdered medium to 1 L of water, autoclave to sterilize.
7. LB-Miller Broth: Mix 10 g of tryptone, 5 g of yeast extract, and 10 g of NaCl in a total of 1 L of water, autoclave to sterilize.
8. BRM medium: Mix 16 g of tryptone, 10 g of yeast extract, 5 g of NaCl, 5.68 g of Na_2HPO_4, 2.64 g of NaH_2PO_4, and 5 g of glucose in a total of 1 L of water, autoclave to sterilize.
9. 4× YT medium: Mix 32 g of tryptone, 20 g of yeast extract, and 5 g of NaCl in a total of 1 L of water, autoclave to sterilize.
10. ZYM-20052: Mix 10 g of N-Z-amine AS (Sigma-Aldrich), 5 g of yeast extract, 3.55 g of Na_2HPO_4, 3.4 g of KH_2PO_4, 2.67 g of NH_4Cl, 0.71 g of Na_2SO_4, 0.241 g of $MgSO_4$, 20 mL of glycerol, 0.5 g of glucose, 2 g of α-lactose, 0.29 g of sodium citrate, and 4.05 g of disodium succinate in a total of 1 liter of water, autoclave to sterilize.
11. Innova-44 incubating, refrigerated shaker with 1″ orbit (New Brunswick Scientific).
12. 250 mL baffled shake flasks (Corning).
13. 0.5 M Isopropyl β-D-1-thiogalactopyranoside (IPTG): Prepared in water in 1 mL aliquots and stored at −80 °C.
14. Genesys 10S UV-VIS spectrophotometer (Thermo Fisher).
15. Micromax RF refrigerated microcentrifuge (Thermo Fisher).
16. Avanti J-20 XP centrifuge (Beckman Coulter).

2.4 Cell Lysis and Analysis

1. 1 M Tris–HCl, pH 7.5: Dissolve 121.1 g of Tris base in 500 mL water, add concentrated HCl to a pH of 7.5, then add water to a final volume of 1 liter.
2. Ready-Lyse Lysozyme Solution (Epicentre).

3. OmniCleave Endonuclease (Epicentre).
4. 1 M $MgCl_2$: Dissolve 2.03 g of $MgCl_2.6H_2O$ in 10 mL water.
5. Stock Lysis Buffer A: Mix 5 mL of 1 M Tris–HCl, pH 7.5 and 0.9 g of dextrose, and bring the volume up to 100 mL with water.
6. Stock Lysis Buffer B: Mix 1 mL of 1 M Tris–HCl, pH 7.5, 0.37 g of KCl, and 0.1 g of sodium deoxycholate, and bring the volume up to 100 mL with water.
7. Micromax RF refrigerated microcentrifuge (Thermo Fisher).

3 Methods

3.1 Construction of Expression Clones

There are a number of options for generation of expression clones, and a detailed discussion of all of these is beyond the scope of this chapter. However, for optimal protein production, our laboratory and others often use a combination of tag elements for solubility enhancement and purification, and a protease cleavage site for removal of tags downstream of initial purification. We will highlight this method here as we commonly apply it for *E. coli* expression. For ease and flexibility of clone construction, we use a recombination-based cloning system such as Gateway cloning [9, 10] to generate an initial vector with the gene of interest flanked at the 5′ end with a protease cleavage site (*see* **Note 1**). If using Gateway cloning, this vector is termed an Entry clone and can be subsequently used to manufacture various expression clones with different purification and solubility tags. We will focus on the most commonly used tag option in this protocol, fusion to a tag which contains a 6× His repeat followed by a maltose-binding protein (MBP) solubility tag [11]. This solubility enhancing tag is used in our laboratory for nearly all heterologous proteins generated in *E. coli*—even in cases where solubility enhancement is not required, the MBP tag also often provides higher yields of protein expression.

1. Starting with a Gateway Entry clone containing an upstream TEV protease site, generate an *E. coli* expression clone by LR recombination of the Entry clone with an appropriate His6-MBP destination vector, for example, pDest-566 which is available from Addgene (*see* **Note 2**). For downstream optimization, this vector must contain a T7 promoter (*see* **Note 3**) and should be relatively low copy number (*see* **Note 4**).
2. Add the following reagents to a microcentrifuge tube in the order given (the total reaction volume should be 10 μL): 1–6 μL of H_2O, 50 ng of Entry clone DNA, 150 ng of Destination Vector DNA, and 2 μL of LR Clonase II.

3. Incubate the reaction mixture for at least 1 h at 30 °C.
4. Add 1 μL of 2 mg/mL Proteinase K to inactivate the LR Clonase and incubate for 15 min at 37 °C (*see* **Note 5**).
5. Transform 1 μL of the LR reaction into high efficiency chemically competent *E. coli* DH10B cells as noted in the manufacturer's instructions (*see* **Note 6**). A good LR cloning result should yield greater than 500 colonies per transformation.
6. After overnight incubation on plates with the appropriate antibiotic selection medium (ampicillin in the case of pDest-566) at 37 °C, pick 2 Expression clone colonies into Falcon 2059 culture tubes containing 2 mL of Circlegrow medium containing 100 μg/mL ampicillin and grow overnight at 37 °C with 200 rpm shaking.
7. Centrifuge 1 mL of the culture in a microcentrifuge to pellet the cells, and isolate plasmid using a Qiagen Miniprep Spin kit, eluting the DNA in 50 μL of elution buffer.
8. LR clones can be verified by agarose gel electrophoresis of supercoiled plasmid DNA, or alternatively, can be confirmed by restriction enzyme analysis (*see* **Note 7**).
9. Expression clones generated by other means than Gateway cloning may require sequencing to validate the proper insertion of the gene of interest, particular if PCR is used to clone the constructs (*see* **Note 8**).

3.2 Generation of E. coli *Expression Strains*

The preferred host strain for T7 promoter-based expression is an *E. coli* BL21 strain which carries the lambda DE3 lysogen engineered to express the T7 RNA polymerase protein. BL21 strains are derived from the *E. coli* B strain and have been specifically constructed for high-level expression of recombinant proteins. These strains have key mutations making them ideal for protein expression, including deletion of several highly active *E. coli* proteases, *lon* and *ompT* [12]. To reduce degradation of mRNA, the strain BL21(DE3) Star can also be used. The Star strain can have higher basal expression of heterologous genes than BL21 strains, due to the increased stability of mRNA [13]. This is achieved by a mutation in the RNaseE gene (*rne*131), which is involved in degradation of unstable messenger RNA molecules. Protein expression in *E. coli* using typical growth conditions with strong promoters produces much higher protein levels than endogenous genes. For this reason, and because heterologous genes often use distinct tRNA recognition codons, it is often the case that endogenous tRNAs for protein synthesis are limiting in *E. coli*. Insufficient tRNA pools can lead to premature translation termination, translation stalling, frameshifting concerns and amino acid misincorporation [14]. To avoid these problems, one can coexpress a plasmid which contains upregulated tRNA genes for these rare codons, such as the pRare plasmid found in *E. coli* Rosetta or Rosetta2 strains (*see* **Note 9**). Other alternative methods exist for dealing with the unusual gene composition of heterologous genes (*see* **Note 10**).

1. To generate competent *E. coli* cells, follow the detailed protocol in [15] (*see* **Note 11**).
2. Thaw one vial of chemically competent cells on ice (*see* **Note 12**).
3. Once thawed, gently mix the competent cells by gently pipetting up and down.
4. Add 50 μL of competent cells to a sterile Eppendorf tube (*see* **Note 13**).
5. Add 10 ng of Expression clone plasmid DNA to the competent cells. Place the mixture on ice for 20 min (*see* **Note 14**).
6. Remove the tube from the ice and immediately heat shock the mixture using a 42 °C water bath for 45 s.
7. Add 80 μL of LB medium (*see* **Note 15**) and shake using a thermomixer at 37 °C for 60 min (*see* **Note 16**).
8. In most cases, add the entire contents of tube to the appropriate solid growth medium that contains an appropriate antibiotic for selection (usually ampicillin or kanamycin for Expression clone selection and chloramphenicol for helper plasmids such as the pRare tRNA vector). In some situations where highly competent cells have been used, a 1:10 or 1:100 dilution plate should also be prepared to ensure single colonies are achieved.
9. Spread the transformation mixture evenly onto the plate (*see* **Note 17**) and allow to dry briefly.
10. Incubate the plate in a 37 °C incubator for 10–16 h.

3.3 E. coli Protein Expression

A number of different scales of expression can be carried out, ranging from 1 mL cultures in a 96-well deep well block, to large 80 or 100 L fermentation vessels. For higher throughput work, the 96-well deep well blocks are an excellent method for carrying out initial screening work. If throughput is lower, as it usually is in our laboratory, we find the optimal scale for initial screening work is usually a 250 mL flask containing 50 mL working volume to ensure proper aeration. This scale often provides enough material to scale-up purification after an initial test expression and still yield hundreds of micrograms or even a milligram of purified protein. The process outlined here is generally scalable in both directions if smaller or larger production is required for your specific application.

1. Using a single colony from a freshly transformed plate (*see* **Note 18**) or an ice chip/scrape from a glycerol stock (*see* **Note 19**), inoculate a 50 mL seed culture of MDAG-135 (*see* **Note 20**) and grow for 14–18 h at 37 °C with 200 rpm shaking.

2. Inoculate 50 mL volumes of medium in a 250 mL baffled shake flask with 2% (v/v) seed culture.
3. For IPTG based expression, grow the culture for 4–5 h at 37 °C with 250 rpm shaking until the OD_{600} reaches ~6–8.
4. Chill the culture to 16 °C (*see* **Note 21**) by placing the culture in a prechilled shaker, and immediately induce with IPTG to a final concentration of 0.5 mM (*see* **Note 22**).
5. For autoinduction, initial cultures from **step 2** in ZYM-20052 medium should be grown for 4–5 h at 37 °C until the OD_{600} reaches 4.5–5.5 (*see* **Note 23**).
6. Chill the culture to 20 °C (*see* **Note 24**) by placing the culture in a prechilled shaker.
7. Grow cultures overnight for an additional 18–20 h.
8. Check OD_{600} and remove 1 OD unit of cells (1 mL/OD_{600}) to a microcentrifuge tube for a test sample.
9. Centrifuge the test sample in a benchtop microfuge at 15,000 × *g* for 5 min, remove supernatant, and freeze the sample at −80 °C.
10. Centrifuge the remaining sample at 5000 × *g* for 30 min.
11. Remove supernatant and freeze the cell pellet at −80 °C (*see* **Note 25**).

When protein expression is problematic under these basic conditions, additional optimization can be carried out on features such as aeration level, time of induction, induction temperature and induction strength.

3.4 Cell Lysis and Analysis

Analysis of expression can be done for both whole-cells (to identify whether protein is being expressed) or for soluble protein (to identify whether protein is in a purifiable fraction). Alternatively, processes for microscale purification have been worked out that will allow direct testing for purification [16]. For the sake of simplicity, we describe here the simplest process for quickly assessing whether soluble protein is being produced in an expression experiment, allowing easy comparison between optimization conditions. This process identifies soluble and insoluble proteins and thus eliminates the need to save multiple samples for whole-cell and soluble protein analysis.

1. Immediately prior to lysing cells, generate working buffer A by adding 1 μL of Ready Lyse (30,000 units) and 2 μL of OmniCleave endonuclease (400 units) to 1 mL of Stock Lysis buffer A (*see* **Note 26**).
2. Generate working buffer B by adding 5 μL of 1 M $MgCl_2$ to 1 mL of Stock Lysis buffer B.

3. Thaw the test sample pellet generated previously at room temperature (*see* **Note 27**).
4. Add 200 μL of working buffer A to the pellet and vortex until cell pellet is completely dissolved.
5. Let the solution sit at room temperature for 5 min.
6. Add 200 μL of working buffer B and vortex to mix.
7. Let the solution sit at room temperature for 5 min.
8. Invert to mix and then remove 50 μL of sample for total protein gel analysis.
9. Centrifuge the remaining 350 μL to remove cellular debris at 20,000 × *g* for 10 min at 4 °C.
10. Carefully remove the supernatant to a clean tube to represent soluble proteins (*see* **Note 28**).
11. Mix appropriate amounts of total and soluble fractions with SDS-PAGE loading buffer, and electrophorese using the manufacturer's protocols for your specific SDS-PAGE system (*see* **Note 29**).

3.5 Optimization of Expression Conditions

Historically, rich media formulations have been used for protein expression experiments to increase both the overall production of cell mass (OD/mL), and the productivity of individual cells (protein/cell). First generation media like Luria Broth or LB were often used to produce proteins prior to the advent of second generation rich media (Terrific Broth, etc.). Utilizing a better understanding of *E. coli* nutrient usage derived from the autoinduction work of Studier, third generation media have been designed to attempt to improve cell yields even more. One such media, Dynamite, was developed by our laboratory and, utilizing the small-scale processes outlined here, we can compare the production of proteins in this medium to that of other first and second generation media, as well as the Studier autoinduction process.

1. A T7 promoter based pDest-566 vector was constructed which expresses a protein of His6-MBP fused to the chloramphenicol acetyltransferase (CAT) gene.
2. Eight different media were purchased or formulated (*see* Table 1) and used in 50 mL test expressions per the protocols in Subheading 3.3. ZYM-20052 was used for autoinduction while the remaining seven media were induced with IPTG as described.
3. OD_{600} values were measured after overnight growth in *E. coli* BL21(DE3) as well as in *E. coli* DH10B (*see* **Note 30**), and are shown in Fig. 1.
4. Equivalent OD samples of cells were taken as described in Subheading 3.4, and total protein was analyzed by SDS-PAGE (*see* Fig. 2).

Table 1
Growth media used in this study along with their approximate cost per liter at the time of publication

Medium	Vendor	Price (per L)
Superior Broth	Athena	$6.00
Terrific Broth	Teknova	$5.67
LB (Miller)	Becton-Dickinson	$3.00
4xYT	In-house	$4.62
Circlegrow	MP Biomedical	$8.25
ZYM-20052	In-house	$4.06
BRM	In-house	$3.63
Dynamite	In-house	$3.76

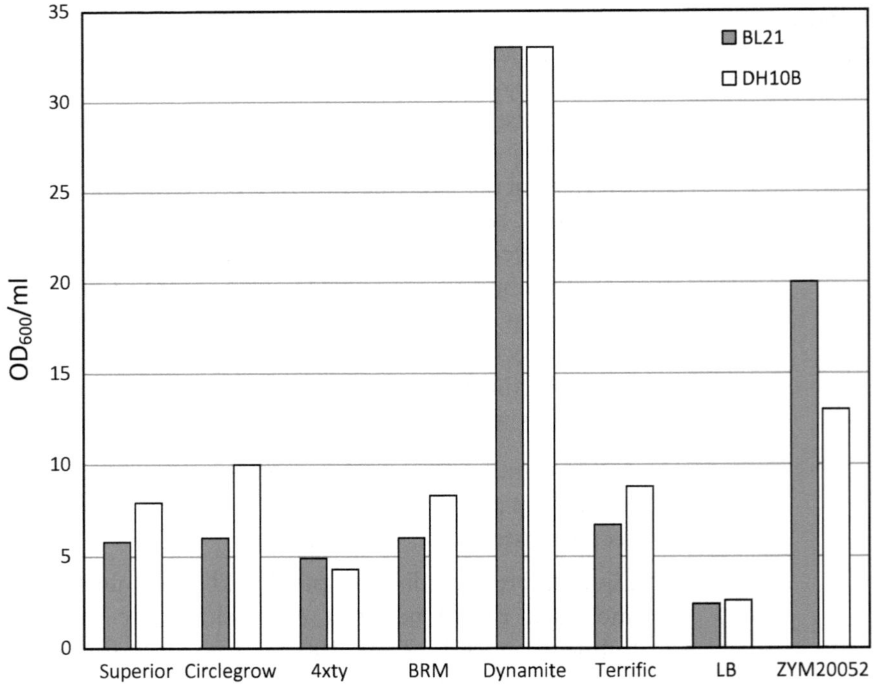

Fig. 1 Growth of two *E. coli* protein expression strains in various growth media. Shown are the production yields of two *E. coli* strains harboring an identical His6-MBP-CAT protein construct grown in 50 mL of different growth media in a 250 mL baffled shake flask at 37 °C. *Dark bars* show the final yield of BL21(DE3) Star cells, while *light bars* show the final yield of DH10B cells

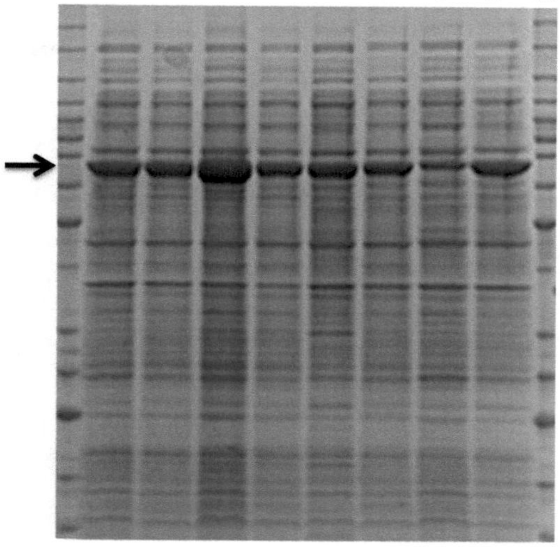

Fig. 2 Whole-cell extracts showing protein expression levels in various growth media. 1 OD unit of cells from expression of His6-MBP-CAT in BL21(DE3) Star in the different media from Fig. 1 were resuspended in detergent lysis buffer as indicated in Subheading 3.4 and samples were analyzed by SDS-PAGE. Molecular weight markers (M) are Benchmark ladder; marker bands flanking the protein of interest represent 60 and 70 kDa. Media are: (S)uperior Broth, (C)irclegrow (4) xYT, (B)RM, (D)ynamite, (T)errific Broth, (L)B-Miller, and (Z)YM-20052

LB medium serves as the benchmark for culture density, with average final OD values around 2.5 OD_{600}/mL. Most of the second generation media peak at levels between 5 and 6 OD_{600}/mL, while autoinduction with ZYM-20052 yields cultures of nearly 20 OD_{600}/mL. Dynamite medium samples finished growth at OD_{600}/mL values greater than 30, representing at least a fivefold increase in cell density over standard second generation media and 50% higher than the autoinduction media. Protein levels, representing per cell productivity since samples are of equal OD loading, are very similar in most of the media (*see* Fig. 2), with the exception of the poor expression often noted in LB. Most notably, Dynamite medium produces similar levels of protein per cell to most second generation media, and only slightly less protein per cell than autoinduction media, making this a highly attractive option. As shown in Table 1, both ZYM-20052 and Dynamite media are also competitive in price with most second generation media.

The choice between autoinduction and IPTG induction in Dynamite medium is not, in our experience, a trivial decision. Solubility and productivity are often protein-specific phenomena, as evidenced by the data in Fig. 3. Here, three different proteins were expressed from the same construct backbone in the same *E. coli* strain using either IPTG induction in Dynamite medium, or

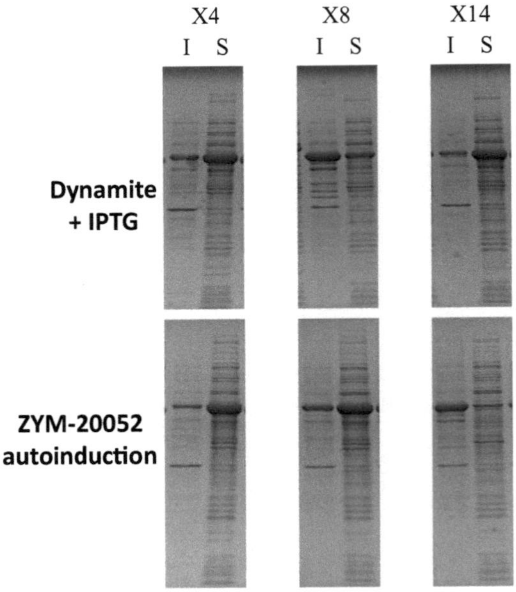

Fig. 3 Solubility analysis of proteins induced with autoinduction or IPTG induction. Three protein expression constructs generated as His6-MBP fusions were expressed in BL21(DE3):pRARE cells and induced either with 0.5 mM IPTG at 16 °C or by autoinduction with a temperature shift to 20 °C. In both cases, expression was carried out for 18–20 h prior to harvest. 1 OD unit of cells were used for detergent lysis as indicated in Subheading 3.4 and samples of soluble (S) and insoluble (I) fractions were analyzed by SDS-PAGE

autoinduction in ZYM-20052. Insoluble and soluble fractions of these samples are analyzed on the SDS-PAGE gel shown, and it is clear that for protein X4, production yields and solubility are nearly identical in the two conditions. However, protein X8 shows greater solubility in autoinduction, while protein X14 shows both higher level expression and much greater solubility in the IPTG-induced Dynamite medium. For this reason, we prefer to test both cases for every protein to ensure that we identify the best option for a given protein. In this case, it is remarkable to note that these three proteins actually share >85% sequence identity, which is a powerful demonstration of how much value can be added by minor optimization of *E. coli* expression. Coupled with an additional round of optimization of solubility tags, we find that we can improve expression of most proteins in *E. coli* dramatically by such parallel screening of expression conditions.

4 Notes

1. Removal of large solubility tags is usually essential for production of proteins of biological and pharmaceutical interest. A number of proteases are available in the literature for this

process, but we find that the best option is the Tobacco Etch Virus protease (TEV) which cleaves at a highly specific amino acid sequence ENLYFQ/G, to leave a single glycine residue at the aminoterminus of the cleaved protein. TEV protease can be generated from readily available clones (pRK793, Addgene) and is easily purified at high yield [17].

2. pDest-566 is based on the common pET series of T7 promoter vectors developed at Novagen. These vectors contain the lacI repressor gene as well as a T7 promoter under the control of the lac operator. This combination ensures strong repression of the promoter under basal conditions prior to induction, and is strongly recommended over other T7 vectors which lack the additional repressor construct and may exhibit leaky expression and subsequent toxicity.
3. While other promoters can be used for protein expression in *E. coli*, the widely used bacteriophage T7 promoter is the best option due to its high level of transcription and tight control via the T7 RNA polymerase system found in most expression strains.
4. It is best to use low copy number vectors with pBR322 origins of replication for protein expression. Higher copy number vectors such as pUC19 can have as many as 500 copies per cell, leading to a huge burden on transcription and translation which often results in toxicity, poor cell growth, lower productivity, and sometimes recombination of DNA leading to production of aberrant proteins.
5. Failure to treat the LR reaction with Proteinase K will result in dramatically reduced colony counts due to the inability of the DNA to transform while coated with Clonase proteins.
6. Any *recA endA* strain (such as DH10B, TOP10, or DH5a) can be used for transformation of Gateway LR reactions. Be sure that the strain being used does not have the F′ episome as it contains the *ccd*A gene which will detoxify the *ccdB* gene resulting in failure of the negative selection. For good LR results, be sure that the competent cells have a transformation efficiency of at least 1×10^8 cfu/μg. Electrocompetent cells can also be used instead of chemically competent cells; however, the only advantage would be in the case of a very low efficiency reaction—usually the number of colonies obtained with standard chemically competent cells is more than sufficient.
7. Gateway reactions are usually so efficient and accurate that further confirmation of Expression clones is not necessary. However, if desired, the Gateway recombination sites can be cleaved with the restriction enzyme *BsrG*I, which will cut out your gene of interest (if it has no additional sites) and allow verification of insert size. Alternately, other restriction sites can be employed.

8. Many other T7 promoter based expression vectors are available, including a large number of pET series vectors from Novagen, as well as many public domain vectors sold through other companies or available at the Addgene repository. In general, the choice of vectors will depend on the type of cloning system which you use and the type of tags that you want to introduce into your proteins.
9. Rosetta strains contain a plasmid that encodes a set of rare tRNAs (*proL*, *leuW*, *argW*, *glyT*, *argU*, and *ileX*) that can theoretically support production of heterologous proteins which are encoded by genes rich in these “rare” *E. coli* codons. In many cases, the presence of some rare codons is not an impediment to good expression, but clusters of rare codons, particularly early in a sequence, can lead to ribosome stalling and premature termination. In our experience, using a rare tRNA plasmid has no negative effects on protein production, and thus we include this in all of our expression work.
10. As an alternative to using rare tRNA expression, one can also optimize a heterologous gene for expression in *E. coli* using other means. Some data suggests that modifications of culture conditions (lowering the temperature, changing media composition, etc.) can shift the codon usage bias enough to alleviate some codon usage-based expression problems, but this has not been clearly demonstrated to be a consistent phenomenon. Several companies have proprietary protocols for optimization of *E. coli* expression constructs that in addition to fixing codon usage issues, remove other potentially deleterious elements such as cryptic splice sites or promoters, poor mRNA secondary structures, or ribosome pause sites. Companies such as DNA2.0 have produced optimized constructs for some human genes which improve expression 5–10 fold over the native sequences, and this may be a good option for heterologous genes which have identifiable issues (poor codon usage, high GC content, etc.)
11. The procedures in the Hanahan paper [15] can easily generate *E. coli* K-12 strains (such as DH10B) at a competence of 1×10^8 cfu/μg and B strains (such as BL21) at a competence of 5×10^7 cfu/μg. These levels of competence are more than sufficient for plasmid transformation under normal circumstances. If higher levels are required, for instance for expression from libraries of constructs, we suggest purchasing high efficiency cells or using electroporation instead of chemical transformation.
12. It is important to thaw the competent cells slowly on ice, as thawing them at room temperature or warmer conditions can dramatically decrease the competency of the cells.

13. In general, competent cells are considered fragile and sensitive to freeze-thaw cycles, but our data suggests that we can refreeze the remainder of a competent cell vial at −80 °C and reuse once with only a minimal loss of competency. Since transformation of Expression clone DNA usually involves a large amount of plasmid DNA and only a few colonies are required, even a tenfold drop in competence can usually be tolerated.
14. In general, most researchers use a 20 min incubation time before heat shock of the competent cells. However, we have found that for most *E. coli* strains used for plasmid propagation and protein expression, this incubation can be reduced to 5 min if time is an issue without any significant loss of competence.
15. Although most publications call for the use of SOC medium for the transformation grow-out, we have found that simple LB medium works just as well as SOC, but at a significantly reduced cost. It is vital that the LB medium be completely sterile, so we routinely make small aliquots of sterile media and discard them after a single use to avoid any potential for contamination.
16. In general, for ampicillin-resistant vectors, the grow-out period after transformation can be reduced significantly without a negative effect on colony numbers. We routinely shake cultures for less than 20 min before plating, and have on some occasions plated directly after addition of medium. However, for other antibiotics which have more rapid cytostatic effects (kanamycin, gentamycin), a full 60 min grow-out is essential for high efficiency transformation.
17. We have used glass beads as well as L-spreaders for distributing cells evenly across the plate during the transformation step. When performing several transformations at once, we routinely use glass beads. The beads can be washed in ethanol, autoclaved, and reused multiple times, but our data suggests that the L-spreaders yield more colonies on average, likely due to cell adherence to the glass beads. While the beads will suffice for normal transformation procedures, we suggest the use of spreaders for problematic transformations or when cells or DNA is present at lower than suggested concentrations.
18. Freshly transformed plates can routinely be stored for up to 2 weeks at 4 °C prior to culture inoculation and glycerol stock preparation. Beyond 2 weeks, we are generally uncomfortable with use of cells on plates due to the likelihood of mutations or cellular damage.
19. Our data suggests no differences in expression using a colony from a freshly transformed plate or an ice chip from a properly made glycerol stock. However, glycerol stocks must be made

from log-phase cells in order to be stable long-term. Glycerol stocks made from saturated overnight cultures often lead to problems with expression upon repeat growths, and in these cases we suggest a retransformation.

20. MDAG-135 is a "non-expressing" medium developed by the Studier lab to ensure extremely tight control of T7 based expression. Use of this medium ensures that no leaky expression of proteins will occur in the overnight cultures which could potentially lead to selection for mutants or high levels of toxicity.
21. We routinely express difficult proteins at 16 °C to promote correct protein folding. This temperature was chosen as the lowest temperature which still provides reasonable metabolic activity and cell stability. Temperatures lower than this have been used for expression of proteins but often require the presence of chaperone proteins from psychrophilic bacteria to permit reasonable levels of bacterial growth, and we have not observed significant changes in protein yield or quality in these systems.
22. We investigated reducing the temperature of the culture by manually by shaking the culture in ice water. This does improve the speed of the temperature drop, but we did not see an improvement in soluble protein with the proteins we tested in our panel. At larger scales (1 L flasks or bioreactors), the time required for temperature shifts of this nature can be excessive (>2 h) at which point expression is likely to have already started, leading to possible higher levels of insoluble protein. In these cases, and when expressing a difficult protein known to have solubility issues, alternative processes such as adding additional cold medium, or utilizing a jacketed bioreactor may improve the temperature shift timing. In general, for 50 mL scale or smaller, the time difference is mostly irrelevant.
23. Autoinduction occurs at cell densities of approximately OD_{600} of 5.0 using our formulation under moderate aeration (250 rpm, 1″ orbit). Our data suggests there is a shift in autoinduction timing under reduced aeration conditions. This can easily occur when using different growth vessels such as non-baffled flasks or poorly aerated blocks. It is suggested to carefully monitor induction timing when using different types of culture vessels.
24. The autoinduction process requires the metabolism of lactose resulting in increased concentrations of allolactose (the actual inducer) and galactose. The metabolic efficiency of *E. coli* cells appears to drop significantly as you decrease the temperature below 20 °C [7], so we prefer to use this temperature for

growth at induction. The slight increase in temperature of our standard 16 °C IPTG induction conditions has not had a noticeable impact on protein folding in our hands.

25. Many researchers prefer to "snap-freeze" the cells in a dry ice–ethanol bath, but we have found no significant difference in using this process versus simply placing the pellets in a −80 °C freezer and allowing them to freeze. The latter is more convenient and eliminates the need for costly amounts of ethanol and dry ice.
26. We have found that the stock Lysis buffer A and Lysis buffer B solutions can be stored for up to 6 months at 4 °C with no effect on the lysis process, but dilutions of the working stocks should be performed immediately before lysis and discarded within 2 h.
27. While fresh pellets can be used for solubility analysis, we have found more consistent lysis and analysis when samples are frozen and thawed first.
28. While a comparison of total protein and soluble protein is usually the most informative analysis, one can also take the insoluble pellet of the detergent lysis test after supernatant removal, and analysis this for the insoluble fraction of proteins. The easiest way to do this is to wash the pellet one or two times with lysis buffer, and then after a repeat centrifugation, dissolve the final pellet in 8 M urea to resolubilize the proteins. This can be a useful process if your protein is known to form inclusion bodies and will need to ultimately be purified from this insoluble fraction.
29. For rapid analysis of smaller-scale samples, we use the detergent lysis process because it is quick and easy. In general, for most of our scale-up work, our lab routinely uses mechanical lysis using a Microfluidizer for protein expression work. Mechanical lysis removes the added variable of detergent in the lysate, which can sometimes mask solubility issues, and also saves the high cost of enzymes and lysis buffer for larger lysates. In general, however, the detergent lysis results are very similar to those seen in the larger scale microfluidic lysis.
30. These growth media were also compared in a typical molecular biology *E. coli* strain, DH10B (also called TOP10). These cells are usually used only for plasmid production, but can in fact also be used for protein expression using non-T7 based expression systems. In this case, the point of the experiment was to demonstrate that the effects of autoinduction and Dynamite media are the same, independent of the *E. coli* strain used. DH10B is a K-12 type strain, which is significantly different from the *E. coli* B strain family.

Acknowledgments

This work has been funded in whole or in part with federal funds from the National Cancer Institute, National Institutes of Health, under contract HHSN261200800001E. The content of this chapter does not necessarily reflect the views or policies of the Department of Health and Human Services, nor does mention of trade names, commercial products, or organizations imply endorsement by the US Government.

References

1. Studier FW, Rosenberg AH, Dunn JJ et al (1990) Use of T7 RNA polymerase to direct expression of cloned genes. Methods Enzymol 185:60–89
2. Esposito D, Chatterjee DK (2006) Enhancement of soluble protein expression through the use of fusion tags. Curr Opin Biotechnol 17:353–358
3. Huang CJ, Lin H, Yang X (2013) Industrial production of recombinant therapeutics in *Escherichia coli* and its recent advancements. J Ind Microbiol Biotechnol 39:383–399
4. Burgess-Brown NA, Mahajan P, Strain-Damerell C et al (2014) Medium-throughput production of recombinant human proteins: protein production in *E. coli*. Methods Mol Biol 1091:73–94
5. Losen M, Frolich B, Pohl M et al (2004) Effect of oxygen limitation and medium composition on *Escherichia coli* fermentation in shake-flask cultures. Biotechnol Prog 20:1062–1068
6. Dyson MR, Shadbolt SP, Vincent KJ et al (2004) Production of soluble mammalian proteins in *Escherichia coli*: identification of protein features that correlate with successful expression. BMC Biotechnol 4:32
7. Studier FW (2005) Protein production by auto-induction in high density shaking cultures. Protein Expr Purif 41:207–234
8. Vera A, Gonzalez-Montalban N, Aris A et al (2007) The conformational quality of insoluble recombinant proteins is enhanced at low growth temperatures. Biotechnol Bioeng 96:1101–1106
9. Hartley JL, Temple GF, Brasch MA (2000) DNA cloning using in vitro site-specific recombination. Genome Res 10:1788–1795
10. Esposito D, Garvey LA, Chakiath CS (2009) Gateway cloning for protein expression. Methods Mol Biol 498:31–54
11. Fox JD, Routzahn KM, Bucher MH et al (2003) Maltodextrin-binding proteins from diverse bacteria and archaea are potent solubility enhancers. FEBS Lett 537:53–57
12. Grodberg J, Dunn JJ (1988) ompT encodes the *Escherichia coli* outer membrane protease that cleaves T7 RNA polymerase during purification. J Bacteriol 170:1245–1253
13. Michel E, Wuthrich K (2012) High-yield *Escherichia coli*-based cell-free expression of human proteins. J Biomol NMR 53:43–51
14. Welch M, Govindarajan S, Ness JE et al (2009) Design parameters to control synthetic gene expression in *Escherichia coli*. PLoS One 4:e7002
15. Hanahan D, Jessee J, Bloom FR (1991) Plasmid transformation of *Escherichia coli* and other bacteria. Methods Enzymol 204:63–113
16. Gillette WK, Esposito D, Taylor TE et al (2010) Purify first: rapid expression and purification of proteins from XMRV. Protein Expr Purif 76:238–247
17. Kapust RB, Tözsér J, Fox JD et al (2001) Tobacco etch virus protease: mechanism of autolysis and rational design of stable mutants with wild-type catalytic proficiency. Protein Eng 14:993–1000

Chapter 6

Optimization of Membrane Protein Production Using Titratable Strains of *E. coli*

Rosa Morra, Kate Young, David Casas-Mao, Neil Dixon, and Louise E. Bird

Abstract

The heterologous expression of membrane proteins driven by T7 RNA polymerase in *E. coli* is often limited by a mismatch between the transcriptional and translational rates resulting in saturation of the *Sec* translocon and non-insertion of the membrane protein. In order to optimize the levels of folded, functional inserted protein, it is important to correct this mismatch. In this protocol, we describe the use of titratable strains of *E. coli* where two small-molecule inducers are used in a bi-variate analysis to optimize the expression levels by fine tuning the transcriptional and translational rates of an eGFP-tagged membrane protein.

Key words Membrane protein, Green fluorescent protein (GFP), Titratable, T7 RNA polymerase, Riboswitch, RiboTite, T7 lysozyme

1 Introduction

The production of recombinant integral membrane proteins (IMPs) in *Escherichia coli* is challenging due to the generally low levels of expression and is often limited by biological effects of the expression on the host cell. This can be due to a number of reasons including the expressed protein may be toxic [1] or insufficient amino acid/amino acyl tRNA availability because of rare codons in the protein [2–4]. In addition, heterologous expression of IMPs often results in expressed but non-inserted protein (i.e., inclusion bodies), due to insufficient membrane capacity [5] and saturation of the membrane insertion machinery [6, 7]. Thus, for IMPs, screening for correctly folded protein is highly important and fusion to eGFP at either the N or C-terminus is now widely used as a reporter [8–10]. In the screening process, the fluorescence can be either monitored using a plate reader as an overall indicator of expression levels or using in-gel fluorescence to report both expression levels and the integrity of the expressed protein [11, 12].

Nicola A. Burgess-Brown (ed.), *Heterologous Gene Expression in E. coli: Methods and Protocols*, Methods in Molecular Biology, vol. 1586, DOI 10.1007/978-1-4939-6887-9_6, © Springer Science+Business Media LLC 2017

In addition, if GFP also incorporates an affinity tag such as octa-histidine, the in-gel GFP fluorescence signal can be used as an indicator for mono-dispersity and integrity of the membrane proteins during affinity tag purification [11–13].

1.1 T7 RNA Polymerase-Driven Expression of Heterologous Proteins

The *E. coli* strain BL21(DE3) and related strains were originally developed by Studier et al. [14]. In its simplest format, the heterologous protein is introduced on a plasmid under the control of T7 promoter while the bacteriophage T7 RNA polymerase is integrated into the bacterial chromosome as a phage λ lysogen and expressed from a mutant *lac*UV5 promoter following induction with Isopropyl β-D-1-thiogalactopyranoside (IPTG) [14]. However, there is often basal transcription of the T7 RNA polymerase and subsequent expression of the heterologous protein. For toxic proteins, this combined with the high induced transcription levels can be sufficient to induce plasmid instability or the heterologous protein being expressed in inclusion bodies [15]. To overcome these problems, a lac operator can be introduced into the T7 promoter in the expression plasmid to reduce the basal transcription levels and the LacI repressor protein can additionally be supplied on a plasmid or the endogenous *E. coli* protein can be used [16]. Alternatively, a compatible pLys plasmid, expressing the T7 lysozyme gene at different levels depending on the promoter configuration, can be used. T7 lysozyme is a natural allosteric inhibitor of the T7 RNA polymerase, and at low levels it will inhibit basal transcription from T7 promoters. At higher levels it will inhibit the fully induced RNA polymerase, reducing the transcription rate, and thus, the translational rate of the heterologous protein (Fig. 1a, [15]). These strains and plasmids are known as the T7 or pET expression system and have become almost ubiquitous in heterologous protein production in *E. coli*. However, despite these modifications to transcriptional regulation, the expression of membrane proteins often results in mis-folding of much of the expressed membrane protein probably due to the saturation of the *Sec* translocon [1, 7, 17].

Fig. 1 (continued) lysozyme produced by the pLysS plasmid, note T7 lysozyme is constitutively expressed at differential levels in pLys plasmids depending on the configuration of the plasmid. (**b**) KRX: L-rhamnose binds to RhaR depressing the *rha*P_{SR} promoter, in addition L-rhamnose also binds to RhaS which induces transcription of T7 RNA polymerase from the *rha*P_{BAD} promoter. In principle, addition of IPTG is also necessary to derepress the PT7/LacO promoter. (**c**) Lemo21 (DE3): IPTG binding to the LacI repressor results in depression of the lacUV5 promoter and expression of T7 RNA polymerase. L-rhamnose binds RhaR depressing the *rha*P_{SR} promoter, L-rhamnose also binds to RhaS, which induces expression of T7 lysozyme (LysY) from the *rha*P_{BAD} promoter. T7 lysozyme allosterically inhibits the activity of T7 RNA polymerase. (**d**) BL21(IL3): The T7 RNA Polymerase is transcribed from the wild-type Lac promoter in response to IPTG binding to LacI. Translation of the T7 polymerase mRNA is in turn activated in the presence of the small-molecule inducer PPDA that binds to the 5′ orthogonal riboswitch (ORS), releasing the ribosome-binding site. The level of T7 RNA polymerase is proportional to the amount of PPDA present. IPTG-dependent derepression of thePT7/LacO promoter results in transcription of YidC-eGFP by T7 RNA polymerase

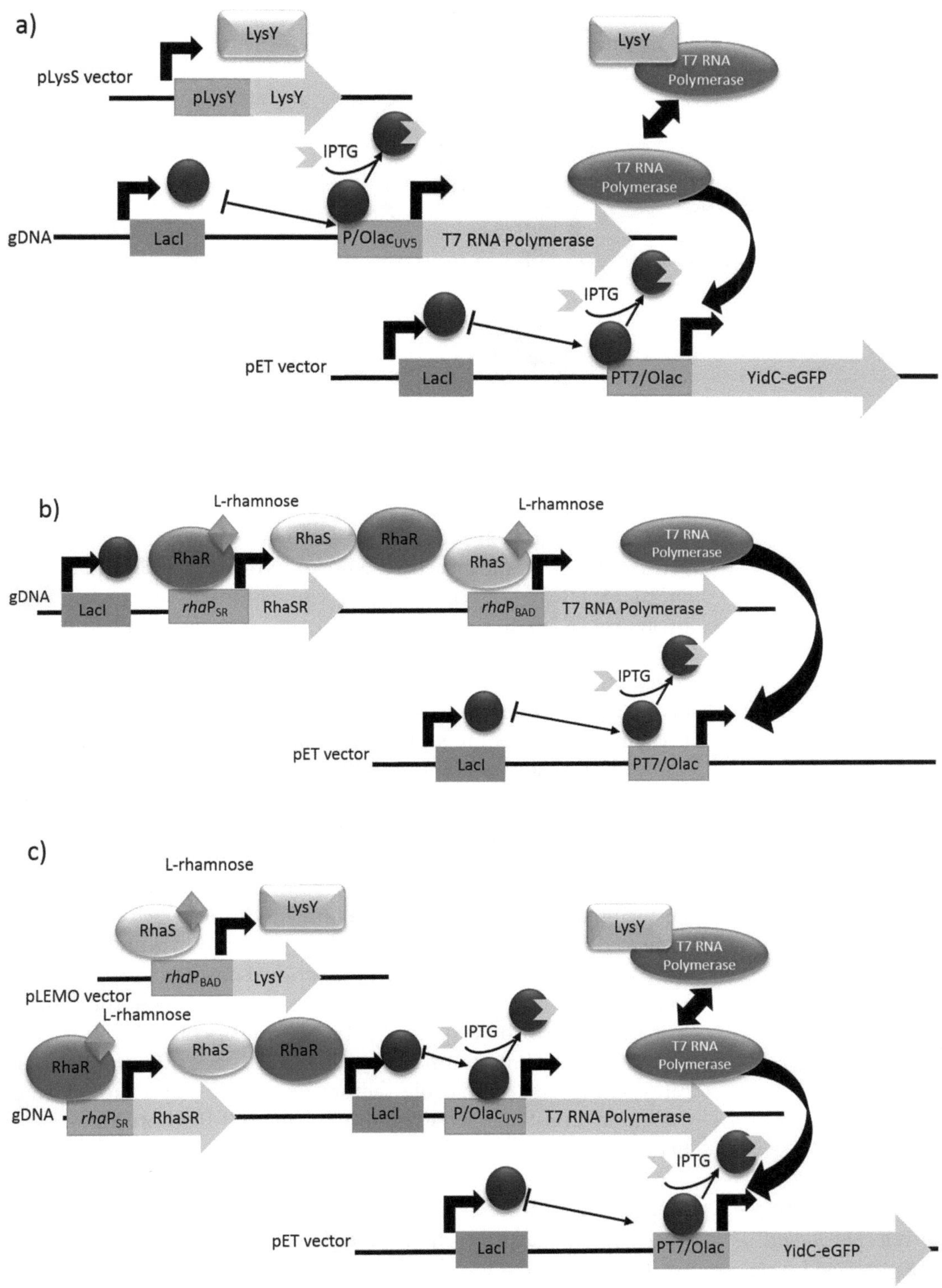

Fig. 1 Schematic diagram of the differential regulation of the T7 promoter in pET44b-YidC-eGFP in representative *E. coli* strains. In all cases, the induced promoters are shown. (**a**) C41 (DE3) pLysS: The inducer IPTG binds to the repressor LacI (LacI is expressed from the genomic copy of the gene and is also on the pET vector). Binding of IPTG decreases the affinity of LacI for the operator, thus derepressing the lacUV5 promoter and PT7/OLac promoter. Expression of T7 RNA polymerase occurs which in turn transcribes YidC-eGFP from the PT7/LacO promoter. The activity of the T7 RNA polymerase is modulated by the level of its allosteric inhibitor T7

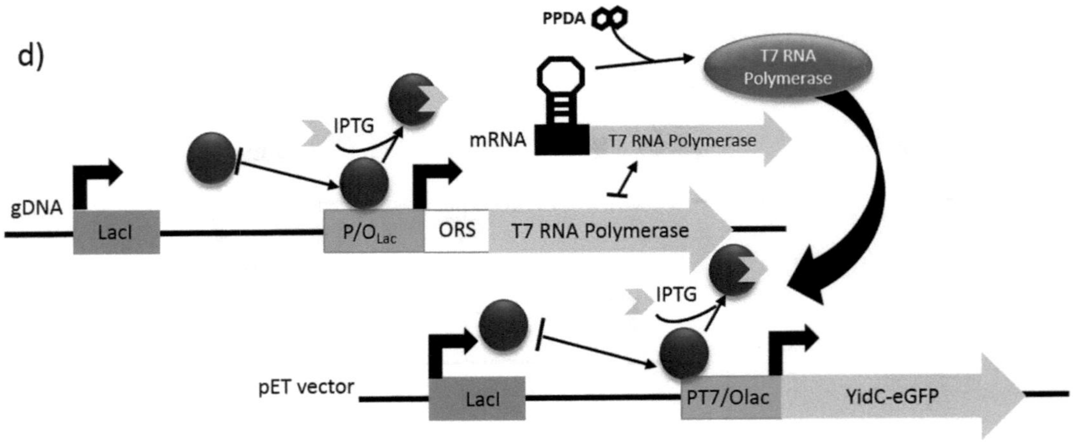

Fig. 1 (continued)

To try and solve the incompatibility between transcription rates and membrane insertion, Miroux and Walker selected isolates of BL21(DE3) that survived the expression of toxic membrane proteins [18]. The strains C41(DE3) and C43(DE3) are known as the Walker strains and are often used to express membrane proteins. However, this is on a protein-by-protein basis, as higher levels of expression of inserted protein are not seen for all membrane proteins in these strains [18, 19]. Analyses of the Walker strains showed that they harbored mutations in the *lac*UV5 promoter and that these are responsible for the often improved membrane protein expression [7]. The mutations found result in lower levels of production of heterologous mRNA due to lower levels of T7 RNA polymerase and hence a slower rate of protein synthesis, thus ensuring that membrane protein translocation machinery is not saturated [6, 7, 20]. These data supported the hypothesis that to maximize the levels of folded, functional inserted protein, it is important to match the rate of transcription and thus, the translational rate with the capacity of the *Sec* translocon. More recently, manipulation of the transcriptional, and indirectly the translational rates, using small-molecule factors in the T7 expression system (often in combination with monitoring GFP fluorescence) has been used to facilitate heterologous expression of IMPs [19–21].

1.2 Small-Molecule-Driven Fine-Tuning of Expression Levels

In Fig. 2a, it can be seen that a YidC-eGFP fusion protein cloned into a pET44 expression plasmid, expressed in C41(DE3) pLysS treated with a range of IPTG concentrations, is induced from a basal value to a maximum value over a relatively small concentration range (e.g., the response is not easily titratable). In order to match the expression levels of a heterologous membrane protein to the capacity of the *Sec* Translocon, inducible recombinant expression should ideally be regulated over a broad dynamic range, i.e., it should have a tunable dose-response curve. Specialized strains of

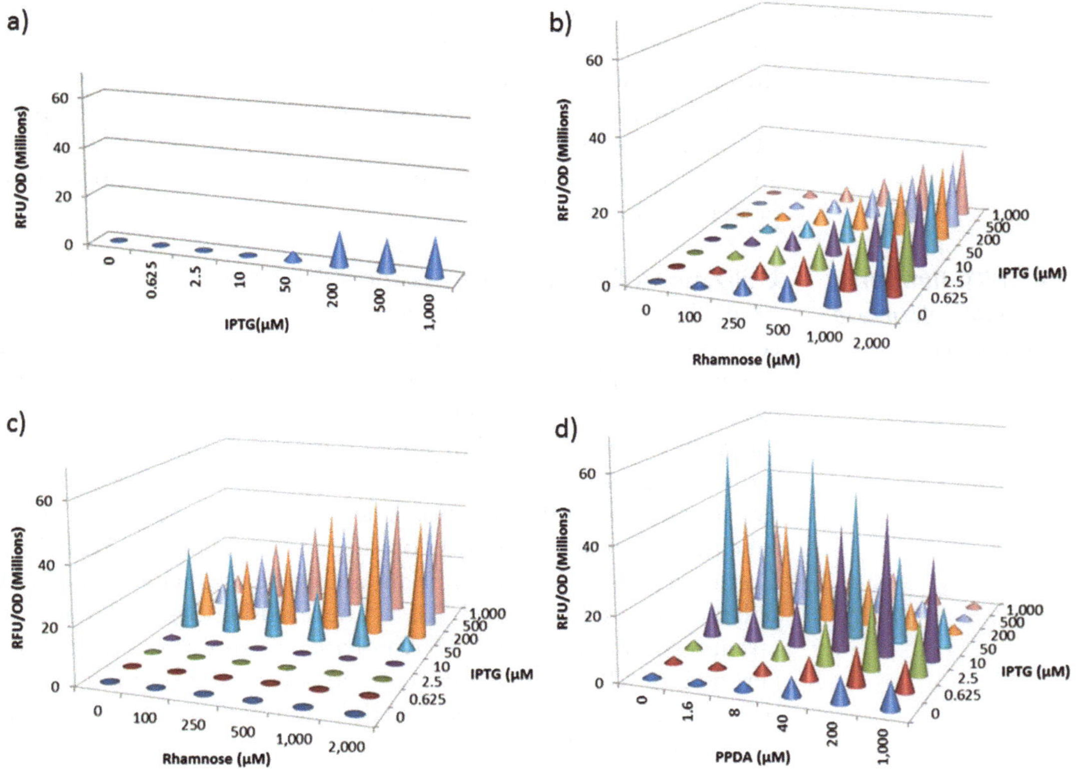

Fig. 2 Expression matrices of YidC-eGFP in response to small-molecule inducers RFU/OD are plotted against small-molecule inducer concentrations. (**a**) C41(DE3) pLysS response to IPTG. (**b**) KRX bi-variate analysis of the response to IPTG and L-rhamnose concentration. (**c**) Lemo21(DE3) bi-variate analysis of the response to IPTG and L-rhamnose. (**d**) BL21(IL3) bi-variate analysis of the response to IPTG and PPDA

E. coli are available in which expression is induced by a second small-molecule inducer in addition to IPTG, permitting fine-tuning of expression levels. These inducers can be used, in combination with IPTG, to either modify the expression rate at either the level of transcription, translation or allosteric control via inhibition of RNA polymerase [19, 20, 22–24].

1.3 Bi-Variate Analysis of Protein Expression

In this type of experiment, two small-molecule inducers are titrated in a matrix against each other to determine the optimal concentration of each molecule. It should be noted that this type of experiment would generally be performed after a primary expression screen in the different strains at one concentration of each small-molecule inducer (e.g., as described by Bird et al. or Lee et al. [10, 25]). In the protocol presented, GFP is used as a reporter of inserted membrane protein and the highest RFU/OD is used as a measure of the maximal expression level. GFP fluorescence/OD readout is used for membrane proteins because as the amount of membrane to be solubilized rises, either the cost rises due to the increased amounts of detergent needed for solubilization as the volume increases, or the detergent solubilization efficiency falls if the resuspension volume is not changed [10]. In this protocol, we describe methods to optimize expression of membrane proteins by a bi-variate analysis in three titratable strains: Lemo21(DE3), KRX, and BL21(IL3), since there can be strain-dependent expression of proteins [19, 25].

1.4 KRX: Titration of T7 RNA Polymerase Transcription

Utilization of a strain that has the T7 RNA polymerase under a titratable promoter allows the level of T7 RNA polymerase to be controlled by regulating its rate of transcription. Controlling the expression level of the T7 RNA polymerase directly affects the expression level of the heterologous protein (Fig. 1b). Two of the most commonly used promoters are the arabinose (P_{BAD}) and L-rhamnose inducible promoters (*rha*P_{BAD}). Both promoters have low levels of basal transcriptional activity and can be used to fine-tune transcriptional levels in a dose-dependent manner [23, 26–28]. Commercially, a strain with an arabinose inducible T7 RNA polymerase is sold as BL21-AI™ (Thermofisher Scientific) and an L-rhamnose inducible T7 RNA polymerase strain is sold as KRX (Promega). KRX is an *E. coli* K12 derivative (BL21-AI is a derivative of BL21 and is a B type strain) that incorporates a chromosomal copy of the T7 RNA polymerase gene in the *rha* locus. The T7 RNA polymerase gene replaces the *rha*BAD cistrons and its expression is thus controlled by the *rha*P_{BAD} promoter. Like many carbohydrate utilization operons, the *rha*P_{BAD} promoter is subject to catabolite repression by glucose and additionally, it is positively activated by a regulatory cascade of two activators, RhaR and RhaS, following addition of L-rhamnose to the medium [27, 29]. Since in KRX the isomerase (RhaA), kinase (RhaB), and aldolase (RhaD) are replaced with the gene for T7 RNA polymerase, the L-rhamnose concentration remains stable during induction as it is not metabolized by the cell [24]. Representative results of a bi-variate analysis of expression pET44-YidC-eGFP with respect to L-rhamnose and IPTG concentrations for KRX are shown in Fig. 2b. KRX shows a good titratable response to L-rhamnose but relatively little response to IPTG at any concentration of L-rhamnose. This observation is similar to the result seen for eGFP alone by Morra et al. [22].

1.5 Inhibition of T7 RNA Polymerase Activity Using Titratable T7 Lysozyme

As already mentioned, T7 lysozyme allosterically inhibits T7 RNA polymerase and this is the basis of the pLys series of plasmids [15]. These plasmids express constitutively T7 lysozyme at a fixed level. A modification of this regulatory system is the Lemo21(DE3) strain where the gene for T7 lysozyme (LysY) is on a plasmid under the control of the highly titratable L-rhamnose promoter *rha*P_{BAD} described above (Fig. 1c, [19, 30]). Wagner et al. showed that by optimizing L-rhamnose levels and thus regulating the activity of the T7 RNA polymerase, a range of membrane proteins were more highly expressed in this strain [19]. Representative results of a bi-variate analysis of expression pET44-YidC-eGFP with respect to L-rhamnose and IPTG concentrations for Lemo21 (DE3) are shown in Fig. 2c. At concentrations of 50 μM IPTG and above, there is expression of the YIDC-eGFP. At 50 μM IPTG, the highest-level of expression is obtained if there is no or a very low level of L-rhamnose. Increasing the concentration of L-rhamnose concentration results in decreased expression, due to inhibition of the T7 RNA polymerase. In contrast, at concentrations of IPTG (≥ 200 μM), increasing the concentration of L-rhamnose results in higher expression. Presumably, over induction (high IPTG, low Rhamnose) results in low levels of membrane-inserted protein as there is too much active T7 RNA polymerase and the YidC-eGFP transcript and/or protein overload the capacity of the host's ribosomal and/or secretion pathways respectively, i.e., forming inclusion bodies. Hence, it would seem that within the optimal induction window, the expression rate/burden of YidC-eGFP are likely balanced relative to insertion capacity of the *Sec* translocon.

1.6 BL21(IL3) Post-transcriptional Control of Heterologous Protein Expression

In bacteria, a number of post-transcriptional control mechanisms that regulate gene expression have been reported. This includes riboswitches [31], tRNA sensing [32, 33], and ribosome pausing [34]. Riboswitches are mostly found in the 5′ untranslated region (UTR) of bacterial mRNAs, and function via a number of mechanisms, including translation initiation, transcription termination, and mRNA cleavage [35–37]. Riboswitches function by changing conformation when they are bound to a specific metabolite [35], thus regulating expression post-transcriptionally; this represents a novel target for use as a genetic control element [38]. Dixon et al. re-engineered the *Vibrio vulnificus* Add-A translational ON riboswitch to no longer bind to its cognate metabolite, but instead to a synthetic inducer [39]. This regulatory element works by sequestration and release of the ribosome-binding site in the absence and presence of the small-molecule inducer respectively. The same authors also demonstrated that the orthogonal riboswitch (ORS) can control heterologous gene expression in vivo [39, 40]. Using these novel translational control elements in combination with a standard transcriptional control element, the RiboTite expression system has recently been developed [22]. The RiboTite system can be used in two different formats, i.e., using the chromosomally engineered expression strain BL21(IL3) alone, or the BL21(IL3) strain in combination with a modified expression plasmid (pETORS). The BL21(IL3) strain contains the T7 RNA polymerase under the dual control of both the lac promoter/operator (t; transcriptional control) and orthogonal riboswitch (T; translational control). This can be combined with a standard pET vector (*t*T/*t*) (Fig. 1d) or a pETORS vector (*t*T/*t*T) [22]. Using expression of eGFP in a pET vector backbone as a test system, Morra et al. showed that both formats had a IPTG (transcriptional inducer) and PPDA (translational inducer)-dependent titratable response; the *t*T/*t* system demonstrated >400-fold induction and dynamic range of ligand response (DRLR = EC_{90}/EC_{10}; in response to PPDA) of 79, while the *t*T/*t*T system displayed around 850-fold induction and a DRLR of 245. The expression levels were responsive to both IPTG concentration and PPDA concentration, and they demonstrated that the system can be precisely tuned to permit dynamic control upon addition of small-molecule inducers [22].

Representative results for the expression of a pET44-YidC-eGFP construct can be seen in Fig. 2d. In this experiment, the BL21(IL3) was tested only with a pET-YidC-eGFP plasmid (e.g., *t*T/t) so that a direct comparison can be made with the other strains tested as an identical plasmid is used. In contrast to the expression of eGFP alone in BL21(IL3) where smooth dose responses were observed with PPDA at all concentrations of IPTG [22], the YidC-eGFP displays a more complex expression profile. Maximal expression is observed with intermediate IPTG concentrations (50 μM) and low PPDA concentrations (1.6–40 μM). As observed for Lemo21, over induction of the T7 RNA polymerase results in a drop in expression levels, presumably outside the optimal induction window the expression rate/burden of YidC-eGFP are overloading the capacity of the *Sec* translocon.

The integrity of the YidC-eGFP recombinant protein was analyzed using in-gel fluorescence (Fig. 3). In all strains, intact fusion protein (upper band) was observed, in addition to a small amount of fusion degradation (lower band). The test experiment demonstrates

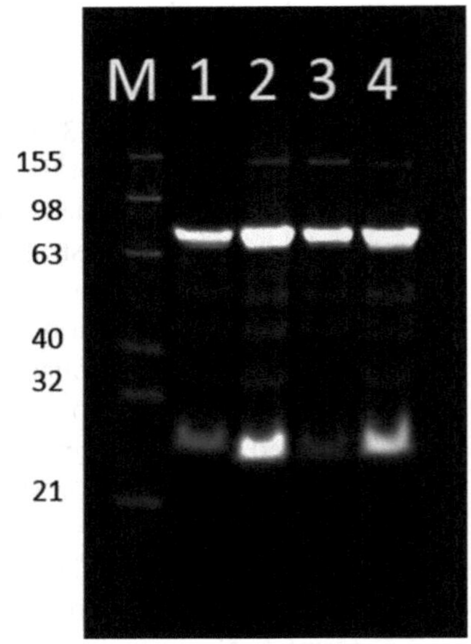

Fig. 3 In-gel fluorescence of the conditions with the highest RFU/OD. M Markers, 1 C41 (DE3) plysS, 2 BL21 (IL3), 3 KRX, 4 Lemo21(DE3)

the utility of carrying out such a bi-variant analysis for the expression of the inner membrane protein YidC-eGFP as the expression level is significantly higher than C41(DE3) pLysS (Fig. 3). In terms of maximal specific expression (RFU/OD), BL21(IL3) produced the greatest amount of protein (Figs. 2 and 3), producing around 25% more YidC-eGFP expression/OD than the next best strain, Lemo 21(DE3), and 163% and 267% more than KRX and C41(DE3) pLysS respectively.

2 Materials

2.1 Vectors

1. An *E. coli T7* promoter expression vector harboring the gene of interest fused to the GFP gene should be used in the expression screening experiment (*see* **Note 1**).
2. An *E. coli T7* promoter expression vector without GFP gene can be used as a negative control if desired.

2.2 Small-Scale *E. coli* Expression Screen

2.2.1 Transformation, Growth and Induction of Cultures

1. *E. coli* expression strains: Chemically competent expression strains of Lemo21(DE3) (NEB), KRX (Promega), and BL21(IL3) [22] are prepared beforehand and stored in 50 μL aliquots at −80 °C [41].
2. Terrific broth (TB): Autoclave 12 g of tryptone, 24 g of yeast extract and 4 mL of glycerol dissolved in 900 mL of water. When the medium has cooled, add 100 mL of 0.17 M KH_2PO_4 and 0.72 M K_2HPO_4 (*see* **Note 2**).
3. SOC: Autoclave 20 g of tryptone, 5 g of yeast extract, 0.585 g of 10 mM NaCl, 0.186 g of KCl, and 0.952 g of anhydrous $MgCl_2$ dissolved in 900 mL of water. When the medium has cooled, add 100 mL of filter sterilized 200 mM Glucose.
4. LB-Agar Plates: The plates are prepared as follows: 5 mL of LB agar supplemented with the appropriate antibiotic/s is poured per well into a 6-well culture plate, and allowed to cool (*see* **Note 3**).
5. 1 M Isopropyl β-D-1-thiogalactopyranoside (IPTG): Dissolve 23.83 g of IPTG in 100 mL of water and filter sterilize through a 0.2 μm filter. Store at −20 °C when not in use.
6. IPTG dilutions: Dilute the 1 M stock to make 10 mL each of working stocks: 0.0625 mM, 0.25 mM, 1 mM, 5 mM, 20 mM, 50 mM, and 100 mM of IPTG. Store at −20 °C when not in use.
7. 1 M and 0.1 M L-rhamnose: Dissolve 18.22 g or 1.822 g of L-rhamnose monohydrate, respectively, in 100 mL of water and filter sterilize through a 0.2 μm filter.
8. L-Rhamnose dilutions: L-rhamnose: Dilute the 1 M or the 0.1 M stock to prepare 5 mM, 12.5 mM, 25 mM, 50 mM, and 100 mM working stocks. Store at −20 °C when not in use.

9. 20 mM Pyrimido[4,5-d]pyrimidine-2,4-diamine. HCl (PPDA) (Peakdale Molecular). Dissolve 234 mg in 50 mL of water, sonicate in a water bath for 30 min, centrifuge at 4500 x *g* for 5 min, filter through a 0.45 μm and then a 0.2 μm filter. Aliquot and store at −20 °C when not in use.
10. PPDA dilutions: Dilute the 20 mM stock to prepare 0.032 mM, 0.16 mM, 0.8 mM, and 4 mM PPDA working stocks. Store at −20 °C when not in use.
11. 96-well deep well plates.
12. Gas adhesive permeable seals.
13. 24-well deep well blocks.
14. 6-well culture plates.
15. 50 mL tubes.
16. Sterile 500 mL and 2 L Erlenmeyer flasks.
17. Refrigerated incubator for flasks and deep well blocks, e.g., Innova 44R which takes both flasks and deep well blocks and Shel lab SSI5R-HS Floor Model Shaking Incubator, 5 Cu. Ft. (144 L) which is more suitable for blocks; and Stuart microtiter plate shaker incubator for blocks only.
18. Laminar flow hood.

2.2.2 Harvesting and Processing of Cultures

1. Phosphate buffered saline (PBS): 137 mM NaCl, 2.7 mM KCl, 10 mM Na_2HPO_4, and 1.8 mM KH_2PO_4. Prepare by adding 8 g of NaCl, 0.2 g of KCl, 1.44 g of Na_2HPO_4, and 0.24 g of KH_2PO_4 to 900 mL of water, adjust the pH to 7.4 with HCl and make the volume up to 1 L. Sterilize by autoclaving (*see* **Note 4**).
2. 96-well deep well plates.
3. Foil seals (e.g., Corning® 96 Well Microplate Aluminum Sealing Tape, these maintain a seal at −80 °C).

2.3 Measurement of Relative Fluorescence Units (RFU) and Optical Density (OD)

1. 96-well clear flat-bottomed microtiter plates.
2. Black walled microtiter plates with an optically clear bottom or solid black microtiter plates.
3. Microplate spectrofluorometer that reads both GFP emission at 512 nm following excitation at 488 nm and absorbance at 595 nm.

2.4 In-Gel Fluorescence

1. Phosphate buffered saline (PBS): 137 mM NaCl, 2.7 mM KCl, 10 mM Na_2HPO_4, 1.8 mM KH_2PO_4. Prepare by adding 8 g of NaCl, 0.2 g of KCl, 1.44 g of Na_2HPO_4, and 0.24 g of KH_2PO_4 to 900 mL of water, adjust the pH to 7.4 with HCl and make the volume up to 1 L. Sterilize by autoclaving (*see* **Note 4**).

2. DNAse I Stock Solutions (40,000 units/mL): To a 200,000 unit bottle of DNAse I, add 5 mL of sterile water. Store 25 μL aliquots at −20 °C.
3. Protease Inhibitor Cocktail: Aliquot a suitable DMSO-based cocktail, mix into 50 μL aliquots, store at −20 °C.
4. DNAseI, Lysozyme, and Protease inhibitor solution (DLPI): Add 1 aliquot of DNAse I and 1 of protease inhibitors to 2 mL of 10 mg/mL Lysozyme solution in PBS.
5. DDM: make a 10% (w/v) solution of n-Dodecyl-β-D-Maltopyranoside in water. Store 100 μL aliquots at −20 °C.
6. 2× SDS-PAGE loading buffer: 100 mM Tris, pH 6.8, 4% (w/v) SDS, 0.2% (w/v) Bromophenol blue, 10% (v/v) β-mercaptoethanol, 20% (v/v) glycerol.
7. 96-well PCR plate.
8. Foil seals.
9. 1.5 mL microcentrifuge tubes.
10. SDS-PAGE gels.
11. SDS-PAGE running buffer.
12. Gel documentation system with blue illumination and a blue-light filter to detect the GFP fusion proteins.

3 Methods

3.1 Transformation into E. coli Expression Strains

1. Thaw an aliquot of competent *E. coli* cells on ice.
2. Use 50–100 ng of mini-prepped GFP expression plasmid (and the expression negative control if desired) to transform the aliquot/s of competent cells (*see* **Note 5**).
3. Incubate on ice for 30 min.
4. Heat-shock the cells for 30 s at 42 °C.
5. Return the cells to ice for 2 min.
6. Add 300 μL of Terrific broth (TB) or SOC per tube.
7. Transfer the cells to a 37 °C static incubator and incubate for 1 h.
8. Transfer 60 μL of cells from the transformation mix onto the LB-Agar plates.
9. Shake the plates in a horizontal plane to spread the cells and dry with the lid off for 15–20 min in a lamina flow hood before replacing the lid and turning upside down.
10. Incubate overnight at 37 °C.

3.2 Preparation of Overnight Cultures

1. Add 10 mL of TB supplemented with 0.2% (w/v) Glucose and the appropriate antibiotics into a 50 mL tube.
2. Pick individual colonies and use to inoculate each tube.
3. Shake the tubes at 250 rpm (in a standard incubator) at 37 °C overnight.

3.3 Growth and Induction of Cultures Using 24-Well Deep Well Blocks (See Note 6)

3.3.1 Lemo21(DE3)

1. Set up six 500 mL flasks each with 100 mL of TB supplemented with 0.2% (w/v) Glucose and appropriate antibiotics.
2. Add the volume required of L-rhamnose from the 100 mM stock to yield one flask each of final concentrations of 0 μM (0 mL), 100 μM (0.1 mL), 250 μM (0.25 mL), 500 μM (0.5 mL), 750 μM (0.75 mL), and 1000 μM (1.0 mL) L-rhamnose.
3. Inoculate each of the six flasks with 1 mL of the overnight starter culture.
4. Allow the flask cultures to grow at 37 °C with shaking at 250 rpm until A595 is 0.4–0.5.
5. Reduce the temperature and allow the flask cultures to grow at 20 °C with shaking at 250 rpm until A595 is 0.6–0.7.
6. While the cells are growing, add 20 μL/well of the IPTG working stocks to four 24-well deep well blocks, as shown in Table 1 (*see* **Note 7**). Duplicates for each condition are set up.
7. Aliquot 2 mL of culture/well.
8. Grow cultures overnight with shaking at 20 °C at 250 rpm (*see* **Note 8**).

3.3.2 KRX (See Note 9)

1. Set up one 2 L flask with 300 mL of TB supplemented with 0.2% (w/v) Glucose and appropriate antibiotics.
2. Add 3 mL of the overnight starter culture to the 2 L flask.
3. Allow the flask cultures to grow at 37 °C with shaking at 250 rpm until A595 is 0.4–0.5.
4. Reduce the temperature and allow the flask cultures to grow at 20 °C with shaking at 250 rpm until A595 is 0.6–0.7.
5. While the cells are growing, add 20 μL/well of the IPTG stocks and 40 μL/well of the L-rhamnose stocks to four 24-well deep well blocks, as shown in Table 2 (*see* **Note 7**).
6. Aliquot 2 mL of culture/well.
7. Grow cultures overnight with shaking at 250 rpm at 20 °C (*see* **Note 8**).

3.3.3 IL3

1. Set up one 2 L flask with 300 mL of TB supplemented with 0.2% (w/v) Glucose and appropriate antibiotics.
2. Add 3 mL of the overnight starter culture to the 2 L.

Table 1

Bi-variate analysis of expression in Lemo21(DE3). The volume of IPTG added to 2 mL of the indicated culture in the analysis of expression in 24-well deep well blocks is described. Block 1 is wells A1 to D6 in the harvested 96-well block, Block 2 is wells E1 to H6, Block 3 is wells A7 to D12, and Block 4 for is wells E7 to H12. The same concentration matrix is used when the screen is carried out in a 96-well block and the volumes are as described in the text of Subheading 3.4

Volume IPTG stock/2 ml culture	Column	0 μM rhamnose culture		100 μM rhamnose culture		250 μM rhamnose culture			500 μM rhamnose culture		750 μM rhamnose culture		1000 μM rhamnose culture	
		1	2	3	4	5	6		1	2	3	4	5	6
20 μL 0 mM	Block 1 A	0 μM IPTG	0 μM IPTG	0 μM IPTG	0 μM IPTG	0 μM IPTG	0 μM IPTG	Block 3 A	0 μM IPTG	0 μM IPTG	0 μM IPTG	0 μM IPTG	0 μM IPTG	0 μM IPTG
20 μL 0.0625 mM	Block 1 B	0.625 μM IPTG	0.625 μM IPTG	0.625 μM IPTG	0.625 μM IPTG	0.625 μM IPTG	0.625 μM IPTG	Block 3 B	0.625 μM IPTG	0.625 μM IPTG	0.625 μM IPTG	0.625 μM IPTG	0.625 μM IPTG	0.625 μM IPTG
20 μL 0.25 mM	Block 1 C	2.5 μM IPTG	2.5 μM IPTG	2.5 μM IPTG	2.5 μM IPTG	2.5 μM IPTG	2.5 μM IPTG	Block 3 C	2.5 μM IPTG	2.5 μM IPTG	2.5 μM IPTG	2.5 μM IPTG	2.5 μM IPTG	2.5 μM IPTG
20 μL 1 mM	Block 1 D	10 μM IPTG	10 μM IPTG	10 μM IPTG	10 μM IPTG	10 μM IPTG	10 μM IPTG	Block 3 D	10 μM IPTG	10 μM IPTG	10 μM IPTG	10 μM IPTG	10 μM IPTG	10 μM IPTG
20 μL 5 mM	Block 2 A	50 μM IPTG	50 μM IPTG	50 μM IPTG	50 μM IPTG	50 μM IPTG	50 μM IPTG	Block 4 A	50 μM IPTG	50 μM IPTG	50 μM IPTG	50 μM IPTG	50 μM IPTG	50 μM IPTG
20 μL 20 mM	Block 2 B	200 μM IPTG	200 μM IPTG	200 μM IPTG	200 μM IPTG	200 μM IPTG	200 μM IPTG	Block 4 B	200 μM IPTG	200 μM IPTG	200 μM IPTG	200 μM IPTG	200 μM IPTG	200 μM IPTG
20 μL 50 mM	Block 2 C	500 μM IPTG	500 μM IPTG	500 μM IPTG	500 μM IPTG	500 μM IPTG	500 μM IPTG	Block 4 C	500 μM IPTG	500 μM IPTG	500 μM IPTG	500 μM IPTG	500 μM IPTG	500 μM IPTG
20 μL 100 mM	Block 2 D	1000 μM IPTG	1000 μM IPTG	1000 μM IPTG	1000 μM IPTG	1000 μM IPTG	1000 μM IPTG	Block 4 D	1000 μM IPTG	1000 μM IPTG	1000 μM IPTG	1000 μM IPTG	1000 μM IPTG	1000 μM IPTG

Table 2
Bi-variate analysis of expression in KRX. The volumes and concentrations of L-rhamnose and IPTG added to 2 mL of culture in the analysis of expression in 24-well deep well blocks are indicated. Block 1 is wells A1 to D6 in the harvested 96-well block, Block 2 is wells E1 to H6, Block 3 is wells A7 to D12, and Block 4 is wells E7-H12. The same concentration matrix is used when the screen is carried out in a 96-well block and the volumes are as described in the text of Subheading 3.4

Volume IPTG stock	Volume rhamnose stock	40 μL 0 mM rhamnose		40 μL 5 mM rhamnose		40 μL 12.5 mM rhamnose	
		1	2	3	4	5	6
20 μL 0 mM	Block 1 A	0 μM IPTG 0 μM rhamnose	0 μM IPTG 0 μM rhamnose	0 μm iptg 100 μm rhamnose	0 μM IPTG 100 μM rhamnose	0 μM IPTG 250 μM rhamnose	0 μM IPTG 250 μM rhamnose
20 μL 0.0625 mM	Block 1 B	0.625 μM IPTG 0 μM rhamnose	0.625 μM IPTG 0 μM rhamnose	0.625 μM IPTG 100 μM rhamnose	0.625 μM IPTG 100 μM rhamnose	0.625 μM IPTG 250 μM rhamnose	0.625 μM IPTG 250 μM rhamnose
20 μL 0.25 mM	Block 1 C	2.5 μM IPTG 0 μM rhamnose	2.5 μM IPTG 0 μM rhamnose	2.5 μM IPTG 100 μM rhamnose	2.5 μM IPTG 100 μM rhamnose	2.5 μM IPTG 250 μM rhamnose	2.5 μM IPTG 250 μM rhamnose
20 μL 1 mM	Block 1 D	10 μM IPTG 0 μM rhamnose	10 μM IPTG 0 μM rhamnose	10 μM IPTG 100 μM rhamnose	10 μM IPTG 100 μM rhamnose	10 μM IPTG 250 μM rhamnose	10 μM IPTG 250 μM rhamnose
20 μL 5 mM	Block 2 A	50 μM IPTG 0 μM rhamnose	50 μM IPTG 0 μM rhamnose	50 μM IPTG 100 μM rhamnose	50 μM IPTG 100 μM rhamnose	50 μM IPTG 250 μM rhamnose	50 μM IPTG 250 μM rhamnose
20 μL 20 mM	Block 2 B	200 μM IPTG 0 μM rhamnose	200 μM IPTG 0 μM rhamnose	200 μM IPTG 100 μM rhamnose	200 μM IPTG 100 μM rhamnose	200 μM IPTG 250 μM rhamnose	200 μM IPTG 250 μM rhamnose

20 µL 50 mM	Block 2 C	500 µM IPTG 0 µM rhamnose	500 µM IPTG 0 µM rhamnose	500 µM IPTG 100 µM rhamnose	500 µM IPTG 100 µM rhamnose	500 µM IPTG 250 µM rhamnose	500 µM IPTG 250 µM rhamnose
20 µL 100 mM	Block 2 D	1000 µM IPTG 0 µM rhamnose	1000 µM IPTG 0 µM rhamnose	1000 µM IPTG 100 µM rhamnose	1000 µM IPTG 100 µM rhamnose	1000 µM IPTG 250 µM rhamnose	1000 µM IPTG 250 µM rhamnose

Volume IPTG stock	Volume Rhamnose Stock	40 µL 25 mM rhamnose		40 µL 50 mM rhamnose		40 µL 100 mM rhamnose	
		1	2	3	4	5	6
20 µL 0 mM	Block 3 A	0 µM IPTG 500 µM rhamnose	0 µM IPTG 500 µM rhamnose	0 µM IPTG 1000 µM rhamnose	0 µM IPTG 1000 µM rhamnose	0 µM IPTG 2000 µM rhamnose	0 µM IPTG 2000 µM rhamnose
20 µL 0.0625 mM	Block 3 B	0.625 µM IPTG 500 µM rhamnose	0.625 µM IPTG 500 µM rhamnose	0.625 µM IPTG 1000 µM rhamnose	0.625 µM IPTG 1000 µM rhamnose	0.625 µM IPTG 2000 µM rhamnose	0.625 µM IPTG 2000 µM rhamnose
20 µL 0.25 mM	Block 3 C	2.5 µM IPTG 500 µM rhamnose	2.5 µM IPTG 500 µM rhamnose	2.5 µM IPTG 1000 µM rhamnose	2.5 µM IPTG 1000 µM rhamnose	2.5 µM IPTG 2000 µM rhamnose	2.5 µM IPTG 2000 µM rhamnose

(continued)

Table 2
(continued)

Volume IPTG stock	Volume Rhamnose Stock	40 µL 25 mM rhamnose		40 µL 50 mM rhamnose		40 µL 100 mM rhamnose	
		1	2	3	4	5	6
20 µL 1 mM	Block 3 D	10 µM IPTG	10 µM IPTG	10 µM IPTG	10 µM IPTG	10 µM IPTG	10 µM IPTG
		500 µM rhamnose	500 µM rhamnose	1000 µM rhamnose	1000 µM rhamnose	2000 µM rhamnose	2000 µM rhamnose
20 µL 5 mM	Block 4 A	50 µM IPTG	50 µM IPTG	50 µM IPTG	50 µM IPTG	50 µM IPTG	50 µM IPTG
		500 µM rhamnose	500 µM rhamnose	1000 µM rhamnose	1000 µM rhamnose	2000 µM rhamnose	2000 µM rhamnose
20 µL 20 mM	Block 4 B	200 µM IPTG	200 µM IPTG	200 µM IPTG	200 µM IPTG	200 µM IPTG	200 µM IPTG
		500 µM rhamnose	500 µM rhamnose	1000 µM rhamnose	1000 µM rhamnose	2000 µM rhamnose	2000 µM rhamnose
20 µL 50 mM	Block 4 C	500 µM IPTG	500 µM IPTG	500 µM IPTG	500 µM IPTG	500 µM IPTG	500 µM IPTG
		500 µM rhamnose	500 µM rhamnose	1000 µM rhamnose	1000 µM rhamnose	2000 µM rhamnose	2000 µM rhamnose
20 µL 100 mM	Block 4 D	1000 µM IPTG	1000 µM IPTG	1000 µM IPTG	1000 µM IPTG	1000 µM IPTG	1000 µM IPTG
		500 µM rhamnose	500 µM rhamnose	1000 µM rhamnose	1000 µM rhamnose	2000 µM rhamnose	2000 µM rhamnose

3. Allow the flask cultures to grow at 37 °C with shaking at 250 rpm until A595 is 0.4–0.5.
4. Reduce the temperature and allow the flask cultures to grow at 20 °C with shaking at 250 rpm until A595 is 0.6–0.7
5. While the cells are growing, add 20 μL/well of the IPTG working stocks and 100 μL/well of the PPDA stocks to four 24-well deep well blocks, as shown in Table 3 (*see* **Note 7**).
6. Aliquot 2 mL of culture/well.
7. Grow cultures overnight with shaking at 250 rpm at 20 °C (*see* **Note 8**).

3.4 Growth and Induction of Cultures Using 96-Well Deep Well Blocks (*See* Note 10)

1. Inoculate individual fresh recombinant colonies from each strain (Lemo21, KRX, and IL3) into 5 mL of TB, supplemented with appropriate antibiotics and 0.2% (w/v) glucose (*see* **Note 11**).
2. Incubate at 37 °C with shaking at 250 rpm.
3. After ~3 h, transfer the 5 mL culture into 25 mL of fresh TB supplemented with appropriate antibiotics, L-rhamnose for Lemo21(DE3), and 0.2% (w/v) glucose.
4. Grow at 37 °C shaking (250 rpm) until A595 ~0.6.
5. For each strain, prepare an inducer concentration matrix in a 96-well microplate (the same matrix as shown in Tables 1–3) by adding 5 μL of 100× IPTG, 10 μL of L-rhamnose, and 25 μL of PPDA working stocks as appropriate. Each inducer concentration matrix can hold duplicates.
6. Add 500 μL of each culture from **step 4** to each well of the matrix 96-well microplate.
7. Grow cells at 20 °C with shaking (1250 rpm) in a suitable microtiter plate shaker incubator for the desired time.

3.5 Harvesting of Cultures

3.5.1 24-Well Blocks

1. Transfer 0.5 mL of culture from each well into 96-well deep-well block (*see* **Note 12**). Taking care to format the block so that the gradients are as shown in Tables 1–3 e.g., blocks 1 and 3 are on the left-hand side and 2 and 4 are on the right-hand side.
2. Seal the block and harvest the cells by centrifugation at 6000 × *g* 4 °C for 10 min.
3. Decant the media from the cell pellets by inverting the block and drain by tapping onto a paper towel.
4. Wash pellets with 1 mL of PBS and spin again as above.
5. Resuspend pellets in 1 mL of PBS.

Table 3
Bi-variate analysis of expression in BL21(IL3). The volumes and concentrations of PPDA and IPTG added to 2 mL of culture in the analysis of expression of BL21 (IL3) in 24-well deep well blocks are indicated. Block 1 is wells A1 to D6 in the harvested 96-well block, Block 2 is wells E1 to H6, Block 3 is wells A7 to D12, and Block 4 is wells E7-H12. The same concentration matrix is used when the screen is carried out in a 96-well block and the volumes are as described in the text of Subheading 3.4

Volume IPTG stock	Volume rhamnose stock	100 µL 0 mM PPDA		100 µL 0.032 mM PPDa		100 µL 0.16 mM PPDA	
		1	2	3	4	5	6
20 µL 0 mM	Block 1 A	0 µM IPTG 0 µM PPDA	0 µM IPTG 0 µM PPDA	0 µM IPTG 1.6 µM PPDA	0 µM IPTG 1.6 µM PPDA	0 µM IPTG 8 µM PPDA	0 µM IPTG 8 µM PPDA
20 µL 0.0625 mM	Block 1 B	0.625 µM IPTG 0 µM PPDA	0.625 µM IPTG 0 µM PPDA	0.625 µM IPTG 1.6 µM PPDA	0.625 µM IPTG 1.6 µM PPDA	0.625 µM IPTG 8 µM PPDA	0.625 µM IPTG 8 µM PPDA
20 µL 0.25 mM	Block 1 C	2.5 µM IPTG 0 µM PPDA	2.5 µM IPTG 0 µM PPDA	2.5 µM IPTG 1.6 µM PPDA	2.5 µM IPTG 1.6 µM PPDA	2.5 µM IPTG 8 µM PPDA	2.5 µM IPTG 8 µM PPDA
20 µL 1 mM	Block 1 D	10 µM IPTG 0 µM PPDA	10 µM IPTG 0 µM PPDA	10 µM IPTG 1.6 µM PPDA	10 µM IPTG 1.6 µM PPDA	10 µM IPTG 8 µM PPDA	10 µM IPTG 8 µM PPDA
20 µL 5 mM	Block 2 A	50 µM IPTG 0 µM PPDA	50 µM IPTG 0 µM PPDA	50 µM IPTG 1.6 µM PPDA	50 µM IPTG 1.6 µM PPDA	50 µM IPTG 8 µM PPDA	50 µM IPTG 8 µM PPDA

20 μL 20 mM	Block 2 B	200 μM IPTG 0 μM PPDA	200 μM IPTG 0 μM PPDA	200 μM IPTG 1.6 μM PPDA	200 μM IPTG 1.6 μM PPDA	200 μM IPTG 8 μM PPDA	200 μM IPTG 8 μM PPDA
20 μL 50 mM	Block 2 C	500 μM IPTG 0 μM PPDA	500 μM IPTG 0 μM PPDA	500 μM IPTG 1.6 μM PPDA	500 μM IPTG 1.6 μM PPDA	500 μM IPTG 8 μM PPDA	500 μM IPTG 8 μM PPDA
20 μL 100 mM	Block 2 D	1000 μM IPTG 0 μM PPDA	1000 μM IPTG 0 μM PPDA	1000 μM IPTG 1.6 μM PPDA	1000 μM IPTG 1.6 μM PPDA	1000 μM IPTG 8 μM PPDA	1000 μM IPTG 8 μM PPDA

Volume IPTG stock	Volume rhamnose stock	100 μL 0.8 mM PPDA		100 μL 4 mM PPDA		100 μL 20 mM PPDA	
		1	2	3	4	5	6
20 μL 0 mM	Block 3 A	0 μM IPTG 40 μM PPDA	0 μM IPTG 40 μM PPDA	0 μM IPTG 200 μM PPDA	0 μM IPTG 200 μM PPDA	0 μM IPTG 1000 μM PPDA	0 μM IPTG 1000 μM PPDA
20 μL 0.0625 mM	Block 3 B	0.625 μM IPTG 40 μM PPDA	0.625 μM IPTG 40 μM PPDA	0.625 μM IPTG 200 μM PPDA	0.625 μM IPTG 200 μM PPDA	0.625 μM IPTG 1000 μM PPDA	0.625 μM IPTG 1000 μM PPDA

(continued)

Table 3
(continued)

Volume IPTG stock	Volume rhamnose stock	100 μL 0.8 mM PPDA		100 μL 4 mM PPDA		100 μL 20 mM PPDA	
		1	2	3	4	5	6
20 μL 0.25 mM	Block 3 C	2.5 μM IPTG	2.5 μM IPTG	2.5 μM IPTG	2.5 μM IPTG	2.5 μM IPTG	2.5 μM IPTG
		40 μM PPDA	40 μM PPDA	200 μM PPDA	200 μM PPDA	1000 μM PPDA	1000 μM PPDA
20 μL 1 mM	Block 3 D	10 μM IPTG	10 μM IPTG	10 μM IPTG	10 μM IPTG	10 μM IPTG	10 μM IPTG
		40 μM PPDA	40 μM PPDA	200 μM PPDA	200 μM PPDA	1000 μM PPDA	1000 μM PPDA
20 μL 5 mM	Block 4 A	50 μM IPTG	50 μM IPTG	50 μM IPTG	50 μM IPTG	50 μM IPTG	50 μM IPTG
		40 μM PPDA	40 μM PPDA	200 μM PPDA	200 μM PPDA	1000 μM PPDA	1000 μM PPDA
20 μL 20 mM	Block 4 B	200 μM IPTG	200 μM IPTG	200 μM IPTG	200 μM IPTG	200 μM IPTG	200 μM IPTG
		40 μM PPDA	40 μM PPDA	200 μM PPDA	200 μM PPDA	1000 μM PPDA	1000 μM PPDA
20 μL 50 mM	Block 4 C	500 μM IPTG	500 μM IPTG	500 μM IPTG	500 μM IPTG	500 μM IPTG	500 μM IPTG
		40 μM PPDA	40 μM PPDA	200 μM PPDA	200 μM PPDA	1000 μM PPDA	1000 μM PPDA
20 μL 100 mM	Block 4 D	1000 μM IPTG	1000 μM IPTG	1000 μM IPTG	1000 μM IPTG	1000 μM IPTG	1000 μM IPTG
		40 μM PPDA	40 μM PPDA	200 μM PPDA	200 μM PPDA	1000 μM PPDA	1000 μM PPDA

3.5.2 96-Well Blocks

1. Seal the block and harvest the cells by centrifugation at 6000 × *g* 4 °C for 10 min.
2. Decant the medium from the cell pellets by inverting the block and drain by tapping onto a paper towel.
3. Wash pellets with 1 mL of PBS and spin again as above.
4. Resuspend pellets in 1 mL of PBS.

3.6 Measurement of Relative Fluorescence Units (RFU) and Optical Density (OD) and Calculation of RFU/OD

1. Transfer 100 μL of the washed resuspended cells to a flat-bottomed black 96-well plate and add 100 μL of PBS (therefore the final dilution factor is ¼, *see* **Note 13**).
2. Use 200 μL of PBS as the blank.
3. Read the fluorescence at 485-20/516-20 (excitation/emission filters) and OD at 595 nm in the Plate reader and calculate normalized fluorescence units as follows:
   (a) Subtract the PBS-only absorbance/fluorescence readings from the wells containing cells (*see* **Note 14**).
   (b) Divide the adjusted fluorescence values by the adjusted absorbance values to normalize for cell density (*see* **Note 15**). This is the RFU/OD.
4. Calculate induction/repression factors by dividing the final normalized fluorescence values by the value for 0 μM inducers.
5. These RFU/OD data can be plotted for a single time-point against ligand concentration (e.g., Fig. 2).
6. If GFP in-gel fluorescence is to be carried out, the 96-well block containing the remainder of the cells should be harvested by centrifugation at 6000 × *g* and 4 °C for 10 min and frozen at −80 °C until required.

3.7 GFP In-Gel Fluorescence

SDS-Page gels can be run on selected samples, e.g., where the highest expression RFU and or RFU/OD is seen to the quality of the expressed protein.

1. Defrost the block from the step above and resuspend the selected samples in 100 μL of PBS.
2. Transfer the selected samples into a microfuge tube.
3. Add 10 μL of DLPI solution and shake for 15 min at 4 °C.
4. Add 10% (w/v) n-Dodecyl β-D-maltoside (DDM) to a final concentration of 1% to lysates.
5. Continue shaking for 60 min at 4 °C.
6. Clear the lysate by centrifuging the microfuge tubes at maximum speed at 4 °C for 10 min.
7. Transfer 10 μL of the cleared lysate to a PCR plate and add 10 μL of SDS PAGE gel loading buffer (*see* **Note 16**).

8. Load 10 μL of this onto SDS-PAGE gel(s) and run at 100–120 V (constant voltage) in a cold room until the dye front reaches the bottom (2–2.5 h).
9. Place the gel onto an imager (e.g., with a blue-light filter to detect the GFP fusion proteins). The exposure time is chosen to ensure that the brightest bands are not saturated.

4 Notes

1. The choice of vector is dictated by the topology of the protein since the GFP needs to be expressed in the cytosol. Thus, for a protein with a cytosolic N-terminus and extracellular C-terminus the fusion to GFP should be at the N-terminus, and for an extracellular N-terminus and cytosolic C-terminus fusion at the C-terminus. Where both termini are cytosolic, then fusion to GFP should be carried out at the C-terminus. Suitable vectors include pOPINN-GFP (fusion at the N-terminus) and pOPINE-3C-GFP or pWALDO (fusion at the C-terminus) [42, 43]. If both termini are extracellular, then a vector such as pWARF+, where fusion to glycophorin-GFP is used to produce a cytosolic C-terminal GFP, can be used [44].
2. The use of phosphate buffer is optional. If phosphate buffer is not used, then add the components to 1 L of water and autoclave.
3. The strains require the following antibiotics in addition to the plasmid antibiotic: IL3—kanamycin during preparation of competent cells but not in the expression screen, KRX—no additional antibiotic, and Lemo 21(DE3)—chloramphenicol.
4. A more convenient way of making PBS solution is to purchase tablets and dissolve them in water according to the manufacturer's directions.
5. Optional: an empty vector can be used as a control for the intrinsic fluorescence of the induced bacteria. If a negative control is used, then cells should be induced (in duplicates) with the gradient of IPTG concentrations and the highest other inducer concentration shown in Tables 1–3.
6. Use this protocol when a conventional flask incubator is used to give sufficient aeration to the cells.
7. Using a repeater pipette (e.g., Ripette genX) increases both the speed and accuracy of dispensing.
8. If the 24-well blocks are transferred to a plate incubator, then rate should be increased to ensure sufficient aeration (e.g., 600 rpm in the Shellab incubators).

9. Since there is relatively little difference between the expression levels at different concentrations of IPTG, the experiment can be performed if desired using a single concentration of IPTG in the range 50–200 μM.
10. Use this protocol when plate incubators are used. The rotation rate should be increased to a high rate so that the cells receive appropriate aeration.
11. Steps 1 and 2 can be replaced by an overnight culture step if desired and 0.25 mL of overnight added to 25 mL of TB.
12. Pipettes that can be expanded and contracted to transfer between 24 and 96-well pitches are useful. We use E1-ClipTip™ Electronic Adjustable Tip Spacing Multichannel Equalizer Pipettes (ThermoFisher Scientific) and 6-channel Rainin Pipet-Lite™ XLS Adjustable Spacer pipettors.
13. The protocol is described for bottom reading machines and the black plates should have a clear bottom; for top reading machines a completely black plate is used and a separate reading for OD is taken using a clear microtiter plate.
14. Exporting these data to a spreadsheet allows easy manipulation of the data.
15. If a vector negative control has been used, then the control RFU/OD values should be subtracted at this point.
16. DO NOT BOIL THE SAMPLES.

Acknowledgments

The Oxford Protein Production Facility-UK is supported by the UK Medical Research Council (MR/K018779/1) and the Manchester Group are supported by the Biotechnology and Biology Research Council, ND holds a BBSRC David Phillips Fellowship [BB/K014773/1].

References

1. Wagner S, Baars L, Ytterberg AJ et al (2007) Consequences of membrane protein overexpression in *Escherichia coli*. Mol Cell Proteomics 6:1527–1550
2. Bill RM, Henderson PJ, Iwata S et al (2011) Overcoming barriers to membrane protein structure determination. Nat Biotechnol 29:335–340
3. Marreddy RK, Geertsma ER, Permentier HP et al (2010) Amino acid accumulation limits the overexpression of proteins in *Lactococcus lactis*. PLoS One 5:e10317
4. Angov E, Hillier CJ, Kincaid RL et al (2008) Heterologous protein expression is enhanced by harmonizing the codon usage frequencies of the target gene with those of the expression host. PLoS One 3:e2189
5. Arechaga I, Miroux B, Karrasch S et al (2000) Characterisation of new intracellular membranes in *Escherichia coli* accompanying large scale over-production of the b subunit of F(1)F(o) ATP synthase. FEBS Lett 482: 215–219

6. Loll PJ (2003) Membrane protein structural biology: the high throughput challenge. J Struct Biol 142:144–153
7. Wagner S, Bader ML, Drew D et al (2006) Rationalizing membrane protein overexpression. Trends Biotechnol 24:364–371
8. Drew DE, von Heijne G, Nordlund P et al (2001) Green fluorescent protein as an indicator to monitor membrane protein overexpression in *Escherichia coli*. FEBS Lett 507: 220–224
9. Sonoda Y, Newstead S, Hu NJ et al (2011) Benchmarking membrane protein detergent stability for improving throughput of high-resolution X-ray structures. Structure 19: 17–25
10. Lee C, Kang HJ, Hjelm A et al (2014) MemStar: a one-shot *Escherichia coli*-based approach for high-level bacterial membrane protein production. FEBS Lett 588:3761–3769
11. Drew D, Lerch M, Kunji E et al (2006) Optimization of membrane protein overexpression and purification using GFP fusions. Nat Methods 3:303–313
12. Drew D, Slotboom DJ, Friso G et al (2005) A scalable, GFP-based pipeline for membrane protein overexpression screening and purification. Protein Science: A Publication of the Protein Society 14:2011–2017
13. Mus-Veteau I (2010) Heterologous expression of membrane proteins for structural analysis. Methods Mol Biol 601:1–16
14. Studier FW, Moffatt BA (1986) Use of bacteriophage T7 RNA polymerase to direct selective high-level expression of cloned genes. J Mol Biol 189:113–130
15. Studier FW (1991) Use of bacteriophage T7 lysozyme to improve an inducible T7 expression system. J Mol Biol 219:37–44
16. Dubendorff JW, Studier FW (1991) Controlling basal expression in an inducible T7 expression system by blocking the target T7 promoter with lac repressor. J Mol Biol 219:45–59
17. Klepsch MM, Persson JO, de Gier JW (2011) Consequences of the overexpression of a eukaryotic membrane protein, the human KDEL receptor, in *Escherichia coli*. J Mol Biol 407:532–542
18. Miroux B, Walker JE (1996) Over-production of proteins in *Escherichia coli*: mutant hosts that allow synthesis of some membrane proteins and globular proteins at high levels. J Mol Biol 260:289–298
19. Wagner S, Klepsch MM, Schlegel S et al (2008) Tuning *Escherichia coli* for membrane protein overexpression. Proc Natl Acad Sci U S A 105:14371–14376
20. Schlegel S, Lofblom J, Lee C et al (2012) Optimizing membrane protein overexpression in the *Escherichia coli* strain Lemo21(DE3). J Mol Biol 423:648–659
21. Hjelm A, Schlegel S, Baumgarten T et al (2013) Optimizing *E. coli*-based membrane protein production using Lemo21(DE3) and GFP-fusions. Methods Mol Biol 1033: 381–400
22. Morra R, Shankar J, Robinson CJ et al (2016) Dual transcriptional-translational cascade permits cellular level tuneable expression control. Nucleic Acids Res 44:e21
23. Chao YP, Chiang CJ, Hung WB (2002) Stringent regulation and high-level expression of heterologous genes in *Escherichia coli* using T7 system controllable by the araBAD promoter. Biotechnol Prog 18:394–400
24. Hartnett J, Gracyalny J, Slater MR (2006) The Single Step (KRX) Competent Cells: Efficient Cloning and High Protein Yields. Promega Notes vol. 96, Promega Corp
25. Bird LE, Rada H, Verma A et al (2015) Green fluorescent protein-based expression screening of membrane proteins in *Escherichia coli*. J Vis Exp 6:e52357
26. Guzman LM, Belin D, Carson MJ et al (1995) Tight regulation, modulation, and high-level expression by vectors containing the arabinose PBAD promoter. J Bacteriol 177:4121–4130
27. Haldimann A, Daniels LL, Wanner BL (1998) Use of new methods for construction of tightly regulated arabinose and rhamnose promoter fusions in studies of the *Escherichia coli* phosphate regulon. J Bacteriol 180:1277–1286
28. Terpe K (2006) Overview of bacterial expression systems for heterologous protein production: from molecular and biochemical fundamentals to commercial systems. Appl Microbiol Biotechnol 72:211–222
29. Egan SM, Schleif RF (1993) A regulatory cascade in the induction of rhaBAD. J Mol Biol 234:87–98
30. Giacalone MJ, Gentile AM, Lovitt BT et al (2006) Toxic protein expression in *Escherichia coli* using a rhamnose-based tightly regulated and tunable promoter system. Biotechniques 40:355–364
31. Serganov A, Nudler E (2013) A decade of riboswitches. Cell 152:17–24
32. Chen G, Yanofsky C (2003) Tandem transcription and translation regulatory sensing of uncharged tryptophan tRNA. Science 301: 211–213

33. Grundy FJ, Henkin TM (1993) tRNA as a positive regulator of transcription antitermination in *B. subtilis*. Cell 74:475–482
34. Li GW, Oh E, Weissman JS (2012) The anti-Shine-Dalgarno sequence drives translational pausing and codon choice in bacteria. Nature 484:538–541
35. Barrick JE, Breaker RR (2007) The distributions, mechanisms, and structures of metabolite-binding riboswitches. Genome Biol 8:R239
36. Mandal M, Breaker RR (2004) Gene regulation by riboswitches. Nat Rev Mol Cell Biol 5:451–463
37. Winkler WC, Breaker RR (2003) Genetic control by metabolite-binding riboswitches. Chembiochem: Eur J Chem Biol 4:1024–1032
38. Link KH, Breaker RR (2009) Engineering ligand-responsive gene-control elements: lessons learned from natural riboswitches. Gene Ther 16:1189–1201
39. Dixon N, Duncan JN, Geerlings T et al (2010) Reengineering orthogonally selective riboswitches. Proc Natl Acad Sci U S A 107: 2830–2835
40. Dixon N, Robinson CJ, Geerlings T et al (2012) Orthogonal riboswitches for tuneable coexpression in bacteria. Angewandte Chemie 51:3620–3624
41. Bird LE (2011) High throughput construction and small scale expression screening of multi-tag vectors in *Escherichia coli*. Methods 55: 29–37
42. Bird LE, Rada H, Flanagan J et al (2014) Application of In-Fusion cloning for the parallel construction of *E. coli* expression vectors. Methods Mol Biol 1116:209–234
43. Waldo GS, Standish BM, Berendzen J et al (1999) Rapid protein-folding assay using green fluorescent protein. Nat Biotechnol 17: 691–695
44. Hsieh JM, Besserer GM, Madej MG et al (2010) Bridging the gap: a GFP-based strategy for overexpression and purification of membrane proteins with intra and extracellular C-termini. Protein Sci 19:868–880

Chapter 7

Optimizing *E. coli*-Based Membrane Protein Production Using Lemo21(DE3) or pReX and GFP-Fusions

Grietje Kuipers, Markus Peschke, Nurzian Bernsel Ismail, Anna Hjelm, Susan Schlegel, David Vikström, Joen Luirink, and Jan-Willem de Gier

Abstract

Optimizing the conditions for the production of membrane proteins in *E. coli* is usually a laborious and time-consuming process. Combining the Lemo21(DE3) strain or the pReX T7-based expression vector with membrane proteins C-terminally fused to Green Fluorescent Protein (GFP) greatly facilitates the optimization of membrane protein production yields. Both Lemo21(DE3) and pReX allow precise regulation of expression intensities of genes encoding membrane proteins, which is critical to identify the optimal production condition for a membrane protein. The use of GFP-fusions allows direct monitoring and visualization of membrane proteins at any stage during the production optimization process.

Key words Membrane protein, Production, *E. coli*, Lemo21(DE3), pReX, Fluorescence, GFP

1 Introduction

The natural abundance of most helical membrane proteins, hereafter referred to as membrane proteins, is usually too low for the isolation of sufficient material for functional and structural studies. The use of natural sources also excludes the possibility of genetically modifying proteins to improve their stability and to facilitate their detection and purification. Despite tremendous efforts, there are very few examples of membrane proteins that have been successfully refolded upon extraction from inclusion bodies under denaturing conditions (e.g., *see* [1]). Therefore, it is preferred to produce membrane proteins in a membrane, from which they can be purified upon detergent extraction. However, the production of membrane proteins in a membrane is usually toxic to the production host [1]. In general, optimizing the conditions for the production of membrane proteins is a laborious and time-consuming process. In this chapter, we have updated

Nicola A. Burgess-Brown (ed.), *Heterologous Gene Expression in E. coli: Methods and Protocols*, Methods in Molecular Biology, vol. 1586, DOI 10.1007/978-1-4939-6887-9_7, © Springer Science+Business Media LLC 2017

and extended a previously published protocol for optimizing the production of membrane proteins using the *E. coli* Lemo21(DE3) strain and GFP-fusions [2].

1.1 The *Escherichia coli* Lemo21(DE3) Strain and the pReX Expression Vector

The bacterium *E. coli* is a widely used host to produce both pro- and eukaryotic membrane proteins [1, 3]. In order to drive the production of the membrane protein of interest in *E. coli*, we utilize the widely used bacteriophage T7-based pET/T7-RNA polymerase (T7-RNAP) expression system, in which expression of the gene encoding the target protein is governed by the T7-RNAP [4]. As production host, we use the BL21(DE3)-derived strain Lemo21(DE3). BL21(DE3) and its derivatives harbor a chromosomal copy of the gene encoding T7-RNAP under the control of the isopropyl β-D-1-thiogalactopyranoside (IPTG)-inducible *lac*UV5 promoter [5]. This promoter is a stronger variant of the *lac* promoter [4]. Upon addition of IPTG, repression of the *lac*UV5 promoter is relieved, leading to expression of the gene encoding the target protein. In BL21(DE3), T7-RNAP activity is very high and fixed. In contrast, in Lemo21(DE3), the activity of the T7-RNAP can be precisely tuned by co-production of its natural inhibitor T7-lysozyme, from the pLemo plasmid (Fig. 1a) [6]. This plasmid is derived from pACYC184 and the expression of the gene encoding T7-lysozyme is governed by the rhamnose promoter [6]. Notably, we used a variant of T7-lysozyme (K128Y) (LysY) that has no amidase activity but retains full inhibition of T7-RNAP [6]. The rhamnose promoter is extremely well titratable and covers a broad range of expression intensities [7]. Recently, we have simplified the Lemo21(DE3) setup by introducing *lysY* under control of the rhamnose promoter onto a Km^R-based pET vector, resulting in the pReX expression vector (Fig. 1b). ReX stands for Regulatable eXpression. The capital X refers to Xbrane Biopharma AB. Notably, pReX can in principle be used in combination with any T7-RNAP-based *E. coli* protein production strain and enables the use of a second plasmid for, e.g., the co-expression of genes encoding chaperones. Using both Lemo21(DE3) and pReX, the production of membrane proteins can be optimized. By tuning the expression intensity of the gene encoding a membrane protein, its production rate can easily be harmonized with the membrane protein biogenesis capacity of the production host [6, 8, 9]. The harmonization of membrane protein production with membrane protein biogenesis capacity alleviates the toxic effects of membrane protein production [6, 9]. This leads to the formation of more biomass resulting in increased membrane protein production yields. It should be noted that for a small number of membrane proteins, we have observed that the membrane protein biogenesis capacity is sufficient without any inhibition of T7-RNAP activity by T7-lysozyme [6, 8, 9]. In these cases, plain BL21(DE3) can be used as production host [6].

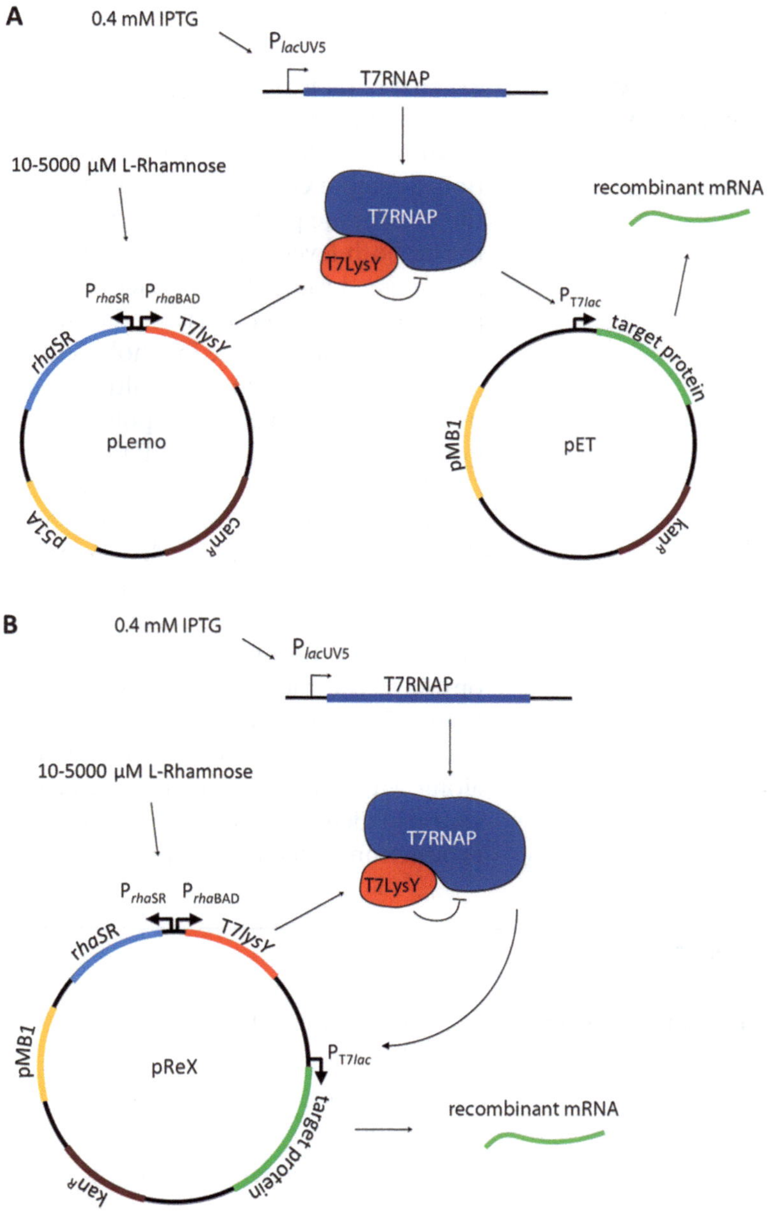

Fig. 1 The Lemo21(DE3) strain and the pReX expression vector. (**a**) Schematic representation of the Lemo21(DE3) strain. Expression of the chromosomally located gene encoding the T7 RNA polymerase (T7-RNAP) is governed by the not well titratable, IPTG inducible *lac*UV5 promoter. Expression of the gene encoding the natural inhibitor of the T7-RNAP, T7-lysozyme (T7 LysY), is governed by the exceptionally well titratable rhamnose promoter from the pLemo plasmid. The pLemo plasmid has a p15A origin of replication and contains a chloramphenicol resistance marker. The gene encoding the target membrane protein is located on a pET vector and its expression is governed by the T7*lac* promoter. The pET vector has a pMB1 origin of replication. For the production of membrane proteins, pET vectors with a kanamycin resistance marker are used. (**b**) Schematic representation of the pReX expression vector. pReX allows for regulated expression of a target gene from a T7*lac* promoter by varying T7 lysozyme levels. pReX is a "merger" between pLemo and pET. pReX has a pMB1 origin of replication and a kanamycin resistance marker. Here, pReX is used in combination with the BL21(DE3) strain

1.2 Membrane Protein GFP-Fusions

To enable rapid screening for the optimal production conditions of membrane proteins, we use membrane protein Green Fluorescent Protein (GFP)-fusions (Fig. 2a) [10–12]. The exceptionally stable GFP-moiety, which is attached to the C-terminus of the membrane protein, can easily be visualized. This allows monitoring both levels and integrity of the target membrane protein at any stage during the production process (Fig. 2b). Membrane protein production levels in the membrane can be estimated by measuring fluorescence in whole cells with a detection limit as low as 10 μg GFP per liter of culture. GFP-fluorescence can also be detected in standard SDS polyacrylamide gels with a detection limit of less than 5 ng of GFP [12]. This in-gel fluorescence allows rapid assessment of the integrity of membrane protein GFP-fusions and can also be used for quantification [12]. In addition, the GFP-moiety allows determining the ratio of membrane inserted to non-inserted membrane protein using an SDS-PAGE/immuno-blotting-based assay [9, 13]. Thus, using membrane protein-GFP-fusions provides information that is very helpful to optimize the membrane protein production yields in the membrane. The generic applicability and simplicity of the Lemo21(DE3) and pReX-based solutions for membrane protein production along with the use of membrane protein GFP-fusions guarantees the rapid identification of the optimal conditions for the *E. coli*-based production of membrane proteins.

2 Materials

2.1 Culturing of Cells

1. Airpore Tape sheets (to cover 24-well uniplates, *see* **item 12**) (Qiagen).
2. 34 mg/mL stock solution of Chloramphenicol, prepared in ethanol.
3. 1 M Isopropyl β-D-1-thiogalactopyranoside (IPTG), filter sterilized.
4. 50 mg/mL stock solution Kanamycin monosulfate, filter sterilized.
5. Lysogeny Broth (LB medium). LB medium is usually referred to as Luria Bertani broth.
6. The pLemo plasmid of the Lemo21(DE3) strain and the pReX expression vector can be obtained from Xbrane Biopharma AB (www.xbrane.com, info@xbrane.com). Lemo21(DE3) can also be obtained from New England Biolabs as competent cells (http://www.neb.com/nebecomm/products/productC2528.asp). For pReX-based protein production, the BL21(DE3) strain was used [5].
7. 0.1 and 0.5 M solutions of L-rhamnose, filter sterilized.

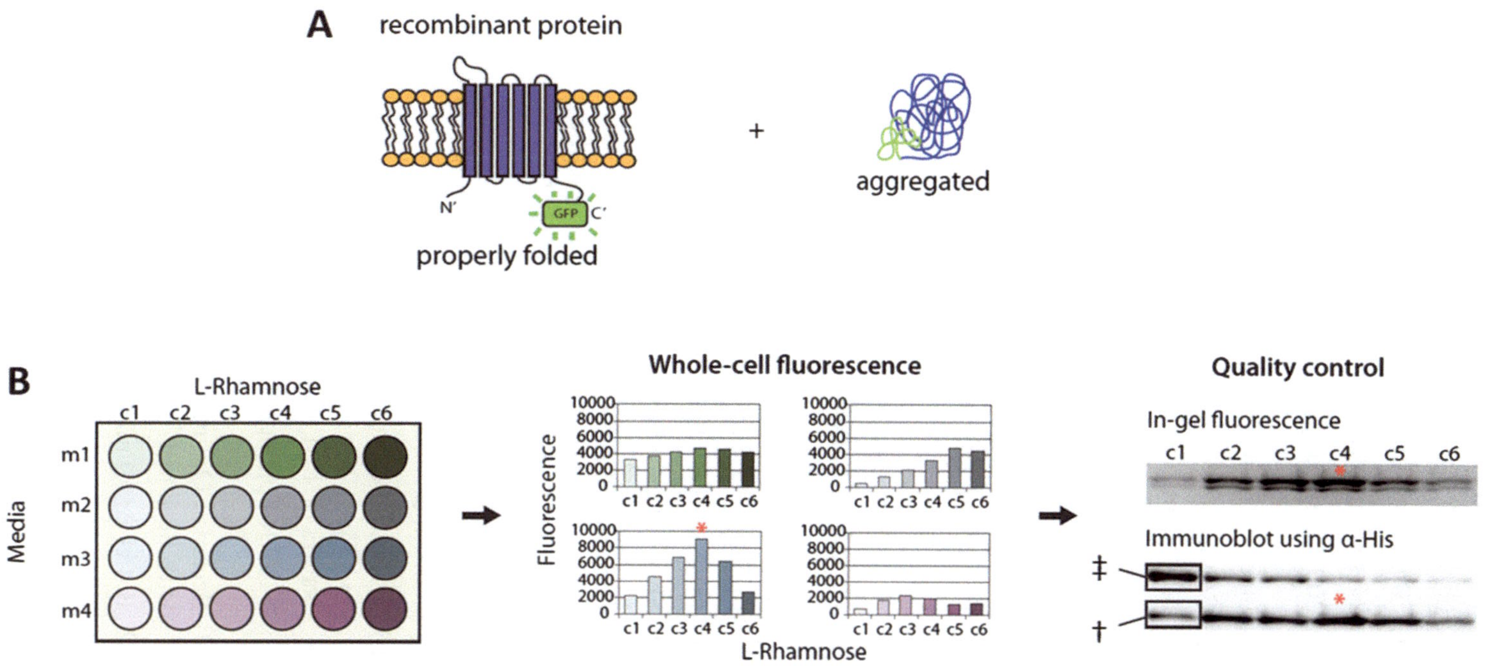

Fig. 2 Monitoring membrane protein production using GFP-fusions. (**a**) Membrane proteins are produced as C-terminal GFP-fusions. The GFP-moiety only folds properly and becomes fluorescent when the membrane protein GFP-fusion is inserted in the membrane. (**b**) Lemo21(DE3) cells harboring a pET vector or BL21(DE3) cells harboring pReX are cultured in the presence of different concentrations of L-rhamnose. The production of membrane protein-GFP-fusions is induced with 0.4 mM IPTG. Whole cell fluorescence is used to monitor the production of membrane protein GFP-fusions in the membrane. Subsequently, in-gel fluorescence is used to assess the integrity of the produced membrane protein GFP-fusions. The ratio of the membrane-inserted membrane protein (single cross) to non-inserted membrane protein (double cross) is monitored using an SDS-PAGE/immuno-blotting-based assay

8. Shaking incubator with temperature control.
9. Tunair 2.5 L baffled shaker flasks.
10. UV-1601 UV-VIS Spectrophotometer.
11. 50 mL Falcon tubes.
12. 24-well uniplates (for the cover, *see* **item 1**).
13. 200 mL shaker flasks.

2.2 Monitoring Production of Membrane Protein GFP-Fusions

1. Blocking buffer: 5% (w/v) milk powder in TBS-T (*see* **item 15**).
2. Coomassie staining solution: 0.4% (w/v) Coomassie Brilliant R250 (Fluka), 50% (v/v) ethanol, 5% (v/v) acetic acid.
3. Destaining solution: 30% (v/v) ethanol, 12% (v/v) acetic acid.
4. ECL-Western Blotting Detection Reagents.
5. Fixing solution: 50% (v/v) ethanol and 12% (v/v) acetic acid.
6. Fuji LAS-1000 charge coupled device (CCD) camera.
7. HRP-conjugated anti-His antibody.
8. Nunc 96-well optical bottom plate, black.
9. Phosphate-buffered saline (PBS): 1.44 g of $Na_2HPO_4{\cdot}2H_2O$ (8.1 mM Phosphate), 0.25 g of KH_2PO_4 (1.9 mM Phosphate), 8.00 g of NaCl, and 0.2 g of KCl in 1000 mL H_2O. Adjust pH to 7.4 using 1 M NaOH or 1 M HCl.
10. Polyvinylidene fluoride (PVDF) membrane (pore size 0.45 μm).
11. SDS-PAGE/Blotting setup: your own choice.
12. Solubilization Buffer (SB): 75 mM Tris–HCl (pH 6.8), 1.5% (v/v) Glycerol, 0.025% (w/v) Bromophenol blue. Before use, add a final concentration of 10% (v/v) β-mercaptoethanol.
13. Microplate spectrofluorometer.
14. Thermomixer equipped with a thermoblock for 1.5 mL microcentrifuge tubes.
15. Tris-buffered saline with 0.05% Tween (TBS-T): 8.4 g of NaCl and 3 g of Tris in 1000 mL H_2O. Adjust pH to 8.0 using 1 M NaOH or 1 M HCl and add 0.05% (v/v) Tween.
16. 5417 R Eppendorf Table top centrifuge.

2.3 Membrane and GFP Isolation

1. Beckman Optima Max XP benchtop ultracentrifuge equipped with Beckman TLA 100.3 rotor.
2. Buffer A: 50 mM Tris (pH 7.5), 300 mM NaCl.
3. Buffer B: 500 mM Imidazole in Buffer A.
4. Buffer C: 20 mM Tris–HCl (pH 7.5), 5 mM EDTA (pH 8), 100 mM NaCl.

5. 1 mg/mL stock solution of Deoxyribonuclease I from bovine pancreas Type IV lyophilized powder.
6. Disposable syringe (10 mL) with a 21-gauge needle.
7. 0.5 M EDTA stock solution, autoclaved.
8. Emulsiflex (C3, Avestin)/French Press/Sonicator.
9. 5 mg/mL stock solution of Lysozyme.
10. 1 M $MgCl_2$ stock solution.
11. Ni-NTA Superflow resin (Qiagen).
12. 100 mg/mL stock solution of Pefabloc SC.
13. Pierce BCA protein assay kit (Thermo Fischer Scientific).
14. Poly-Prep Chromatography Columns (Bio-Rad).

3 Methods

3.1 Determination of Membrane Protein Topology

GFP only folds correctly and becomes fluorescent in the cytoplasm of *E. coli* whereas it does not in the periplasm [14, 15]. Therefore, using a membrane protein GFP-fusion to optimize membrane protein production requires that the membrane protein of interest has a C_{in} topology. If the topology of the protein of interest is not known, use a topology predictor (e.g., MEMSAT3 [16], SCAMPI [17], TOPCONS [18], or all in parallel). If the membrane protein has a C_{out} topology it can be extended with one transmembrane segment at the C-terminus, thereby positioning the GFP-moiety on the cytoplasmic side of the membrane (*see* Subheading 3.2). Fusing GFP to the N-terminus of N_{in}/C_{out} membrane proteins is not recommended since the GFP-moiety may interfere with the targeting of the protein to the membrane and its folding [19, 20].

3.2 Selection of Expression Vector

For Lemo21(DE3)-based protein production, we routinely use the pGFPd and -e expression vectors [12]. These vectors are derived from pET28a(+) and code for a Tobacco Etch Virus (TEV) protease cleavage site between the multiple cloning site and the sequence encoding the GFP-His_8 moiety. In case a membrane protein has a C_{out} topology, use pGFPd and -e derived vectors that have the genetic information encoding the transmembrane segment of glycophorin A between the TEV protease cleavage site and the GFP-His_8 moiety [21]. The pReX vector is also derived from pET28a (Fig. 1b). Notably, pReX does not contain the GFP and TEV sequences. Thus, the complete gene encoding a membrane protein GFP-fusion has to be cloned into pReX. Here, we used pReX in combination with the BL21(DE3) strain [5].

All three pGFPd, -e and pReX, confer resistance to kanamycin. Kanamycin resistance is preferred to ampicillin resistance for membrane protein production. This is due to the different site of action of these two antibiotic resistance markers. Kanamycin targets the 30S subunit of the prokaryotic ribosome and the kanamycin resistance gene encodes a cytoplasmic protein. Ampicillin interferes with the biogenesis of the peptidoglycan layer in the periplasm and is neutralized by the protein encoded by the ampicillin resistance gene, β-lactamase. To reach the periplasm, β-lactamase is translocated through the same protein-conducting channel in the membrane that is also involved in mediating the biogenesis of membrane proteins into the membrane, the so-called Sec-translocon. Thus, the use of the ampicillin resistance marker during membrane protein production may unnecessarily consume Sec-translocon capacity, thereby lowering membrane protein production yields.

3.3 Identification of the Optimal Production Conditions

The first step to produce a membrane protein using Lemo21(DE3) or pReX in combination with BL21(DE3) is to identify the concentration of L-rhamnose that gives highest production yields. We define the highest production yield as the highest amount of full-length membrane protein GFP-fusion that is inserted in the membrane per mL of culture. It should be noted, however, that the isolation of the target protein may be facilitated by choosing a condition that yields the highest amount of target per OD_{600} (per cell) instead, even if the overall yield may be lower in that case. We perform the initial screen at 30 °C since we have noticed that using different temperatures usually does not lead to improved production yields, and we routinely use Lysogeny Broth (LB) medium (e.g., [9]). However, other culture media, including terrific broth medium and autoinduction-based media, and temperatures other than 30 °C can be used [9, 22]. Notably, defined autoinduction media conveniently allow labeling of proteins for NMR and crystallography experiments [23]. The optimal concentration of L-rhamnose strongly depends on the medium and the culture conditions (e.g., oxygen levels) used. In Fig. 3, an example is shown of the use of pReX combined with BL21(DE3) to optimize the production of a model membrane protein C-terminally fused to GFP using whole cell and in-gel fluorescence. The following steps refer to the screening of one target, but multiple targets can be screened in parallel.

3.3.1 Transformation and Culturing

For Lemo21(DE3), transform the expression vector harboring the gene encoding the membrane protein GFP-fusion of interest into Lemo21(DE3). For pReX, transform pReX containing the gene encoding the membrane protein GFP-fusion of interest into BL21(DE3) (*see* **Notes 1–3**). Always use fresh transformants (not older than 4–5 days). The use of glycerol stocks of transformed

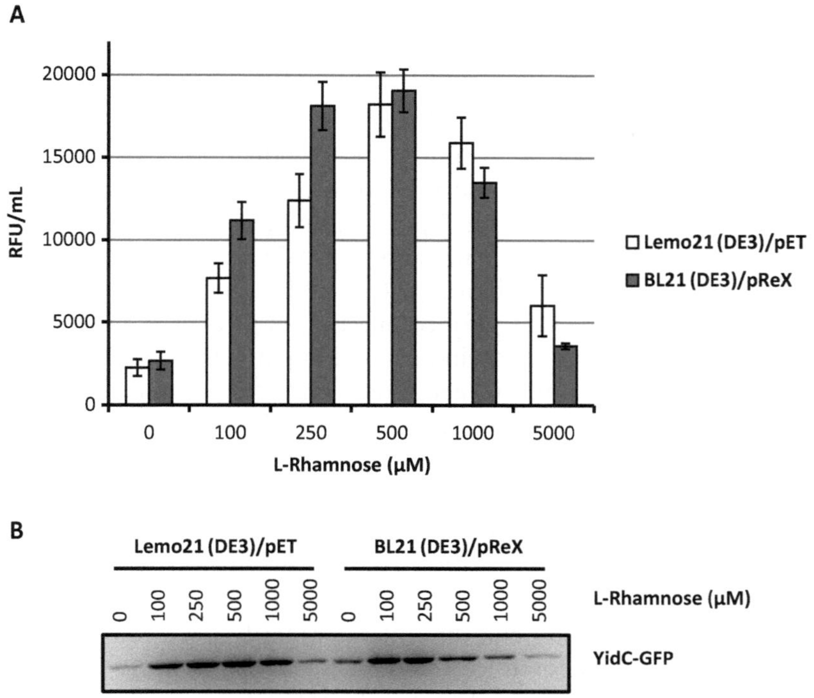

Fig. 3 Comparing the pReX and Lemo21(DE3)-based optimization of the production of the membrane protein YidC. The pReX-based production of the *E. coli* membrane protein YidC C-terminally fused to GFP in BL21(DE3) cells was compared to Lemo21(DE3)/pET-based production of YidC-GFP. Cells were cultured in LB medium at 30 °C, and production of YidC-GFP was induced with 0.4 mM IPTG for 8 h. (**a**) The effect of varying gene expression levels on YidC-GFP production yields as monitored by whole cell fluorescence (RFU/mL). (**b**) The integrity of the produced YidC-GFP was monitored using in-gel fluorescence

cells as starting material can lead to severe reduction of production yields and is not recommended.

1. Set up, using a single colony from a transformation plate, an overnight (o/n) culture in a 15 mL Falcon tube containing 3 mL of LB medium supplemented with the appropriate antibiotic for the expression vector used (50 μg/mL kanamycin for pGFPd/e and pReX) and if Lemo21(DE3) is used, also 34 μg/mL chloramphenicol (for maintaining pLemo). Incubate in a shaking incubator at 30 °C, 220 rpm (*see* **Note 2**).
2. Prepare nine 50 mL Falcon tubes with 12 mL of LB medium each, containing the appropriate antibiotics. Add L-rhamnose to eight of the Falcon tubes to a final concentration of 10, 50, 100, 250, 500, 1000, 2500, and 5000 μM (*see* **Notes 4–6**). The ninth Falcon tube does not contain any L-rhamnose (0 μM).
3. Inoculate each Falcon tube with a 50-fold dilution of the o/n culture. Incubate at 30 °C, 220 rpm and monitor the OD_{600} of the cultures.

4. At an OD_{600} of 0.4–0.5 (this OD_{600} will be reached approximately 2–2.5 h after inoculation), induce expression of the gene encoding the membrane protein GFP-fusion by adding IPTG to a final concentration of 0.4 mM (*see* **Note 7**).
5. Take, 4, 8, and 24 h after induction, 1 mL of culture for whole cell fluorescence measurements using a plate reader (*see* Subheading 3.3.2). Simultaneously, take 100 μL for OD_{600} measurements and approximately 500 μL for SDS-PAGE(/ immunoblotting) (*see* Subheadings 3.3.3 and 3.3.4). Measure the OD_{600} for comparison of growth and calculations of membrane protein production per OD_{600} (*see* **Note 7**). The whole cell and in-gel fluorescence measurements allow determining the optimal concentration of L-rhamnose for the production of a membrane protein (*see* **Note 8**).

3.3.2 Whole Cell Fluorescence

1. Transfer 1 mL of culture volume to a 1.5 mL Eppendorf tube and collect the cells by centrifugation for 2 min at 15,700 × *g*. Carefully remove the supernatant.
2. Resuspend the pellet in 100 μL of ice-cold PBS and leave it on ice for at least 30 min. This will allow the GFP-moiety to fold. Alternatively, wash the cell pellet once in between using 1 mL of ice-cold PBS and repeat the centrifugation step.
3. Transfer the 100 μL suspension into to a black Nunc 96-well optical bottom plate and measure GFP-fluorescence (emission: 512 nm, excitation: 485 nm, cut off: 495 nm) in a microtiter plate spectrofluorometer. For maximal sensitivity, select the option "bottom read" (*see* **Note 9**). To estimate membrane protein production yields from whole cell fluorescence using purified GFP, refer to Subheading 3.4.

3.3.3 In-Gel Fluorescence

Measuring whole cell fluorescence does not allow discrimination between the full-length fusion protein and degradation products. GFP is an exceptionally stable molecule and remains fluorescent even if the membrane protein of interest has been degraded. The integrity of the produced material can be rapidly assessed using in-gel fluorescence.

1. Harvest the cells from 500 μL of culture volume by centrifugation for 2 min at 15,700 × *g*. Carefully remove the supernatant.
2. Resuspend the cell pellets in PBS to an equal OD_{600} (we dilute to a final concentration of 0.2 OD_{600} units (ODU)/10 μL PBS). Add an equal volume of SB buffer to each suspension (final concentration 0.1 ODU/10 μL solution). Ensure homogeneity of the cell suspension. Different concentrations of purified GFP of a known concentration (see Subheading 3.4) may be included in the analysis from here on. This will allow to

accurately estimate production yields and to discriminate between the full-length membrane protein GFP-fusion and degradation products (*see* **Note 10**).

3. Incubate all samples (cell suspensions and purified GFP) for 5–10 min at 37 °C (*see* **Note 11**).
4. Analyze a fraction of each sample corresponding to 0.05–0.2 ODU of sample by means of standard SDS-PAGE including an appropriate molecular weight marker. The amount of ODU to be analyzed depends on the setup used for SDS-PAGE (e.g., gel type, pocket size, etc.).
5. Rinse the gel with distilled water and detect in-gel fluorescence with a CCD camera system. Expose the gel to blue-light at 460 nm and capture images with increasing exposure time until the desired band intensity is reached. Fluorescence intensities can be quantified using appropriate software and production yields estimated by comparing intensities to a GFP reference sample of known concentration (*see* **Note 12**).
6. Optionally, to control sample loading, the gel can be stained for 2 h in Coomassie staining solution after fixation. Subsequently, the gel is incubated in destaining solution.

3.3.4 Monitoring the Ratio of Membrane Inserted to Non-inserted Membrane Protein

An assay based on SDS-PAGE/immuno-blotting of GFP-fused membrane proteins allows distinguishing between membrane proteins that are properly inserted in the membrane and incorrectly folded membrane proteins, which are not inserted in the membrane [9, 13]. If a membrane protein-GFP-fusion is not inserted in the membrane and ends up in aggregates, its GFP-moiety does not fold properly (Fig. 2). Only if the membrane protein GFP-fusion is inserted in the membrane, the GFP-moiety folds properly and becomes fluorescent. Correctly folded GFP is not denatured in SDS-PAGE solubilization buffer (SB) at temperatures below 37 °C. As a consequence, a membrane protein GFP-fusion that has been inserted in the membrane will migrate faster in a gel than a non-inserted fusion.

1. Follow the in-gel fluorescence protocol up until **step 5** and run an additional gel in case the one used to monitor in-gel fluorescence has been stained with Coomassie (*see* **Note 13**).
2. Transfer the proteins from the SDS-gel to a PVDF-membrane using a wet-based western-blotting setup of your choice.
3. For the detection of the membrane protein-GFP-fusions, use an antibody recognizing the C-terminal His-tag of the GFP-moiety (*see* **Note 14**). We routinely use a HRP-conjugated anti-His antibody. Briefly, block unspecific binding sites by incubating the membrane for 1 h in blocking buffer (5% milk or BSA in TBS-T), then rinse the membranes three times with TBS-T, and incubate with anti-His antibody in TBS-T for

45–60 min. Wash again with excessive amounts of TBS-T (three times 10 min) to remove nonspecifically bound antibody.

4. Detect the membrane protein GFP-fusions using the detection method of your choice. We use a chemiluminescence-based assay (ECL Western blotting detection kit from Amersham) and detect the signal using a CCD camera.

3.4 GFP-His$_8$ as a Reference for Whole Cell Fluorescence and In-Gel Fluorescence Measurements

Both whole cell fluorescence and in-gel fluorescence can be used to estimate production yields and require purified GFP-His$_8$ as a reference. Here, we briefly describe how GFP-His$_8$ is produced. For a more detailed protocol, see ref. [12].

1. Transform Lemo21(DE3) with a plasmid encoding GFP fused to a His$_8$ purification tag. We use pET20bGFP-His$_8$ (Amp^R) as a standard [12]. Since the produced GFP-His$_8$ is soluble and located in the cytoplasm, the use of the ampicillin resistance marker does not interfere with production yields.
2. When using Lemo21(DE3) to produce GFP-His$_8$, add L-rhamnose to a final concentration of 750 μM, 34 μg/mL chloramphenicol, and 100 μg/mL ampicillin [6].
3. For the production of GFP-His$_8$, induce the expression of the gene encoding the protein for approx. 4 h as described in Subheading 3.5.1 and process the cells according to Subheading 3.5.2. Proceed with the supernatant rather than the pellet.
4. Isolate GFP-His$_8$ using immobilized metal affinity chromatography (IMAC). Pack a column with the volume of Ni-NTA resin appropriate for the amount protein that will be isolated. Equilibrate Ni-NTA column with five column volumes (CVs) of Buffer A. Add imidazole to a final concentration of 10 mM to the sample, and load it onto the Ni-NTA column. Wash the Ni-NTA column with 20 CVs of 10% buffer B and elute with 50% buffer B.
5. Pool the major GFP-His$_8$ containing fractions (as determined by fluorescence) and dialyze o/n in buffer C. As soluble GFP-His$_8$ is produced to very high yields and since it serves a reference purpose only, it is not essential to retain all, but the protein should contain as few contaminants as possible.
6. Determine the protein concentration using a BCA assay according to the instructions of the manufacturer and measure GFP-fluorescence from 0.01 to 0.3 mg/mL GFP-His$_8$. Check the purity of GFP-His$_8$ by using standard SDS-PAGE followed by Coomassie staining/destaining.
7. Plot the GFP-fluorescence versus the protein concentration and use the plot to convert the GFP-fluorescence from any sample to mg/mL of GFP-His$_8$.

8. Estimate production yields by dividing the molecular weight of the produced membrane-protein GFP-fusion by 28 kDa (MW of GFP-His_8) and multiply the obtained value with the amount of GFP-His_8 as determined in the previous step.
9. When using purified GFP-His_8 as a reference to estimate production yields from whole-cell fluorescence (*see* Subheading 3.3.2) and when assessing the efficiency of a detergent [12], keep in mind that GFP-fluorescence is dampened by whole cells/membranes. Whole cells dampen the GFP-fluorescence by a factor of approximately 1.5 and membranes dampen the GFP-fluorescence by a factor of 1.3 [12].

3.5 Scaling Up Membrane Protein Production

After production screening, the optimal condition is used to set up cultures that can be used for the isolation of membrane protein material suitable for functional and structural studies. Here, we have used LB medium for scaling up the production of the target protein, but other media can be used as well (*see* Subheading 3.3). For more information on detergent screening and the isolation of membrane proteins, we refer to, e.g., [2, 12].

3.5.1 Scaling Up of Cultures

1. We use 2.5 L baffled shaker flasks for scaling up the production of membrane proteins. However, it is also possible to use fermenters (15 L fermenters have successfully been used for scaling up Lemo21(DE3)-based membrane protein production).
2. Set up an o/n culture in a 200 mL shaker flask containing 20 mL of LB medium with the appropriate antibiotics (*see* Subheading 3.3.1, **step 1**). Incubate at 30 °C, 220 rpm in a shaking incubator.
3. Inoculate 1 L of LB medium (with appropriate antibiotics and the optimal concentration of L-rhamnose) with the o/n culture in a 2.5 L baffled shaker flask and incubate at 30 °C, 220 rpm.
4. Induce expression of the gene encoding the target membrane protein as described before, at an OD_{600} of approximately 0.4–0.5, with 0.4 mM IPTG (final concentration) for the time determined to be optimal by the production screen. Before harvesting the cells, take 1 mL of culture for measuring whole cell fluorescence and quality control of the produced material (*see* Subheadings 3.3.2–3.3.4).

3.5.2 Isolation of Membranes

From here on, all steps should be carried out on ice or at 4 °C. Centrifugation steps are also performed at 4 °C.

1. Harvest the cells by centrifugation for 20 min at 6200 × *g*. Discard the supernatant and carefully resuspend the pellet in 50 mL of ice-cold PBS.
2. Pellet the cells according to the previous step, discard the supernatant, and resuspend the pellet in 10 mL of ice-cold

PBS. If needed, the pellet or the suspension can be frozen in liquid nitrogen and stored at −80 °C up to 6 months. Freezing and thawing may facilitate breaking the cells.

3. Add 1 mg/mL of Pefabloc SC (or another protease inhibitor mix of your choice). Add 1 mM EDTA and 0.5 mg/mL lysozyme (final concentration) and incubate on ice for 30–60 min (*see* **Note 15**). For improved efficiency of lysis, the suspension may be stirred slowly. Add 5 μg/mL DNaseI and 2 mM $MgCl_2$ and incubate for approx. 15 min. Break the cells using an Emulsiflex (15,000 p.s.i., 3–5 cycles), or a method of your choice (e.g., French press, sonication). Most cells are broken when the suspension has turned translucent.
4. Clear the suspension of unbroken cells/debris by centrifugation at 24,000 × *g* for 20 min. Transfer the supernatant (containing the membranes) to a clean tube and repeat the centrifugation step.
5. To collect the membranes, centrifuge the supernatant for 45 min at 150,000 × *g*. Discard the supernatant and resuspend the membrane pellet in 10 mL of ice-cold PBS using a disposable 10 mL syringe with a 21-gauge needle.
6. Fill up the centrifugation tube with ice-cold PBS and harvest the membranes once more for 45 min at 150,000 × *g*. This step will remove residual EDTA which otherwise would interfere with the IMAC step later.
7. Resuspend the membrane pellet in 5 mL of ice-cold PBS, essentially as described before. If desired, membrane suspensions may be frozen in liquid nitrogen and stored at −80 °C for up to 6 months. However, some membrane protein crystallographers avoid freezing and storing of membranes and continue with purification immediately as repeated freezing/thawing may negatively affect the material.

3.6 mNeonGreen as a Fluorescent Fusion Tag for Monitoring Membrane Protein Production

This protocol is based on GFP as a fluorescent fusion tag. However, other fluorescent fusion tags may also be used. Recently, monomeric yellow-green fluorescent protein, mNeonGreen (mNG), was derived from a tetrameric fluorescent protein originating from the cephalochordate *Branchiostoma lanceolatum* [24]. mNG has only a low homology to GFP. Notably, mNG has a 1.5–3 times enhanced brightness compared to the GFP variant EGFP, and a similar photostability. We have successfully used mNG as a fusion partner to monitor the production of a derivative of the *E. coli* tail-anchored membrane protein DjlC (mNG-DjlC-TMD) in Lemo21(DE3) (Fig. 4). When using mNG as a fusion partner, the protocols above can be used with only a minor modification. For whole cell fluorescence, the following settings should be used: emission at 517 nm, excitation at 506 nm, cutoff at 515 nm.

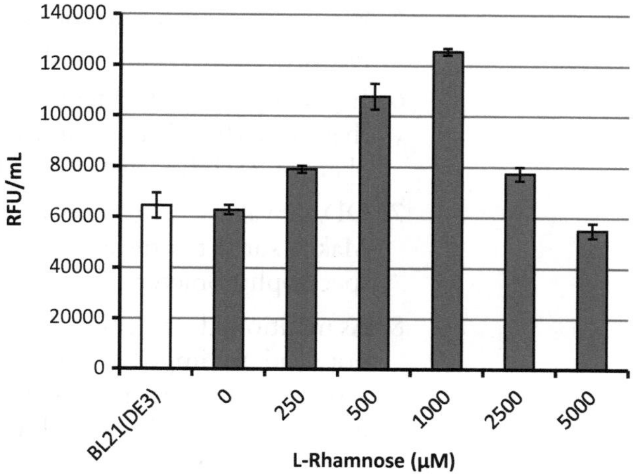

Fig. 4 The use of mNeonGreen as a fusion tag to monitor membrane protein production. The production of a truncated version of the *E. coli* tail-anchored membrane protein DjlC N-terminally fused to mNeonGreen (mNG) was optimized using Lemo21(DE3). As a reference, the protein was also produced in BL21(DE3). The gene encoding the mNG-DjlC fusion protein was expressed from a standard pET vector. Cells were cultured in LB medium at 30 °C, and expression of the target gene was induced with 0.4 mM IPTG for 24 h. Whole cell fluorescence (RFU/mL) was used to monitor protein production

4 Notes

1. An empty expression vector may be included as control.
2. If "leaky" production of the target membrane protein is toxic, the addition of L-rhamnose to the plates used for the transformation and the o/n culture will reduce the toxicity of the "leaky" production.
3. pReX can in principle be used in combination with any T7-RNAP-based protein production strain. This includes strains containing a with pReX compatible plasmid for e.g., the co-expression of genes encoding chaperones. Notably, pReX has a pMB1 origin of replication, which is compatible with, e.g., the widely used p15A and SC101 origins of replication.
4. When many different conditions and/or many targets are included in a membrane protein production screen, one may consider using 24-well plates rather than Falcon tubes. However, the maximum culture volume of a well is 5 mL, which strongly limits sampling options. Furthermore, it should be kept in mind that after taking samples the aeration of the cultures can change, which may influence growth/protein production characteristics. Therefore, when using 24-well plates, we advise running multiple cultures in parallel to allow sampling from separate but identical cultures.

5. We use these concentrations of L-rhamnose by default for the production screening but they may of course be adapted.
6. If the optimal production yield is reached without any L-rhamnose (0 μM) in Lemo21(DE3), consider continuing with plain BL21(DE3) instead.
7. OD_{600} values vary between different spectrophotometers. Make sure to measure within the linear range of the spectrophotometer.
8. As mentioned, screening at different temperatures usually does not lead to improved production yields in Lemo21(DE3). However, if severe degradation of the membrane protein GFP-fusion is observed, a switch to lower production temperatures or a shortening of induction times may be considered. Production at lower temperatures usually results in a different optimal L-rhamnose concentration.
9. The level of background fluorescence is usually quite low but can lead to overestimation of membrane protein production yields, especially if protein production levels are very low. Measure whole cell fluorescence of cells harboring an empty expression vector to account for background fluorescence. If production yields are lower than 200 μg/L, the signal-to-noise ratio may be improved by increasing the amount of cells analyzed. Use 5 mL of culture for fluorescence measurements in such a case. Alternatively, mNG may be used (*see* Subheading 3.6).
10. Instead of adjusting the cell suspensions to the same OD_{600}, they may also be adjusted to the same fluorescence levels (useful if various constructs with different production levels as determined by whole cell fluorescence are screened simultaneously). That way, weak bands can be detected easily without interference from neighboring, stronger bands.
11. Incubation at temperatures higher than 37 °C is not recommended as this can lead to aggregation of membrane proteins and loss of GFP-fluorescence. In addition, incubation at higher temperatures does not allow discrimination between the membrane inserted and non-inserted version of the target protein (*see* Subheading 3.3.4). For a small number of membrane protein fluorescent protein fusions, we have observed that incubating at 37 °C results in the formation of aggregates. In these cases, we advise incubating the samples at 30 °C or room temperature. If frozen cells are used, add $MgCl_2$ to a final concentration of 1 mM and DNaseI (1–5 units/10 μL) to the samples and incubate for 15 min on ice before adding SB buffer to yield a homogenous suspension. Alternatively, benzonase can be used. The samples can also be centrifuged for 10 min at 7000 × *g* to "pellet" the DNA that can make samples slimy and viscous. Notably, after centrifugation, take the sample to be

loaded on a gel as much as possible from the top of the sample in the tube.

12. GFP-His$_8$ has a molecular weight of approx. 28 kDa; however, GFP remains folded in SDS and the apparent molecular weight in SDS-gels is lower (approx. 20 kDa). For quantification, use an image without any saturated signals.
13. In order to determine the best possible ratio of membrane-inserted to non-inserted protein, we advise not only to investigate the sample that yielded the highest fluorescence/mL but also some samples of adjacent rhamnose concentrations.
14. We have experienced that antisera recognizing GFP can bind differently to folded and unfolded GFP. Therefore, we use an antibody recognizing the His-tag. When evaluating the results, keep in mind that the binding behavior varies between different antibodies and that binding is not necessarily linear.
15. The T7-lysozyme variant, LysY, which is produced by Lemo21(DE3) and cells harboring pReX is not lytic [6]. Adding lysozyme is not essential; however, it tremendously facilitates breaking the cells.

Acknowledgments

This work was supported by grants from the Swedish Research Council, the Carl Tryggers Stiftelse, the Marianne and Marcus Wallenberg Foundation, NIH grant 5R01GM081827-03, and the SSF supported Center for Biomembrane Research to JWdG and the People Programme (Marie Curie Actions) of the European Union's Seventh Framework Programme FP7/2007-2013/under REA grant agreement n°607072 (NK and MK) and a SystemsX Transition Postdoc Fellowship (SusS).

References

1. Wagner S, Bader ML, Drew D et al (2006) Rationalizing membrane protein overexpression. Trends Biotechnol 24:364–371
2. Hjelm A, Schlegel S, Baumgarten T et al (2013) Optimizing *E. coli*-based membrane protein production using Lemo21(DE3) and GFP-fusions. Methods Mol Biol 1033:381–400
3. Schlegel S, Hjelm A, Baumgarten T et al (2014) Bacterial-based membrane protein production. Biochim Biophys Acta 1843:1739–1749
4. Samuelson JC (2011) Recent developments in difficult protein expression: a guide to *E. coli* strains, promoters, and relevant host mutations. Methods Mol Biol 705:195–209
5. Studier FW, Moffatt BA (1986) Use of bacteriophage T7 RNA polymerase to direct selective high-level expression of cloned genes. J Mol Biol 189:113–130
6. Wagner S, Klepsch MM, Schlegel S et al (2008) Tuning *Escherichia coli* for membrane protein overexpression. Proc Natl Acad Sci U S A 105:14371–14376
7. Giacalone MJ, Gentile AM, Lovitt BT et al (2006) Toxic protein expression in *Escherichia coli* using a rhamnose-based tightly regulated and tunable promoter system. Biotechniques 40:355–364
8. Wagner S, Baars L, Ytterberg AJ et al (2007) Consequences of membrane protein overexpression in *Escherichia coli*. Mol Cell Proteomics 6:1527–1550

9. Schlegel S, Löfblom J, Lee C et al (2012) Optimizing membrane protein overexpression in the *Escherichia coli* strain Lemo21(DE3). J Mol Biol 423:648–659
10. Drew DE, von Heijne G, Nordlund P et al (2001) Green fluorescent protein as an indicator to monitor membrane protein overexpression in *Escherichia coli*. FEBS Lett 507:220–224
11. Drew D, Slotboom DJ, Friso G et al (2005) A scalable, GFP-based pipeline for membrane protein overexpression screening and purification. Protein Sci 14:2011–2017
12. Drew D, Lerch M, Kunji E et al (2006) Optimization of membrane protein overexpression and purification using GFP fusions. Nat Methods 3:303–313
13. Geertsma ER, Groeneveld M, Slotboom DJ et al (2008) Quality control of overexpressed membrane proteins. Proc Natl Acad Sci U S A 105:5722–5727
14. Feilmeier BJ, Iseminger G, Schroeder D et al (2000) Green fluorescent protein functions as a reporter for protein localization in *Escherichia coli*. J Bacteriol 182:4068–4076
15. Drew D, Sjöstrand D, Nilsson J et al (2002) Rapid topology mapping of *Escherichia coli* inner-membrane proteins by prediction and PhoA/GFP fusion analysis. Proc Natl Acad Sci U S A 99:2690–2695
16. Nugent T, Jones DT (2009) Transmembrane protein topology prediction using support vector machines. BMC Bioinformatics 10:159
17. Peters C, Tsirigos KD, Shu N et al (2016) Improved topology prediction using the terminal hydrophobic helices rule. Bioinformatics 32:1158–1162
18. Tsirigos KD, Peters C, Shu N et al (2015) The TOPCONS web server for consensus prediction of membrane protein topology and signal peptides. Nucleic Acids Res 43:W401–W407
19. Kawate T, Gouaux E (2006) Fluorescence-detection size-exclusion chromatography for precrystallization screening of integral membrane proteins. Structure 14:673–681
20. Luirink J, Yu Z, Wagner S et al (2012) Biogenesis of inner membrane proteins in *Escherichia coli*. Biochim Biophys Acta 1817:965–976
21. Hsieh JM, Besserer GM, Madej MG et al (2010) Bridging the gap: a GFP-based strategy for overexpression and purification of membrane proteins with intra and extracellular C-termini. Protein Sci 19:868–880
22. Löw C, Jegerschöld C, Kovermann M et al (2012) Optimisation of over-expression in *E. coli* and biophysical characterisation of human membrane protein synaptogyrin 1. PLoS One 7:e38244
23. Studier FW (2005) Protein production by auto-induction in high density shaking cultures. Protein Expr Purif 41:207–234
24. Shaner NC, Lambert GG, Chammas A et al (2013) A bright monomeric green fluorescent protein derived from *Branchiostoma lanceolatum*. Nat Methods 10:407–409

Chapter 8

High Yield of Recombinant Protein in Shaken *E. coli* Cultures with Enzymatic Glucose Release Medium EnPresso B

Kaisa Ukkonen, Antje Neubauer, Vinit J. Pereira, and Antti Vasala

Abstract

Expression of recombinant proteins in sufficient quantities is essential for protein structure-function studies. The most commonly used method for recombinant protein production is overexpression in *E. coli* cultures. However, producing high yields of functional proteins in *E. coli* can be a challenge in conventional shaken cultures. This is often due to nonoptimal growth conditions, which result in low cell yields and insoluble or incorrectly folded target protein. To overcome the shortcomings of shake flask cultivation, we present a cultivation method based on enzymatic glucose delivery. This system mimics the fed-batch principle used in bioreactor cultivations and provides high yields of biomass and recombinant proteins in shaken cultivations.

Key words Recombinant proteins, Protein expression, Bacterial culture, Microbial growth media, Fed-batch culture, EnBase, EnPresso, LB medium, TB medium

1 Introduction

Many organisms are used today as hosts or "factories" for producing recombinant proteins by overexpression. *E. coli* is one of the microbial factories that is widely used. The reasons attributed to the popularity of *E. coli* include its inherent ability to grow rapidly within a span of days on an easily constituted growth medium and its ability to express heterologous proteins with sufficient quantities. However, the yield of active protein is highly influenced by the growth characteristics and nutritional requirements of the host strain. Numerous optimization strategies have been developed to address the challenge of improving the specific and volumetric protein yields [1].

The cultivation of *E. coli* for protein expression is commonly performed in liquid growth media composed of complex nutrient sources, typically yeast extract and peptones. The most commonly

Nicola A. Burgess-Brown (ed.), *Heterologous Gene Expression in E. coli: Methods and Protocols*, Methods in Molecular Biology, vol. 1586, DOI 10.1007/978-1-4939-6887-9_8, © Springer Science+Business Media LLC 2017

used variant of such media is the simple Lysogeny Broth (LB), even though LB was not originally designed for recombinant protein production. Attempts to develop improved media for high cell density cultivation have applied richer compositions, buffer components, and glucose or glycerol as a supplemental carbon source [2, 3]. However, the problems encountered with the commonly applied cultivation practices indicate that ideal conditions for successful recombinant protein synthesis are difficult to obtain in shaken cultures.

Limitations of LB and related rich medium formulations arise from the fact that they promote fast, exponential growth of the culture, after which the culture stagnates to a relatively low cell density. Reasons behind this include exhaustion of nutrients, unfavorable pH changes, exhaustion of oxygen due to high respiration rate, and production of growth-inhibitory by-products through overflow and anaerobic metabolism [4–8]. The resulting low cell density limits the protein productivity per volume, and protein expression in the exponential growth phase may lead to problems in protein folding and solubility.

Growth rate of the culture is a key parameter for successful protein production [9]. Uncontrolled growth in LB and related media is associated with uncontrolled protein synthesis rate, which may overload the protein folding machinery and lead to formation of inclusion bodies or incorrectly folded proteins [10]. Despite recently developed special expression strains and vectors for improved protein expression, it is commonly acknowledged that research on membrane proteins and toxic or disulfide bond containing proteins is hampered by inadequate protein yields. For most eukaryotic proteins, growth rate of 0.2–0.3 divisions per hour would be optimal. However, steady maintenance of such growth rates mostly requires controlled feeding of a limiting nutrient. This is typically achieved by a controlled feed pump [11] and has therefore been difficult to obtain in shaken cultures. In practice, good productivity would require both high induction cell density and slow growth during the expression. The most commonly applied approach to slow down the growth and expression in shaken cultures is the use of low cultivation temperature; this method however compromises the cell yield [12]. Alternatively, some researchers have used an approach whereby bacteria cultivated in LB medium are harvested at mid-logarithmic phase, changed to fresh medium, and then induced for recombinant protein production [13]. The autoinduction method developed by Studier provides both relatively high cell yields (OD_{600} = 15) and (auto)induction at a cell density that is typically five times higher than the induction cell density using LB medium [14]. Lactose autoinduction media often provide high protein yields, but the expression level can be compromised by changes in aeration capacity [15, 16]. Moreover, the use of autoinduction medium is only applicable with *E. coli* strains that encode functional lactose permease, e.g., BL21 derivatives.

Studies by Neubauer et al. aimed to transfer the high yields of recombinant proteins obtained in bioreactors to shake flask cultures [17]. This led to the development of the enzyme-based (EnBase) glucose release technology that allows for control of growth rate in microbial cultures so as to achieve very high cell densities [18]. EnBase technology works on the basis of glucose-limited fed-batch cultivation, which is routinely applied by pumps and sophisticated control systems in bioreactors. The core of the EnBase technology comprises an amylase enzyme that breaks down a polysaccharide in a controlled manner to gradually release glucose into the growth medium over the extended period of the cultivation process. The continuous and gradual release of glucose and the presence of pH stabilising constituents in the growth medium enable high cell densities (OD_{600} of 20–50) and consequently high yields of recombinant protein in shake flasks [19]. The complete growth medium comprising the EnBase technology and optimized composition of other nutrient sources for *E. coli* cultivation is called EnPresso B. Compared to LB and other conventional growth media, EnPresso B can provide multi-fold improved yields of recombinant proteins and biocatalysis products per culture volume (*see* Table 1 for representative case studies). While this chapter focuses on the expression of unlabeled

Table 1
Comparison of protein and other bioproduct yields obtained by *E. coli* cultivations with traditional growth medium compared to EnPresso growth system

Target product	Traditional growth medium	EnPresso growth system	Improvement	Reference
Chimeric-truncated form of tissue-type plasminogen activator (t-PA)	0.9 mg/mL of total protein 35.8 IU/mg specific activity with LB	2.5 mg/mL of total protein 46.6 IU/mg specific activity	2.8-fold 30 % higher	[20]
Single-domain antibody 7C12	13 mg/L with LB	130 mg/L	Tenfold	[21]
Single-domain antibody EG2	37 mg/L with LB	200 mg/L	5.4-fold	[21]
Valinomycin	0.3 mg/L with TB	appr. 10 mg/L	33-fold	[22]
21-Acetoxy derivatives by conversion with growing cells	41 ± 3.5 mg/L with TB	50 ± 1.3 mg/L	22 % higher conversion	[23]
Nanobodies	22 mg/L with LB	45 mg/L	Twofold	[24]
Rat lactate dehydrogenase A	93 mg/L with LB	280 mg/L	Threefold	[25]

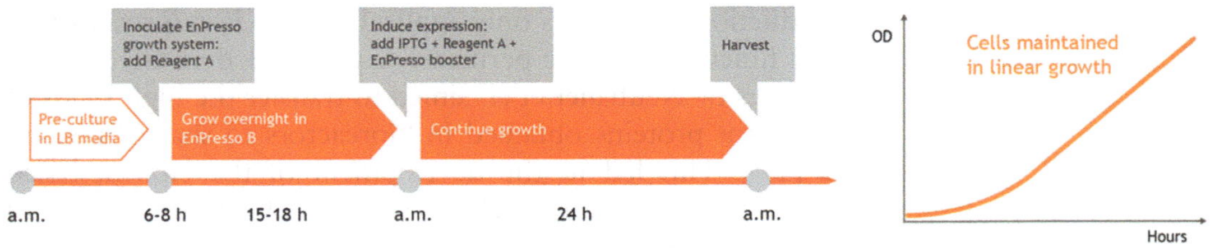

Fig. 1 Workflow showing EnPresso B cultivation with schematic of microbial OD_{600} over time

recombinant proteins by the use of EnPresso B medium, it should be noted that the EnBase technology is also applicable to high-yield expression of ^{15}N [26] and selenomethionine labeled proteins (unpublished results). Protein labeling can be accomplished in the related EnPresso B Defined Nitrogen-free medium, which allows for supplementation of a selected nitrogen source [26].

Protein folding should be regarded as a cotranslational rather than a posttranslational event [27]. Long production phase and slow protein synthesis provided by EnPresso B allow sufficient time for the protein folding machinery to operate. Thus, EnPresso B can enhance the production of large and difficult-to-fold proteins, and especially the production of recombinant proteins that require the synthesis of several subunits or helper proteins. Recently, EnPresso B was reported to provide a significant benefit when applied with the CyDisCo (Cytoplasmic Disulfide bond formation in *E. coli*) expression technology. The *E. coli* CyDisCo strains, which encode helper proteins for eukaryotic disulfide bond formation, produced excellent yields of disulfide bonded recombinant proteins when grown in EnPresso B medium [28].

In this chapter, we describe how to generate high yields of recombinant proteins by cultivating *E. coli* in shaken cultures. We describe the complete protein expression method that applies to the EnPresso B growth system (summarized in Fig. 1). The protocol includes preparation of the medium, preparation of a preculture and inoculation, considerations for the choice of cultivation vessel and conditions, induction of protein expression, and harvest of the cells.

2 Materials

Prepare all solutions using ultrapure water. All solutions used in the cultivation should be sterilized by either autoclaving or filtration and thereafter handled under a laminar flow hood to prevent contamination.

1. LB medium: 10 g of tryptone, 5 g of yeast extract, 5 g of NaCl, pH 7.0. Dissolve in about 900 mL of water in a glass beaker.

Let the solution cool to room temperature and adjust pH to 7.0. Transfer into a 1 L glass bottle and fill up to 1 L. Sterilize by autoclaving and store at 4 °C.

2. 20 % (w/v) Glucose solution: Add 20 g of D-glucose in 70 mL of water in a glass beaker. Dissolve by mixing in a magnetic stirrer with heating. Transfer into a 100 mL glass bottle and fill up to 100 mL. Sterilize by autoclaving and store at room temperature.

3. 0.1 M Isopropyl β-D-1-thiogalactopyranoside (IPTG): Weigh 0.48 g of IPTG into a 50 mL centrifuge tube and add 20 mL of water. Dissolve by mixing and sterilize by filtering through a 0.2 μm syringe filter. Aliquot the filtered solution aseptically into sterile microcentrifuge tubes and store at −20 °C.

4. Antibiotic solution: 100 mg/mL of ampicillin or 50 mg/mL of kanamycin (depending on the resistance provided by the recombinant vector). Weigh 1.0 g of ampicillin sodium salt or 0.5 g of kanamycin monosulfate into a centrifuge tube and dissolve in 10 mL of water. Sterilize the solution by filtering through a 0.2 μm syringe filter. Aliquot the filtered solution aseptically into sterile microcentrifuge tubes and store at −20 °C (*see* **Note 1**).

5. EnPresso B medium (sterile; *see* **Note 2**): Add 50 mL of autoclaved water to a sterilized conical 500 mL Erlenmeyer flask (*see* **Notes 3** and **4**). Carefully open one EnPresso B tablet bag (white) and aseptically transfer the two tablets from the bag into the Erlenmeyer flask (*see* **Note 5**). Close the flask opening tightly with aluminum foil. Place the aluminum covered flask onto an orbital shaker at 30–37 °C and shake at 200–250 rpm until the tablets have dissolved (*see* **Note 6**).

6. EnPresso B Booster tablet: provided within EnPresso B kit.

7. Reagent A: provided within EnPresso B kit.

8. AirOtop membrane seals (Thomson Instrument Company) (*see* **Notes 7–9**).

3 Methods

3.1 Preculture

1. Add 2 mL of preculture medium (LB), 2 μL of antibiotic solution, and 20 μL of glucose solution into a sterile 50 mL tube (*see* **Notes 10** and **11**).

2. Inoculate a colony of *E. coli* cells transformed with the recombinant plasmid into the preculture tube (*see* **Note 12**).

3. Incubate preculture at 37 °C with vigorous shaking (200–250 rpm) for 6 h or until OD_{600} in the culture is 2.5–5.0.

3.2 Protein Expression Culture

1. Add 50 μL of antibiotic solution to the dissolved EnPresso B medium in the Erlenmeyer flask (*see* **Note 13**).
2. Add 2 mL of the preculture to the EnPresso B medium.
3. Add 25 μL of Reagent A to the EnPresso B medium (*see* **Note 14**).
4. Close the flask with AirOtop membrane seal before removing the flask from laminar flow bench.
5. Place the flask onto an orbital shaker and incubate at 30 °C with 200–250 rpm shaking (*see* **Notes 15** and **16**).
6. Incubate the flask overnight (16–18 h).
7. Take the flask to the laminar flow bench and open the seal. Add aseptically one Booster tablet (black bag) (*see* **Note 17**), 25 μL of Reagent A (*see* **Note 18**), and 0.2 mL of 0.1 M IPTG (*see* **Note 19**).
8. Close the flask with a fresh AirOtop seal and place the flask back onto an orbital shaker.
9. Continue incubation at 30 °C with 200–250 rpm shaking (*see* **Note 15**).
10. Harvest the culture after 6–24 h incubation (*see* **Note 20**): transfer the culture into two 50 mL centrifuge tubes. Centrifuge at high speed (e.g., 10,000 × *g*) and 4 °C until the cells and supernatant are separated (*see* **Note 21**).
11. Discard supernatant and store the pellet at −20 °C (*see* **Note 22**).

4 Notes

1. Ampicillin and kanamycin are the most commonly used selection markers for protein expression in *E. coli*, but also other antibiotics such as tetracycline and chloramphenicol are compatible with EnPresso B medium.
2. Prepare EnPresso B medium fresh before use, or store at 4 °C until the next day.
3. It is recommended to select a culture volume that is a multiple of 50 mL, as one EnPresso B tablet bag (containing 2 tablets) is sufficient for preparing 50 mL of medium. Select a sufficiently large Erlenmeyer flask for the cultivation; volume of the medium should be not more than 10 % of the flask nominal volume (e.g., 500 mL flask for 50 mL of medium). Fill volumes higher than 10 % do not support proper oxygen transfer rates in conical and non-baffled flasks.
4. Alternatively to conical flasks, baffled flasks such as Ultra Yield Flasks™ (Thomson Instrument Company) can be used. Ultra Yield Flasks™ generally provide higher cell densities and

product yields compared to Erlenmeyer flasks [29]. Use 20 % fill volume (e.g., 250 mL flask for 50 mL of medium) with the Ultra Yield Flasks™.

5. EnPresso B tablets and the Booster tablets are pre-sterilized. Thus, no further sterilization of the medium is needed when the tablets are handled aseptically and dissolved in sterilized water.
6. EnPresso B tablets dissolve in water in 10–15 min at the given temperatures. The dissolution may take longer if a tablet sticks to the bottom of the flask and does not move around when shaken. In this case, the tablet can be detached either by fast and vigorous manual shaking, or by nudging with a sterile pipette. A good way to prevent the tablets from sticking to the middle of the flask bottom is to add the tablets into an empty flask, and tilt the flask (approximately 20°) so that the tablets move to one edge. Keep the flask tilted in this way while pouring the sterile water in and while carrying the flask to the incubator to prevent the tablets from moving from the edge to the middle. As the liquid will circulate around the edges of the flask bottom, it will wash the tablets along and they will not stick to the glass as easily as when residing in the middle.
7. An optimal flask closure allows for efficient gas exchange while keeping the culture sterile. Aluminum foil or cotton plugs should not be used as closures, since they severely limit the supply of oxygen into the flask and transfer of carbon dioxide out of the flask. Decreasing partial pressure of oxygen in the flask headspace leads to a decrease in oxygen transfer into the liquid culture. The AirOtop membrane seals, or corresponding adhesive membranes from other suppliers, are strongly recommended. Thin silicone sponge closures (Sigma-Aldrich) are also suitable.
8. If using baffled flasks, spillage of the liquid medium can cause wetting of the flask closure. In this case, the best closure option is the AirOtop seal, which is made of hydrophobic material and is able to stay dry when exposed to moderate spillage. A wet membrane or sponge closure blocks the airflow and reduces oxygen supply. A wet closure should always be replaced with a new dry closure.
9. When using adhesive membrane seals as flask closures, the seals may leave sticky residues of the adhesive glue on the flask mouth. When cleaning the flasks, this residue can be removed by scrubbing with acetone.
10. Do not fill more than 10 mL of LB medium into one 50 mL tube. Overfilling will result in insufficient oxygen supply and lower cell density. Severe oxygen limitation during preculture may also affect the main culture adversely. If the expression

culture is larger (e.g., 500 mL) it is preferable to prepare multiple 2–10 mL precultures in separate tubes rather than one larger preculture in a single tube.

11. The preculture is supplemented with 2 g/L of glucose to control leaky expression. LB medium often contains traces of lactose, which may be sufficient to induce some premature expression from lactose-inducible promoters [30]. Presence of glucose in the medium represses the uptake of lactose into the cell and thus inhibits expression of the recombinant genes controlled by lactose-inducible promoters.
12. Alternatively, preculture can be inoculated with a glycerol stock of transformed *E. coli* cells.
13. If using baffled flasks, it is important to prevent excessive foam formation. To eliminate foaming, add Antifoam 204 (Sigma-Aldrich) to 1:10,000 v/v at the beginning of cultivation. Antifoam 204 is autoclavable and can also be diluted to allow for easier pipetting. Dilution may cause turbidity of the antifoam solution, but this does not affect its properties.
14. Add the glucose-releasing reagent (Reagent A) as the last component into the culture to avoid any glucose accumulation before the cells are inoculated.
15. Use 200 rpm in a device with 50 mm (2 in.) shaking diameter, and 250 rpm in a device with 25 mm (1 in.) shaking diameter. The maximum rate of oxygen transfer is dependent not only on the shaking speed (rpm) but also on the diameter of orbital shaking. At a fixed shaking speed, larger shaking diameter contributes to higher maximum oxygen transfer rate. If using Ultra Yield Flasks™, the shaking diameter must not exceed 25 mm (1 in.). In our experience, shaking of Ultra Yield Flasks™ with 50 mm diameter will result in heavy spillage of the medium out of the flasks.
16. It is not advisable to attach Ultra Yield Flasks™ or other plastic shake flasks to a sticky mat-shaking table. In our experience, plastic flasks do not adhere to the sticky mat table sufficiently but may detach at higher shaking speeds. Clamps should be used instead.
17. If culture volume is lower than 50 mL, the Booster tablet cannot be added directly to the culture. In this case, dissolve the Booster tablet into 5 mL of sterile water to prepare a 10× concentrated solution, and add an aliquot of this solution to the culture in 1:10 v/v. For dissolving, add one Booster tablet into 5 mL of sterile water in a 50 mL microcentrifuge tube. Place the tube onto an orbital shaker in upright or slightly tilted position, and shake at 30–37 °C until the tablet has dissolved. This may take up to 1 h. To speed up the work in the morning

of induction, Booster tablet can be dissolved the previous day and the solution stored overnight at 4 °C before use.

18. If using Ultra Yield Flasks™, add 75 μL of Reagent A.
19. Cultivation can alternatively be performed by an autoinduction method. Autoinduction by IPTG is applicable to *E. coli* strains that express functional lactose permease. In such strains, IPTG uptake at low concentrations (0.02–0.04 mM) is regulated by the lactose permease. This mechanism enables glucose-mediated repression of lac-inducible promoters at low IPTG concentrations, and induction at the time when the initial batch of glucose has been consumed [31]. The slow glucose feed provided by the EnBase technology does not prevent lactose permease activity. For autoinduction, supplement the medium (50 mL) with 0.01 mL of 0.1 M IPTG, one Booster tablet, 0.06 mL of 20 % glucose, and 50 μL of Reagent A at the time of inoculation (Subheading 3, **steps 2** and **3**). Incubate for 18–24 h until the next day under the specified conditions (skip Subheading 3, **steps 7–9**) and continue from Subheading 3, **step 10**.
20. Optimal harvest time is dependent on the target protein. Proteins that are toxic to the host cells should be harvested after about 6 h expression. Nontoxic proteins can be expressed overnight, and they are usually harvested at maximum yields the next morning. Clones carrying the pLysS or pLysE plasmid are more susceptible to lysis during prolonged expression periods, and some of the product may be lost to lysis if cultivation is continued overnight. Therefore, the optimal harvest time for pLysS and pLysE clones is likely 6–15 h, again depending on the target protein.
21. If the recombinant protein is fused with maltose-binding protein (MBP), wash the cell pellet to remove residues of the supernatant. Without washing, traces of the EnPresso medium polysaccharide may be carried over to the cell lysate and may be bound by the MBP tag. To wash the pellet, first remove the supernatant and add corresponding volume of 0.9 % NaCl on the top of the pellet. Mix carefully on a vortex or using a pipette to resuspend the pellet. Centrifuge the sample at high speed (e.g., 10,000 × *g*) at 4 °C until the cells and supernatant are separated. Discard the supernatant after centrifugation, and repeat the resuspension and centrifugation steps one more time. Washing must always be done before freezing the pellet as frozen and thawed cells may lyse during the procedure.
22. Prolonged cultivation at a low growth rate may sometimes render *E. coli* cells more resistant to lysis. To enable efficient lysis of the cells, freeze (−20 °C) and thaw the pellet before proceeding to lysis.

References

1. Graslund S, Nordlund P, Weigelt J et al (2008) Protein production and purification. Nat Methods 5:135–146
2. Tartoff KD, Hobbs CA (1987) Improved media for growing plasmid and cosmid clones. Bethesda Res Lab Focus 9:12
3. Duttweiler HM, Gross DS (1998) Bacterial growth medium that significantly increases the yield of recombinant plasmid. Biotechniques 24:438–444
4. Losen M, Frolich B, Pohl M et al (2004) Effect of oxygen limitation and medium composition on *Escherichia coli* fermentation in shake-flask cultures. Biotechnol Prog 20:1062–1068
5. Luli GW, Strohl WR (1990) Comparison of growth, acetate production, and acetate inhibition of *Escherichia coli* strains in batch and fed-batch fermentations. Appl Environ Microbiol 56:1004–1011
6. Warnecke T, Gill RT (2005) Organic acid toxicity, tolerance, and production in *Escherichia coli* biorefining applications. Microb Cell Fact 4:25
7. Vemuri GN, Altman E, Sangurdekar DP et al (2006) Overflow metabolism in *Escherichia coli* during steady-state growth: transcriptional regulation and effect of the redox ratio. Appl Environ Microbiol 72:3653–3661
8. Sezonov G, Joseleau-Petit D, D'Ari R (2007) *Escherichia coli* physiology in Luria-Bertani broth. J Bacteriol 189:8746–8749
9. Neubauer P, Winter J (2001) Expression and fermentation strategies for recombinant protein production in *Escherichia coli*. Springer, Dordrecht, pp 195–258
10. Fahnert B, Lilie H, Neubauer P (2004) Inclusion bodies: formation and utilisation. Adv Biochem Eng Biotechnol 89:93–142
11. Lee SY (1996) High cell-density culture of *Escherichia coli*. Trends Biotechnol 14:98–105
12. Schein CH, Noteborn MH (1988) Formation of soluble recombinant proteins in *Escherichia coli* is favored by lower growth temperature. Nat Biotechnol 6:291–294
13. Sivashanmugam A, Murray V, Cui C et al (2009) Practical protocols for production of very high yields of recombinant proteins using *Escherichia coli*. Protein Sci 18:936–948
14. Studier FW (2005) Protein production by auto-induction in high-density shaking cultures. Protein Expr Purif 41:207–234
15. Blommel PG, Becker KJ, Duvnjak P et al (2007) Enhanced bacterial protein expression during auto-induction obtained by alteration of lac repressor dosage and medium composition. Biotechnol Prog 23:585–598
16. Ukkonen K, Mayer S, Vasala A et al (2013) Use of slow glucose feeding as supporting carbon source in lactose autoinduction medium improves the robustness of protein expression at different aeration conditions. Protein Expr Purif 91:147–154
17. Neubauer P, Panula-Perälä J, Siurkus J et al (2009) EnBase™: Novel high-cell-density-culture based screening platform. Chem Ing Tech 81:1247–1248
18. Neubauer P, Neubauer A, Vasala A (2015) Enzyme-based fed-batch technique in liquid cultures. US Patent 9,127,261
19. Krause M, Ukkonen K, Haataja T et al (2010) A novel fed-batch based cultivation method provides high cell-density and improves yield of soluble recombinant proteins in shaken cultures. Microb Cell Fact 9:11
20. Mahboudi F, Barkhordari F, Godarzi RM et al (2013) A fed-batch based cultivation mode in *Escherichia coli* results in improved specific activity of a novel chimeric-truncated form of tissue plasminogen activator. J Appl Microbiol 114:364–372
21. Zarschler K, Witecy S, Kapplusch F et al (2013) High-yield production of functional soluble single-domain antibodies in the cytoplasm of *Escherichia coli*. Microb Cell Fact 12:1
22. Li J, Jaitzig J, Hillig F et al (2014) Enhanced production of the nonribosomal peptide antibiotic valinomycin in *Escherichia coli* through small-scale high cell density fed-batch cultivation. Appl Microbiol Biotechnol 98:591–601
23. Mosa A, Hutter MC, Zapp J et al (2015) Regioselective acetylation of C21 hydroxysteroids by the bacterial chloramphenicol acetyltransferase I. Chembiochem 16:1670–1679
24. Ta DT, Redeker ES, Billen B et al (2015) An efficient protocol towards site-specifically clickable nanobodies in high yield: cytoplasmic expression in *Escherichia coli* combined with intein-mediated protein ligation. Protein Eng Des Sel 28:351–363
25. Nowicki MW, Blackburn EA, McNae IW et al (2015) A streamlined, automated protocol for the production of milligram quantities of untagged recombinant rat lactate dehydrogenase a using ÄKTAxpress TM. PLoS One 10:e0146164
26. Rubino JT, Martinelli M, Cantini F et al (2016) Structural characterization of zinc-bound Zmp1, a zinc-dependent metalloprotease

secreted by *Clostridium difficile*. J Biol Inorg Chem 21:185–196

27. Fedorov AN, Baldwin TO (1997) Cotranslational protein folding. J Biol Chem 272:32715–32718
28. Nguyen VD, Hatahet F, Salo KE et al (2011) Pre-expression of a sulfhydryl oxidase significantly increases the yields of eukaryotic disulfide bond containing proteins expressed in the cytoplasm of *E. coli*. Microb Cell Fact 10:1
29. Ukkonen K, Vasala A, Ojamo H et al (2011) High-yield production of biologically active recombinant protein in shake flask culture by combination of enzyme-based glucose delivery and increased oxygen transfer. Microb Cell Fact 10:107
30. Nair R, Salvi P, Banerjee S et al (2009) Yeast extract mediated autoinduction of lacUV5 promoter: an insight. N Biotechnol 26:282–288
31. Faust G, Stand A, Weuster-Botz D (2015) IPTG can replace lactose in auto-induction media to enhance protein expression in batch-cultured *Escherichia coli*. Eng Life Sci 15: 824–829

Part II

Protein Purification and Solubility Enhancement

Chapter 9

A Generic Protocol for Purifying Disulfide-Bonded Domains and Random Protein Fragments Using Fusion Proteins with SUMO3 and Cleavage by SenP2 Protease

Hüseyin Besir

Abstract

Recombinant expression of heterologous proteins in *E. coli* is well established for a wide range of proteins, although in many cases, purifying soluble and properly folded proteins remains challenging (Sorensen and Mortensen, J Biotechnol 115:113–128, 2005; Correa and Oppezzo, Methods Mol Biol 1258:27–44, 2015). Proteins that contain disulfide bonds (e.g., cytokines, growth factors) are often particularly difficult to purify in soluble form and still need optimizing of protocols in almost every step of the process (Berkmen, Protein Expr Purif 82:240–251, 2012; de Marco, Microb Cell Fact 11:129, 2012). Expression of disulfide bonded proteins in the periplasm of *E. coli* is one approach that can help to obtain soluble protein with the correct disulfide bridges forming in the periplasm. This offers the appropriate conditions for disulfide formation although periplasmic expression can also result in low expression levels and incorrect folding of the target protein (Schlapschy and Skerra, Methods Mol Biol 705:211–224, 2011). Generation of specific antibodies often requires a specific antigenic sequence of a protein in order to get an efficient immune response and minimize cross-reactivity of antibodies. Larger proteins like GST (Glutathione-S-transferase) or MBP (maltose binding protein) as solubilizing fusion partners are frequently used to keep antigens soluble and immunize animals. This approach has the disadvantage that the immune response against the fusion partner leads to additional antibodies that need to be separated from the antigen-specific antibodies. For both classes of proteins mentioned above, a protocol has been developed and optimized using the human version of small ubiquitin-like modifier 3 (SUMO3) protein and its corresponding protease SenP2. This chapter describes the experimental steps for expression, purification, refolding, and cleavage that are applicable to both disulfide-bonded proteins with a defined structure and random protein fragments for antibody generation or larger peptides with defined sequence that are difficult express on their own.

Key words SUMO fusion protein, Disulfide bonded proteins, Growth factors, Cytokines, Protein refolding, Insoluble protein, His-tag purification, SenP2 protease, Antigen purification

1 Introduction

A number of methods and tools have been described to overcome limitations of recombinant protein expression in *E. coli* [1, 2]. Disulfide bonded proteins like are particularly difficult to produce in soluble form in *E. coli* [3, 4]. Periplasmic expression is one

Nicola A. Burgess-Brown (ed.), *Heterologous Gene Expression in E. coli: Methods and Protocols*, Methods in Molecular Biology, vol. 1586, DOI 10.1007/978-1-4939-6887-9_9, © Springer Science+Business Media LLC 2017

approach that can help to obtain soluble protein with the correct disulfide bridges but can also result in low expression levels and incorrect folding of the target protein [5]. Using solubility enhancing fusion proteins is one option to increase the expression level of a challenging target protein. Among these fusion proteins, the small ubiquitin-like modifier (SUMO) proteins shows significant advantages compared to other fusion proteins. SUMO proteins are small proteins of approximately 100 amino acids found in all eukaryotes but not in bacteria and involved in a number of cellular processes [6]. SUMO proteins from yeast (SMT3) and human (SUMO-1, -2, and -3) have been used as N-terminal fusion partners for recombinant proteins and found to offer some advantages over traditional fusion proteins. Its small size and generally high expression levels in *E. coli* result in a high fraction of the recombinant product being the target protein, while larger fusion proteins like MBP (~45 kDa), GST(~28 kDa), or NusA (~53 kDa) make up a significant part of the product.

The most beneficial property of the SUMO proteins, however, is clearly the specificity and robustness of their corresponding protease [7, 8] which, in contrast to commonly used TEV or 3C proteases, allow the cleavage under partially denaturing conditions in up to 3 M urea and, most importantly, cleaving before the first amino acid of the target protein without leaving any additional amino acids at the N-terminus. This property is especially useful for proteins that need a defined N-terminus for activity or binding, or avoiding of unwanted antigenic sequences by remaining amino acids. In the experiments described in this chapter, the human SUMO3 protein and its corresponding protease SenP2 were used.

Numerous studies have been published that describe the use of the SUMO fusion proteins in *E. coli* [9–11]. Lu et al. [12] have published a protocol also using SUMO fusion and refolding similar to the one described below which they used for generating a number of disulfide bonded cytokines. The following protocol allows the generation of a wide range of folded or unfolded proteins and large peptides with defined N-termini, where the N-terminal SUMO fusion protein has been cleaved off (Fig. 1).

For the expression in *E. coli* and purification of SUMO-fusion proteins, a synthetic human *sumo3* gene was generated and cloned into the pETM11 vector [13] containing an N-terminal His6-sequence for purification via Ni-columns and a C-terminal *egfp* gene which is replaced by the target sequence (Fig. 1). Two restriction sites for *Age*I and *Bam*HI have been added for cloning without changing the amino acid sequence of SUMO3. An alternative vector was generated for using the sequence and ligation-independent cloning (SLIC) approach and described in Scholz et al. [14], essentially leading to the same final expression construct as this one after cloning.

```
         XbaI
         |
caattcccctctagaaataattttgatttaactttaagaaggagatataccATGaaacat  < 120
Q  F  P  S  R  N  N  F  D  L  T  L  R  R  Y  T  M  K  H
         70        80        90        100       110
     His-tag                                           >TEV-site
     |                                                 |
caccatcaccatcaccccatgagcgattacgacatccccactactgagaatctttatttt  < 180
H  H  H  H  H  P  M  S  D  Y  D  I  P  T  T  E  N  L  Y  F
         130       140       150       160       170
       NcoI
       | |-> sumo3
cagggcgccatgggcaacgatcacattaacctgaaagtggccggt....SUMO3......
Q  G  A  M  G  N  D  H  I  N  L  K  V  A  G  ....300bp......
         190       200       210       220
                                         Cleavage site
   AgeI     BamHI                        (after GlyGly)
   |        |  |-> egfp
cagaccggtggatccgtgagcaagggcgaggagctgttcaccggggtggtgcccatcctg  < 480
Q  T  G  G  S  V  S  K  G  E  E  L  F  T  G  V  V  P  I  L
         430       440       450       460       470
                                                      SacI
                                                      |
......EGFP.....ctggagttcgtgaccgccgccgggatcactctcggcatggacgag  < 1140
......714bp....L  E  F  V  T  A  A  G  I  T  L  G  M  D  E
                   1100      1110      1120      1130
                   EagI    XhoI
         HindIII   NotI    SciI    His-tag (C-term.)
         |         ||      |
ctctacaagtgaaagcttgcggccgcactcgagcaccaccaccaccaccactgagatccg  < 1200
L  Y  K  *  K  L  A  A  A  L  E  H  H  H  H  H  H  *
         1150      1160      1170      1180      1190
```

Fig. 1 Sequence map of the pETM11-SUMO3GFP expression vector. Two restriction sites (AgeI, BamHI) are available for the 5′-end of a target gene which will replace the gene for EGFP using one of the restriction sites downstream of the EGFP gene

In the following section, the expression and purification of insoluble SUMO fusion proteins are described in detail, as well as the purification of SenP2 protease needed for cleavage of the fusion proteins. The initial purification step of the SUMO fusion protein is done by centrifugation and separating the soluble cell extract from the insoluble recombinant protein. After washing and subsequent solubilization of the insoluble material in buffer containing 6 M guanidinium, the protein is purified under denaturing conditions using Ni-NTA columns. Guanidinium is only used for solubilizing the protein, while the purification is done in buffers with 6 M urea which is more economic and, more importantly, necessary for the activity of the SenP2 protease under the conditions for the cleavage reaction described below. SenP2 activity is inhibited by guanidinium in concentrations above ~0.5 M. Following the Ni-NTA purification, the urea concentration is lowered by dialysis against a buffer with 2.5 M urea and a redox buffer system [reduced glutathione (GSH)/oxidized glutathione (GSSG), 2 mM:0.5 mM] which will allow most disulfide bonded proteins to stay in solution and refold, and the protease to have enough activity to efficiently cleave the SUMO3 fusion protein in a few hours or overnight. In a second Ni-NTA purification step, the His-tagged SUMO3 and uncleaved fusion protein will be separated from the cleaved target protein which can be further purified by an anion-exchange or gel filtration step. Quality control and confirmation of the protein mass by intact mass spectrometry is essential, especially in case of disulfide bonded proteins where the mass reduction of 2 Da compared to the theoretical molecular weight would indicate each disulfide bond that is formed. Subsequently, the protein can be used directly for immunization or activity assays.

2 Materials

Prepare all solutions using ultrapure water (prepared by purifying deionized water to attain a sensitivity of 18 M Ω cm at 25 °C) and analytical grade reagents. Sterilize all solutions either by autoclaving or filtering (0.22 μm filter). For cooling tubes or flasks, use an ice–water bath where the vessel is in contact with water to enable more efficient cooling than just placing the vessel on or in crushed ice. Prepare and store all reagents at room temperature (unless indicated otherwise). Follow all waste disposal regulations when disposing of waste materials.

2.1 Expression of HisSenP2 Protease and SUMO3-Fusion Proteins

1. BL21(DE3) CodonPlus RIL (Stratagene) *E. coli* cells transformed with expression vectors pETM11-HisSenP2 or pETM11-SUMO3-[gene of interest] and plated on LB agar plates containing 33 μg/mL kanamycin (kan) and 10 μg/mL chloramphenicol (cam) (*see* **Note 1**).
2. LB Kan-33/Cam-10 medium: Luria–Bertani (LB) broth supplemented with 33 mg/mL kanamycin and 10 μg/mL chloramphenicol just before use.
3. 1 M isopropyl β-D-1-thiogalactopyranoside (IPTG) solution.
4. 1 M L-glutathione reduced (GSH): Prepare only 5–10 mL and adjust pH with 10 M NaOH to 7–8. Measure pH with indicator strips and not a pH meter because of the small volume and thiol compound (can be harmful for pH electrodes, *see* **Note 2**).
5. 1 M L-glutathione oxidized (GSSG): Prepare only 5–10 mL and adjust pH with 10 M NaOH to 7–8. Measure pH with indicator strips, not pH meter because of the small volume and thiol compound (can be harmful for pH electrodes, *see* **Note 2**).
6. Phosphate-buffered saline (PBS): 10 mM Na_2HPO_4, pH 7.4, 2 mM KH_2PO_4, 2.7 mM KCl, 140 mM NaCl.
7. 10 mg/mL DNAseI: Prepare in PBS.
8. 100 mg/mL lysozyme: Prepare in PBS.
9. Complete protease inhibitor cocktail tablets (EDTA-free, Roche): 1 tablet for 50 mL of lysate.

10. Bacterial shaking incubators (refrigerated).
11. 1.5-mL tubes.
12. 50-mL tubes.
13. 1.5- and 50-mL tube rack.
14. 3000 mL flask clamps.
15. Preparative centrifuge with rotor suitable for 500 or 1000 mL bottles.
16. Centrifuge bottles (500 or 1000 mL).

2.2 Purification of His-Tagged Proteins

All solutions for protein purification are stored at 4 °C.

1. Lysis/loading/wash buffer for native purification conditions: 50 mM Tris–HCl, pH 8.0, 300 mM NaCl, 20 mM imidazole. For cell lysis, add 1 tablet of Complete protease inhibitor cocktail tablet to 50 mL of lysate.
2. Elution buffer for Ni-NTA column: 50 mM Tris–HCl pH 8.0, 300 mM NaCl, 300 mM imidazole.
3. Anion-exchange running buffer: 50 mM Tris–HCl pH 7.0, 50 mM NaCl.
4. Anion-exchange elution buffer: 50 mM Tris–HCl pH 7.0, 1 M NaCl.
5. Cation-exchange running buffer: 50 mM Tris–HCl pH 6.0, 50 mM NaCl.
6. Cation-exchange elution buffer: 50 mM Tris–HCl pH 6.0, 1 M NaCl.
7. Storage buffer for HisSenP2: 100 mM Tris–HCl pH 8.0, 100 mM NaCl, 1 mM DTT, 2 mM EDTA, 50 % (v/v) glycerol.
8. Washing buffer for insoluble protein: Phosphate-buffered saline (PBS): 10 mM Na_2HPO_4, pH 7.4, 2 mM KH_2PO_4, 2.7 mM KCl, 140 mM NaCl, 0.1 % (v/v) Triton X-100.
9. Solubilization buffer for insoluble protein: Phosphate-buffered saline (PBS): 10 mM Na_2HPO_4, pH 7.4, 2 mM KH_2PO_4, 2.7 mM KCl, 140 mM NaCl, 6 M Guanidinium.
10. For 1 L, use 100 mL of 10× PBS buffer and add 573.18 g of guanidinium chloride, fill up to 900 mL and adjust the pH to 7.5 with 10 M NaOH solution (Guanidinium is acidic and will lower the pH.).
11. Running buffer for denaturing purification: 6 M urea, 50 mM Tris–HCl pH 8.0, 250 mM NaCl, 20 mM imidazole.
12. Elution buffer for denaturing purification: 6 M urea, 50 mM Tris–HCl pH 8.0, 250 mM NaCl, 300 mM imidazole.

13. Dialysis buffer for refolded protein: 2.5 M urea, 50 mM Tris–HCl pH 7.9, 250 mM NaCl, 20 mM imidazole, 2 mM GSH, 0.5 mM GSSG.
14. Anion-exchange running buffer for refolded protein: 50 mM Tris–HCl pH 7.0, 50 mM NaCl (optional: +2.5 M urea, *see* **Note** 7).
15. Anion-exchange elution buffer for refolded protein: 50 mM Tris–HCl pH 7.0, 1 M NaCl (optional: +2.5 M urea, *see* **Note** 7).
16. Cation-exchange elution buffer for refolded protein: 50 mM Tris–HCl pH 6.0, 50 mM NaCl (optional: +2.5 M urea, *see* **Note** 7).
17. Cation-exchange elution buffer for refolded protein: 50 mM Tris–HCl pH 6.0, 1 M NaCl (optional: +2.5 M urea, *see* **Note** 7).
18. Sonicator (e.g., Branson sonifier 450) with standard disruptor horn.
19. Ultracentrifuge with rotor for 30 or 60 mL ultracentrifuge tubes (e.g., Beckman Ti45 or Ti70 rotor).
20. 30 or 60 mL ultracentrifuge tubes.
21. Ni-NTA columns (prepacked 5 mL, e.g., Ni-NTA Sepharose, Qiagen).
22. Anion-exchange (prepacked 5 mL, e.g., HiTrapQ, GE Healthcare) and cation-exchange (prepacked 5 mL, e.g., HiTrapSP, GE Healthcare) columns.
23. 0.45 μm sterile filter units with Luer lock for syringe filtration.
24. Dialysis tubing with 3 or 10 kDa molecular weight cutoff (MWCO) (pretreated dialysis tubing with no trace metal ions, e.g., Spectra/Por® 7 Pretreated dialysis tubing). Use dialysis tubing according to MW of the target protein (for proteins with MW <20 kDa, a tubing with MWCO of 3 kDa should be used; for MW >20 kDa, a MWCO of 10 kDa is appropriate).
25. FPLC system with a superloop, if available (e.g., ÄKTA, GE Healthcare or NGC, Bio-Rad).
26. UV/Vis spectrophotometer with a quartz or UV-transparent plastic cuvettes.
27. Electrophoresis system for SDS-PAGE with power supply, electrophoresis chamber, polyacrylamide gels, SDS sample buffer, MW marker with appropriate size range.
28. Coomassie staining/destaining solution: [e.g., 45 % (v/v) methanol, 10 % (v/v) acetic acid, 0.25 % (w/v) Brilliant Blue G-250 in water for staining and 10 % (v/v) methanol and 10 % (v/v) acetic acid in water for destaining].
29. Access to mass spectrometry instrument or service facility for intact mass determination.

3 Methods

All procedures after harvesting the bacterial culture are carried out at 4 °C in a cold room or by placing the tubes in an ice–water bath unless indicated otherwise.

3.1 Expression and Purification of HisSenP2 Protease (4 L Total Culture Vol.)

1. Freshly transform the construct pETM11-HisSenP2 into BL21(DE3) CodonPlus RIL and plate it on LB-agar containing 33 μg/mL kanamycin and 10 μg/mL chloramphenicol.
2. Inoculate 50 mL of LB medium containing 33 μg/mL kanamycin (kan) and 10 μg/mL chloramphenicol (cam) and incubate overnight at 37 °C and 250 rpm.
3. Use 10 mL of pre-culture to inoculate 1 L of LB medium containing 33 μg/mL kan and 10 μg/mL cam.
4. Grow the culture until OD600 reaches 0.4–0.6. Take 200 μL of the culture before induction and store the cell pellet after centrifuging in a 1.5-mL tube at 10,000 × *g*.
5. Reduce the temperature in the shaker to 18 °C and induce expression 30–60 min later with 0.2 mM IPTG. Continue incubation overnight at 18 °C and 180 rpm.
6. Harvest the cells at 6000 × *g* for 20 min, discard the supernatant and store cell pellets at −20 °C or −80 °C until purification.
7. Resuspend a 4 L-pellet of HisSenP2 in ~40 mL of lysis buffer in a 50 mL tube by pipetting up and down slowly with a 10 mL disposable pipette until the cell suspension is homogeneous.
8. Add 0.1 mg/mL DNase (final conc.), 5 mM $MgCl_2$, and 1 mg/mL Lysozyme (final conc.).
9. Incubate at room temperature for ~15 min, then cool on ice for 15 min.
10. Lyse the cells by sonication (medium intensity, 50 % duty cycle, 5× 30 s, 60 s, pausing after each cycle).
11. Centrifuge supernatant at ~100,000 × *g* for 45 min at 4 °C and filter the supernatant through a 0.45 μm syringe filter into a new tube placed in an ice–water bath.
12. Load the supernatant on a 5 mL Ni-NTA column (1 column per 4 L pellet), charged with Ni^{2+} and equilibrated in lysis buffer. Recommended flow rate for binding is 0.5 mL/min. Collect 10 mL flowthrough fractions.
13. Wash column(s) with running buffer until 280 nm absorption goes down. Collect 10 mL wash fractions.
14. Elute with gradient from 0 to 100 % elution buffer (in 10 CVs), collect 5 mL fractions and check 20 μL of peak fractions

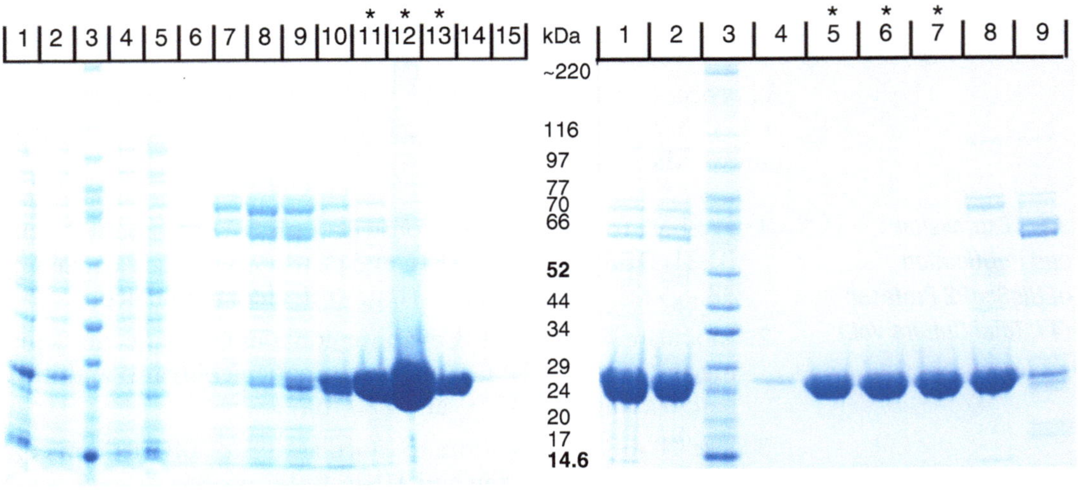

Fig. 2 Purification of HisSenP2, SDS-PAGE images after Ni-NTA purification (*left*) and anion-exchange step (*right*). Lanes with * indicate fractions used for further purification (*left*) and storage (*right*), respectively

by SDS-PAGE. Pool fractions with highest SenP2 (*see* Fig. 2, left) concentration and dialyze overnight at 4 °C against 1 L anion-exchange buffer (*see* **Note 3**).

15. Load the dialyzed sample on an anion-exchange column (5 mL HiTrapQ) and collect 5 mL fractions. Check 20 μL of each fraction by SDS-PAGE and pool fractions with high HisSenP2 concentration. HisSenP2 does not bind to HiTrapQ but most of the contaminant protein bands and nucleic acid will. The flowthrough contains a highly purified preparation of HisSenP2 (*see* Fig. 2, right).
16. Dialyze pooled flowthrough fractions with HisSenP2 overnight in the cold room against 2× Storage buffer without glycerol (200 mM Tris–HCl pH 8.0, 200 mM NaCl, 2 mM DTT, 4 mM EDTA).
17. Measure the concentration of HisSenP2 (e.g., by Bradford or Lowry assay), dilute to 1 mg/mL by adding an appropriate volume 2× Storage buffer and add glycerol 1:1 to reach 50 % (v/v) glycerol in 1× Storage buffer.
18. Prepare aliquots of 0.5 or 1 mL in 1.5 mL tubes, freeze in liquid N_2. HisSenP2 can be stored at −80 °C for at least 12 months.

3.2 Expression and Purification of SUMO-Fusion Protein

1. Freshly transform the construct pETM11-SUMO3-[gene of interest] into BL21(DE3) CodonPlus RIL and plate it on LB-agar containing 33 μg/mL kan and 10 μg/mL cam.
2. Inoculate 50 mL of LB medium containing 33 μg/mL kan and 10 μg/mL cam and incubate overnight at 37 °C and 250 rpm.

3. Use 10 mL of pre-culture to inoculate 1 L of LB medium containing 33 µg/mL kan and 10 µg/mL cam.
4. Grow the culture until OD600 reaches 0.4–0.6. Take 200 µL of the culture before induction and store the cell pellet after centrifuging in a 1.5 mL tube at 10,000 × *g*.
5. Reduce the temperature in the shaker to 18 °C and induce expression 30–60 min later with 0.2 mM IPTG. Continue incubation overnight at 18 °C and 180 rpm.
6. Harvest the cells at 6000 × *g* for 20 min, discard the supernatant and store cell pellets at −20 °C or −80 °C until purification.
7. Resuspend a pellet of SUMO3 fusion protein (usually from 3 L culture) in ~40 mL of lysis buffer in a 50 mL tube by pipetting up and down slowly with a 10 mL disposable pipet until the cell suspension is homogeneous.
8. Add 0.1 mg/mL DNase (final conc.), 5 mM $MgCl_2$ and 1 mg/mL Lysozyme (final conc.).
9. Incubate at room temperature for ~15 min, then cool it on ice for 15 min.
10. Lyse the cells by sonication (medium intensity, 50 % duty cycle, 5× 30 s, 60 s pausing after each cycle).
11. Centrifuge lysate at ~100,000 × *g* for 45 min at 4 °C.
12. Pipet the supernatant into a fresh tube (store at 4 °C) and add 30 mL of PBS +0.1 % (v/v) Triton X-100 to the pellet. Resuspend pellet by pipetting up and down and centrifuge at 20,000 × *g* for 20 min.
13. Remove the supernatant into a fresh 50 mL tube and keep for on ice for checking by SDS-PAGE.
14. Add another 30 mL of PBS +0.1 % (v/v) Triton X-100 to the pellet and repeat the washing step twice, keeping the supernatant in a new tube each time (*see* **Note 4**).
15. Resuspend (usually almost white) pellet in PBS +6 M guanidinium overnight at room temperature on a rolling device or gel shaker.
16. Centrifuge solubilized sample at ~100,000 × *g* for 45 min at 4 °C, carefully pipet the supernatant without remaining particles from the pellet and filter the supernatant through a 0.45 µm syringe filter into a new tube placed in an ice–water bath (*see* **Note 5**).
17. Load the solubilized sample on a 5 mL Ni-NTA column (1 column per 3 L culture), charged with Ni^{2+} and equilibrated in denaturing running buffer. Recommended flow rate for binding is 0.5 mL/min. Collect 10 mL flowthrough fractions.
18. Wash column(s) with denaturing running buffer until 280 nm absorption goes back to baseline. Collect 10 mL wash fractions.

19. Elute with gradient from 0 to 100 % denaturing elution buffer (in 10 CVs), collect 5 mL fractions and check 20 μL of peak fractions by SDS-PAGE.

20. Pool fractions with highest protein amount and measure protein concentration, add pH-adjusted GSH and GSSG to a final concentration of 2 and 0.5 mM, respectively, and dialyze overnight at 4 °C against 1 L of redox refolding buffer with the same GSH/GSSG concentration ratio and 2.5 M urea.

21. Centrifuge the dialyzed sample for 30 min at 4 °C with 10,000 × *g*. In case of heavy precipitation of the sample, measure the protein concentration of the supernatant after centrifugation.

22. Transfer the supernatant carefully into a fresh tube and add HisSenP2 at a ratio 1:20 or 1:50 (w/w), incubate the digestion reaction overnight at 4 °C while dialyzing against the ion-exchange buffer without urea (*see* **Notes 6** and 7).

23. Centrifuge the digestion sample for 30 min at 4 °C with 10,000 × *g* and filter the supernatant through a 0.45 μm syringe filter before loading onto a 5 mL HiTrapQ cation-exchange column, then collect 5 mL fractions. Check 20 μL of each fraction by SDS-PAGE. For disulfide bonded proteins, the presence of intermolecular disulfide bonds can be checked by non-reducing SDS-PAGE where gels, SDS-PAGE running and sample loading buffer are prepared without a reducing agent like DTT or β-Mercaptoethanol (as an example, *see* Fig. 3, right, with PDGF dimer bands).

24. Check the identity of the protein in each fraction by mass spectrometry of the intact protein to confirm expected disulfide bond formation (*see* **Note 8**). Examples of three SUMO-

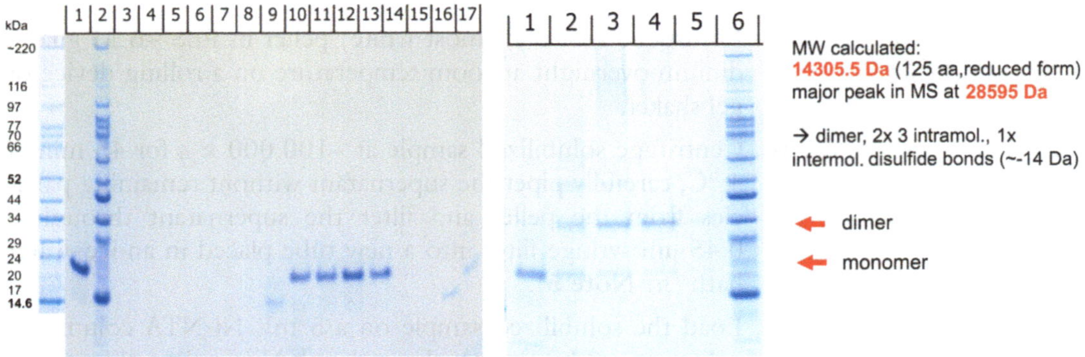

Fig. 3 Purified PDGF-A after cation-exchange on HiTrapSP. Fractions in *lanes 10–13* of the reducing SDS-PAGE gel were eluted at different elution buffer concentration and also analysed by non-reducing SDS-PAGE (gel on the right, *lane 1–4*). Each fraction contained clearly different PDGF-A species. Sample in *lane 4* was analyzed by intact mass spectrometry

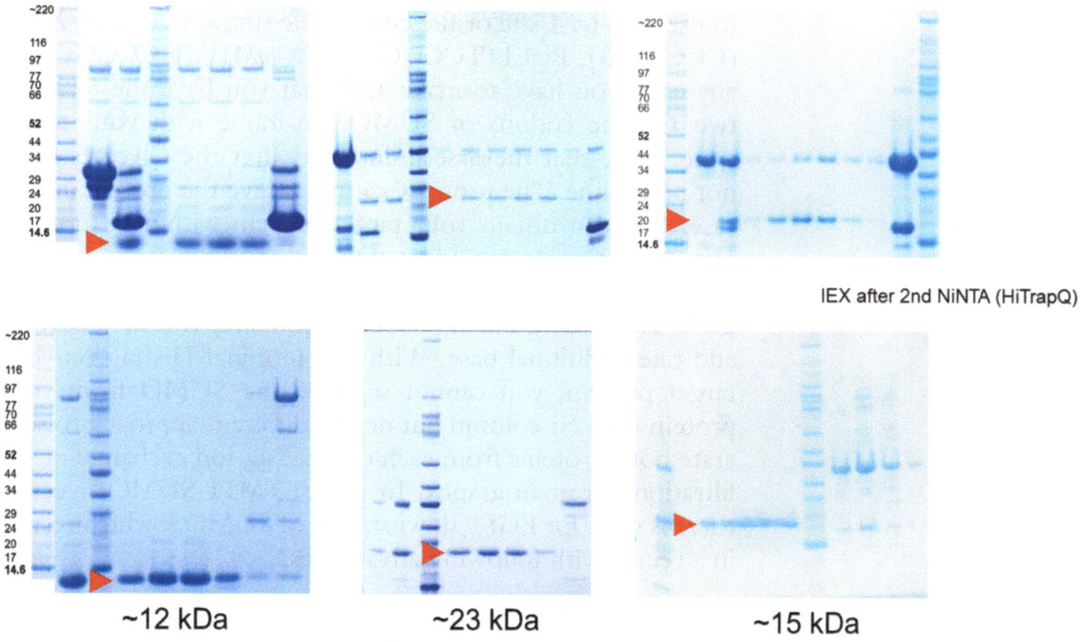

Fig. 4 Results of purification of three example SUMO3 fusion proteins using the described protocol after refolding and final ion-exchange step

fusion proteins prepared according to this protocol are shown in Fig. 4.

25. Dialyze the purified and pooled fractions against 1 L of PBS overnight at 4 °C.
26. Filter through a 0.45 μm syringe filter and measure the concentration of the protein. Adjust the final concentration by dilution with PBS or by concentrating in a membrane filter device with appropriate MW cutoff.
27. Store the fractions with the pure target protein in small aliquots (e.g., 0.1–0.5 mL) at −80 °C after snap-freezing in liquid N2.

4 Notes

1. For cloning of target genes into the pETM11-SUMO3 vector, the target gene can be inserted via the *Bam*HI-site and one of the restriction sites downstream of it. In this case, you have to add an N-terminal Ser due to the TCC-codon in the *Bam*HI-site (if you use the *Bam*HI-compatible *Bgl*II-site AGATCT in your insert, it is again a Ser coded by TCT). This Ser will be left in your protein after cleavage with SenP2. If no additional amino acids are wanted at the N-terminus, one has

to use the *Age* I-site or the compatible sites for *Ava*I or *Xma*I (CCCGGG), *Bsp*EI (TCCGGA) or *Ngo*MIV (GCCGGC). In any case, you have to make sure that you have the terminal two Glycine codons of SUMO3 in-frame with your target gene. Note that the first amino acid after the Glycines must not be Proline. The protease cannot cleave the protein in that case. You can obtain your target protein with a C-terminal His_6-tag if you do not add a stop codon to your gene. In this case, make sure that the second His_6-tag is in-frame with your gene. Depending on the 3′-restriction site, you may have to add one additional base. With a C-terminal His-tag on your target protein, you cannot separate the SUMO from your protein via a Ni-column but need a different approach to separate both proteins from each other, e.g., ion exchange or gel filtration chromatography. In the pETM11-SUMO3 vector, there is gene for EGFP downstream of SUMO3, which results in a vector with following advantages:

- Easy confirmation of double digest of the vector due to release of GFP fragment which is replaced subsequently by your target gene.
- Optional N-terminal tagging of your target protein with EGFP by using *Sac* I and one of the 3′-restriction sites downstream of the EGFP gene (optional with or without C-terminal His_6-tag).
- Optional C-terminal tagging of your target protein with EGFP by using *Age*I and *Bam*HI.

2. Sometimes the washing step with PBS +0.1 % (v/v) Triton X-100 can lead to solubilization of significant amounts of the insoluble protein, so discarding the supernatant after each washing step could lead to loss of most of the valuable target protein. In those cases, the supernatant can be used for His-tag purification under native conditions. A refolding step under in a GSH/GSSG redox buffer should be done for such proteins too in order to allow formation of correct disulfide bonds.
3. Instead of dialysis between chromatography runs, one could use buffer exchange columns (also called desalting columns) as a quicker and more convenient way of changing the buffer conditions. Especially in diluted samples, dialysis can result in significant loss of protein due to adsorption and precipitation in the dialysis tubing. In such cases, desalting columns should be used if available.
4. It is crucial to adjust the pH value of 1 M L-glutathione (reduced and oxidized) solution to 7–8 in order to maximize the refolding efficiency. Otherwise, addition of L-glutathione at high concentration lowers the pH of the buffer and can lead to protein aggregation when added directly to the protein solution before dialysis. Using pH-strips is accurate enough for

this step, for the small volumes of reduced and oxidized GSH stock solutions, using a pH-meter is inconvenient and not recommended.

5. If at this point the filter is clogged rapidly, the clear sample could be loaded onto the Ni-NTA column equilibrated in denaturing running buffer with 6 M urea. If there are visible particles in the sample, the centrifugation step should be repeated and the supernatant pipetted carefully into a fresh tube.
6. Disulfide bonded proteins should be dialyzed against cation-exchange buffer and loaded on a cation-exchange column (e.g., HiTrapSP) while other protein fragments should be purified by an anion-exchange. If the anion-exchange separation is not satisfying, buffer exchange and a cation-exchange step should be performed alternatively.
7. Most cytokines and growth factors are stable during the SenP2 digest and dialysis against buffer without urea and remain properly folded in solution, while in case of random protein fragments, precipitation of the protein is not unlikely. Therefore, ion-exchange buffer with 2.5 M urea could be used instead of urea-free buffer for dialysis and the last purification step. For immunizations, the presence of urea is acceptable to some extent.
8. In case of intramolecular disulfide bonds, each one would result in −2 Da smaller MW than the theoretical MW. If intermolecular disulfide bonds are expected (e.g., PDGF) the MW of the dimer will be reduced by −2 Da, corresponding to −1 Da compared to the MW of one monomer.

References

1. Sorensen HP, Mortensen KK (2005) Advanced genetic strategies for recombinant protein expression in *Escherichia coli*. J Biotechnol 115:113–128
2. Correa A, Oppezzo P (2015) Overcoming the solubility problem in *E. coli*: available approaches for recombinant protein production. Methods Mol Biol 1258:27–44
3. Berkmen M (2012) Production of disulphide-bonded proteins in *Escherichia coli*. Protein Expr Purif 82:240–251
4. de Marco A (2012) Recent contributions in the field of the recombinant expression of disulphide bonded proteins in bacteria. Microb Cell Fact 11:129
5. Schlapschy M, Skerra A (2011) Periplasmic chaperones used to enhance functional secretion of proteins in *E. coli*. Methods Mol Biol 705:211–224
6. Panavas T, Sanders C, Butt TR (2009) SUMO fusion technology for enhanced protein production in prokaryotic and eukaryotic expression systems. Methods Mol Biol 497: 303–317
7. Reverter D, Lima CD (2006) Structural basis for SENP2 protease interactions with SUMO precursors and conjugated substrates. Nat Struct Mol Biol 13:1060–1068
8. Mukhopadhyay D, Dasso M (2007) Modification in reverse: the SUMO proteases. Trends Biochem Sci 32:286–295
9. Marblestone JG, Edavettal SC, Lim Y et al (2006) Comparison of SUMO fusion technology with traditional gene fusion systems: enhanced expression and solubility with SUMO. Protein Sci 15:182–189
10. Peciak K, Tommasi R, Choi J et al (2014) Expression of soluble and active interferon

consensus in SUMO fusion expression system in *E. coli*. Protein Expr Purif 99:18–26

11. Hoffmann A, Müller MQ, Gloser M et al (2010) Recombinant production of bioactive human TNF-alpha by SUMO-fusion system—high yields from shake-flask culture. Protein Expr Purif 72:238–243
12. Lu Q, Burns MC, McDevitt PJ et al (2009) Optimized procedures for producing biologically active chemokines. Protein Expr Purif 65:251–260
13. Dümmler A, Lawrence A-M, de Marco A (2005) Simplified screening for the detection of soluble fusion constructs expressed in *E. coli* using a modular set of vectors. Microb Cell Fact 4:34
14. Scholz J, Besir H, Strasser C et al (2013) A new method to customize protein expression vectors for fast, efficient and background free parallel cloning. BMC Biotechnol 13:12

Chapter 10

A Strategy for Production of Correctly Folded Disulfide-Rich Peptides in the Periplasm of *E. coli*

Natalie J. Saez, Ben Cristofori-Armstrong, Raveendra Anangi, and Glenn F. King

Abstract

Recombinant expression of disulfide-reticulated peptides and proteins is often challenging. We describe a method that exploits the periplasmic disulfide-bond forming machinery of *Escherichia coli* and combines this with a cleavable, solubility-enhancing fusion tag to obtain higher yields of correctly folded target protein than is achievable via cytoplasmic expression. The protocols provided herein cover all aspects of this approach, from vector construction and transformation to purification of the cleaved target protein and subsequent quality control.

Key words *E. coli*, Periplasm, Recombinant expression, Purification, Liquid chromatography, TEV protease cleavage, Disulfide-rich peptide (DRP), Disulfide-rich protein, Venom peptide

1 Introduction

Disulfide bonds are found in the vast majority of secreted proteins and peptides, including human hormones, growth factors, antibodies, as well as most protein therapeutics [1]. Animal venoms represent an extreme example of disulfide-rich peptide (DRP) secretion, as they often contain hundreds or even thousands of secreted DRPs [2, 3]. While some of these venom peptides are promising leads for the development of therapeutics [4–6] or bioinsecticides [7–9], their production is often less than straightforward.

The major challenge for production of recombinant DRPs is obtaining the native disulfide-bond isomer in reasonable yield. A DRP containing three disulfide bonds (six cysteines) can theoretically form 15 different disulfide-bond isomers, and the number of possible isomers rapidly increases to 105 for four disulfide bonds and 945 for five disulfide bonds.

Nicola A. Burgess-Brown (ed.), *Heterologous Gene Expression in E. coli: Methods and Protocols*, Methods in Molecular Biology, vol. 1586, DOI 10.1007/978-1-4939-6887-9_10, © Springer Science+Business Media LLC 2017

Intracellular (cytoplasmic) expression in *E. coli* has been widely used for heterologous protein expression because it has the potential to produce high protein yields in a fast and cost-effective manner using simple plasmid constructs [1, 10]. However, cytoplasmic expression of DRPs is challenging because the reducing environment (i.e., low redox potential) of the intracellular space makes formation of disulfide bonds virtually impossible [10]. Misfolded DRPs often aggregate into inclusion bodies within the cytoplasm and their recovery requires reduction/denaturation, solubilization, and refolding [10], a time-consuming and laborious process that has to be optimized for each peptide. Strategies to improve folding of DRPs in the *E. coli* cytoplasm include: (1) the use of genetically modified versions of *E. coli* (Origami strains) with mutations that disable the thioredoxin and glutathione reductive pathways, thus making the cytoplasm more conducive to disulfide-bond formation; (2) SHuffle strains that additionally express the periplasmic disulfide-bond isomerase DsbC within the cytoplasm [1, 10]. Although these strains allow disulfide-bond formation within the *E. coli* cytoplasm, they generally do not improve the yield of soluble folded peptide beyond that obtained in unmodified *E. coli* strains, where disulfide-bond formation occurs ex vivo following cell lysis [11].

An alternative approach to production of DRPs is to express them in the periplasm of *E. coli*, where the molecular machinery for disulfide-bond formation is located [10, 12, 13] (Fig. 1). Additional advantages of this approach include the smaller repertoire of proteases in the periplasm compared with the cytoplasm and the

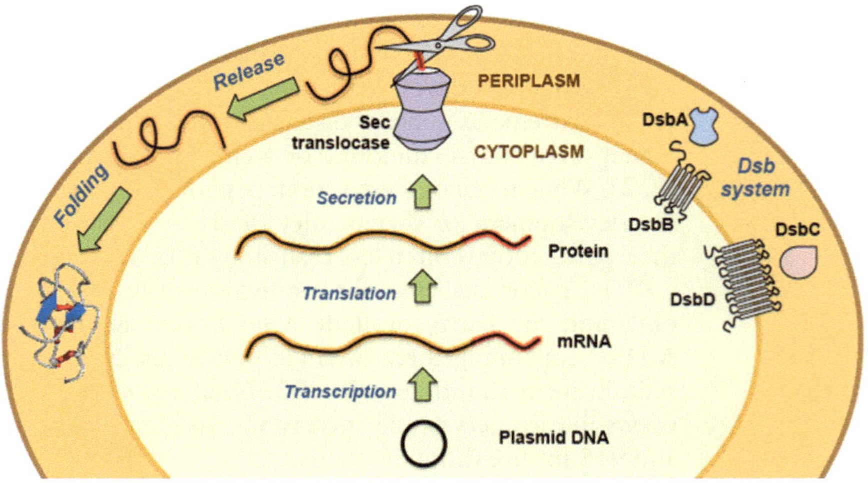

Fig. 1 Schematic of the pathway for production of secreted proteins in *E. coli*. After translation of the encoding mRNA in the cytoplasm, the protein is translocated into the periplasm. During translocation, the signal sequence is removed to release the mature protein. The periplasmic Dsb system, comprised of the DsbA, DsbB, DsbC, and DsbD proteins, subsequently aids in disulfide-bond formation. Adapted from ref. [19]

potential for simpler purification of the recombinant DRP via isolation of the less protein-rich periplasmic compartment as a first step in the purification procedure [10]. We have used this approach extensively to express venom peptides containing between two and six disulfide bonds [14–20]. For periplasmic expression, we use the pLIC-MBP vector (available from Addgene), which encodes a periplasmic signal sequence (which is removed during protein translocation to the periplasm), followed by an N-terminal His_6 tag to aid purification, maltose binding protein (MBP) to enhance solubility, and a tobacco etch virus (TEV) protease recognition site immediately preceding the target peptide sequence (Fig. 2).

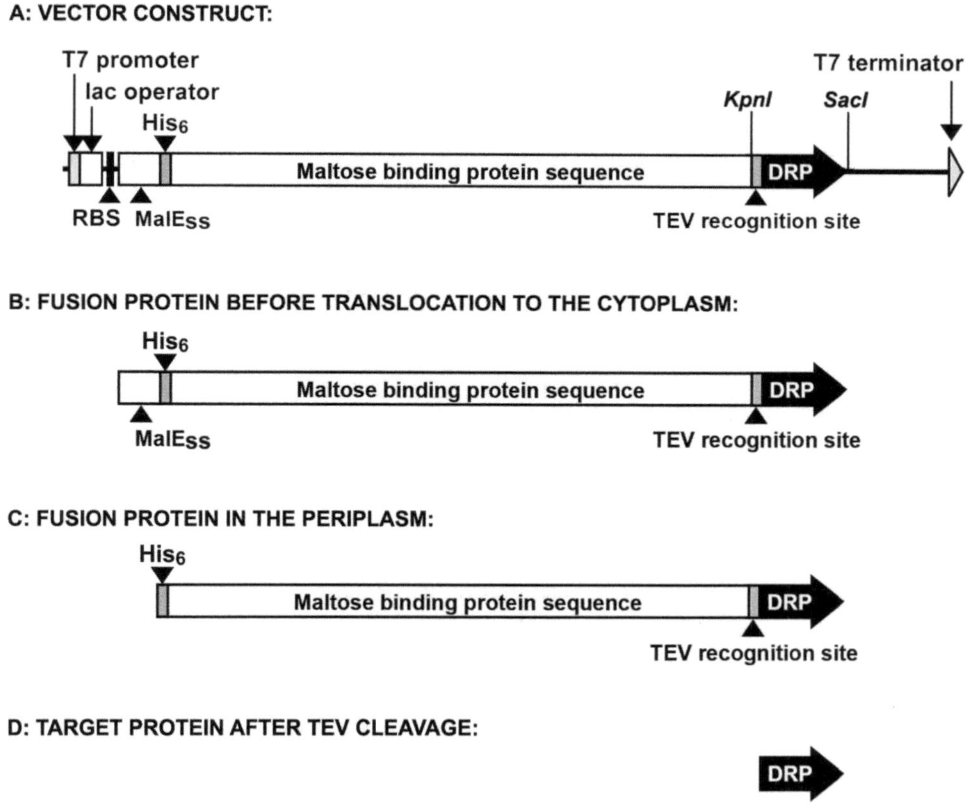

Fig. 2 Schematic representation of the vector construct, expressed fusion protein, and final target DRP. (**a**) The vector-coding region includes a MalE signal sequence (MalEss) for periplasmic export, a His_6 affinity tag, a solubility-enhancing maltose binding protein (MBP) fusion tag, and the coding sequence of the selected disulfide-rich peptide or protein (DRP), with a TEV protease recognition site inserted between the fusion tag and the peptide coding regions. The location of key elements of the vector are shown, including the ribosome binding site (RBS) and key restriction sites. (**b**) The fusion protein produced in the cytoplasm contains a MalE signal sequence that directs the nascent protein to the periplasm. (**c**) As the fusion protein translocates to the periplasm, the MalE signal sequence is removed to yield the mature fusion protein that will be purified from the soluble *E. coli* lysate. (**d**) After affinity purification of the fusion protein, it is cleaved with TEV protease to yield the target disulfide-rich peptide or protein (DRP)

This vector was originally designed for ligation-independent cloning (LIC) [21–23] but LIC would leave several nonnative residues at the N-terminus of the target DRP following TEV protease cleavage. To circumvent this problem, we cloned the TEV recognition site (ENLYFQG) immediately in front of the target sequence and ligated it back into the vector using traditional methods (explained further in Subheading 3.1 and 4). Although TEV protease cleavage is highly specific and robust, one potential disadvantage is that it cleaves between the penultimate and last residues (P1 and P1′ positions, respectively) of its recognition site, leaving a nonnative Gly residue at the N-terminus of the target peptide that might affect its activity. However, TEV protease is relatively insensitive to the nature of the amino acid residue in the P1′ position, and therefore the P1′ Gly can be replaced by short-chain amino acids such as Ser, Ala, Cys, Asn, and Asp [24]. Cleavage efficiency is even reasonable when the P1′ Gly is replaced by bulkier amino acids such as Tyr, Phe, Met, Lys, His, and Gln. Relatively few amino acids in the P1′ position such as Pro, Val, Ile, and Glu lead to poor cleavage efficiency [24]. Thus, in most cases, one can use the first amino acid of the target peptide in the P1′ position of the TEV protease cleavage site, leading to native peptide after cleavage from the fusion protein. However, if the target peptide begins with a non-preferred residue, we recommend using Gly, Ala, or Ser in the P1′ site.

Here, we describe protocols for the production of recombinant disulfide-rich peptides in the periplasm of *E. coli*, ranging from construct design and transformation, inoculation of cultures from glycerol stocks, induction of expression and fusion-protein purification, TEV cleavage and peptide purification, to quantification and quality control. After initial production of glycerol stocks, the entire procedure can be performed as required in as little as one week. These protocols provide the basis for producing DRPs for functional and structural characterization. Example results for the expression of a typical spider venom peptide are shown in Fig. 3. While these protocols are derived from our experience with disulfide-rich venom peptides (*see* Table 1), they can also be applied to nonvenom DRPs or disulfide-rich proteins.

2 Materials

NB: Ensure the appropriate personal protective equipment (PPE) is used when handling solvents, acids, bases, irritants, or other hazardous chemicals. Suggested PPE and safety precautions can be found online in the relevant Material Safety Data Sheet for each chemical.

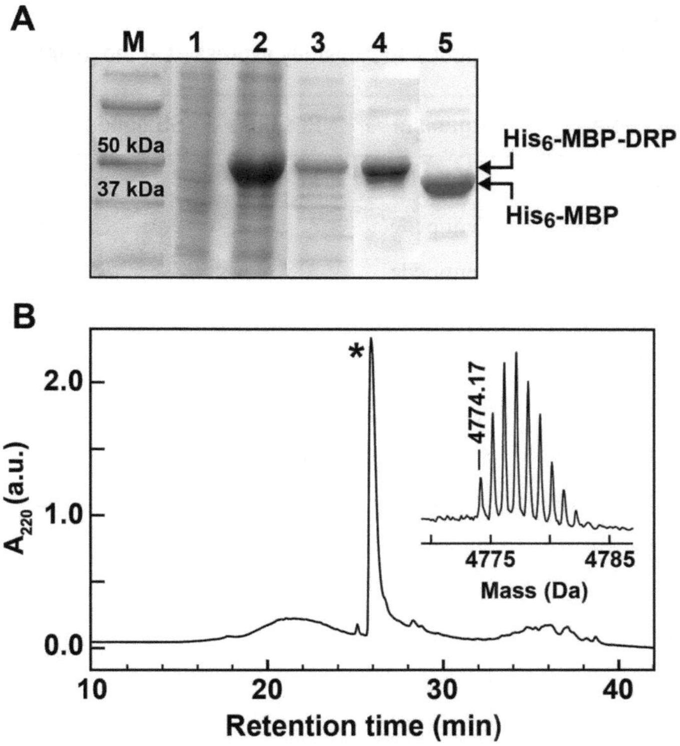

Fig. 3 Typical outcome as exemplified by the disulfide-rich venom peptide PcTx1 (π-TRTX-Pc1a). (**a**) SDS-PAGE gel showing key samples from expression and purification. *Lanes*: (*1*) cell extract before IPTG induction; (*2*) extract from IPTG-induced cells; (*3*) soluble extract after cell disruption; (*4*) fusion protein eluted from Ni-NTA column before TEV protease cleavage; (*5*) post-cleavage sample. His_6-MBP-DRP denotes the PcTx1 fusion protein, while His_6-MBP denotes the cleaved fusion tag. (**b**) RP-HPLC chromatogram showing the final step in the purification of PcTx1; the *asterisk* denotes the peak corresponding to correctly folded recombinant peptide. The *inset* is a MALDI-TOF MS spectrum showing the $M+H^+$ ion for the purified recombinant peptide (observed, 4774.17 Da; calculated, 4774.21 Da). Adapted from ref. [14]

2.1 Plasmid Construction

1. Expression plasmids (*see* **Note 1**).

2.2 Transformation

1. 90 × 16 mm sterile petri dish.
2. Competent BL21(DE3) *E. coli* cells can be purchased or produced in-house. Competent cells are flash frozen (using liquid nitrogen) in 50 μL aliquots and stored at −80 °C.
3. Ampicillin: Prepare in advance at a stock concentration of 100 mg/mL in water. Store stocks in 1 mL aliquots at −20 °C.
4. LB broth: Prepare in advance and store at room temperature. Dissolve 10 g of tryptone, 5 g of yeast extract, and 10 g of NaCl in 950 mL water. Adjust the pH of the medium to 7.0

Table 1
Disulfide-rich venom peptides successfully expressed in the periplasm of *E. coli* (updated from [19])

Toxin name	Organism	Residues	SS bonds	Yield (mg/L)	Target	Ref.
TxIA	Cone snail	17	2	>5	nAChR	[25]
μ-SLPTX-Ssm1a	Centipede	32	2	0.1–1	Na_v	[26]
κ-SLPTX-Ssm2a	Centipede	31	3	1–5	K_v	[26]
U_3-SYTX-Sth1a	Spider	31	3	2–3	Not known	[27]
U_3-SYTX-Sth1h	Spider	32	3	2–3	Not known	[27]
β-TRTX-Cm1a	Spider	33	3	1–5	Na_v	[28]
β-TRTX-Cm1b	Spider	33	3	1–5	Na_v	[28]
μ-TRTX-Hhn2a	Spider	33	3	1–5	Na_v	[29]
μ-TRTX-Hhn2b	Spider	33	3	1–5	Na_v	[30]
OAIP1	Spider	33	3	1–5	Not known	[31]
OAIP2	Spider	33	3	1–5	Not known	[32]
OAIP3	Spider	34	3	1–5	Not known	[32]
U_1-TRTX-Pc1a	Spider	33	3	1–5	Not known	[33]
U_2-TRTX-Pc1a	Spider	33	3	1–5	Not known	[33]
μ-TRTX-Tp1a	Spider	33	3	0.4	Na_v	[18]
β-TRTX-Ps1a	Spider	34	3	1–5	Na_v	[28]
κ-TRTX-Gr3a	Spider	34	3	>5	K_v	[34]
μ-TRTX-Hhn1a	Spider	35	3	0.1–1	Na_v	[35]
μ-TRTX-Hd1a	Spider	35	3	0.5–1	Na_v	[20]
U_5-SYTX-Sth1a	Spider	36	3	1–2	Not known	[27]
π-TRTX-Pc1a	Spider	40	3	>5	ASIC1a	[14]
APETx2	Sea anemone	42	3	0.5–1.0	ASIC3	[15]
μ-SPLTX$_3$-Ssm6a	Centipede	46	3	2.5	Na_v	[36]
U_1-AGTX-Ta1a	Spider	51	3	>5	Not known	[37]
κ-SLPTX-Ssm1a	Centipede	52	3	>5	K_v	[26]
U_2-CUTX-As1a	Spider	37	4	1–5	Not known	[38]
μ-SGTX-Sf1a	Spider	46	4	>5	Insect Na_v	[17]
μ-DGTX-Dc1a	Spider	56	4	1–5	Insect Na_v	[16]
ω-CNTX-Pn4a	Spider	55	6	1–5	Ca_v	[39]
DkTx	Spider	76	6	0.1–1	TRPV1	[40]

using 1 M NaOH and make up to 1 L. Autoclave for 20 min at 121 °C in volumes of less than 500 mL (*see* **Note 2**).

5. LB agar: Dissolve 10 g of tryptone, 5 g of yeast extract, and 10 g of NaCl in 950 mL water. Adjust the pH of the medium to 7.0 using 1 M NaOH and make up to 1 L. Autoclave for 20 min at 121 °C in volumes of less than 350 mL.
6. LB agar plates are made by melting LB agar in a microwave, then allowing the solution to cool to 55 °C (*see* **Note 3**). Add appropriate antibiotics and swirl to mix, then immediately pour ~20 mL into a petri dish (*see* **Note 4**). Place lids on the plates, leave to cool for 30–60 min (until solidified), then invert and store in a plastic bag at 4 °C (*see* **Note 5**).
7. Shaking/rolling incubator set to 37 °C.
8. Plate incubator set to 37 °C.

2.3 5 mL Starter Culture

1. LB broth: Prepare in advance and store at room temperature. Suspend 10 g of tryptone, 5 g of yeast extract, and 10 g of NaCl in ~950 mL water. Adjust pH of medium to 7.0 using 1 M NaOH and make up to 1 L (*see* **Note 2**). Autoclave in volumes of less than 500 mL.
2. Ampicillin: Prepare in advance at a stock concentration of 100 mg/mL in water. Store stocks in 1 mL aliquots at −20 °C.
3. Sterile culture tube.

2.4 Glycerol Stocks

1. Sterile 80% glycerol: Add 2 mL of water to 8 mL of 100% glycerol in a 50 mL Falcon tube. Roll gently on a roller until evenly mixed. Autoclave.
2. Sterile Eppendorf tubes.

2.5 50 mL Starter Culture

1. LB broth: Prepare in advance and store at room temperature. Suspend 10 g of tryptone, 5 g of yeast extract, and 10 g of NaCl in ~950 mL water. Adjust the pH of the medium to 7.0 using 1 M NaOH and make up to 1 L (*see* **Note 2**). Autoclave in volumes of less than 500 mL.
2. Ampicillin: Prepare in advance at a stock concentration of 100 mg/mL in water. Store stocks in 1 mL aliquots at −20 °C.
3. Sterile 250 mL Erlenmeyer flask.
4. Shaking incubator set to 37 °C and 180 rpm.

2.6 1 L Expression Culture

1. LB broth (1 L): Prepare in advance and store at room temperature. Suspend 10 g of tryptone, 5 g of yeast extract, and 10 g of NaCl in ~950 mL water. Adjust pH of medium to 7.0 using 1 M NaOH and make up to 1 L (*see* **Note 2**).
2. Ampicillin: Prepare in advance at a stock concentration of 100 mg/mL in water. Store stocks in 1 mL aliquots at −20 °C.

3. Sterile 5 L baffled flask (*see* **Note 6**).
4. Bunsen burner or biosafety cabinet.
5. Aluminum foil.
6. Refrigerated shaking incubator.
7. UV cuvette and spectrophotometer suitable for measuring at 600 nm wavelength.
8. 3× SDS-PAGE loading dye: 188 mM Tris–HCl pH 6.8, 6% (w/v) SDS, 15% (v/v) β-mercaptoethanol, 30% (v/v) glycerol, 0.02% (w/v) bromophenol blue. Store stocks in 1 mL aliquots at −20 °C.
9. 1.5 mL Eppendorf tubes.
10. 1 M IPTG stock: Prepare in advance at a stock concentration of 1 M in water. Sterile filter (0.45 μm filter). Store stocks in 1 mL aliquots at −20 °C.
11. Large capacity centrifuge: suitable for centrifuging volumes of 1 L, at a speed of 5000 × *g*, at 4 °C.
12. 50 mL Falcon tube.
13. Spatula.

2.7 Cell Lysis

1. TN buffer: 20 mM Tris, 200 mM NaCl, pH 8.0.
2. Magnetic stirrer and stirrer bar.
3. Cell disruptor (*see* **Note 7**).
4. DNase (optional): 10 mg/mL stock concentration in water, stored in 10 μL aliquots at −20 °C.
5. Protease inhibitor cocktail (optional): Use according to manufacturer's instructions.
6. 3× SDS-PAGE loading dye: 188 mM Tris–HCl pH 6.8, 6% (w/v) SDS, 15% (v/v) β-mercaptoethanol, 30% (v/v) glycerol, 0.02% (w/v) bromophenol blue. Store stocks in 1 mL aliquots at −20 °C.
7. 1.5 mL Eppendorf tubes.
8. Centrifuge: suitable for centrifuging 40 mL tubes, at a speed of 41,000 × *g*, at 4 °C.

2.8 Nickel Affinity Purification

1. Column to hold resin: empty column with a total volume of 50–100 mL.
2. Ni-NTA resin: any nickel affinity resin can be used at the user's discretion.
3. Nickel Sulfate ($NiSO_4$) (*see* **Note 8**): 100 mM stock stored at room temperature.
4. TN buffer: 20 mM Tris, 200 mM NaCl, pH 8.0.

5. Magnetic stirrer and stirrer bar (optional).
6. 1 M Imidazole, pH 8.0: stored at room temperature. Imidazole stock can be added directly to TN buffer to give TN buffer containing the desired concentration of imidazole.
7. 3× SDS-PAGE loading dye: 188 mM Tris–HCl pH 6.8, 6% (w/v) SDS, 15% (v/v) β-mercaptoethanol, 30% (v/v) glycerol, 0.02% (w/v) bromophenol blue. Store stocks in 1 mL aliquots at −20 °C.
8. 1.5 mL Eppendorf tubes.

2.9 TEV Cleavage

1. Millipore Amicon Ultra-15 30 K concentrator/centrifugal filter.
2. TN buffer (± imidazole): 20 mM Tris, 200 mM NaCl, pH 8.0 (± the appropriate volume of imidazole stock solution).
3. 1 M Imidazole, pH 8.0: stored at room temperature. Imidazole stock can be added directly to TN buffer to give TN buffer containing the desired concentration of imidazole.
4. Centrifuge: suitable for above concentrator (equivalent to a 50 mL Falcon tube) and 3000 × *g*, 4 °C.
5. 2× Redox buffer: Prepare fresh. For 5 mL, add 1.8 mg of reduced glutathione (GSH) to 2.4 mg of oxidized glutathione (GSSG) and suspend in TN buffer (without imidazole) to a final volume of 5 mL. Use immediately.
6. 3× SDS-PAGE loading dye: 188 mM Tris–HCl pH 6.8, 6% (w/v) SDS, 15% (v/v) β-mercaptoethanol, 30% (v/v) glycerol, 0.02% (w/v) bromophenol blue. Store stocks in 1 mL aliquots at −20 °C.
7. 1.5 mL Eppendorf tubes.
8. TEV protease: TEV protease is made in-house and stored at a concentration of 1 mg/mL at −80 °C. Commercial TEV protease can also be used.

2.10 SDS-PAGE Analysis

1. Heat block set to 95 °C.
2. Protein Molecular weight markers.
3. SDS-PAGE system: any system can be used at the user's discretion.
4. SDS-PAGE gel: Gels can be purchased or made in-house. Generally, we run samples on a 12.5% Tris-glycine resolving gel (with a 4% stacking gel).
5. Gel stain: 40% methanol, 10% acetic acid in water, with 0.25% Coomassie Blue R-250.
6. Gel destain: 10% methanol, 10% acetic acid in water.

2.11 Peptide Purification by Liquid Chromatography

1. HPLC solvent delivery system with binary high-pressure gradient delivery unit coupled to a UV-VIS detector. We use a Shimadzu Prominence system.
2. Semi-preparative reversed-phase C4 column. We use a Phenomenex Jupiter 300 column (10 × 250 mm, 5 μm, 300 Å).
3. Analytical reversed-phase C18 column. We use an Agilent Zorbax SB300 column (4.6 × 250 mm, 5 μm, 300 Å).
4. Solvent A: 0.05% (v/v) trifluoroacetic acid (TFA) in water (*see* **Note 9**).
5. Solvent B: 0.043% (v/v) TFA, 90% acetonitrile (HPLC grade) (*see* **Note 10**) and 10% water (*see* **Note 11**).
6. Solvent filtration apparatus with a 0.22 μm filter membrane.
7. Sample syringe filters, 0.22 μm porosity.
8. Manual HPLC injection syringe.
9. Electrospray ionization mass spectrometer (ESI-MS). We use an API-2000 ESI-QdQ triple quadrupole system.
10. Freeze dryer, or vacuum concentrator.

2.12 Quality Control and Quantitation

1. HPLC solvent delivery system with binary high-pressure gradient delivery unit and column oven (optional) coupled to a UV-VIS detector.
2. Narrow bore reversed-phase C18 column. We use a Thermo Aquasil column (2.1 × 50 mm, 5 μm, 120 Å).
3. Matrix-assisted laser desorption ionization-time of flight mass spectrometer (MALDI-TOF MS). We use a Model 4700 Proteomics Bioanalyzer system.
4. Ground steel target plate.
5. 5 mg/mL α-cyano-4-hydroxy-cinnamic acid (CHCA) MALDI matrix: prepared in 50/50 acetonitrile/water (*see* **Note 12**).
6. Matrix solvent: 0.1% (v/v) TFA, 50% acetonitrile (HPLC grade), and 50% water.
7. NanoDrop™ spectrophotometer (ThermoFisher).

3 Methods

NB: For sterility, Subheadings 3.1–3.5 should be performed over a Bunsen burner or in a biosafety cabinet to avoid contamination whenever the culture tubes or flasks are open.

3.1 Plasmid Construction

1. A synthetic gene encoding the peptide of interest, with codons optimized for *E. coli* expression, can be constructed in-house or purchased from a number of companies, including GeneArt

(Regensburg, Germany). The peptide gene is then cloned into a variant of the pLIC-MBP expression vector (*see* **Note 13**).

2. If the plasmid is prepared commercially, it will arrive as lyophilized material, and should be reconstituted in DNase-free water to a final concentration of 50-100 ng/μL.
3. Plasmid is stored at −20 °C until needed for transformation.

3.2 Transformation

1. Thaw a tube containing 50 μL of competent BL21(DE3) *E. coli* cells on ice for up to 30 min (or until all ice crystals disappear) (*see* **Note 14**).
2. Add 0.5–3 μL (50 ng/μL) of plasmid DNA to the cell mixture (*see* **Note 15**). Mix the cells and DNA by carefully flicking the tube four to five times.
3. Incubate on ice for 30 min (*see* **Note 16**).
4. Heat shock at 42 °C for 45 s, then put back on ice for 2–5 min (*see* **Note 17**).
5. Add 1 mL of room-temperature LB broth and incubate at 37 °C for 60 min (shake at 200 rpm or rotate) (*see* **Note 18**).
6. While cells are incubating, warm LB agar plates containing 100 μg/mL ampicillin to 37 °C (*see* **Note 19**).
7. After incubation, spin the transformation culture at 5000 × *g* for 5 min to pellet cells, and discard most of the medium. Resuspend cells in the remaining medium (50–100 μL) and spread onto the top one-third of the plate. Streak the cells across the remaining two-thirds of the plate (*see* **Note 20**).
8. Incubate the plate upside down at 37 °C overnight (not longer than 18 h).

3.3 5 mL Starter Culture

1. Dispense 5 mL of LB broth and 5 μL of 100 mg/mL ampicillin into a sterile culture tube.
2. Inoculate by picking a single colony from the transformation plate (*see* **Note 21**) and swirl it inside the LB broth.
3. Incubate at 37 °C with shaking at 180 rpm for 16 h, or overnight.

3.4 Glycerol Stocks

1. To the 5 mL overnight starter culture (from Subheading 3.3) dispense 1.5 mL of 80% (v/v) glycerol (final concentration of ~20% (v/v) glycerol).
2. Pipette gently to ensure the glycerol is evenly distributed throughout the culture.
3. Aliquot the culture/glycerol suspension into 250 μL or 1 mL lots in Eppendorf tubes for long-term storage at −80 °C.

3.5 50 mL Starter Culture

1. Dispense 50 mL of LB broth and 50 μL of 100 mg/mL ampicillin into a sterile 250 mL Erlenmeyer flask.
2. Defrost a glycerol stock on ice, then inoculate the 50 mL culture using 250 μL of glycerol stock.
3. Incubate at 37 °C with shaking at 180 rpm for 16 h, or overnight.

3.6 1 L Expression Culture

1. The next morning, dispense 1 L of LB broth (*see* **Note 22**) and 1 mL of 100 mg/mL ampicillin into a sterile 5 L baffled flask (*see* **Note 6**).
2. Over a bunsen burner, or in a biosafety cabinet, pour 25 mL of the 50 mL starter culture into the expression flask and cover with foil or an appropriate lid, then incubate at 37 °C with shaking at 120 rpm (*see* **Note 23**) for 3 h.
3. Add 1 mL sample of culture to a cuvette and measure optical density at 600 nm (OD600) in a spectrophotometer. Continue growing, if necessary, until the OD600 reaches 0.8–1.0, then incubator temperature can be reduced to 16 °C if required (*see* **Note 24**).
4. Take a sample of the un-induced culture for SDS-PAGE analysis. Place 40 μL into an Eppendorf tube and mix with 20 μL of 3× SDS-PAGE loading dye. Save until analysis.
5. Defrost 0.25 mL of 1 M IPTG, dispense into the expression culture, and incubate the flask for a further 16 h at 16 °C (or whatever temperature is optimal for DRP expression) (*see* **Note 24**).
6. The following morning, place 40 μL of culture into an Eppendorf tube and mix with 20 μL of 3× SDS-PAGE loading dye. Save until SDS-PAGE analysis.
7. Centrifuge the culture at 5000 × *g* for 10 min at 4 °C, discard the supernatant, and collect the cell pellet (by scraping with a spatula) into a 50 mL Falcon tube (*see* **Note 25**) and freeze at −80 °C until purification (*see* **Note 26**).

3.7 Cell Lysis

NB: Subheading 3.7 and **steps 1–6** of Subheading 3.8 can be performed concurrently.

1. Dissolve the bacterial cell pellet in an appropriate volume (*see* **Note 27**) of TN buffer.
2. Use a magnetic stirrer bar to obtain a well-dispersed solution (*see* **Note 28**).
3. Lyse cells using a cell disruptor (*see* **Note 7**); we use 25–27 kPa (TS Series Cell Disrupter, Constant Systems Ltd.) and a temperature below 6 °C (*see* **Notes 29** and **30**). Other methods of cell lysis such as sonication or French press could be used if preferred.

4. Make a sample of whole-cell lysate for SDS-PAGE analysis. Place 40 μL of whole cell lysate into an Eppendorf tube and mix with 20 μL of 3× SDS-PAGE loading dye. Save until SDS-PAGE analysis.
5. Centrifuge the whole-cell lysate using 40 mL centrifuge tubes at 41,000 × *g* for 30 min at 4 °C.
6. Collect the supernatant (soluble lysate) for protein purification.
7. Make a sample of soluble lysate for SDS-PAGE. Put 40 μL of soluble lysate into an Eppendorf tube and mix with 20 μL of 3× SDS-PAGE loading dye. Save until SDS-PAGE analysis.

3.8 Nickel Affinity Purification

1. Place 10 mL of 50% Ni-NTA resin slurry into an empty column to give a column resin (bed) volume of 5 mL.
2. Drain off excess liquid (*see* **Note 31**).
3. Wash the 5 mL of Ni-NTA resin with 50 mL (or 10 bed volumes) of filtered distilled water.
4. Add 5 bed volumes of 100 mM $NiSO_4$ (*see* **Note 8**) to the Ni-NTA resin and let it stand for at least 30 min (in order for Ni to bind to resin).
5. Wash the excess $NiSO_4$ with 5 bed volumes of filtered distilled water. Collect all Ni-contaminated waste separately and dispose appropriately (*see* **Note 8**).
6. Equilibrate the Ni-NTA resin with 10 bed volumes of TN buffer.
7. Add the soluble lysate (from Subheading 3.7) to charged Ni-NTA resin and allow it to flow slowly by gravity (*see* **Notes 32** and **33**).
8. Wash column with 5 bed volumes of TN buffer containing 15 mM imidazole to remove bacterial proteins that are bound nonspecifically to the column.
9. Make a sample of resin before elution for SDS-PAGE analysis. Place 40 μL of resin/buffer slurry into an Eppendorf tube and mix with 20 μL of 3× SDS-PAGE loading dye. Save until SDS-PAGE analysis.
10. Elute and collect His-tagged protein in two lots of 10 mL TN buffer containing ~250 mM imidazole (*see* **Note 34**).
11. Make a sample of resin after elution for SDS-PAGE analysis. Place 40 μL of resin/buffer slurry into an Eppendorf tube and mix with 20 μL of 3× SDS-PAGE loading dye. Save until SDS-PAGE analysis.

3.9 TEV Protease Cleavage

1. Prewash a Millipore Amicon Ultra-15 30 K concentrator/centrifugal filter device with 10 mL TN buffer. Spin at 3000 × *g* for 10 min at 4 °C.
2. Concentrate the 20 mL of eluted fusion protein (from Subheading 3.8) down to ~5 mL and dilute back with TN buffer *without imidazole* to 15 mL. Concentrate to 5 mL.
3. Collect flow-through from Millipore concentrator in a separate Falcon tube in case any fusion protein passes through the filter.
4. Make a sample of concentrator flow-through for SDS-PAGE analysis. Place 40 μL of concentrator flow-through into an Eppendorf tube and mix with 20 μL of 3× SDS-PAGE loading dye. Save until SDS-PAGE analysis.
5. Prepare 5 mL of 2× Redox buffer.
6. Remove the 5 mL of fusion protein from the concentrator and add 5 mL of 2× Redox buffer.
7. Make a "before-cleavage" sample for SDS-PAGE analysis. Place 40 μL of fusion protein/redox buffer solution into an Eppendorf tube and mix with 20 μL of 3× SDS-PAGE loading dye. Save until SDS-PAGE analysis.
8. To the 10 mL of fusion protein/redox buffer add 100 μL of 1 mg/mL TEV protease and incubate overnight at room temperature with gentle rolling (*see* **Note 33**).
9. The following morning, make a "post-cleavage" sample for SDS-PAGE analysis. Place 40 μL of cleavage mixture into an Eppendorf tube and mix with 20 μL of 3× SDS-PAGE loading dye. Save until SDS-PAGE analysis.
10. The cleavage mixture can be frozen at −20 °C until peptide purification, allowing time for the SDS-PAGE analysis to be performed, or you can proceed immediately to Subheading 3.11.

3.10 SDS-PAGE Analysis of DRP Expression, Purification, and Cleavage

1. Heat the SDS-PAGE samples at 95 °C for 10 min.
2. After heating, centrifuge the samples at 13,000 × *g* for 2–3 min.
3. Load 20 μL of sample into each well (use only 10 μL for uninduced and induced culture samples to avoid overloading the gel). Add 7 μL of molecular size markers into a separate well (*see* **Note 35**).
4. Run gel in an electrophoresis apparatus with a voltage of 100 V for 10 min and then increase to 150 V for 60 min.
5. After completion of electrophoresis, carefully remove the gel from the plates, wash briefly in water, and then stain in gel stain for a minimum of 1 h (up to 4 h).
6. Remove gel stain, wash with water, and then destain gels in destaining solution for a minimum of 1 h.

7. Check for the presence of the appropriate sized bands in each sample (also *see* Fig. 3a) and ensure cleavage was successful (*see* **Note 36**).

3.11 Peptide Purification by Liquid Chromatography

1. Filter solvents through a 0.22 μm filter before use. This ensures no particulate matter is present that could block solvent lines or the column, and also acts to degas the solvents (*see* **Note 37**).
2. The sample from the TEV protease cleavage reaction is prepared for HPLC by removing the His_6-MBP and His_6-TEV protease. These proteins are precipitated by the addition of 1% (final, v/v) TFA to the mixture, followed by centrifugation at 20,000 × *g* for 20 min (*see* **Note 38**). The supernatant is then passed through a 0.22 μm syringe filter to remove any remaining particulates.
3. Connect semi-preparative C4 column (*see* **Note 39**) to HPLC system according to the manufacturer's instructions and equilibrate column in 5% solvent B.
4. Once a stable baseline has been obtained, perform a blank run by injecting the same volume/composition as your test sample via a HPLC injection syringe. If required, repeat until no contaminants are eluted in the blank run. A useful gradient profile for peptide purification is 5–50% solvent B in solvent A over 45 min to elute the peptide of interest, followed by 50–80% solvent B over 5 min to elute any remaining MBP, TEV protease, or other contaminating proteins in the mixture (*see* **Note 40**). Set flow rate at 3–5 mL/min depending on back pressure (*see* **Note 41**).
5. Inject 5 mL of the processed TEV protease cleavage-reaction mixture using the above method, and monitor elution at 214 and 280 nm (*see* **Note 42**).
6. Collect peak fractions manually in 5 mL tubes (*see* **Note 43**).
7. Fractions are analyzed to identify the desired product based on mass and to assess purity. Electrospray mass spectrometry is the preferred method due to its speed and mass accuracy (*see* **Note 44**).
8. In some cases, further purification will be required to obtain >95% peptide purity. In these cases, an analytical C18 column can be used as it gives improved resolution over a semi-preparative C4 column (*see* **Note 45**). The system is set up as per the initial fractionation, but the flow rate and solvent gradient are changed. The flow rate is set to 0.8–1 mL/min depending on the back pressure and the gradient adjusted to match the elution time of the peptide of interest (*see* **Note 46**). An example HPLC chromatogram is shown in Fig. 3b.
9. Pool the pure peaks of interest and lyophilize (vacuum concentration could also be used) (*see* **Note 47**).

3.12 Quality Control and Quantitation

Confirmation of the purity and integrity of peptides after lyophilization is important for downstream functional and structural studies as some peptides are sensitive to the drying processes (*see* **Note 48**). We recommend performing a final HPLC run using a 1% per min gradient with temperature control (*see* **Note 49**). Using a small i.d. C18 column will allow for good resolution and sensitivity; only small amounts of peptide will be needed (~2 μg for each HPLC run). The final peptide product should also be analyzed by mass spectrometry (MS). We recommend using MALDI-TOF MS (*see* **Note 50**) using the following procedure. An example MALDI-TOF MS result is shown in Fig. 3b (inset).

1. Dissolve 5 mg/mL of CHCA in the matrix solvent and sonicate for 5 min (*see* **Note 51**).
2. Apply 0.4 μL of the matrix mix to a target spot. Overlay this with 0.4 μL of the peptide sample dissolved in 0.1% TFA in water (*see* **Note 52**).
3. Allow samples to air dry.
4. Perform MALDI-TOF MS as per individual instrument instructions.

Accurate quantitation of the peptide after quality control is also critical. There are many approaches to quantitating peptides (*see* **Note 53**); we use a combination of HPLC peak area at 214 nm and a NanoDrop that estimates concentration from UV absorption at 280 nm (A_{280}). The gold standard for determining peptide concentration is quantitative amino acid analysis; however, this procedure often requires outsourcing to specialized laboratories.

HPLC peak area at 214 nm

1. Prepare a stock solution of a "standard peptide" that is known to be pure and determine the concentration by amino acid analysis (*see* **Note 54**).
2. Perform HPLC of the standard peptide and your peptide of interest under the same conditions. Do this in duplicate (*see* **Note 55**).
3. Calculate the area under the HPLC peak (at 214 nm) for the peptide of interest and compare that to the area from the known amount of the standard peptide to calculate the concentration.

NanoDrop quantitation

1. Obtain the A_{280} measurement using the NanoDrop, blanking with the solution in which the peptide has been constituted (*see* **Note 56**).
2. Perform the above step in triplicate measurements for at least three dilutions of the peptide (*see* **Note 57**).
3. Calculate the concentration of peptide using the molar extinction coefficient (ε) and Beer-Lambert's law ($C_{\text{peptide}} = A_{280}/(\varepsilon \times l_{\text{path length}})$) (*see* **Note 58**).

4 Notes

1. The pLIC-MBP plasmid (containing a target gene encoding APETx2, a 42-residue DRP derived from a sea anemone) can be obtained from Addgene (https://www.addgene.org/72668/). The vector is derived from a pET-21 background and should not be used commercially.
2. If you are using ready-made LB broth and/or agar powder, use the suggested amount of powder mixture for the appropriate amount of water.
3. The melted LB agar can be cooled by placing in a water bath or oven set to 55 °C; this allows the temperature to be held and the solution to be left unattended. If the solution is too warm when the antibiotic is added, it may degrade the antibiotic. If the solution is too cool, the agar may solidify before you are able to add antibiotic or pour the plates.
4. If bubbles are produced when pouring the LB agar into the petri dish, these can be removed by quickly passing the flame of an inverted Bunsen burner over the top of the plate.
5. If the plates are kept sterile, they can be stored for ~1 month and still retain ampicillin selection.
6. Using baffled flasks increases aeration of the cultures, which improves cell growth and protein yield. If baffled flasks are not available, regular non-baffled flasks can be used. If using regular flasks, note that the speed of shaking must be increased during culture growth and induction of expression.
7. If a cell disruptor is not available, other conventional lysis methods can be substituted. For example, lysis by sonication, French press, freeze-thaw cycles, or periplasmic extraction by osmotic shock.
8. Nickel is toxic. Collect nickel waste separately. Use appropriate personal protective equipment, and dispose of unused/waste material according to the relevant Material Safety Data Sheet.
9. TFA is the most commonly used additive in HPLC solvents due to its volatility and compatibility with downstream applications such as mass spectrometry. Although the amount of TFA used in the solvents can be altered, adding more than 0.1% TFA will in most cases not improve resolution but will increase in the baseline absorbance at 214 nm. Alternative acid additives such as phosphoric acid, formic acid, hydrochloric acid, and acetic acid can also be used and may be useful to provide different separation profiles of peptides.
10. The peptide retention time and peak resolution can be manipulated by changing the mobile phase solvent composition. We most commonly use acetonitrile as it provides the lowest

viscosity solvent mixture, allowing for higher flow rates to be used during separation. Methanol and 2-propanol are two other commonly used organic solvents, which along with acetonitrile exhibit high optical transparency at the commonly used detection wavelengths of 214 and 280 nm.

11. Recipes for solvent mixtures are most commonly presented as volume ratios (v/v). Care must be taken when preparing solvent mixtures, as there is a contraction in volume when mixing water with an organic solvent, and temperature effects on solvent volume. For example, mixing 50 mL of water and 50 mL of acetonitrile (both at room temperature) will result in a total volume slightly less than 100 mL (~95 mL) due to shrinkage upon mixing. Furthermore, this mixture is endothermic, causing the mixture to cool and further reduce in volume (note, the mixture of water and methanol is exothermic). Although there are several procedures for making solvents, we use the easiest and most widely practiced method. For example, if preparing 1 L of solvent B: (a) measure 100 mL of water in a graduated cylinder; (b) measure 900 mL of acetonitrile in a separate graduated cylinder; (c) place both liquids in a bottle and shake thoroughly, then (d) add 430 μL of TFA to the mixture. It is important to consistently prepare solvents in the same manner as this will lead to the most reproducible results.

12. MALDI matrices are recrystallized reagents, of which the most commonly used for peptides are CHCA, 2,5-dihydroxybenzoic acid (DHB), and 3,5-dimethoxy-4-hydroxycinnamic acid (SA, or sinapinic acid). CHCA is a good first choice matrix for peptides <10 kDa, providing a good signal-to-noise ratio.

13. The variant pLIC-MBP vector we use is an IPTG-inducible plasmid that produces a $MalE_{SS}$-His_6-MBP-DRP fusion protein with a TEV protease cleavage site directly preceding the peptide-coding region. The vector, which was originally designed for ligation-independent cloning (LIC) [21–23], contains a plasmid-encoded TEV protease recognition site within the LIC region. However, this site was originally introduced for high-throughput cloning and results in several nonnative residues inserted N-terminal to the gene of interest. We do not use this plasmid-encoded protease recognition site, but instead introduce a TEV recognition site into the synthetic peptide-coding DNA region before cloning. Cloning is then performed with restriction enzymes (*KpnI* and *SacI*) to remove the entire LIC region and plasmid-encoded TEV recognition site. The MalE signal sequence ($MalE_{SS}$) directs the fusion protein to the periplasm, MBP enhances solubility, and the His_6-tag enables purification of the fusion protein via nickel affinity chromatography.

14. Cells can be thawed by hand or on the bench to save time; however, thawing above 0 °C will decrease the transformation efficiency.
15. Sometimes less is more with the amount of DNA used during a transformation. When using highly competent cells, using less DNA often increases the transformation efficiency. The volume of DNA added to the cells also affects transformation efficiency. We recommend using a DNA volume of not more than 10% of the competent cell volume (the lower the better).
16. The incubation time on ice can be reduced to as low as 2 min; however, the transformation efficiency will decrease roughly twofold for every 10 min the incubation time is shortened.
17. The optimal time for heat shock will vary between cell strains as well as the tubes used. We find that 45 s works well when using the 2.0 mL tube provided with commercial cells, and standard 1.5 mL Eppendorf tubes.
18. To save time the outgrowth step can be eliminated when using ampicillin-resistant plasmids, but not for other antibiotics. This step allows the cells to recover and express the antibiotic resistance proteins encoded on the transformed plasmid, allowing them to grow once plated on an antibiotic containing agar plate. We find incubation for 60 min to be optimal; decreasing the incubation time decreases transformation efficiency. Shaking or rolling the tubes during incubation increases transformation efficiency; however, it is not essential.
19. Preheating the agar plates is not required; however, it allows for more rapid colony formation and facile spreading of cells during **step** 7.
20. Several different strategies can be used to plate cells with the final objective to achieve individual colonies. Many protocols suggest plating two different concentrations of cells on separate agar plates. Cells can be spread evenly across the agar plate using a cell spreader or plating beads. We recommend the spread-and-streak technique in **step** 7 to save time and LB agar plates.
21. Colonies can be picked using an inoculation loop, sterile stick, or pipette tip.
22. LB broth can be replaced with M9 minimal medium to produce labeled protein for subsequent structural characterization using NMR spectroscopy.
23. If not using a baffled flask, shaking speed can be increased to 180 rpm.

24. Some target proteins can be expressed with reasonable yields of soluble material at 37 °C and therefore the temperature does not always need to be reduced to 16 °C for DRP expression. However, more difficult DRPs may require slower rates of production to achieve their correct fold. For this reason, 16 °C is used as a more generally applicable protocol. If 37 °C is used for expression, then expression needs to only proceed for a maximum of 4–5 h after induction in **step 5**.
25. If desired, the 50 mL falcon can be pre-weighed, so that the final weight of the pellet can be accurately determined (useful for calculating volume of lysis buffer to be used in Subheading 3.7).
26. Alternatively, purification can continue immediately, without freezing.
27. The volume is usually 15–20 mL of TN buffer per gram of cell pellet. If the pellet weight is unknown, an arbitrary volume of 100 mL can be used (for 1 L of culture).
28. It is important that there are no lumps of cell pellet remaining and that the cells are well dispersed in solution to prevent any blockage when using the cell disruptor in **step 3**, and to achieve the most efficient lysis.
29. Lysis will release various cellular enzymes and proteases. If the target fusion protein is particularly vulnerable to proteolytic cleavage, it is recommended that a protease inhibitor cocktail is added to the lysate.
30. The cell lysate can be quite viscous. To make it easier to handle, DNase can be added (100 μg per 100 mL of lysate, from 10 mg/mL stock). After the addition of DNase and stirring for 15 min on a magnetic stirrer, the lysate should be nonviscous and the protocol can be continued.
31. Ensure that the resin does not dry out to maintain optimal binding capacity.
32. Alternatively, the resin can be added to the soluble lysate and stirred/mixed for 15–30 min for optimal binding, before proceeding with the fusion protein purification.
33. If the target DRP is temperature sensitive, then purification and TEV protease cleavage can be performed in a cold room or at 4 °C to maintain protein integrity.
34. The exact amount of imidazole required to elute the DRP of interest will need to be determined by initial small-scale trials, but will generally be between 100 and 500 mM imidazole.
35. Pick an appropriate molecular weight marker based on the sizes of the fusion protein and target DRP.
36. If cleavage is incomplete, additional TEV protease can be added and cleavage left to proceed for longer. Note, however,

that cleavage efficiency is <90% for some fusion proteins regardless of the amount of TEV protease added and the amount of time allowed for the cleavage reaction.

37. An alternative to using a 0.22 μm filter is a Whatman No. 4 filter paper, which acts as a 1 μm filter that can be more cost efficient. Many HPLC systems have online degassing capabilities; check your specific instrument to determine whether further degassing is required.

38. Several approaches can be used to remove the His_6-MBP and His_6-TEV protease from the cleavage mixture. We use TFA precipitation, as it is simple and quick. However, there is a possibility of co-precipitation of the DRP of interest, which could decrease the final yield. An alternative strategy is re-passage of the cleavage mixture over a Ni-NTA column: the His_6-tagged proteins will be bound while the DRP will be eluted. Although this works well in some instances, many hydrophobic peptides will be retained on the Ni-NTA beads due to their backbone having similar properties to that of a C4 RP-HPLC column. We have also used syringe-driven solid-phase extraction (SPE) column in a similar manner; however, similar problems can occur with retention of DRP in the column and the need for high organic solvent concentrations to elute the peptide (resulting in the MBP and TEV also eluting). A centrifugal filter with a molecular weight cutoff of 15 kDa can also be used to allow the peptide to pass through, while retaining the MBP and TEV protease. However, we find that some DRPs bind to the filter membrane and are not retrievable.

39. The choice of HPLC column used will depend on the properties of the DRP being produced. We use a semi-preparative column (10 mm internal diameter (i.d.)) for this step due to the sample size loaded per run. The selection of column i.d. will depend on the sample capacity and detection sensitivity desired. We use column lengths between 150 and 250 mm, as longer columns result in increased resolution. However, a longer column has the disadvantage that larger proteins (such as MBP and TEV protease) may bind irreversibly to the column, a major problem if these proteins are not efficiently removed prior to the HPLC step. Retention of analytes depends on a number of column factors including the phase and ligand chain length, flexibility, and exposure of surface silanols. We find using a common C4 alkyl chain chemistry allows for good resolution of the peptide of interest, as well as removal of residual large proteins loaded on the column. The geometry of the column packing, both the particle diameter and pore size, are also important factors in optimizing resolution. A smaller particle diameter can improve resolution; 5 μm is most commonly used for peptides. The selection of pore size will also be determined

by the analytes being studied. Although the sizes of our target DRP, TEV protease, and MBP are quite different, we are most interested in improving resolution of the peptide. Given the complex mixture of the loaded sample, we suggest using a 300 Å column.

40. The choice of gradient conditions will largely depend on the elution time of the peptide of interest. A longer gradient will often result in better separation; however, this may not be needed in all applications. It is important to ramp the gradient up to around 80% acetonitrile at the end of the separation to remove any residual large or highly hydrophobic proteins bound to the column.

41. The flow rate used will depend on the i.d. of the column. The flow rates generally used are 3–5 mL/min for semi-preparative columns (10 mm i.d.), 0.8–1 mL/min for analytical columns (4.6 mm i.d.), and 0.15–0.3 mL/min for narrow-bore columns (2.1 mm i.d.). A higher flow rate generally provides better resolution. Specific columns and HPLC systems will indicate the back pressure that can be used for each application; however, as a general rule, we aim to maintain the back pressure below ~18 MPa (~2600 psi).

42. UV detection of peptides and proteins is most commonly performed at 210–220 nm, which detects absorbance by the peptide backbone, and 280 nm, which detects absorbance by the side chains of tryptophan and tyrosine residues. When a dual wavelength detector is not available, a single wavelength between 210 and 220 nm is advised. Photodiode array detectors can be extremely powerful to enhance detection capabilities but they are not necessary for this application.

43. We use manual collection as it provides the most flexibility; however, an array of sophisticated automated fraction collectors can also be programmed if preferred. The choice of collection-tube size will depend on the flow rate and volume desired per tube. It is important to use tubes with good recovery and low levels of contamination. The dead volume between the detector flow cell and collection point is another important factor to consider, especially when working with low flow rates (<0.3 mL/min).

44. Mass spectrometry can be used to identify the correct component within several closely eluting peaks, and can also be used as an indicator of purity. However, it is important to note that good mass spectral data alone does not guarantee peptide purity. This is particularly important when producing disulfide-rich peptides that can be folded into different disulfide isomers yet still have the same mass. The RP-HPLC peak shape should also be considered when judging the purity of compounds.

45. In most cases, the improved resolution of an analytical C18 column will suffice to purify the peptide of interest. In rare circumstances an orthogonal chromatography step is required, for example ion exchange chromatography, or changing the RP-HPLC column chemistry or mobile phase. However, separating disulfide isomers can sometimes be difficult due to the similar physiochemical properties of the isomers. In most cases, it is best to optimize the growth and purification conditions (before liquid chromatography) to produce the correctly folded isomer.
46. As is the case with the initial semi-preparative HPLC separation, the gradient should be adjusted once the elution time of a peptide under a given set of conditions is known. This will be more time and cost efficient.
47. Alternatively, sensitive target peptides or proteins can be buffer exchanged and concentrated for downstream applications.
48. Peptides are susceptible to oxidative processes during lyophilization. The most common issues are oxidation of methionine residues, and alterations to cysteine residues that can dimerize or internally shuffle disulfide bonds.
49. Controlling temperature with a column oven during HPLC eliminates temperature variability that can affect retention time and peak shape. Temperature influences solvent viscosity, and in some cases it can alter the structure or conformation of peptides during HPLC. Although separation of peptides using HPLC is often performed at ambient temperature, it is recommended to control temperature (at 40 °C) during the final quality control and quantitation steps. Increasing temperature during HPLC is often used as a way to improve the peak shape of analytes; however, this does not occur for all peptides (e.g., PcTx1).
50. MALDI-TOF MS is preferred over ESI MS in this step as MALDI produces mostly singularly charged ions that are easier to interpret. It is also more sensitive and can easily be automated.
51. We recommend preparing fresh matrix every time as this gives the best crystal formation; however, the matrix can be prepared and frozen for later use.
52. This technique allows the matrix and peptide to mix in solution and the peptide to embed into the matrix crystals, resulting in the formation of co-crystals.
53. A_{280} measurements can be made using traditional cuvette-based instruments rather than a NanoDrop. Although the use of cuvettes is nondestructive and sample can be recovered, the NanoDrop is easier to use and requires much smaller volumes

(~2 μL compared to ~300 μL for standard quartz cuvettes). Dye-based assays such as the Bradford protein assay or bicinchoninic acid (BCA) protein assay, as well as the Folin-reducing agent-based Lowry protein assay, are also commonly used for peptide/protein quantitation. The major disadvantage of these assays is that the amount of dye bound or Folin reagent reduced depends on the sequence of the DRP. The DRP is then quantitated against a reference standard that can have significantly different properties. Thus, an appropriate standard must be found for each peptide that is being measured, and a standard curve acquired each time to accurately estimate unknown peptide concentrations.

54. Once a reference "standard peptide" has been quantitated by amino acid analysis, this peptide can be used as the known concentration standard for all subsequent HPLC quantifications. It is therefore more cost efficient to determine the concentration of a large amount of a stock solution of standard peptide. Care must be taken to monitor any potential breakdown or alterations to the standard peptide over time.
55. Analyzing each sample in duplicate will ensure there is no variability between samples due to differences in the amount of peptide injected.
56. We find an accurate linear range for the NanoDrop to be 0.05–2.0 absorbance units.
57. The A_{280} absorbance is based almost entirely on the presence of tyrosine and tryptophan residues in the DRP. Not all peptides are rich in these amino acids, making it important to measure dilutions of the stock solution to ensure readings are within the linear range. For this same reason, it is hard to give a concentration range for peptides that can be reliably measured using the NanoDrop.
58. Estimates of the molar extinction coefficient (ε) in solution based on the peptide's amino acid sequence can be obtained using online tools such as ProtParam (http://web.expasy.org/protparam/).

Acknowledgments

This work was supported by the Australian Research Council (DP130103813) and the Australian National Health and Medical Research Council (Principal Research Fellowship and Project Grant APP1063798 to G.F.K.). B.C.-A. is supported by an Australian Postgraduate Award from the Australian Government.

References

1. Lobstein J, Emrich CA, Jeans C et al (2012) SHuffle, a novel *Escherichia coli* protein expression strain capable of correctly folding disulfide bonded proteins in its cytoplasm. Microb Cell Fact 11:56
2. Escoubas P, Sollod BL, King GF (2006) Venom landscapes: mining the complexity of spider venoms via a combined cDNA and mass spectrometric approach. Toxicon 47:650–663
3. Dutertre S, Jin AH, Kaas Q et al (2013) Deep venomics reveals the mechanism for expanded peptide diversity in cone snail venom. Mol Cell Proteomics 12:312–329
4. Lewis RJ, Garcia ML (2003) Therapeutic potential of venom peptides. Nat Rev Drug Discov 2:790–802
5. King GF (2011) Venoms as a platform for human drugs: translating toxins into therapeutics. Expert Opin Biol Ther 11:1469–1484
6. Zambelli VO, Pasqualoto KF, Picolo G et al. (2016) Harnessing the knowledge of animal toxins to generate drugs. Pharmacol Res 112:30–36
7. Schwartz EF, Mourao CB, Moreira KG et al (2012) Arthropod venoms: a vast arsenal of insecticidal neuropeptides. Biopolymers 98: 385–405
8. King GF, Hardy MC (2013) Spider-venom peptides: structure, pharmacology, and potential for control of insect pests. Annu Rev Entomol 58:475–496
9. Smith JJ, Herzig V, King GF et al (2013) The insecticidal potential of venom peptides. Cell Mol Life Sci 70:3665–3693
10. Salinas G, Pellizza L, Margenat M et al (2011) Tuned *Escherichia coli* as a host for the expression of disulfide-rich proteins. Biotechnol J 6:686–699
11. Nozach H, Fruchart-Gaillard C, Fenaille F et al (2013) High throughput screening identifies disulfide isomerase DsbC as a very efficient partner for recombinant expression of small disulfide-rich proteins in *E. coli*. Microb Cell Fact 12:37
12. Baneyx F, Mujacic M (2004) Recombinant protein folding and misfolding in *Escherichia coli*. Nat Biotechnol 22:1399–1408
13. Choi JH, Lee SY (2004) Secretory and extracellular production of recombinant proteins using *Escherichia coli*. Appl Microbiol Biotechnol 64:625–635
14. Saez NJ, Mobli M, Bieri M et al (2011) A dynamic pharmacophore drives the interaction between Psalmotoxin-1 and the putative drug target acid-sensing ion channel 1a. Mol Pharmacol 80:796–808
15. Anangi R, Rash LD, Mobli M et al (2012) Functional expression in *Escherichia coli* of the disulfide-rich sea anemone peptide APETx2, a potent blocker of acid-sensing ion channel 3. Mar Drugs 10:1605–1618
16. Bende NS, Dziemborowicz S, Mobli M et al (2014) A distinct sodium channel voltage-sensor locus determines insect selectivity of the spider toxin Dc1a. Nat Commun 5:4350
17. Bende NS, Dziemborowicz S, Herzig V et al (2015) The insecticidal spider toxin SFI1 is a knottin peptide that blocks the pore of insect voltage-gated sodium channels via a large β-hairpin loop. FEBS J 282:904–920
18. Cardoso FC, Dekan Z, Rosengren KJ et al (2015) Identification and characterization of ProTx-III [μ-TRTX-Tp1a], a new voltage-gated sodium channel inhibitor from venom of the tarantula *Thrixopelma pruriens*. Mol Pharmacol 88:291–303
19. Klint JK, Senff S, Saez NJ et al (2013) Production of recombinant disulfide-rich venom peptides for structural and functional analysis via expression in the periplasm of *E. coli*. PLoS One 8:e63865
20. Klint JK, Smith JJ, Vetter I et al (2015) Seven novel modulators of the analgesic target $Na_V1.7$ uncovered using a high-throughput venom-based discovery approach. Br J Pharmacol 172:2445–2458
21. Cabrita LD, Dai W, Bottomley SP (2006) A family of *E. coli* expression vectors for laboratory scale and high throughput soluble protein production. BMC Biotechnol 6:12
22. Eschenfeldt WH, Lucy S, Millard CS et al (2009) A family of LIC vectors for high-throughput cloning and purification of proteins. Methods Mol Biol 498:105–115
23. Stols L, Gu M, Dieckman L et al (2002) A new vector for high-throughput, ligation-independent cloning encoding a tobacco etch virus protease cleavage site. Protein Expr Purif 25:8–15
24. Kapust RB, Tozser J, Copeland TD et al (2002) The P1′ specificity of tobacco etch virus protease. Biochem Biophys Res Commun 294:949–955
25. Dutertre S, Ulens C, Buttner R et al (2007) AChBP-targeted α-conotoxin correlates distinct binding orientations with nAChR subtype selectivity. EMBO J 26:3858–3867
26. Yang S, Liu Z, Xiao Y et al (2012) Chemical punch packed in venoms makes centipedes

excellent predators. Mol Cell Proteomics 11: 640–650

27. Ariki NK, Muñoz LE, Armitage EL et al. (2016) Characterization of three venom peptides from the spitting spider *Scytodes thoracica*. PLoS One 11:e0156291
28. Bosmans F, Rash L, Zhu S et al (2006) Four novel tarantula toxins as selective modulators of voltage-gated sodium channel subtypes. Mol Pharmacol 69:419–429
29. Xiao Y, Liang S (2003) Inhibition of neuronal tetrodotoxin-sensitive Na^+ channels by two spider toxins: hainantoxin-III and hainantoxin-IV. Eur J Pharmacol 477:1–7
30. Li D, Xiao Y, Hu W et al (2003) Function and solution structure of hainantoxin-I, a novel insect sodium channel inhibitor from the Chinese bird spider *Selenocosmia hainana*. FEBS Lett 555:616–622
31. Hardy MC, Daly NL, Mobli M et al (2013) Isolation of an orally active insecticidal toxin from the venom of an Australian tarantula. PLoS One 8:e73136
32. King GF, Hardy MC (2013) Pest-controlling agents isolated from spider venom and uses thereof. Australian Patent WO 2013026105 A1
33. Choi SJ, Parent R, Guillaume C et al (2004) Isolation and characterization of Psalmopeotoxin I and II: two novel antimalarial peptides from the venom of the tarantula *Psalmopoeus cambridgei*. FEBS Lett 572: 109–117
34. Jung HJ, Lee JY, Kim SH et al (2005) Solution structure and lipid membrane partitioning of VSTx1, an inhibitor of the KvAP potassium channel. Biochemistry 44:6015–6023
35. Xiao YC, Liang SP (2003) Purification and characterization of Hainantoxin-V, a tetrodotoxin-sensitive sodium channel inhibitor from the venom of the spider *Selenocosmia hainana*. Toxicon 41:643–650
36. Yang S, Xiao Y, Kang D et al (2013) Discovery of a selective $Na_V1.7$ inhibitor from centipede venom with analgesic efficacy exceeding morphine in rodent pain models. Proc Natl Acad Sci U S A 110:17534–17539
37. Undheim EA, Grimm LL, Low CF et al (2015) Weaponization of a hormone: convergent recruitment of hyperglycemic hormone into the venom of arthropod predators. Structure 23:1283–1292
38. Bende NS, Kang E, Herzig V et al (2013) The insecticidal neurotoxin Aps III is an atypical knottin peptide that potently blocks insect voltage-gated sodium channels. Biochem Pharmacol 85:1542–1554
39. Cordeiro MN, de Figueiredo SG, Valentim Ado C et al (1993) Purification and amino acid sequences of six Tx3 type neurotoxins from the venom of the Brazilian 'armed' spider *Phoneutria nigriventer* (Keys). Toxicon 31:35–42
40. Bohlen CJ, Priel A, Zhou S et al (2010) A bivalent tarantula toxin activates the capsaicin receptor, TRPV1, by targeting the outer pore domain. Cell 141:834–845

Chapter 11

Split GFP Complementation as Reporter of Membrane Protein Expression and Stability in *E. coli*: A Tool to Engineer Stability in a LAT Transporter

Ekaitz Errasti-Murugarren, Arturo Rodríguez-Banqueri, and José Luis Vázquez-Ibar

Abstract

Obtaining enough quantity of recombinant membrane transport proteins with optimal purity and stability for structural studies is a remarkable challenge. In this chapter, we describe a protocol to engineer SteT, the amino acid transporter of *Bacillus subtilis*, in order to improve its heterologous expression in *Escherichia coli* and its stability in detergent micelles. We built a library of 70 SteT mutants, combining a random mutagenesis protocol with a split GFP assay as reporter of protein folding and membrane insertion. Mutagenesis was restricted to residues situated in the transmembrane domains. Improved versions of SteT were successfully identified after analyzing the expression yield and monodispersity in detergent micelles of the library's members.

Key words Split GFP, Membrane transport proteins, Heterologous expression, SteT, LAT, FSEC

1 Introduction

Membrane transport proteins (MTPs) are integral membrane proteins indispensable for cell viability as they selectively mediate the uptake of essential molecules across cellular membranes. Despite their biological and pharmacological importance, MTPs (and integral membrane proteins in general) are poorly represented in the Protein Data Bank, comprising less than 1% of all entries.

In general, obtaining sufficient amount of a highly pure, stable, and functional MTP in heterologous expression hosts like *Escherichia coli* (*E. coli*) is often the initial obstacle toward their structural determination [1]. Different *E. coli* strains with special phenotypes have been engineered [2, 3] along with numerous plasmid expression vectors with the aim of improving both the production and the quality of the obtained protein (e.g., [4]).

Nicola A. Burgess-Brown (ed.), *Heterologous Gene Expression in E. coli: Methods and Protocols*, Methods in Molecular Biology, vol. 1586, DOI 10.1007/978-1-4939-6887-9_11, © Springer Science+Business Media LLC 2017

However, optimization of the gene encoding the MTP target appears to be necessary as well. Protein mutagenesis has demonstrated its remarkable utility not only to increase protein production, but also to build stability in MTPs [5–8]. Moreover, combining mutagenesis with the use of the green fluorescent protein (GFP) as fusion tag of MTPs has enormously contributed on speeding up their structural determination [7, 9, 10]. The fluorescence of the GFP is a remarkable reporter of protein expression yield, membrane localization, and stability after detergent solubilization, being also very useful in high-throughput approaches [11, 12]. In this chapter, we describe a protocol that uses these two tools to optimize SteT, the L-Serine/L-Threonine exchanger of *Bacillus subtilis* and a prokaryotic paradigm of the mammalian L-amino acid transporter (LAT) family [13, 14]. Attempts to determine the 3D structure of SteT have been unsuccessful due to its poor expression in *E. coli* (even after codon optimization) and poor stability in detergent micelles. The protocol is divided into different steps (Fig. 1). Using error-prone PCR, we generated a

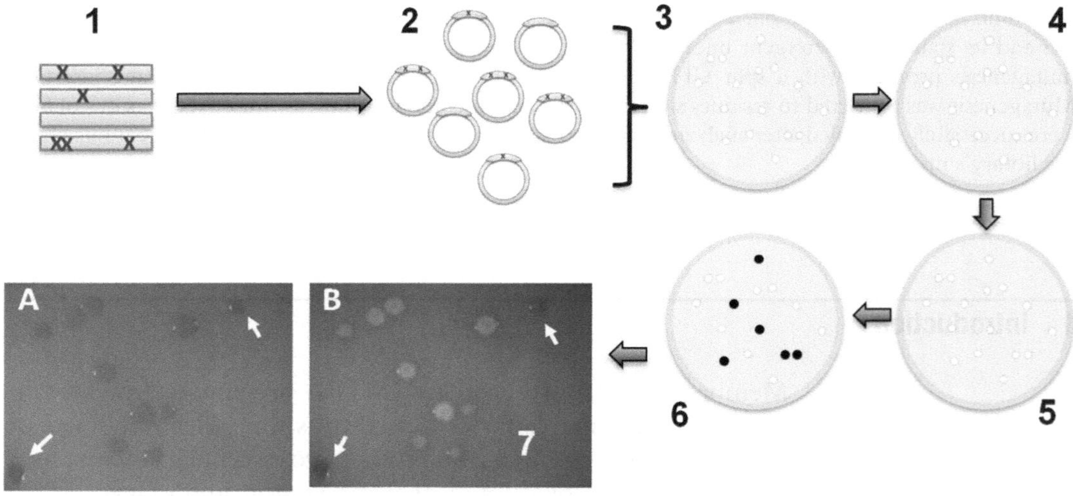

Fig. 1 Scheme of the protocol used for SteT random mutants selection using the split GFP assay. (*1*) Error-prone PCR is used to generate random amino acid replacements. (*2*) A second PCR reaction is used to clone the SteT mutants into the pTET-GFP_{11} vector. (*3*) The resulting pool of pTETSteTvariants-GFP_{11} is transformed in *E. coli* BL21(DE3) cells harboring the pETGFP_{1-10} vector and cells are plated on nitrocellulose membranes placed on the top of LB plates. *E. coli* colonies grow during 16 h at 37 °C. (*4*) After colonies appear, SteT-GFP_{11} is induced by transferring the membrane to a second LB plate containing ANTET and incubating for 3 h at 30 °C. (*5*) Excess of ANTET is removed by transferring the membrane to a third LB plate with no inducing agent (resting plate) and incubating for 1 h at 30 °C. (*6*) GFP_{1-10} is induced by transferring the membrane to another LB plate containing IPTG and incubating for 3 h at 30 °C. (*7*) *E. coli* colonies that emit GFP fluorescence due to the complementation of GFP_{1-10} and GFP_{11} are observed under UV or blue light. The panels of this scheme show *E. coli* colonies expressing sequentially SteT-GFP_{11} and GFP_{1-10} without excitation light (**a**), and after shading with UV light emission (**b**). The two pictures show exactly the same colonies. *White arrows* represent the clones that show no fluorescence and therefore no SteT mutants have expressed and/or inserted into the cytoplasmic membrane. This protocol is adapted from [21]

pool of SteT random mutants containing single or double amino acid replacements. Well-folded mutants that reached the inner membrane of *E. coli* were identified immediately after cloning using an assay based on the "in vivo" molecular complementation of the GFP (Figs. 1 and 2) [15]. This simple assay discards before DNA sequencing, SteT mutants that did not retain proper folding as wild-type. In addition, by only fusing a small portion of the GFP to SteT, any eventual impact of the full-length GFP during translation and membrane insertion of the SteT mutants is minimized. Next, only mutants presenting amino acid replacements within the transmembrane regions were considered for the final library of SteT mutants. Intramolecular interactions between transmembrane domains play major roles on MTPs stability and function. Indeed, point mutations within these regions are able to increase substantially protein stability in detergent [16, 17], although sometimes the effect is just the opposite. In addition, single replacements are also prone to stabilize (or lock) specific structural conformers of a MTP [18, 19], lowering their conformational heterogeneity in detergent micelles, thus increasing the probability of forming well-ordered crystals for X-ray diffraction.

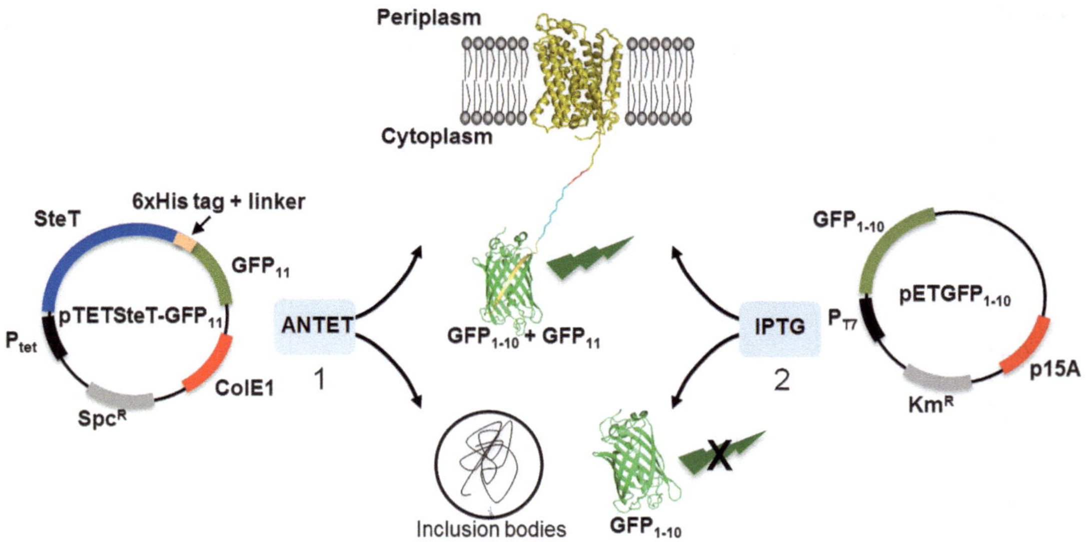

Fig. 2 Schematic of the split GFP assay as reporter of membrane protein expression and stability in the membrane. The split GFP assay consists of two plasmids: pTET-GFP_{11}, and pET–GFP_{1-10} [21]. pTET carries the gene encoding the target membrane protein (SteT) fused to a small part (15 amino acids) of GFP (GFP_{11}) at its C terminus, and pET carries the gene encoding the rest of the GFP molecule (GFP_{1-10}, 215 amino acids). Plasmids are compatible, containing the ColE1 and the p15A origins of replication, respectively. They also encode two antibiotic resistance genes: spectinomycin (SpcR) in pTET and kanamycin (KmR) in pET. Protein expression is controlled by two promoters: Ptet (ANTET inducible) in pTET and PT7 (IPTG inducible) in pET. Inducing sequentially these two genes, first SteT-GFP11 (adding ANTET) and second, GFP1–10 (adding IPTG), allows the identification of SteT variants that are expressed and inserted into the cytoplasmic membrane of *E. coli* since the two fragments of the GFP complements and fluorescence are emitted. SteT variants that do not reach the membrane or are poorly stable are confined into inclusion bodies and no complementation occurs

To identify evolved SteT variants, we screened the SteT random library measuring two properties: (1) the expression yield in *E. coli* and (2) the degree of aggregation (or monodispersity) after solubilizing the protein in detergent. In both cases, the fluorescence of the GFP was used as reporter. Fluorescence size-exclusion chromatography (FSEC) allows a quick analysis of protein monodispersity using non-purified GFP-fused samples by evaluating the shape and elution time of the protein elution peak [10]. Using this protocol, we succeeded on finding several SteT mutants presenting a better expression yield than wild-type and, most importantly, showing as well a notable improvement of stability in detergent micelles with respect to wild-type (Fig. 3) [7]. Two of these mutants were already submitted for crystallization trials.

2 Materials

2.1 Cloning SteT into the pTET-GFP$_{11}$ Vector and Generation of SteT Random Mutants

1. cDNA encoding SteT (Genscript) with optimized codons for *E. coli* transcription and translation.
2. Primers to amplify the SteT coding region for error-prone PCR.
3. pTETGFP$_{11}$ and pETGFP$_{1-10}$ plasmid vectors, generously provided by G.S. Waldo (Bioscience Division, Los Alamos National Laboratory, Los Alamos, NM, USA).

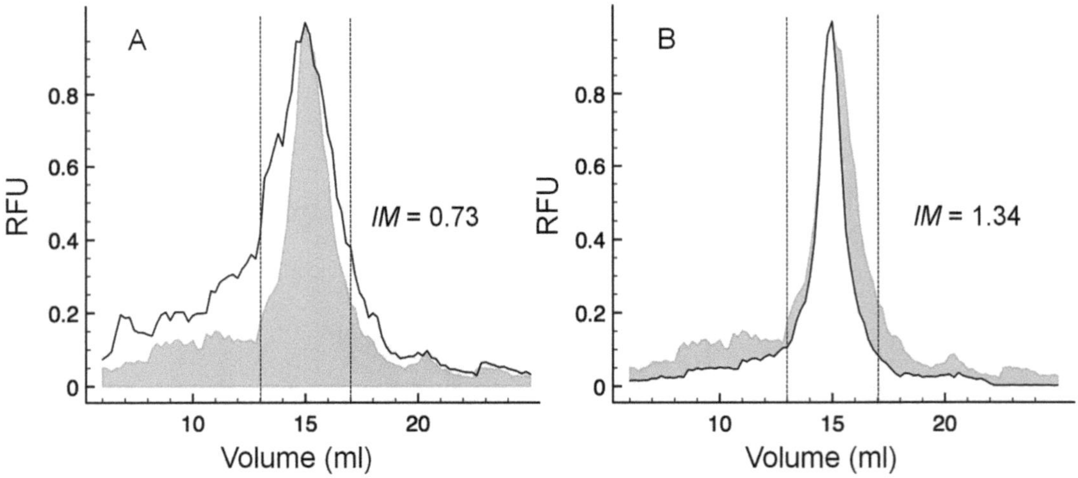

Fig. 3 Examples of normalized FSEC profiles of SteT-GFP variants. In order to compare the monodispersity in DDM of each SteT mutant, FSEC chromatograms were normalized and overlapped versus wild-type (*shadow area*). Panel *A* shows a typical result of a double mutation (G35R/G55R) that causes a decrease of monodispersity in DDM with respect to wild-type. Panel *B* displays the improving effect of the double mutation I134V/A377T on SteT monodispersity in DDM. To evaluate more quantitatively changes in monodispersity of each mutant versus wild-type, we assigned to each FSEC a numeric value named index of monodispersity (*IM*, *see* Subheading 3.2, **step 2**), calculated from the area under the main elution peak of SteT (delimited in the graph between the horizontal dashed lines). *IM* values < 1 indicate less monodispersity (or stability) than wild-type, whereas mutants with *IM* values > 1 more display better monodispersity in DDM micelles than wild-type [7]. RFU stands for relative fluorescent units

4. *NcoI*, *NdeI*, and *BamHI* DNA restriction enzymes (NEB).
5. High fidelity DNA-polymerase (Roche).
6. QIAquick PCR Purification Kit (Qiagen) for PCR purification.
7. T4 DNA ligase, 400 U/μL (NEB), and T4 ligase buffer for ligation of digested PCR products (NEB).
8. *E. coli* BL21(DE3) (Thermofisher) and *E. coli* XL1Blue (Agilent Technologies) competent cells.
9. QIAprep Spin Miniprep Kit (Qiagen) for plasmid DNA purification.
10. GeneMorph II EZClone Domain Mutagenesis Kit (Agilent Technologies) provides all the enzymes for the error-prone PCR reaction and subsequent cloning of the random library into the destination vector: pTETGFP$_{11}$.
11. Luria-Bertani (LB) medium and LB-agar plates with the appropriate antibiotic concentration as indicated: 35 μg/mL of kanamycin for pTETGFP$_{11}$ and 75 μg/mL of spectinomycin for pETGFP$_{1-10}$. A stock solution of 1000× of each antibiotic is prepared in Milli-Q water, sterilized by passing through 0.22 μm sterile filters and stored at −20 °C until use.
12. Solutions to prepare *E. coli* BL21(DE3) competent cells harboring the pET-GFP$_{1-10}$ plasmid: (a) 0.1 M $CaCl_2$ and (b) 0.1 M $CaCl_2$ + 15% (v/v) glycerol. Both solutions are prepared with autoclaved Milli-Q water, filtered through a 0.22 μm sterile filter and stored at 4 °C.

2.2 "In Colony" Split GFP Complementation Assay

1. *E. coli* BL21(DE3) cells co-transformed with pTET-SteT-GFP$_{11}$ and pETGFP$_{1-10}$.
2. 0.3 μg/mL Anhydrous tetracycline (ANTET; ACROS organics): A 1000× solution is prepared in ethanol and stored at −20 °C.
3. 0.4 mM Isopropyl-β-D-thiogalactopyranoside (IPTG): A stock solution of 1 M in sterile Milli-Q water is prepared, sterilized by passing through a 0.22 μm sterile filter and stored at −20 °C.
4. Phosphate buffered saline (PBS): 0.01 M K^+-Phosphate, 150 mM NaCl.
5. Luria-Bertani (LB) media and LB-agar plates with the appropriate antibiotic concentration as indicated: 35 μg/mL of kanamycin for pTETGFP$_{11}$ and 75 μg/mL of spectinomycin for pETGFP$_{1-10}$. A stock solution of 1000× of each antibiotic was prepared in Milli-Q water, sterilized by passing through 0.22 μm sterile filters, and stored at −20 °C until use.
6. Nitrocellulose filter membrane (Amersham Hybond-N). These filters are autoclaved before use.

2.3 Split GFP Complementation Assay in Liquid Cultures

1. *E. coli* BL21(DE3) competent cells harboring the pET-GFP_{1-10} vector.
2. pTET-SteTmutants-GFP_{11} plasmid vectors coding for the final library of SteT mutants. These mutants contain only one or two amino acid substitutions exclusively localized in the transmembrane regions of SteT.
3. Luria-Bertani (LB) medium supplemented with 35 μg/mL of kanamycin and 75 μg/mL of spectinomycin.
4. 0.3 μg/mL Anhydrous tetracycline (ANTET; ACROS organics). A 1000× solution is prepared in ethanol and stored at −20 °C.
7. 0.4 mM Isopropyl β-D-thiogalactoside (IPTG). Stock solutions of 1 M in sterile Milli-Q water are stored at −20 °C.
5. Phosphate buffered saline (PBS): 0.01 M K^+-Phosphate, 150 mM NaCl.
6. FLx800 Fluorescence plate Reader (Biotek).

2.4 Stability of SteT Random Mutants After Detergent Solubilization

2.4.1 Isolation of E. coli Plasma Membranes

1. Cell lysis buffer: 20 mM Tris-Base (pH 8.0), 350 mM NaCl, 1 tablet Complete protease inhibitor cocktail (Roche) per 25 mL of buffer, 1 mM Pefabloc serine protease inhibitor (Roche) (*see* **Note 1**).
2. Membrane resuspension buffer: 20 mM Tris-Base (pH 8.0), 150 mM NaCl, 10% (v/v) glycerol. Protease inhibitors are also added (*see* **Note 1**).
3. Cell disruptor (Constant Systems).
4. 10 μg/mL DNAse (Roche).
5. Benchtop ultracentrifuge, Beckman Coulter Optima MAX series with TLA-55 and TLA 120.1 rotors (Beckman).
6. Potter-homogenizer.
7. Bicinchoninic acid (BCA) protein assay Kit (Pierce).

2.4.2 FSEC Analysis

1. n-Dodecyl-β-D-maltopyranoside (DDM) solgrade (Anatrace).
2. ÄKTA Purifier FPLC system (GE Healthcare) equipped with a Frac-950 fraction collector (GE Healthcare).
3. Superose 6 10/300 GL Tricorn gel filtration column (GE Healthcare).
4. FSEC buffer: 20 mM Tris-Base (pH 7.6), 150 mM NaCl, 0.05% (w/v) DDM.
5. Benchtop ultracentrifuge, Beckman Coulter Optima MAX series with TLA-55 and TLA 120.1 rotors (Beckman).
6. 1.5 mL polyallomer microcentrifuge tubes (Beckman).
7. Round-bottom 96-well black plates (ThermoFisher).
8. FLx800 Fluorescence plate Reader (Biotek).

3 Methods

3.1 Generating the Library of SteT Random Mutants

A quite successful approach when optimizing integral membrane proteins for structural studies consists of generating libraries of mutated versions of these targets and identifying among them evolved (and functional) versions presenting two basic characteristics: (1) good yield of production and (2) stability in detergent micelles [17, 20]. Following this principle, we describe here the protocol used to build stability in SteT [7], the best-characterized prokaryotic paradigm of LATs; a mammalian amino acid transporter family for which no high-resolution 3D structure has been determined. We decided to build a SteT library of mutants using random mutagenesis, but taking into account two constraints: (1) a maximum of two amino acid substitutions per SteT sequence and (2) all the replacements must be localized in the transmembrane domains of the protein.

Random amino acid replacements along the SteT sequence and cloning of the resulting SteT random mutants into the expression vector pTETGFP$_{11}$ (Fig. 1) were carried out using error-prone PCR optimized in our laboratory (following the manufacturer's instructions) to obtain the desired mutational rate. In the final construct, each SteT mutant contains a 15-amino acid fragment of the GFP (GFP$_{11}$) fused after the C-terminal end (Fig. 2). The remaining 215-amino acid fragment of the GFP (GFP$_{1\text{-}10}$) is encoded in a second and compatible plasmid vector (pETGFP$_{1\text{-}10}$) (Fig. 2). In addition, both vectors encode two different antibiotic resistance markers (spectinomycin for pTETGFP$_{11}$ and kanamycin for pETGFP$_{1\text{-}10}$), as well as two different inducible promoters (ANTET for pTETGFP$_{11}$ and IPTG for pETGFP$_{1\text{-}10}$) [21]. After co-transforming these two vectors into *E. coli* BL21(DE3) cells, an assay based on the molecular complementation of the GFP was then conducted to quickly identify SteT random mutants expressing and inserting in the *E. coli* inner membrane [15] (Fig. 2). In this assay, sequential coexpression of the SteT mutant-GFP$_{11}$ followed by GFP$_{1\text{-}10}$ resulted in fluorescence emission when the two GFP fragments complement; a process that only occurs if the SteT mutant is properly inserted into the inner membrane of *E. coli* [7, 15] (Fig. 2). Also, misfolded or poorly stable proteins as result of mutagenesis are quickly removed from the membrane and confined as aggregates in inclusion bodies where GFP complementation does not occur (Fig. 2) [15]. This approach was described initially to analyze the "in vivo" solubility of recombinant soluble proteins [21, 22] and further optimized in our laboratory for membrane proteins [15].

3.1.1 Cloning SteT into the pTET-GFP$_{11}$ Vector and Generation of SteT Random Mutants by Error-Prone PCR

1. Amplify by high-fidelity PCR the gene encoding SteT using primers that introduce *NcoI* and *BamHI* restriction enzyme sites for subsequent cloning into the pTETGFP$_{11}$ vector (Fig. 2).
2. Double digest the SteT PCR fragment (insert) and the pTET-SteT-GFP$_{11}$ vector with *NcoI* and *BamHI* restriction enzymes.
3. Purify the SteT PCR digested fragment by running a 1% agarose gel and isolating the band using a DNA gel purification kit.
4. Ligate the resulting isolated fragment into the digested pTET-GFP$_{11}$ vector using T4-DNA ligase, transform the reaction into XL1-blue competent cells, plate them onto an LB-plate supplemented with 75 μg/mL of spectinomycin, and incubate overnight at 37 °C.
5. After *E. coli* colonies appear, screen for the correct clone containing pTET-SteT-GFP$_{11}$ by enzymatic digestion and finally verify by DNA sequencing.
6. To introduce random mutations in the SteT coding region of pTET-SteT-GFP$_{11}$, use the GeneMorph II EZClone Domain Mutagenesis Kit. The first reaction consists of an error-prone PCR reaction using the Mutazyme II DNA polymerase and two primers that anneal the N- and C-terminal regions of SteT, respectively (Fig. 1). Perform the mutagenesis according to the kit instructions but always using 500 ng of the template pTET-SteT-GFP$_{11}$. The settings used during thermocycling are shown in Table 1.
7. The resulting mutated PCR products from the error-prone PCR reaction are purified using the PCR purification kit. Thereafter, the purified product composed of SteT random mutants is cloned into the pTETGFP$_{11}$ vector using a second PCR reaction (EZclone reaction) where the same purified PCR products will serve as *megaprimers* and the pTET-SteT-GFP$_{11}$ vector as template (Fig. 1). The EZclone reaction uses a high-fidelity enzyme to avoid unwanted secondary mutations and is performed following the kit instructions.

Table 1
Settings used during the error-prone PCR reaction (Subheading 3.1.1)

Segment	Temperature, °C	Time	No. cycles
1	95	2 min	1
	95	30 s	
2	55	30 s	28
	70	2 min 30 s	
3	72	10 min	1

8. Digest the EZclone reaction with 1 μL of *DpnI* (included in the GeneMorph II EZClone Domain Mutagenesis Kit) for 3 h at 37 °C. Transform the digested reaction containing the pool of pTET-SteTmutants-GFP_{11} into *E. coli* BL21(DE3) competent cells harboring the $pETGFP_{1-10}$ plasmid (*see* Subheading 3.1.2) (Fig. 1). Before moving forward in the protocol, screen by DNA sequencing a few colonies from the previous transformation to verify that you are generating random mutations, homogeneously distributed along the sequence and with the desired frequency (*see* **Note 2**).

3.1.2 Making *E. coli* BL21(DE3) Competent Cells Harboring $pETGFP_{1-10}$

1. Transform the $pETGFP_{1-10}$ plasmid into *E. coli* BL21(DE3) competent cells and plate onto an LB-agar plate containing 35 μg/mL of kanamycin.
2. Inoculate one colony into 2 mL of LB medium supplemented with kanamycin and incubate at 37 °C for 16 h, with shaking at 220 rpm.
3. In a 500 mL flask, dilute 1 mL of the overnight culture into 100 mL of LB medium containing kanamycin.
4. Incubate this medium at 37 °C and 220 rpm until OD_{600} ~0.3.
5. Chill the culture on ice for 15 min.
6. Centrifuge the cells for 10 min at 3300 × *g* and 4 °C. Discard the medium and resuspend the cell-pellet in 30–40 mL of ice-cold 0.1 M $CaCl_2$ and keep the cells on ice for 30 min.
7. Centrifuge the cells for 10 min at 3300 × *g* and 4 °C. Remove the supernatant and resuspend the cell-pellet in 6 mL of ice-cold 0.1 M $CaCl_2$, 15% (v/v) glycerol solution. Aliquot the cell suspension into sterile 1.5 mL micro-centrifuge tubes (200 μL aliquots). Freeze the tubes in liquid nitrogen and store at – 80 °C.

3.1.3 "In Colony" Identification of Mutants Expressing and Inserting in the *E. coli* Inner Membrane Using the Split GFP Complementation Assay

1. Put a sterile nitrocellulose filter paper on the top of an LB-agar plate containing spectinomycin and kanamycin and plate on top the transformed *E. coli* BL21(DE3) cells with both pTET-SteTmutants-GFP_{11} and $pETGFP_{1-10}$ (Fig. 1). Grow the transformation for 16 h at 37 °C or until the colonies appear on the top of the filter. Note that each colony contains a unique SteT random mutant.
2. To start the induction of the SteT random mutants, transfer the filter paper containing the *E. coli* colonies onto a new LB-agar plate supplemented with 0.3 μg/mL of ANTET and spectinomycin and kanamycin and incubate for 3 h at 30 °C (Figs. 1 and 2) (*see* **Note 3**).
3. After the incubation, move the filter paper to a new LB-agar plate containing no inducing agent and the two antibiotics (spectinomycin and kanamycin), for 1 h at 30 °C (Fig. 1).

4. Induce GFP_{1-10} by transferring the filter paper onto a new LB-agar plate supplemented with 0.4 mM of IPTG and both antibiotics (spectinomycin and kanamycin) and incubate at 30 °C for 3 h (Figs. 1 and 2).
5. After incubation, observe the GFP fluorescence in the colonies using either a fluorescence microscope or an imagine device (both blue or UV light can be used as excitation source) (Fig. 1). Green colonies indicate GFP complementation and therefore, SteT mutant expression and membrane insertion.
6. Inoculate selected green colonies into 4 mL of LB medium containing 75 μg/mL of spectinomycin, grow overnight at 37 °C, and isolate the plasmid DNA of each clone. Send the clones for DNA sequencing to identify the nature of mutations and their localization in the SteT sequence (*see* **Note 4**).
7. For the final library and further screening, select only SteT mutants containing amino acid replacements exclusively located in the transmembrane regions.

3.2 Expression Yield of the Mutant Library Quantified Using the Split GFP Assay in Liquid Cultures

1. Grow 20 mL of *E. coli* BL21(DE3) harboring each pTET-SteTmutant-GFP_{11} and pETGFP$_{1-10}$ (include the antibiotics spectinomycin and kanamycin) and induce expression of SteTmutant-GFP_{11} by adding 0.3 μg/mL of ANTET and incubating at 30 °C overnight (*see* **Note 5**). Thereafter, induce GFP_{1-10} expression in the same culture by adding 0.4 mM IPTG and incubate for 1 h at 30 °C. In parallel, perform the same protocol with cells harboring either pTET-SteTwild-type-GFP_{11} and pETGFP$_{1-10}$ or pETGFP$_{1-10}$ alone. For the latter, include only kanamycin as antibiotic.
2. Pellet the cells at 5000 × *g* for 15 min at 4 °C and wash twice with PBS. Resuspend in PBS, but adjusting the cell density to OD_{600} = 0.2.
3. Transfer 200 μL of each sample into a 96-well black plate and measure the GFP fluorescence (emission 512 nm and excitation 488 nm) in a fluorescence plate reader. The relative expression yield of each mutant with respect to wild-type is calculated by dividing the fluorescent emission of each mutant by the fluorescent emission of wild-type. The fluorescence background measured from cells harboring only pETGFP$_{1-10}$ is subtracted from each measurement.

3.3 Stability of SteT Random Mutants After Detergent Solubilization

Stability after detergent solubilization is another fundamental property when selecting optimal MTPs candidates for structural determination. There are different methods to evaluate stability in detergent micelles, some of them based on the resistance of the solubilized proteins to thermal denaturation (or thermostability) [6, 23]. Perhaps the most exploited one is the analysis of the elution profile after size-exclusion chromatography (SEC). A monodisperse elution profile given by single and symmetric elution peaks is a direct proof of protein stability in a given detergent, whereas multiple and wide peaks (polydispersity) indicate protein tendency to aggregate and, therefore, instability. In fact, there is a direct relationship between monodispersity measured in a SEC assay and probability of forming well-ordered crystals for X-ray diffraction [1]. Consequently, we used SEC to evaluate the degree of monodispersity of the library of 70 SteT random mutants after solubilizing them in DDM, a detergent known to solubilize in a functional state wild-type SteT [13, 14]. In addition, SEC chromatograms were constructed by chasing the fluorescence of the complemented GFP fused at the C-terminal end of each mutant (FSEC) (Fig. 3) [7]; consequently, no protein purification was needed.

3.3.1 Isolation of *E. coli* Plasma Membranes

1. Grow 200 mL of *E. coli* BL21(DE3) cells harboring a given pTET-SteTmutant-GFP_{11} and pETGFP$_{1–10}$. Induce sequentially both proteins with ANTET and IPTG as described in Subheading 3.2. In parallel, perform the same experiment with cells co-transformed with pTET-SteTwild-type-GFP_{11} and pETGFP$_{1–10}$ that will be used as control.
2. Harvest the cells by centrifugation at 5000 × *g* for 15 min at 4 °C. Decant the supernatant and resuspend the cell pellet in lysis buffer (1 mL per 0.2 g of cell-pellet). If desired, cell suspension can be frozen in liquid nitrogen and stored at −80 °C.
3. Add the DNAse to the cell suspension to reduce viscosity and pass it three times at 25,000 psi and 4 °C in the cell disruptor (*see* **Note 6**). After cell lysis, keep a small aliquot on ice.
4. Centrifuge the lysed cells at 15000 × *g* for 30 min at 4 °C to remove the unbroken cells and cell debris. Collect the supernatant containing the inner membranes. Take an aliquot and store it on ice.
5. Transfer the two aliquots from **steps 3** and **4** into a black 96-well plate and measure the GFP fluorescence (emission 512 nm and excitation 488 nm) in a microplate spectrofluorometer. Comparison of the fluorescence from these two samples indicates the efficiency of cell lysis.
6. Pellet the inner membranes from **step 4** by ultracentrifugation at 200,000 × *g* for 2 h and 4 °C. Discard the supernatant and resuspend the cell membranes in resuspension buffer using the potter-homogenizer and attain a final concentration of 10 mg/mL of total protein. Protein concentration is measured using the BCA protein assay kit following the manufacturer's instructions.
7. If desired, cell membranes can be frozen in liquid nitrogen and stored at −80 °C for up to 6 months.

3.3.2 FSEC Analysis of SteT Random Mutants After DDM Solubilization

1. Transfer 900 μL of membrane suspension of a given SteT-GFP variant into 1.5 mL Beckman polyallomer microcentrifuge tubes.
2. Add 1% (w/v) of DDM (final concentration) to the membrane suspension and incubate for 1 h and 4 °C in an orbital mixer. After incubation, take an aliquot to store on ice and centrifuge the rest at 120,000 × *g* for 1 h and 4 °C.
3. After ultracentrifugation, discard the pellet, take a small aliquot, and transfer it together with the one from **step 2** into a black 96-well plate and measure the GFP fluorescence (emission at 512 nm and excitation at 488 nm) in a microplate spectrofluorometer. Comparing the fluorescence of both samples allows the calculation of the solubilization efficiency.
4. Inject 500 μL of the supernatant obtained from the previous step (DDM-solubilized membranes) onto a Superose 6 10/300 GL column equilibrated with FSEC buffer (*see* **Note 7**).
5. After the first 6 mL of elution of each FSEC, start collecting 200 μL eluting-fractions in a 96-well plate. At the end of the run, measure the GFP fluorescence of each fraction in a microplate spectrofluorometer (*see* **Note 8**) and plot fluorescence values versus elution volume to obtain the FSEC profile of each SteT mutant (Fig. 3).
6. To evaluate and compare the degree of monodispersity of each mutant with respect to wild-type, a numeric value was assigned to each FSEC chromatogram. This value, named index of monodispersity (*IM*), compares the wideness of the elution peak of each mutant versus the corresponding wideness of the wild-type elution peak (Fig. 3). Therefore, mutants with values of *IM* > 1 show better monodispersity than wild-type, whereas mutants with values of *IM* < 1 present a worse monodispersity behavior than wild-type (Fig. 3). The expression used for calculating *IM* is the following:

$$IM = \frac{\int_{13\,ml}^{17\,ml} \text{N FSEC}_{\text{wild type}}}{\int_{13\,ml}^{17\,ml} \text{N FSEC}_{\text{SteT variant}}} \tag{1}$$

where ∫ refers to the area under the curve of the normalized FSEC chromatogram (N-FSEC) between the 13 and the 17 mL of elution (Fig. 3). This interval of elution volume includes the main elution peak of SteT wild-type monomer (around 15 mL).

4 Notes

1. If further purification using Immobilized Metal Affinity Chromatography (IMAC) will be performed, it is important to take into account that Complete protease inhibitor cocktail

tablets (Roche) are available both with or without EDTA. Nickel-NTA resin allows a low concentration of EDTA in the purification buffer while cobalt-based (TALON) resins do not.

2. It is always necessary to do preliminary tests of the error-prone PCR reaction obtained from the EZclone mutagenesis kit. Although the kit is designed to provide a random and uniform mutational spectrum along the sequence delimited with the primers, a preliminary screening of, for example, different template concentrations should be tested.
3. To optimize membrane protein expression, it is recommended to test other induction temperatures as well as different concentrations of ANTET using always a positive control in parallel (in our case, wild-type SteT).
4. Since the goal is to select only mutations within the transmembrane domains, it is recommended to have a good topological model of the protein.
5. Before starting the screening of mutants' expression, it is advised to carry out some preliminary tests using wild-type to find the optimal temperature, induction time, and ANTET concentration. For example, test the expression with 0.3, 0.5, or 1 μg/mL of ANTET during different times (1, 4, or 16 h) at 20, 25, 30, or 37 °C.
6. It is recommended to start cell disruption at lower pressures as the high viscosity of the cell suspension could result in the blockage of the inlet valve of the cell disruptor. If breakage efficiency is low, four passes at incremental pressures (e.g., 25, 30, 35, and 37 Kpsi) at 4 °C are recommended.
7. The use of 0.05% (w/v) of DDM does not rescue aggregated membrane proteins even if they are solubilized in a different detergent. This fact allows assaying multiple detergent-solubilized membrane proteins in a single column without changing the running buffer.
8. Fractionation starts after 6 mL of elution because the void volume of the Superose 6 10/300 GL column is around 7–8 mL. In this way, we avoid collecting irrelevant fractions and all the fractions of one single experiment can be collected in a single plate. In case of having a fluorescence detector associated with the FPLC system for in-line detection, sample fractionation is not necessary.

Acknowledgments

We are very grateful to Geoffrey S. Waldo and members of his lab for sending us the split GFP expression plasmids. We also thank Manuel Palacín for his support and critical advice. This work was supported in part by the Spanish Ministry of Science and Innovation, grants BFU2008-04637 (J.L.V.I.). E. Errasti-Murugarren is a recipient of a Sara Borrell contract from the Instituto de Salud Carlos III (Spain). IRB Barcelona is the recipient of a Severo Ochoa Award of Excellence from MINECO (Government of Spain). The authors declare no competing financial interests.

References

1. Sonoda Y, Newstead S, Hu N-J et al (2011) Benchmarking membrane protein detergent stability for improving throughput of high-resolution X-ray structures. Structure (London, UK: 1993) 19:17–25
2. Hjelm A, Schlegel S, Baumgarten T et al (2013) Optimizing *E. coli*-based membrane protein production using Lemo21(DE3) and GFP-fusions. Methods Mol Biol (Clifton, NJ) 1033:381–400
3. Hattab G, Warschawski DE, Moncoq K et al (2015) *Escherichia coli* as host for membrane protein structure determination: a global analysis. Sci Rep 5:12097
4. Leviatan S, Sawada K, Moriyama Y et al (2010) Combinatorial method for overexpression of membrane proteins in *Escherichia coli*. J Biol Chem 285:23548–23556
5. Bill RM, Henderson PJF, Iwata S et al (2011) Overcoming barriers to membrane protein structure determination. Nat Biotechnol 29:335–340
6. Serrano-Vega MJ, Magnani F, Shibata Y et al (2008) Conformational thermostabilization of the beta1-adrenergic receptor in a detergent-resistant form. Proc Natl Acad Sci U S A 105:877–882
7. Rodríguez-Banqueri A, Errasti-Murugarren E, Bartoccioni P et al (2016) Stabilization of a prokaryotic LAT transporter by random mutagenesis. J Gen Physiol 147:353–368
8. Schlinkmann KM, Plückthun A (2013) Directed evolution of G-protein-coupled receptors for high functional expression and detergent stability. Methods Enzymol 520:67–97
9. Drew D, Lerch M, Kunji E et al (2006) Optimization of membrane protein overexpression and purification using GFP fusions. Nat Methods 3:303–313
10. Kawate T, Gouaux E (2006) Fluorescence-detection size-exclusion chromatography for precrystallization screening of integral membrane proteins. Structure 14:673–681
11. Goehring A, Lee C-H, Wang KH et al (2014) Screening and large-scale expression of membrane proteins in mammalian cells for structural studies. Nat Protoc 9:2574–2585
12. Listwan P, Terwilliger TC, Waldo GS (2009) Automated, high-throughput platform for protein solubility screening using a split-GFP system. J Struct Funct Genomics 10:47–55
13. Reig N, del Rio C, Casagrande F et al (2007) Functional and structural characterization of the first prokaryotic member of the L-amino acid transporter (LAT) family: a model for APC transporters. J Biol Chem 282:13270–13281
14. Bartoccioni P, del Rio C, Ratera M et al (2010) Role of transmembrane domain 8 in substrate selectivity and translocation of SteT, a member of the L-amino acid transporter (LAT) family. J Biol Chem 285:28764–28776
15. Rodríguez-Banqueri A, Kowalczyk L, Palacín M et al (2012) Assessment of membrane protein expression and stability using a split green fluorescent protein reporter. Anal Biochem 423:7–14
16. Smirnova IN, Kaback HR (2003) A mutation in the lactose permease of *Escherichia coli* that decreases conformational flexibility and increases protein stability. Biochemistry 42: 3025–3031
17. Tate CG, Schertler GFX (2009) Engineering G protein-coupled receptors to facilitate their structure determination. Curr Opin Struct Biol 19:386–395
18. Abramson J, Smirnova I, Kasho V et al (2003) Structure and mechanism of the lactose permease of *Escherichia coli*. Science 301:610–615
19. Kowalczyk L, Ratera M, Paladino A et al (2011) Molecular basis of substrate-induced permeation by an amino acid antiporter. Proc Natl Acad Sci U S A 108:3935–3940
20. Scott DJ, Kummer L, Tremmel D et al (2013) Stabilizing membrane proteins through protein engineering. Curr Opin Chem Biol 17: 427–435

21. Cabantous S, Waldo GS (2006) In vivo and in vitro protein solubility assays using split GFP. Nat Methods 3:845–854
22. Cabantous S, Terwilliger TC, Waldo GS (2005) Protein tagging and detection with engineered self-assembling fragments of green fluorescent protein. Nat Biotechnol 23:102–107
23. Miller JL, Tate CG (2011) Engineering an ultra-thermostable β(1)-adrenoceptor. J Mol Biol 413:628–638

Chapter 12

Acting on Folding Effectors to Improve Recombinant Protein Yields and Functional Quality

Ario de Marco

Abstract

Molecular and chemical chaperones/foldases can strongly contribute to improve the amounts and the structural quality of recombinant proteins. Several methodologies have been proposed to optimize their beneficial effects. This chapter presents a condensed summary of the biotechnological opportunities offered by this approach followed by a protocol describing the method we use for expressing disulfide bond-dependent recombinant antibodies in the cytoplasm of bacteria engineered to overexpress sulfhydryl oxidase and DsbC isomerase. The system is based on the possibility to trigger the foldase expression independently and before the induction of the target protein. As a consequence, the recombinant antibody synthesis starts only after enough foldases have accumulated to promote correct folding of the antibody.

Key words Molecular chaperones, Disulfide isomerases, Sulfhydryl oxidase, Osmolytes, Protein soluble aggregates, Protein quality assessment, Secretion efficiency

1 Introduction

According to the classical protein folding theory [1], the amino acid sequence should already possess the whole information necessary to guide the polypeptide to its final native conformation. The process is a progressive rearrangement toward the thermodynamically most convenient structure. However, effective folding can be impaired because some intermediate conformations may be kinetically trapped into states that need high activation energy for jumping to the successive. There are classes of enzymes such as disulfide isomerases, sulfhydryl oxidases, or prolyl isomerases that enable to overcome the rate-limiting steps of the protein folding process. These catalytic foldases work in combination with molecular chaperones to promote the development of protein folding intermediates into mature proteins. Such intermediates often present hydrophobic patches on their surface and some molecular chaperones prevent their aggregation by holding them. Chaperones can also contribute to active folding by stretching the intermediate

Nicola A. Burgess-Brown (ed.), *Heterologous Gene Expression in E. coli: Methods and Protocols*, Methods in Molecular Biology, vol. 1586, DOI 10.1007/978-1-4939-6887-9_12, © Springer Science+Business Media LLC 2017

molecules into less structured forms and in this way they enable the substrate folding development into alternative and more productive structural rearrangements. Even though the contribution of chaperones is more relevant under metabolic stress conditions such as high temperature or high protein expression rate, now it is known that several proteins are substrates of foldases even under physiological conditions [2] and that they rely on foldase activities to reach their native conformation. This is true also for relatively simple proteins, as demonstrated by the results that indicated that 5% of *E. coli* bacteria are obliged clients of GroEL/ES chaperonins and that this interaction is absolutely necessary to complete their folding [3]. Many other bacterial proteins need to interact with at least one chaperone during the folding process to be able to complete it [3].

Consequently, in the absence of sufficient amounts of foldases, several proteins tend to aggregate and finally precipitate. Misfolding becomes even more critical during recombinant expression in heterologous systems because this method increases the molecular crowding, whereas the host folding machinery can be not adequate for the heterologous substrates. Despite the possibility to isolate the aggregate proteins from inclusion bodies and to induce their in vitro refolding, the conventional approach for obtaining functional recombinant proteins is based on methods which allow for their recovery in soluble form. The strategies to counterbalance protein aggregation are multiple and can be complementary. Low growing temperatures, for instance, should slow down the host metabolism and assure a better ratio between new synthetized polypeptides and available foldases. The concentration of chaperones and foldases can be then increased by overexpressing these molecules and the positive effect on recombinant protein folding can be increased by the combined accumulation of chemical chaperones that minimize unproductive interactions between partially unfolded protein intermediates [4]. Of course, at the same time it is necessary to limit the energetic cost that the high-rate synthesis of chaperones can represent for the host cells. In this perspective, the optimal vector design for molecular chaperone overexpression should consider the biological context (for instance, polymeric holders should accumulate at significantly higher rates than catalytic molecules) and the technical cunnings to implement it. For instance, the success of the strategy illustrated in our previous work [5] strongly depended on the completeness of the chaperone network that was possible to express by means of a set of eight vectors [6]. These had three different antibiotic resistances to enable their co-expression, were characterized by different origin of replication to assure variable copy numbers in each cell, and the promoters regulating the expression rate of each chaperone were selected to modulate their final accumulation with the aim of reproducing the wild type ratios. In most of the experiments reported in the literature, the natural foldases already present in the cell type used for recombinant expression were overexpressed. Hybrid combinations in which heterologous foldases are used are less investigated. In theory, the overexpression of eukaryotic foldases in bacteria in combination with the overexpression of eukaryotic target proteins could be profitable because they could have higher binding specificity for the substrates. This approach is however often difficult to be implemented because—like in the case of chaperonins—the eukaryotic counterparts are by far more complex in terms of number of involved molecules and consequently the experiment becomes technically too demanding.

Osmolytes such as amino acids, sugars and methylamines that can act as chemical chaperones have been used to stabilize protein structure and to impair aggregation processes. Their accumulation is usually triggered by stress factors that can be mimicked under controlled conditions [7]. As an alternative, host bacteria can be engineered to overexpress osmolytes [8]. Apart from their direct stabilizing effect even on constitutionally unstable protein mutants [9], osmolytes can contribute to the increase of recombinant protein yields by assisting or directly stabilizing molecular chaperones [4, 10]. The strategy for osmolyte choice and their applications in protein biotechnology have been reviewed recently [11].

Tags and fusion partners can improve the solubility of folding intermediates as well and, by such a way, indirectly increase the yields of native recombinant proteins because the molecules have a longer useful time to rearrange their structure until the native conformation is reached [12]. Therefore, as it can be appreciated by this short and not exhaustive summary, there is an extremely high number of molecular, chemical, and even physical elements that can be beneficial for the folding of recombinant proteins. Their contribution can be cumulative or interfere with that of other factors and, consequently, a rational screening of this multifactorial system should be performed to identify the optimal combination for any given target protein. The systematic use of Design-of-Experiment and combinatorial approaches [13–15] will probably offer the only reliable opportunity to exploit the whole potentiality that folding helpers can represent for the success of the recombinant protein technology.

In the following protocol,a particular application is described, namely how to produce monodispersed and functionally active recombinant proteins, and specifically recombinant antibodies (Fig. 1), in the cytoplasm of *E. coli*. The advantage of cytoplasmic production is due to the larger volume and more efficient folding machinery with respect to the periplasmic space. The method is suitable for antibody fragments such as VHHs and scFvs but resulted extremely efficient also to produce IgG-like antibodies reconstituted by fusing variable domains to Fc domains [16]. Particularly valuable is the opportunity to exploit this system for

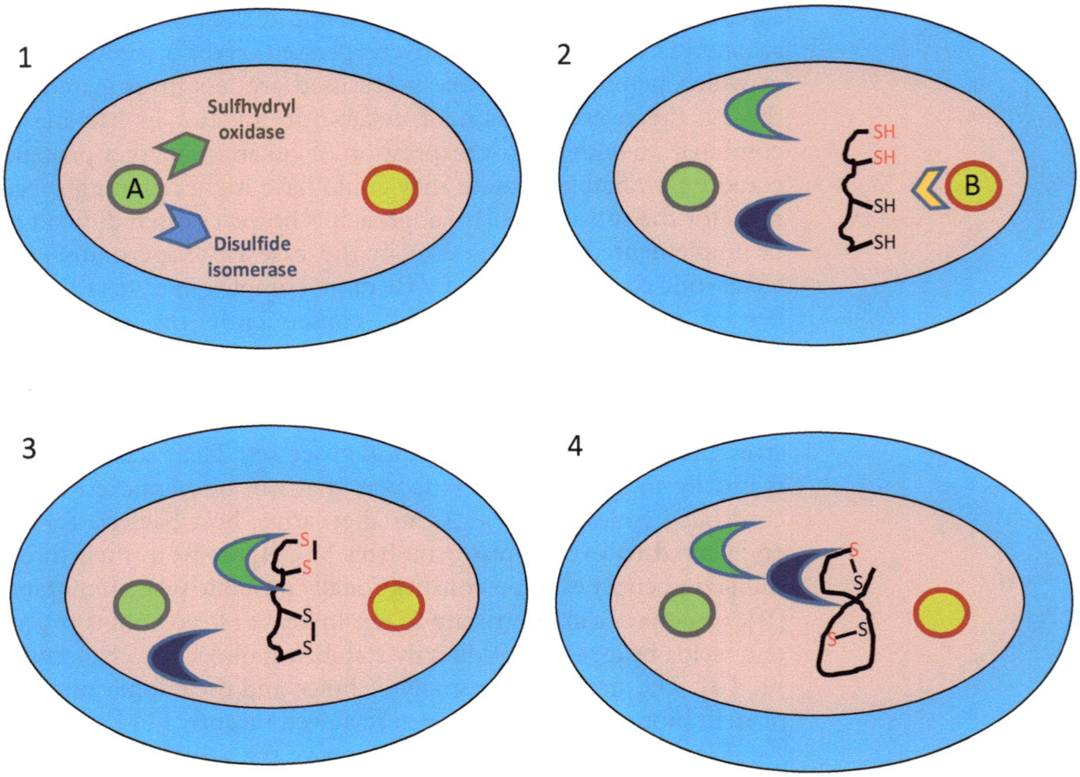

Fig. 1 Four-step formation of correct disulfide bonds in recombinant antibodies expressed in bacterial cytoplasm. (*1*) Arabinose-dependent cytoplasmic expression of sulfhydryl oxidase and disulfide isomerase from plasmid A. (*2*) The two enzymes accumulate before inducing the IPTG-dependent cytoplasmic expression of the recombinant antibody from plasmid B. (*3*) Sulfhydryl oxidase-dependent disulfide bond formation. (*4*) Disulfide isomerase-dependent disulfide bond isomerization

producing fusion immune-reagents in which the partners have different redox condition requisites for achieving their native folding [16, 17].

2 Materials

2.1 Vectors

1. An IPTG-inducible vector for the cytoplasmic expression of recombinant polypeptides (ampicillin or kanamycin resistance) with at least a His-tag.
2. An arabinose-inducible cytoplasmic expression vector for the expression of sulfhydryl oxidase (SO) and disulfide isomerase (DIso) carrying a further resistance (*see* **Note 1**).

2.2 Small-Scale Protein Production

1. Culture medium (*see* **Note 2**). Luria–Bertani (LB) medium is prepared by dissolving 10 g of tryptone, 5 g of yeast extract, and 10 g of NaCl in 950 mL of deionized water. Adjust the pH

to 7.0 using 1 N NaOH and bring the volume up to 1 L. Autoclave and add the required antibiotic(s) after the medium has cooled down to 55 °C.

2. 1 M IPTG in H_2O.
3. 40% (w/v) glucose.
4. BL21(DE3) bacteria (competent cells already co-transformed with the vector for the arabinose-dependent SO and DIso expression and the IPTG-dependent vector for the expression of the antibody fragment).
5. 1 mg/mL DNase I.
6. 100 mg/mL lysozyme.
7. Ni-NTA magnetic beads.
8. Lysis buffer: 50 mM Tris–HCl, pH 8.0, 500 mM NaCl, 5 mM $MgCl_2$.
9. Washing buffer: 50 mM Tris–HCl, pH 8.0, 500 mM NaCl, 15 mM imidazole, 0.02% (v/v) Triton.
10. PBST: Dissolve 8 g of NaCl, 0.2 g of KCl, 1.44 g of Na_2HPO_4, 0.24 g of KH_2PO_4, 2 mL of Tween 20 in 800 mL of distilled H_2O, adjust pH to 7.2, and add distilled H_2O to 1000 mL before sterilization.
11. Water bath sonication equipment.
12. 2 mL Eppendorf tubes.
13. Tubes for microbiology (12–15 mL) (*see* **Note 3**).
14. 100 mg/mL ampicillin or 50 mg/mL kanamycin stock solutions in H_2O (according to the vector used for expressing the target protein).
15. 34 mg/mL chloramphenicol stock solution in 100% ethanol.
16. 12.5 mg/mL arabinose in H_2O.
17. 8 M urea in H_2O.

2.3 Small-Scale Production Output Analysis

1. Equipment for running polyacrylamide gels.
2. Buffers for SDS-PAGE (sample buffer, running buffer).
3. Staining cuvette.
4. Colloidal blue for polyacrylamide gel staining (*see* **Note 4**).

2.4 Cytoplasmic Large-Scale Protein Production of Antibody Fragments

1. Culture medium (as described in Subheading 2.2, **item 1**).
2. 1 M IPTG in H_2O.
3. 40% (w/v) glucose.
4. BL21(DE3) bacteria co-transformed with the suitable constructs.
5. 1 mg/mL DNase I.
6. 100 mg/mL lysozyme.

7. Ni-NTA chromatographic column.
8. Lysis buffer: 50 mM Tris–HCl, pH 8.0, 500 mM NaCl, 5 mM $MgCl_2$.
9. Washing buffer: 50 mM Tris–HCl, pH 8.0, 500 mM NaCl, 20 mM imidazole.
10. Elution buffer: 50 mM Tris–HCl, pH 8.0, 150 mM NaCl, 200 mM imidazole.
11. Falcon conical centrifuge tubes (50 mL) (*see* **Note 5**).
12. Cooled benchtop centrifuge with adaptors for 50 mL Falcon tubes.
13. Cooled centrifuge.
14. Centrifuge tubes (50 mL).
15. FPLC equipment.
16. 100 mg/mL ampicillin or 50 mg/mL kanamycin stock solutions in H_2O (according to the vector used for expressing the target protein).
17. 34 mg/mL chloramphenicol stock solution in 100% ethanol.
18. Arabinose in powder.
19. SEC buffer: 50 mM Tris–HCl, pH 8.0, 150 mM NaCl.
20. HiTrap Desalting column.
21. Sonication device.
22. Microbiology filter unit (pore diameter, 0.2 μm).

2.5 Protein Structural Quality Evaluation

1. Equipment for running polyacrylamide gels.
2. Staining tray.
3. Colloidal blue for polyacrylamide gel staining.
4. Superdex 75 5/150 GL (GE Healthcare).
5. Column for size-exclusion chromatography (SEC).
6. SEC buffer: 50 mM Tris–HCl, pH 8.0, 150 mM NaCl.
7. ÄKTA-FPLC (GE Healthcare).
8. Spectrofluorometer.
9. Spectrofluorometer cuvette.

3 Methods

3.1 Vector Preparation

1. Antibody fragments from phage display libraries are directly subcloned by cut-and-paste from the original phagemid into an IPTG-inducible and His-tagged bacterial vector for cytoplasmic expression. Further tags such as GFP, Fc-domain, biotinylation sequence, alkaline phosphatase can be present to produce a fusion immune-reagent (*see* **Note 6**).

3.2 Small-Scale Protein Production (See Note 7)

1. Use BL21(DE3) cells transformed with the two plasmids to inoculate a microbiology (12–15 mL) tube containing 2 mL of LB medium, 1% (w/v) glucose, 2 μL of antibiotic stock solutions for selecting both vectors, the one expressing the recombinant antibody (ampicillin or kanamycin) and the one expressing SO and DIso (chloramphenicol) (*see* **Notes 8** and **9**).
2. Grow overnight at 30 °C in an inclined rack (30°) inside a shaker (180 rpm).
3. The day after, add 3 μL of the pre-cultures to three 12–15 mL tubes each one filled with 3 mL of LB (or TB), 3 μL of ampicillin/kanamycin stock solution and 3 μL of chloramphenicol stock solution.
4. Let the bacteria grow at 37 °C in an orbital shaker (210 rpm) until the OD_{600} reaches 0.4.
5. Add 60 μL of arabinose stock solution and switch the temperature to 20 °C.
6. Induce the recombinant antibody expression after 30 min (the OD_{600} of the culture will have reached approximately the value of 0.6) by adding 0.2 mM IPTG.
7. Let the culture grow for 18 h at 20 °C.
8. Harvest the pellet from single tubes by centrifuging the culture medium (15 min × 11,000 × *g* at 4 °C) after 3, 6, and 18 h (*see* **Note 10**).
9. Remove the medium and store the pellet at −20 °C.
10. Add 20 μL of magnetic bead slurry to a 2 mL Eppendorf tube.
11. Set the tube into a magnetic rack and carefully remove the solution.
12. Transfer the tube into a standard rack and resuspend the beads in 400 μL of PBST.
13. Set the tube to the magnetic rack and remove the buffer.
14. Transfer the tube in a standard rack and collect the beads in 50 μL of lysis buffer.
15. Resuspend the bacteria pellet in 350 μL of lysis buffer.
16. Sonicate for 5 min in a water bath at room temperature.
17. Add lysozyme to a final concentration of 1 mg/mL and DNase I to a final concentration of 50 μg/mL and incubate for 30 min at room temperature by continuous rocking. No viscous material indicating the presence of indigested nucleic acids should be detectable at the end of this step.
18. Separate the supernatant fractions by centrifugation (5 min at 16,100 × *g*).
19. Remove 10 μL of supernatant and add it to 10 μL of 2× SDS sample buffer (total soluble fraction).

20. Resuspend the pellet in 40 μL of 8 M urea, recover 10 μL of the solubilized fraction and add it to 10 μL of 2× SDS sample buffer (insoluble fraction).
21. Add the rest of the supernatant to the tubes with the pretreated beads.
22. Incubate the tubes for 30 min under constant rotation.
23. Separate the beads from the supernatant by means of the magnetic rack and discard the supernatant.
24. Remove the magnet, resuspend the beads in 400 μL of washing buffer and incubate for 30 min under constant rotation.
25. Repeat **steps 22** and **23**.
26. Separate the beads from the supernatant by means of the magnetic rack and carefully discard the supernatant (*see* **Note 11**).
27. Remove the magnet and incubate the beads for 10 min in the presence of 25 μL of 50 mM Tris–HCl, pH 8.0, 250 mM NaCl, and 250 mM imidazole to elute the purified antibody.
28. Separate the beads by means of the magnet and recover the supernatant in an Eppendorf tube.
29. Remove 10 μL of supernatant and add it to 10 μL of 2× SDS sample buffer (elution fraction).
30. Use the three fractions collected during the purification for running an SDS polyacrylamide gel.

3.3 Small-Scale Production Output Analysis: SDS-PAGE (See Note 12)

1. Run a 12% SDS-PAGE gel, remove it from the glass and place it on the tray.
2. Rinse the gel 3 × 5 min in H_2O.
3. Cover the gel with colloidal blue and incubate for 30–60 min at room temperature (*see* **Note 13**).
4. Destain by washing in H_2O (*see* **Note 14**).

3.4 Cytoplasmic Large-Scale Recombinant Antibody Production

1. Use an aliquot of frozen BL21(DE3) cells transformed with the two plasmids to inoculate a 15 mL flask containing 10 mL of culture medium, 1% (w/v) glucose, and 10 μL of antibiotic stock solutions for selecting both vectors, the one expressing the recombinant antibody (ampicillin or kanamycin) and the one expressing SO and DIso (chloramphenicol).
2. Grow overnight at 30 °C in a shaker (180 rpm).
3. The day after, add 2.5 mL of the pre-cultures to a 2000 mL flask filled with 500 mL of LB (or TB), 500 μL of ampicillin/kanamycin stock solution and 500 μL of chloramphenicol stock solution.
4. Let the bacteria grow at 37 °C in an orbital shaker (210 rpm) until the OD_{600} reaches 0.4.

5. Add arabinose to a final concentration corresponding to 0.5% and switch the temperature to 20 °C.
6. Induce the recombinant antibody expression after 30 min (the OD_{600} of the culture will have reached approximately the value of 0.6) by adding 0.2 mM IPTG.
7. Let the culture grow at 20 °C for the time defined by the small scale production test.
8. Harvest the pellet by centrifuging the medium (15 min × 11,000 × *g* at 4 °C).
9. Remove the medium and resuspend the bacteria in 12 mL of cooled PBS.
10. Fill the resuspension fraction in a 50 mL Falcon tubes, centrifuge them 10 min at 4500 × *g* in a benchtop centrifuge precooled at 4 °C and discard the supernatant.
11. Freeze the pellet (*see* **Note 15**).
12. Resuspend the bacteria pellet in 10 mL of lysis buffer.
13. Sonicate on ice (*see* **Note 16**).
14. Add lysozyme to a final concentration of 1 mg/mL and DNase I to a final concentration of 50 μg/mL and incubate for 30 min at room temperature by continuous rocking. No viscous material indicating the presence of indigested nucleic acids should be detectable at the end of this step.
15. Fill the homogenate in centrifuge tubes and separate the supernatant by centrifugation (15 min at 16,100 × *g*).
16. Recover the supernatant and remove any debris using a filter unit with pores of 0.2 μm.
17. Load the supernatant onto a Ni-NTA column operated by a FPLC system and pre-equilibrated with washing buffer (*see* **Note 17**).
18. Wash extensively the column with washing buffer (signal back to the background).
19. Elute the recombinant antibodies by adding elution buffer and recover 0.5 mL fractions in separate tubes.
20. Analyze protein content and quality by SDS-PAGE and pool the content from protein-enriched tubes.
21. Desalt using a desalting column and SEC buffer before proceeding to protein quality evaluation.
22. Evaluate the protein quality by SDS-PAGE analysis, as described in Subheading 3.3.

3.5 Analytical Size Exclusion Chromatography (SEC)

1. Pre-equilibrate the SEC column in 5 volumes of SEC buffer.
2. Load the sample in a mini-loop of 15 μL of volume.

3. Run the gel filtration at a flow-rate of 0.2 mL/min and collect the absorbance signal at 280 nm.
4. Analyze the peak distribution (*see* **Note 18**).

3.6 Fluorimetric Aggregation Index (AI)

1. Set up the spectrofluorometero selecting the excitation at 280 nm and the emission at 340 nm. Use a scan rate of 5 and record the emission signal between 260 and 400 nm (*see* **Note 19**).
2. Select 600 V for the sensitivity and shift to lower values when the signal is saturated.
3. Scan and record the emission values at 280 and 340 nm.
4. Calculate the ratio between the values at 280 and 340 nm (*see* **Note 20**).

3.7 Functional Evaluation

1. When possible, comparative functional evaluation against reliable controls (experimental or reported in the literature) should be performed to complete the quality analysis of the purified proteins. In contrast, the simple demonstration that the sample has activity does not rule out that it is partially damaged through partial misfolding or degradation. A recent article indicates what standard analyses should be performed to thoroughly evaluate the structural quality of a purified protein [18].

4 Notes

1. Although vectors with different characteristics and enzymes of different origin could be probably effectively used for this application, we suggest to refer to those developed in Ruddock's group because they have been widely validated for the production of both protein and antibody fragments [16, 17, 19]. Two versions are available: Erv1p sulfhydryl oxidase from *S. cerevisiae* is paired to either the *E. coli* DsbC or to the human PDI since isomerases of evolutionary different organisms could be better suited for substrates with variable folding needs.
2. Luria–Bertani—LB—medium represents probably the standard for bacterial growth in most of the labs. However, *E. coli* can grow at much higher densities in richer media such as Terrific Broth—TB. Given the small volumes used for screening purification, it can happen that the amount of purified protein remains below the limit of detection in blue-stained gels. Increasing the bacterial volume will allow for the recovery of 2–3 times more protein and can help in yield comparison.
3. The reported protocol is conceived for labs that have basic equipment. When available, 96 deep-well plates filled with

0.6–1 mL culture medium can represent a valuable alternative. However, because of the difficult aeration of these miniaturized cultures, good productions can be obtained only at the elevated rotation rates enabled by ad hoc shakers that can reach 600–1000 rpm. The operator must also pay attention to the spill over that can happen when the paper filter is removed from the plate top. Medium drops can easily contaminate near wells. In standard recombinant protein productions, the use of self-induction media limits the manipulation steps and consequently reduces the contamination danger. In this specific case the option is more difficult to implement because the expression of the helper enzymes is arabinose-dependent and should anticipate the induction of the T7 polymerase activity.

4. Colloidal blue dyes allow detection of proteins at lower concentrations than standard Coomassie blue (up to 10 ng) and are water soluble. Their use avoids having to operate with and to stock toxic products.
5. Any tube with similar characteristics will fit as well, such as Sarstedt tubes.
6. A large variety of expression vectors are available that enable the production of ready-to-use immune-reagents suitable for the final specific application. For instance, they promote the expression of fusion proteins with the SNAP-tag, toxins, Fc-domains of different origins or interleukin-2 [16, 20, 21].
7. The aim of this step is to verify the clone expression and to identify the optimal growth conditions for the recovery of full-length constructs before moving to large-scale production.
8. The protocol is based on the capacity of accumulating SO and DIso in the bacterial cytoplasm before the induction of the recombinant antibodies, the native structure of which needs the formation of correct disulfide bonds. Therefore, the expression leakage of the IPTG-dependent vector must be absolutely avoided because any synthetized polypeptide could aggregate in the absence of SO and DIso and could generate aggregation seeds that could compromise the correct folding also of the recombinant antibodies expressed after SO and DIso accumulation.
9. At this time is useful to prepare frozen stocks of transformed bacteria by adding glycerol to a final concentration of 20%. Store the bacteria at −80 °C.
10. The method is suitable for parallel analysis of different clones. Of course, the 96-well format (*see* **Note 4**) will result more convenient when several samples need to be analyzed.
11. It is important to avoid wasting part of the beads during liquid removal. It is suggested to use a P1000 pipette for removing

the bulk of the buffer but a P200 with small tips to remove the last drops without touching the magnetic beads.

12. Here it is reported the minimal analysis (an SDS-PAGE) necessary to evaluate the output of the expression test. When the resources are available, an analytical SEC could be performed to assess the protein monodispersity [22].

13. The commercially available products can request variable condition for optimal staining. Apart from the manufacturers' suggestions, several tips are available on the web to increase the sensitivity or reduce the incubation time.

14. The presence of a single band of the expected mass in the elution fraction will indicate the successful expression and IMAC purification of the target protein/antibody. A calibration curve with known concentrations of a standard protein such a BSA could help in estimating the yields/L culture. The presence of bands with mass smaller than expected usually indicate that the protein was (partially) degraded or was not expressed correctly as a full-length protein. Growth time, specificities of the fusion tags, and linker fragility can all contribute to this result. The comparison among the three growth times should indicate the optimal range for expressing that specific construct, namely the accumulation period that maximizes the yields of full-length polypeptide with respect to by-products. In the absence of any purified protein, the analysis of the total soluble protein will allow for evaluating the actual presence of soluble target protein (the protein could be present in the supernatant but does not bind to the beads, for instance because forms soluble aggregates). If no soluble target protein is detectable, it could have accumulated in the pellet as insoluble aggregates. Different growth conditions or the use of alternative tags can improve the solubility (see Introduction). If no target protein is detectable even in the pellet, probably something went wrong during cloning.

15. Pellets from 500 mL of bacterial culture have usually a volume between 3 and 8 mL, according to the used medium.

16. Sonication devices may possess very variable characteristics and it is not meaningful to indicate general operating conditions. Nevertheless, it is important that the lysate remains cool (better shorter and repeated pulses followed by rest on ice than long treatments at high intensity because they heat the solution and induce aggregation) and that at the end no apparent solution viscosity is visible.

17. If possible, run the purification steps at 4 °C.

18. This fast analysis allows for the identification of polymerization or aggregation products as well as the presence of degradation products. It can be coupled with in-line multi-angle light scattering.

19. The cuvette has a minimal volume of 140 μL.
20. The AI index has been initially proposed by Nominé et al. and has been later validated using independent approaches [23, 24]. The method does not provide an absolute "quality score" but the ratio between the signal of light scattering (caused by aggregates) at 280 nm and the emission signal of the aromatic residues (at 340 nm, possible only when they are accessible and not buried into aggregates) offers a reliable, fast, and inexpensive indication of the sample monodispersity. Values of the 280/340 ratio below 0.2 correlate with elevated monodispersity and indicate an excellent protein structural quality. No sample is wasted during the measurement and the effect of the buffer modification can be easily monitored.

References

1. Anfinsen CB (1972) The formation and stabilization of protein structure. Biochem J 128: 737–749
2. Taipale M, Krykbaeva I, Koeva M et al (2012) Quantitative analysis of HSP90-client interactions reveals principles of substrate recognition. Cell 150:987–1001
3. Kerner MJ, Naylor DJ, Ishihama Y et al (2005) Proteome-wide analysis of chaperonin-dependent protein folding in *Escherichia coli*. Cell 122:209–220
4. de Marco A, Vigh L, Diamant S et al (2005) Native folding of aggregation-prone recombinant proteins in *Escherichia coli* by osmolytes, plasmid- or benzyl alcohol-overexpressed molecular chaperones. Cell Stress Chaperones 10:329–339
5. de Marco A, Deuerling E, Mogk A et al (2007) Chaperone-based procedure to increase yields of soluble recombinant proteins produced in *E. coli*. BMC Biotechnol 7:32
6. de Marco A (2007) Protocol for preparing proteins with improved solubility by co-expression with molecular chaperones in *Escherichia coli*. Nat Protoc 2:2632–2639
7. Diamant S, Rosenthal D, Azem A (2003) Dicarboxylic amino acids and glycine-betaine regulate chaperone-mediated protein disaggregation under stress. Mol Microbiol 49: 401–410
8. Schultz T, Liu J, Capasso P et al (2007) The solubility of recombinant proteins expressed in *Escherichia coli* is increased by *otsA* and *otsB* co-transformation. Biochem Biophys Res Commun 355:234–239
9. Bandyopadhyay A, Saxena K, Kasturia N et al (2012) Chemical chaperones assist intracellular folding to buffer mutational variations. Nat Chem Biol 8:238–245
10. Diamant S, Eliahu N, Rosenthal D et al (2001) Chemical chaperones regulate molecular chaperones in vitro and in cells under combined salt and heat stresses. J Biol Chem 276:39586–39591
11. de Marco A (2014) Osmolytes as chemical chaperones to use in protein biotechnology. In: Doglia SM, Lotti M (eds) Protein aggregation in bacteria: functional and structural properties of inclusion bodies in bacterial cells. Wiley, Hoboken, NJ, pp 77–92
12. Esposito D, Chatterjee DK (2006) Enhancement of soluble protein expression through the use of fusion tags. Curr Opin Biotechnol 17:353–358
13. Swalley SE, Fulghum JR, Chambers SP (2006) Screening factors effecting a response in soluble protein expression: formalized approach using design of experiments. Anal Biochem 351:122–127
14. Bora N, Bawa Z, Bill RM et al (2012) The implementation of a design of experiments strategy to increase recombinant protein yields in yeast (review). Methods Mol Biol 866: 115–127
15. Nozach H, Fruchart-Gaillard C, Fenaille F (2013) High throughput screening identifies disulfide isomerase DsbC as a very efficient partner for recombinant expression of small disulfide-rich proteins in *E. coli*. Microb Cell Fact 12:37
16. Djender S, Schneider A, Beugnet A et al (2014) Bacterial cytoplasm as an effective cell compartment for producing functional VHH-based affinity reagents and *Camelidae* IgG-like recombinant antibodies. Microb Cell Fact 13:140
17. Veggiani G, de Marco A (2011) Improved quantitative and qualitative production of single-domain intrabodies mediated by the

co-expression of Erv1p sulfhydryl oxidase. Protein Expr Purif 79:111–114

18. Raynal B, Lenormand P, Baron B et al (2014) Quality assessment and optimization of purified protein samples: why and how? Microb Cell Fact 13:180
19. Nguyen VD, Hatahet F, Salo KE et al (2011) Pre-expression of a sulfhydryl-oxidase significantly increases the yields of eukaryotic disulfide bond containing proteins expressed in the cytoplasm of *E. coli*. Microb Cell Fact 10:1
20. Liu JL, Zabetakis D, Walper SA et al (2014) Bioconjugates of rhizavidin with single domain antibodies as bifunctional immunoreagents. J Immunol Methods 411:37–42
21. de Marco A (2015) Recombinant antibody production evolves into multiple options aimed at yielding reagents suitable for application-specific needs. Microb Cell Fact 14:125
22. Sala E, de Marco A (2010) Screening optimized protein purification protocols by coupling small-scale expression and mini-size exclusion chromatography. Protein Expr Purif 74:231–235
23. Nominé Y, Ristriani T, Laurent C et al (2001) A strategy for optimizing the monodispersity of fusion proteins: application to purification of recombinant HPV E6 oncoprotein. Protein Eng 14:297–305
24. Capasso P, Aliprandi M, Ossolengo G et al (2009) Monodispersity of recombinant Cre recombinase correlates with its effectiveness in vivo. BMC Biotechnol 9:80

Chapter 13

Protein Folding Using a Vortex Fluidic Device

Joshua Britton*, Joshua N. Smith*, Colin L. Raston, and Gregory A. Weiss

Abstract

Essentially all biochemistry and most molecular biology experiments require recombinant proteins. However, large, hydrophobic proteins typically aggregate into insoluble and misfolded species, and are directed into inclusion bodies. Current techniques to fold proteins recovered from inclusion bodies rely on denaturation followed by dialysis or rapid dilution. Such approaches can be time consuming, wasteful, and inefficient. Here, we describe rapid protein folding using a vortex fluidic device (VFD). This process uses mechanical energy introduced into thin films to rapidly and efficiently fold proteins. With the VFD in continuous flow mode, large volumes of protein solution can be processed per day with 100-fold reductions in both folding times and buffer volumes.

Key words Protein folding, Vortex fluidics, Continuous flow, Misfolded proteins, Inclusion bodies, Bacterial protein expression

1 Introduction

Protein overexpression provides important tools for a wide range of applications ranging from therapeutics to laboratory reagents [1–3]. Bacterial protein overexpression capitalizes upon the low cost, high growth rate, and versatility of *Escherichia coli* [4, 5]. Although many proteins correctly fold during overexpression in *E. coli*, larger or very hydrophobic proteins often generate misfolded and insoluble aggregates, which are directed into inclusion bodies [6–8]. Recovering such aggregated proteins often involves chemical denaturation of the inclusion bodies, followed by multi-day dialysis to remove the denaturant [9, 10]. An alternative method applies high pressure to drive proteins into their folded states [11]. Illustrative of the wasteful inefficiency of conventional protein folding, liters of buffer are typically used to refold small quantities (mg) of protein [12, 13]. To address this issue, we have developed an approach to

*These authors contributed equally to this work.

Nicola A. Burgess-Brown (ed.), *Heterologous Gene Expression in E. coli: Methods and Protocols*, Methods in Molecular Biology, vol. 1586, DOI 10.1007/978-1-4939-6887-9_13, © Springer Science+Business Media LLC 2017

harness mechanical energy supplied by a VFD to rapidly fold proteins in vitro under standard conditions.

The VFD, a commercially available benchtop device, can drive a range of chemical transformations in its thin films [14–19]. Rapid rotation of a solution in a glass sample tube creates a dynamic thin film with ≈230 μm thickness [16]. Within this film, reagents experience micromixing, shear stress, and vibrational effects [19–24]. At specific rotational speeds, the sample tube enters a harmonic vibration that generates Faraday or pressure waves in the fluid, which can accelerate protein folding [25, 26].

The method described here is demonstrated with the folding of recombinant hen egg white lysozyme (HEWL) overexpressed in *E. coli*. The insoluble, misfolded protein results from overinduction, and a correctly folded form can be obtained with short VFD processing times. Though folding HEWL is illustrated here, the approach is generalizable to include other proteins, as demonstrated for cAMP dependent protein kinase A (PKA) and a truncated form of the membrane protein caveolin-1 [27].

2 Materials

2.1 Buffers

All buffers are prepared at 1 L scale with double-distilled H_2O (ddH_2O) purified to a resistivity of 18 MΩ and filtered through a 0.22 μm filter. Buffers are autoclaved, and then stored at room temperature unless otherwise indicated. All reagents should be at least reagent grade and ACS-certified.

1. Buffer 1: 50 mM NaH_2PO_4, 500 mM NaCl, pH 8.0. To a 1 L Erlenmeyer flask, add 6.00 g of NaH_2PO_4 and 29.22 g of NaCl. Add approximately 600 mL of ddH_2O, and dissolve by magnetic stirring. Adjust the pH to 8.0 using 6.0 M NaOH and transfer the solution to a graduated cylinder before bringing the volume to 1 L with ddH_2O (*see* **Note 1**).
2. Buffer 2 (Denaturing buffer): 20 mM Tris–HCl, 10 mM NaCl, 8.0 M urea, pH 8.0. To a 1 L Erlenmeyer flask, add 3.15 g of Tris–HCl, 0.58 g of NaCl, and 480.48 g of urea. Add approximately 300 mL of ddH_2O and dissolve the powders into solution using magnetic stirring. Adjust the pH of the solution to 8.0 using 6.0 M NaOH and transfer to a graduated cylinder before bringing the final volume to 1 L with ddH_2O (*see* **Note 2**).
3. Buffer 3 (Elution buffer): 20 mM Tris–HCl, 400 mM NaCl, 8.0 M urea, pH 7.8. To a 1 L Erlenmeyer flask, add 3.15 g of Tris–HCl, 23.38 g of NaCl, and 480.48 g of urea. Add approximately 300 mL of ddH_2O and dissolve the powders in solution using magnetic stirring. Adjust the pH of the solution to 8.0 using 6.0 M NaOH and transfer to a graduated cylinder before bringing the final volume to 1 L with ddH_2O (*see* **Note 3**).

4. Buffer 4 (Phosphate-buffered saline): 10 mM Na_2HPO_4, 2.0 mM KH_2PO_4, 2.7 mM KCl, 137 mM NaCl, pH 7.2. To a 1 L Erlenmeyer flask, add 1.42 g of Na_2HPO_4, 0.27 g of KH_2PO_4, 0.20 g of KCl, and 8.0 g of NaCl. Add approximately 600 mL of ddH_2O and dissolve the powders into solution using magnetic stirring. Adjust the pH of the solution to 7.2 using 6.0 M NaOH, and transfer to a graduated cylinder before bringing the final volume to 1 L with ddH_2O.

2.2 VFD Components and Equipment

1. The VFD, sample tube (20 mm external diameter and 17.7 mm inner diameter) and delivery jet feeds are purchased from Vortex Fluidic Technologies (www.vortexfluidictechnologies.com).
2. A laboratory syringe pump with the appropriate tubing (e.g., WPI model AL-1000).
3. VFD-sample tube caps: B19 Suba-Seals® (Sigma).
4. Nachi 6005NR open bearings (Nachi-Fujikoshi) (*see* **Note 4**).
5. VFD sample tube lubricant (Dow Corning 976 V High Vacuum Grease).
6. All-purpose machine oil (Singer).

2.3 Protein Overexpression, Purification, and Analysis

All reagents for protein overexpression should be molecular biology grade and the appropriate measures taken to maintain sterile working conditions.

1. Super optimal broth with catabolite repression (SOC): 2% (w/v) tryptone, 0.5% (w/v) yeast extract, 10 mM NaCl, 2.5 mM KCl, 10 mM $MgCl_2$, 10 mM $MgSO_4$, 20 mM glucose. To a 1 L autoclavable bottle, add 20 g of tryptone, 5 g of yeast extract, 0.584 g of NaCl, 0.186 g of KCl, 0.952 g of $MgCl_2$, and 2.467 g of $MgSO_4 \cdot 7H_2O$. Bring the solution to ≈1 L with ddH_2O and swirl to dissolve the solid contents. Autoclave the solution at 121 °C for 20 min. Allow the solution to cool to room temperature, add 3.60 g of glucose and invert the bottle to mix the contents before use.
2. Lysogeny Broth (LB): 1% (w/v) tryptone, 0.5% (w/v) yeast extract, 1% (w/v) NaCl. To a 1 L autoclavable bottle, add 10 g of tryptone, 5 g of yeast extract, and 10 g of NaCl. Bring the final volume to ≈1 L with ddH_2O and autoclave at 121 °C for 20 min. Allow the solution to cool to room temperature and invert the bottle to mix the contents before use.
3. 40 mg/mL kanamycin sulfate: To a 50 mL plastic conical tube, dissolve 1.60 g of kanamycin sulfate in approximately 35 mL of ddH_2O. Bring the total volume to ≈40 mL with ddH_2O. Filter the solution with a 0.22 μm syringe filter and divide the solution into several 1 mL aliquots in 1.5 mL centrifuge tubes. Store the aliquots at −20 °C.

4. Sterile petri dish.
5. LB/Kanamycin agar plate: 1% (w/v) tryptone, 0.5% (w/v) yeast extract, 1% (w/v) NaCl, 1.5% (w/v) agar. To a 1 L autoclavable bottle, add 10 g of tryptone, 5 g of yeast extract, 10 g of NaCl, and 15 g of agar. Bring the final volume to ≈1 L with ddH_2O and autoclave at 121 °C for 20 min. Allow the solution to cool to approximately 55 °C and add 1 mL of the 40 mg/mL kanamycin sulfate from above (giving a final concentration of 40 μg/mL) and swirl to mix. Add 20 mL of this mixed solution to a petri dish (1 L of solution is sufficient for 50 petri dishes). Allow the dishes to cool until the agar has solidified (approximately 1 h) and store at 4 °C.
6. 1.0 M isopropyl β-D-1-thiogalactopyranoside (IPTG): In a 50 mL plastic conical tube, dissolve 5 g of IPTG in 15 mL ddH_2O. Transfer to a graduated cylinder and add ddH_2O to a final volume of 21.0 mL with ddH_2O. Filter the solution with a 0.22 μm syringe filter and divide this into several 1 mL aliquots in 1.5 mL centrifuge tubes. Store the aliquots at −20 °C.
7. The gene encoding for HEWL was purchased from Addgene, and was cloned into a pET28c vector (GE Healthcare).
8. $CaCl_2$ competent *E. coli* BL21(DE3) cells (EMD Millipore).
9. UNOSphere S cation exchange media (Bio-Rad).
10. Kimble-Chase Kontes FlexColumn (Fisher).
11. Pierce™ BCA Protein Assay Kit (Thermo Fisher).
12. Circular Dichroism quartz cuvette (Hellma).
13. EnzChek® Lysozyme Assay Kit (Thermo Fisher).
14. 10 kDa cutoff protein concentrator (Sartorius).

3 Methods

3.1 Overexpression of HEWL in E. coli

All procedures are performed at room temperature under sterile conditions unless otherwise indicated.

1. Transform a plasmid encoding HEWL into $CaCl_2$ competent *E. coli* BL21(DE3) cells as follows: Add 2 μL of HEWL DNA to 75 μL of competent cells. Incubate the mixture on ice for 30 min. Heat shock the mixture at 42 °C for 45 s and then incubate on ice for 2 min. To rescue the cells, add 250 μL of SOC medium to the mixture and incubate in an incubator shaker at 37 °C and 225 rpm for 1 h. Remove from the shaker and plate cells (150 μL) onto an LB/Kanamycin agar plate. Incubate the plate for 10–12 h in an incubator at 37 °C.
2. Pick a single colony of the transformed cells containing the HEWL vector, and inoculate 10 mL of LB containing 10 μL of 40 mg/mL kanamycin sulfate (final concentration of 40 μg/mL).

Incubate this mixture for 10–12 h in an incubator shaker at 37 °C and 225 rpm.

3. Transfer this 10 mL solution into a 2 L baffled flask containing 1 L of LB and 1 mL of 40 mg/mL kanamycin sulfate. Grow the culture in an incubator shaker at 37 °C and 225 rpm until the cells reach an optical density with OD_{600} of 0.8 (≈2.5 h). Then, add 1 mL of the 1 M IPTG solution (final concentration of 1 mM) and incubate for another 6 h at 30 °C and 225 rpm.
4. Centrifuge the culture at 5524 × *g* for 20 min at 4 °C. Remove the supernatant and discard into 10% (v/v) bleach. Resuspend the pellet in Buffer 1 (20 mL) by mixing or pipetting the solution over the pellet, but do not vortex the pellet (*see* **Note 5**).
5. Lyse the resuspended cells using probe sonication. For this process, incubate the cells on ice and use eight cycles of 30 s pulses (20 W) with 1 min of cooling (without sonication) between cycles (*see* **Note 6**).
6. Centrifuge the viscous solution at 4 °C for 50 min at 30,996 × *g* with a floor model centrifuge (e.g., Beckman Avanti™ J-25) with a rotor suitable for 50 mL centrifuge tubes (e.g., Beckman JA-17).
7. Discard the supernatant and retain the pellet, which contains HEWL in inclusion bodies (*see* **Note 7**). Add 30 mL of Buffer 2 to the pellet and incubate at 4 °C with shaking overnight to fully dissolve, denature and solubilize any remaining protein (*see* **Note 8**).
8. Centrifuge the dissolved pellet at 33,264 × *g* for 1 h. Retain the supernatant and discard any remaining pelleted cell debris. The supernatant should now contain denatured HEWL.
9. HEWL may then be purified using UNOSphere S-Cation exchange media in a column by hand. For this procedure, add 3 mL of the resin to a Kontes FlexColumn and rinse several times with water (1.0 mL/min) and then 3–4 column volumes of Buffer 2 (1.0 mL/min). Incubate the denatured HEWL with the resin for 2 h at 4 °C, then wash the resin-adhered protein with 50 mL of Buffer 2 (1.0 mL/min). Elute the purified protein from the column using 50 mL of Buffer 3 (1.0 mL/min). SDS-PAGE can be used to determine the purity of the eluted HEWL (*see* **Note 9**).
10. Protein concentration in the eluted solution is then assayed by measuring the absorbance at 280 nm (molar extinction coefficient for HEWL of 37,970 M^{-1} cm^{-1}) or by using a BCA assay kit. The solution is then either concentrated or diluted (Buffer 3) to ≈4.4 mg/mL using a 10 kDa concentrator. This protein solution can next be folded.

3.2 Refolding HEWL Using the VFD

The VFD (Fig. 1) should be set at an angle of 45° relative to the horizontal position and the bearings should be well oiled (*see* **Note 10**).

3.2.1 Small Scale Folding of HEWL by Confined Mode VFD Processing (<3 mL)

1. To decrease the concentration of urea in the HEWL solution in Buffer 3, disperse 10 μL of the solution in 990 μL of Buffer 4, and mix with a 1 mL pipette.
2. Add the diluted solution (≈44 μg/mL) to a 20 mm external diameter glass sample tube, and seal the sample tube with a Suba-Seal® septum.
3. Insert the sample tube into the VFD. To do this, first apply a thin layer of lubricant around the sample tube. Then, insert the sample tube gently through the upper housing unit, until it is firmly mounted in the housing unit of the lower bearing.
4. Place the safety shield on the device and rotate the sample tube at a 5k rpm rotational speed for 5 min (*see* **Note 11**).

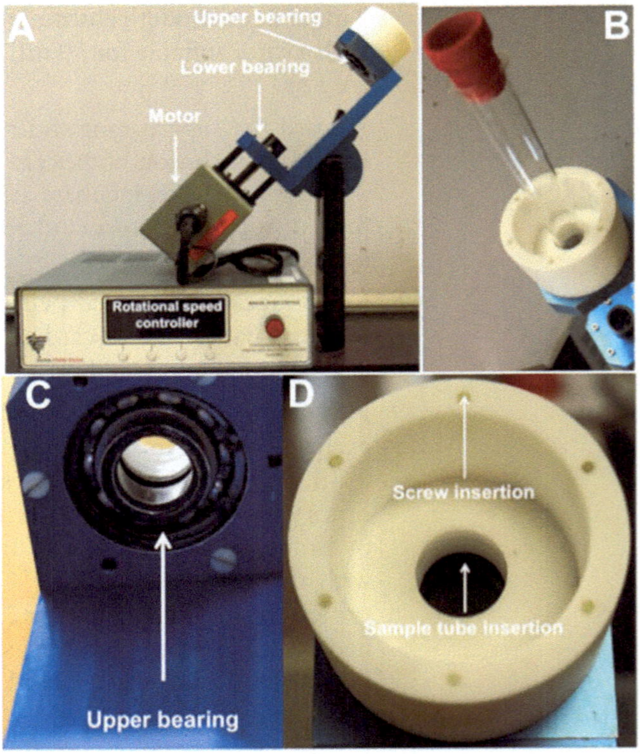

Fig. 1 Schematic diagram of a VFD. (**a**) The VFD and its rotational speed controller unit. The electric motor connects to the controller box and the lower bearing. Both the upper and lower bearings hold the sample tube in place during rotation. (**b**) The sample tube is inserted into the upper bearing. (**c**) The upper bearing of the device showing connection to the housing unit by six screws. (**d**) The housing unit of the VFD. The holder to the continuous flow jet feeds, termed the housing lid, attaches to the housing unit via the screw inserts

5. Remove the contents of the tube and transfer into a 2 mL centrifuge tube containing 100 μL of 100% glycerol. Quickly mix and flash freeze the sample in liquid nitrogen or a mixture of dry ice and ethanol. Store the sample at −80 °C until analysis (*see* **Note 12**).

3.2.2 Large-Scale Folding of HEWL by Continuous Flow VFD Processing

1. The solution of HEWL, currently in Buffer 3, is rapidly diluted into Buffer 4. Disperse 1 mL of the HEWL solution in Buffer 3 in 99 mL of Buffer 4 contained in a 250 mL beaker or flask. Mix this solution rapidly by swirling.
2. As shown in Fig. 2, the VFD can be configured for continuous flow operation.

   Process the solution by flowing through the VFD at a rate of 0.1 mL/min. Collect the solution of folded protein in a beaker or flask on ice.

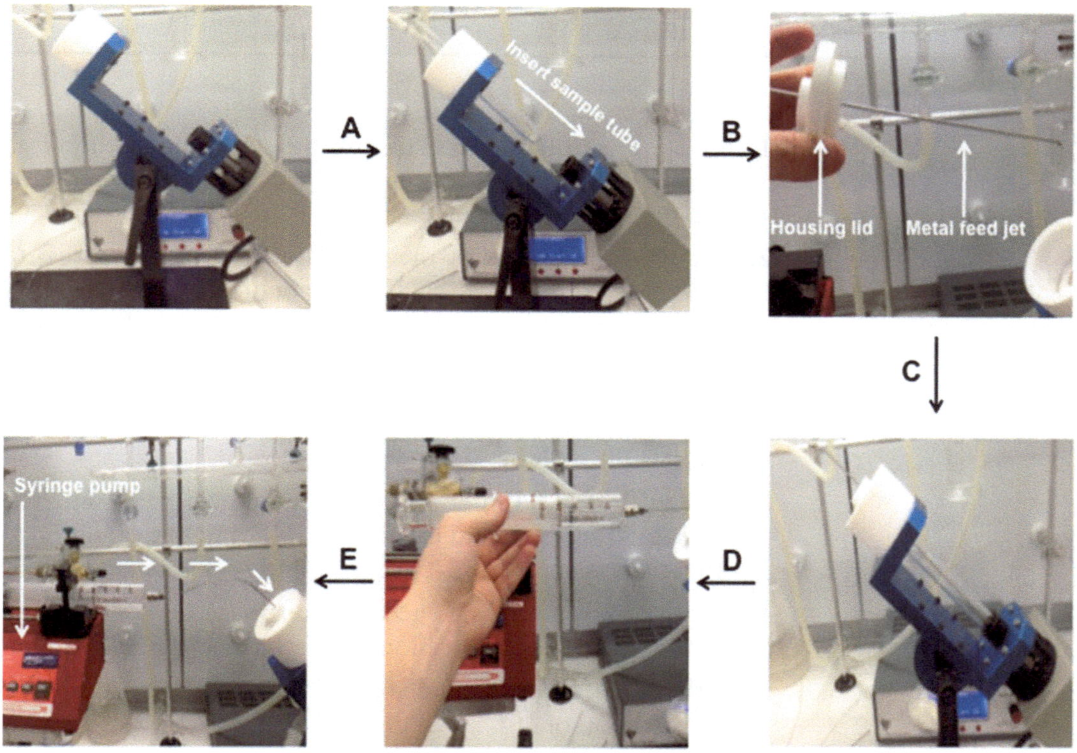

Fig. 2 A step-by-step guide for continuous flow protein folding by VFD. (**a**) Insert the 20 mm external diameter sample tube. For this process, we suggest adding a lubricant to the sample tube before applying any force. (**b**) Insert a metal jet feed through the housing lid until ≈75% of the jet feed has been inserted (*see* **Note 14**). (**c**) Add the housing lid onto the housing unit and fasten with plastic screws. Orient the jet feed with the exit facing the lower side of the sample tube. (**d**) Draw the solution of unfolded HEWL into a 50 mL syringe, and vent any air bubbles. (**e**) Insert the syringe first into the jet feed, and then place the syringe into the syringe pump. Select the required flow rate and start the VFD and syringe pump. For the confined mode of operation (1–3 mL), add the unfolded HEWL solution to the sample tube via a pipette, and cap with a Suba seal

3.2.3 Analysis of Refolded HEWL

Circular Dichroism

1. Circular dichroism (CD) is used to determine the secondary structure of the refolded HEWL samples. The sample (200 or 300 μL) is added directly to a CD cuvette. The spectrum is obtained using a spectropolarimeter such as the Jasco J-810 spectropolarimeter scanning 20 nm/min over four accumulations. Based on previous published results and crystal structures, the expected secondary structure of HEWL is largely α-helical [28] (*see* **Note 13**).

HEWL Activity Assay

2. To verify the protein folding efficacy of the sample, perform a HEWL activity assay. The experiment uses the commercially available Enzcheck Lysozyme Assay Kit. Follow the manufacturer's instructions to determine the activity of the refolded samples. A non-VFD-processed sample provides a direct comparison.

4 Notes

1. NaH_2PO_4 can be difficult to dissolve and requires thorough mixing. We recommend using a stir plate and optionally gentle heating.
2. Add all components of the buffer, apart from the urea, to 300 mL of ddH_2O as this large amount of urea will greatly increase the solution volume. The reaction is endothermic and will require gentle heating and stirring to completely dissolve the salts. The pH of this buffer is temperature dependent; thus, allow the solution to equilibrate at room temperature before adjusting the pH as detailed above.
3. As with Buffer 2, Buffer 3 requires heating with stirring. The pH should again be adjusted after allowing the solution to reach room temperature.
4. The bearings in the VFD need to be replaced every ≈3000 h (*see* Fig. 1 for bearing location during replacement).
5. At this point, the resuspended solution can be stored at −80 °C after addition of 10% (v/v) glycerol.
6. Other methods of lysing cells, such as using a French press, are also acceptable.
7. A layer of cell debris can often be observed on the outside of the pellet. This layer can be removed and discarded along with removal of the supernatant.
8. We find it easier to dissolve the pellet in urea by breaking up the pellet into smaller pieces using a sterile spatula. We then transfer the solution into a plastic conical tube to dissolve overnight.

9. If purity is poor after this step, wash steps can be added. Slowly increase the concentration of NaCl in the binding buffer from 0 to 50 mM in 10 mM increments. Additionally, size exclusion chromatography can be performed to further purify the protein.
10. As shown in several publications [16, 25], any deviation from the 45° tilt angle may drastically reduce the efficiency of the VFD. Additionally, the vibrations responsible for driving protein folding are produced by vibrations inherent to the VFD. Such vibrations can be sensitive to small changes in operational parameters, such as tilt angle, rotational speed, sample tube size, and wear on housing unit and bearings. Therefore, device maintenance is critical. For example, the addition of oil (e.g., machine oil, Singer) to the bearings before each use can increase reproducibility.
11. Several factors contribute to the optimal rotational speed for protein refolding, including quality and wear of both bearings and the housing unit. If refolding is not observed, vary the rotational speed of the sample tube in 50 rpm increments around 5k rpm to optimize folding conditions. Additionally, the housing unit and bearings can be replaced easily. The link below contains a 3D printing file for the housing unit (https://grabcad.com/library/vfd-collar-1, Fig. 1); high density ABS plastic is sufficient for production of this part.
12. Rapid dilution is commonly used to fold proteins. For this comparison between VFD and non-VFD processed conditions, samples are flash frozen after processing to prevent further conformational changes. For the non-VFD control, identical treatments include dilution of the sample and incubation at room temperature for the same time period as VFD processing (5 min); therefore, a direct comparison can be made with the VFD-processed solution.
13. Use the Dichroweb program to analyze the CD spectrum of the proteins, and assign its secondary structure [29].
14. The jet feeds must be held firmly in the housing unit lid. If they are not held firmly, then vibration of the jet feeds against the rotating sample tube will cause the glass sample tube to shatter.

Acknowledgments

J.B. thanks the Taihi Hong Memorial award for support. G.W. gratefully acknowledges support from the National Institute of General Medical Sciences of the NIH (1RO1-GM100700-01). C.R. acknowledges the Australian Research Council and the Government of South Australia for their financial support of this project.

References

1. Andersen DC, Krummen L (2002) Recombinant protein expression for therapeutic applications. Curr Opin Biotechnol 13:117–123
2. Hunt I (2005) From gene to protein: a review of new and enabling technologies for multi-parallel protein expression. Protein Expr Purif 40:1–22
3. Graumann K, Premstaller A (2006) Manufacturing of recombinant therapeutic proteins in microbial systems. Biotechnol J 1: 164–186
4. Hewitt L, McDonnell JM (2004) Screening and optimizing protein production in *E coli*. Methods Mol Biol (Clifton, NJ) 278:1–16
5. Rosano Gá L, Ceccarelli EA (2014) Recombinant protein expression in *Escherichia coli*: advances and challenges. Front Microbiol 5:172
6. Wang L, Maji SK, Sawaya MR et al (2008) Bacterial inclusion bodies contain amyloid-like structure. PLoS Biol 6:e195
7. Chang ESH, Liao T-Y, Lim T-S et al (2009) A new amyloid-like β-aggregate with amyloid characteristics, except fibril morphology. J Mol Biol 385:1257–1265
8. Ventura S, Villaverde A (2006) Protein quality in bacterial inclusion bodies. Trends Biotechnol 24:179–185
9. Singh A, Upadhyay V, Upadhyay AK et al (2015) Protein recovery from inclusion bodies of *Escherichia coli* using mild solubilization process. Microb Cell Fact 14:41
10. Vallejo LF, Rinas U (2004) Strategies for the recovery of active proteins through refolding of bacterial inclusion body proteins. Microb Cell Fact 3:11
11. Qoronfleh MW, Hesterberg LK, Seefeldt MB (2007) Confronting high-throughput protein refolding using high pressure and solution screens. Protein Expr Purif 55:209–224
12. Radford SE (2000) Protein folding: progress made and promises ahead. Trends Biochem Sci 25:611–618
13. Tsumoto K, Ejima D, Kumagai I et al (2003) Practical considerations in refolding proteins from inclusion bodies. Protein Expr Purif 28:1–8
14. Peng W, Chen X, Zhu S et al (2014) Room temperature vortex fluidic synthesis of monodispersed amorphous proto-vaterite. Chem Commun 50:11764–11767
15. Wahid MH, Eroglu E, LaVars SM et al (2015) Microencapsulation of bacterial strains in graphene oxide nano-sheets using vortex fluidics. RSC Adv 5:37424–37430
16. Yasmin L, Chen X, Stubbs KA et al (2013) Optimising a vortex fluidic device for controlling chemical reactivity and selectivity. Sci Rep 3:2282
17. Yasmin L, Eggers PK, Skelton BW et al (2014) Thin film microfluidic synthesis of fluorescent highly substituted pyridines. Green Chem 16:3450–3453
18. Britton J, Chalker JM, Raston CL (2015) Rapid Vortex Fluidics: Continuous Flow Synthesis of Amides and Local Anesthetic Lidocaine. Chem A Eur J 21:10660–10665
19. Britton J, Raston CL (2015) Rapid high conversion of high free fatty acid feedstock into biodiesel using continuous flow vortex fluidics. RSC Adv 5:2276–2280
20. Gandy MN, Raston CL, Stubbs KA (2015) Photoredox catalysis under shear using thin film vortex microfluidics. Chem Commun 51:11041–11044
21. Mo J, Eggers PK, Chen X et al (2015) Shear induced carboplatin binding within the cavity of a phospholipid mimic for increased anticancer efficacy. Sci Rep 5:10414
22. Vimalanathan K, Gascooke JR, Suarez-Martinez I et al (2016) Fluid dynamic lateral slicing of high tensile strength carbon nanotubes. Sci Rep 6:22865
23. Yasmin L, Stubbs KA, Raston CL (2014) Vortex fluidic promoted Diels–Alder reactions in an aqueous medium. Tetrahedron Lett 55: 2246–2248
24. Britton J, Raston CL (2014) Continuous flow vortex fluidic production of biodiesel. RSC Adv 4:49850–49854
25. Britton J, Dalziel SB, Raston CL (2015) Continuous flow Fischer esterifications harnessing vibrational-coupled thin film fluidics. RSC Adv 5:1655–1660
26. Britton J, Dalziel SB, Raston CL (2016) The synthesis of di-carboxylate esters using continuous flow vortex fluidics. Green Chem 18:2193–2200
27. Yuan TZ, Ormonde CFG, Kudlacek ST et al (2015) Shear-stress-mediated refolding of proteins from aggregates and inclusion bodies. Chembiochem 16:393–396
28. Tanaka FF, Leslie P, Pranab K et al (1975) The circular dichroism of lysozyme. J Biol Chem 250:6977–6982
29. Whitmore L, Wallace BA (2004) DICHROWEB, an online server for protein secondary structure analyses from circular dichroism spectroscopic data. Nucleic Acids Res 32(Web Server Issue):668–673

Chapter 14

Removal of Affinity Tags with TEV Protease

Sreejith Raran-Kurussi, Scott Cherry, Di Zhang, and David S. Waugh

Abstract

Although affinity tags are highly effective tools for the expression and purification of recombinant proteins, they generally need to be removed prior to structural and functional studies. This chapter describes a simple method for overproducing a soluble form of a stable variant of tobacco etch virus (TEV) protease in *Escherichia coli* and a protocol for purifying it to homogeneity so that it can be used as a reagent for removing affinity tags from recombinant proteins by site-specific endoproteolysis. Further, we cleave a model substrate protein (MBP-NusG) in vitro using the purified TEV protease to illustrate a protease cleavage protocol that can be employed for simple pilot experiments and large-scale protein preparations.

Key words Affinity chromatography, Affinity tag, Fusion protein, His-tag, IMAC, Immobilized metal affinity chromatography, Maltose-binding protein, MBP, TEV protease, Tobacco etch virus protease

1 Introduction

The use of fusion technology has become a widespread practice in the production of recombinant proteins for various applications. Although it was originally designed to facilitate the detection and purification of proteins, subsequently it has become clear that some fusion tags offer extra benefits like improving the yield of their fusion partners, protecting them from intracellular proteolysis, enhancing their solubility, and even facilitating their folding [1]. However, all tags, whether large or small, have the potential to interfere with the structure and function of purified proteins [2–5]. For this reason, it is generally advisable to remove the tag(s) at some stage.

The stringent specificity of viral proteases makes them attractive tools for removing affinity tags. The nuclear inclusion protease from tobacco etch virus (TEV) is probably the best-characterized enzyme of this type. TEV protease recognizes the amino acid sequence ENLYFQ/G with high efficiency and cleaves between Q and G. Its stringent sequence specificity, ease of production, and ability to tolerate a variety of residues at the P1′ position of

Nicola A. Burgess-Brown (ed.), *Heterologous Gene Expression in E. coli: Methods and Protocols*, Methods in Molecular Biology, vol. 1586, DOI 10.1007/978-1-4939-6887-9_14,

its recognition site have contributed to its popularity as an endoproteolytic reagent [6, 7].

Here, we describe a method for the large-scale production of a highly active and stable variant (L56V, S135G, S219V mutant) of TEV protease in *E. coli* and its purification to homogeneity. The protease is initially produced as a fusion to the C-terminus of MBP, which causes it to accumulate in a soluble and active form rather than in inclusion bodies. The fusion protein cleaves itself in vivo to remove the MBP moiety, yielding a soluble TEV protease catalytic domain with an N-terminal polyhistidine tag. The His7-tagged TEV protease can be purified in two steps using immobilized metal affinity chromatography (IMAC) followed by gel filtration. An S219V mutation in the protease reduces its rate of autolysis by approximately 100-fold and also yields an enzyme with greater catalytic efficiency than the wild-type protease [8]. The L56V and S135G mutations enhance the stability and solubility of the protease [9]. The presence of a polyhistidine (His7-tag) on the N-terminus of the protease facilitates not only its purification but also its separation from the digestion products of a His-tagged fusion protein in a subtractive IMAC procedure [10]. We also describe a simple and rapid method to test the solubility of proteins after removing their N-terminal fusion tags in a crude cell lysate.

2 Materials

2.1 Overproduction of His7-TEV (L56V, S135G, S219V) Protease in E. coli

1. A glycerol stock of *E. coli* BL21(DE3) CodonPlus-RIL cells containing the TEV protease expression vector pDZ2087 (*see* **Note 1**).
2. LB medium and LB agar plates containing 100 μg/mL ampicillin (for pDZ2087 selection) and 30 μg/mL chloramphenicol (for pRIL selection). LB medium: Add 10 g of Bacto tryptone, 5 g of Bacto yeast extract, and 5 g of NaCl to 1 L of H_2O and sterilize by autoclaving. For LB agar, also add 12 g of bacto agar before autoclaving. To prepare plates, allow medium to cool until flask or bottle can be held in hands without burning, then add 1 mL of ampicillin stock solution (100 mg/mL in H_2O, filter sterilized), mix by gentle swirling, and pour or pipet ca. 30 mL into each sterile petri dish (100 mm dia.).
3. Isopropyl-thio-β-D-galactopyranoside (IPTG), dioxane-free. Prepare a stock solution of 200 mM in H_2O and filter sterilize. Store at −20 °C.
4. Shaker/incubator.
5. Sterile baffled-bottom flasks.

6. A high speed centrifuge (e.g., Sorvall refrigerated centrifuge).
7. A spectrophotometer and cuvette that can measure absorbance at 600 nm.

2.2 Purification of His7-TEV (L56V, S135G, S219V) Protease

1. Cell lysis/IMAC equilibration buffer: 50 mM sodium phosphate (pH 8.0), 200 mM NaCl, 25 mM imidazole. Filter through a 0.22 μm polyethersulfone membrane and store at 4 °C.
2. A mechanical device to disrupt *E. coli* cells (e.g., a sonicator, French press, or cell homogenizer) (*see* **Note 2**).
3. Polyethersulfone filtration unit (0.22 and 0.45 μm).
4. A solution of 5% (w/v) polyetheleneimine, pH 8.0. Mix 50 mL of 50% (w/v) polyethylenimine with H_2O to a volume of 450 mL. Adjust the pH to 8.0 with concentrated HCl, and let cool to room temperature. Adjust the volume to 500 mL with H_2O and check the pH. Adjust if necessary. Filter through a 0.22 μm polyethersulfone filtration unit. The solution is stable for at least 3 years when stored at 4 °C.
5. A spectrophotometer and cuvette that can measure absorbance at 280 nm.
6. ÄKTA Explorer chromatography system or the equivalent.
7. Ni-NTA Superflow resin (Qiagen Incorporated).
8. Column XK 26/20 (Amersham Biosciences).
9. IMAC equilibration buffer: 50 mM sodium phosphate (pH 8.0), 200 mM NaCl, 25 mM imidazole. Filter through a 0.22 μm polyethersulfone membrane and store at 4 °C.
10. IMAC elution buffer: 50 mM sodium phosphate (pH 8.0), 200 mM NaCl, 250 mM imidazole. Filter through a 0.22 μm polyethersulfone membrane and store at 4 °C.
11. 0.5 M ethylenediaminetetraacetic acid (EDTA), pH 8.0 stock solution.
12. 1 M stock solution of 1,4-dithio-dl-threitol (DTT). Prepare 10 mL by mixing 1.55 g of DTT with H_2O to a final volume of 10 mL. Place solution on ice. Use immediately or store at −20 °C.
13. An Amicon Stirred Ultrafiltration Cell concentrator and YM10 ultrafiltration membranes (Millipore Corporation).
14. A HiPrep 26/60 Sephacryl S-100 HR column (GE Healthcare Life Sciences).
15. Gel filtration buffer: 25 mM sodium phosphate (pH 7.5), 100 mM NaCl. Filter through a 0.22 μm polyethersulfone membrane and store at 4 °C.
16. A Dewar flask filled with liquid nitrogen.

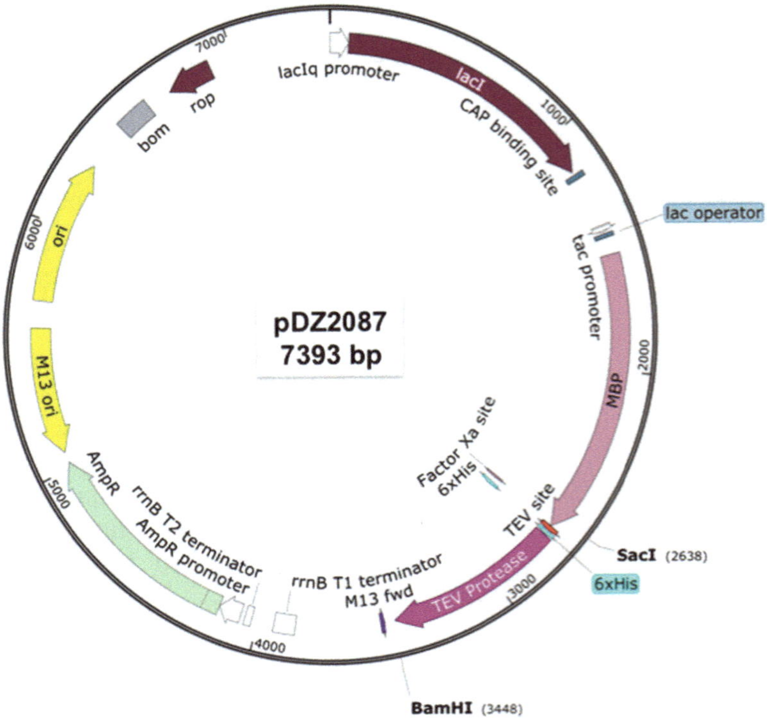

Fig. 1 A schematic map of the modified TEV protease expression vector pDZ2087 that produces a fusion protein product with the configuration MBP-ENLYFQ/G-His_7-TEV. Self-cleavage of the MBP fusion protein by TEV protease generates His7-TEV protease in vivo. (The TEV site is *underlined* and the site of cleavage is marked by a *forward slash* in the text above)

3 Methods

3.1 Overproduction of Soluble His7-TEV Protease in E. coli

The induction of pDZ2087 with IPTG produces an MBP fusion protein (Fig. 1) that self-cleaves in vivo to generate a soluble His7-TEV (L56V, S135G, S219V) protease. Virtually all the protease remains soluble after intracellular processing if the temperature is reduced from 37 to 30 °C after the addition of IPTG.

1. Inoculate 50–150 mL of LB broth containing 100 μg/mL ampicillin and 30 μg/mL chloramphenicol in a 500 mL baffle-bottom shake flask from a glycerol stock of pDZ2087 transformed *E. coli* BL21(DE3) CodonPlus-RIL cells. Place in an incubator and shake overnight at 250 rpm and 37 °C.
2. Add 25 mL of the saturated overnight culture to each 1 L of fresh LB broth containing 100 μg/mL ampicillin, 30 μg/mL chloramphenicol, and 0.2% (w/v) glucose in a 4 L baffle-bottom shake flask. To ensure that there will be an adequate yield of pure protein at the end of the process, we ordinarily grow 4–6 L of cells at a time.

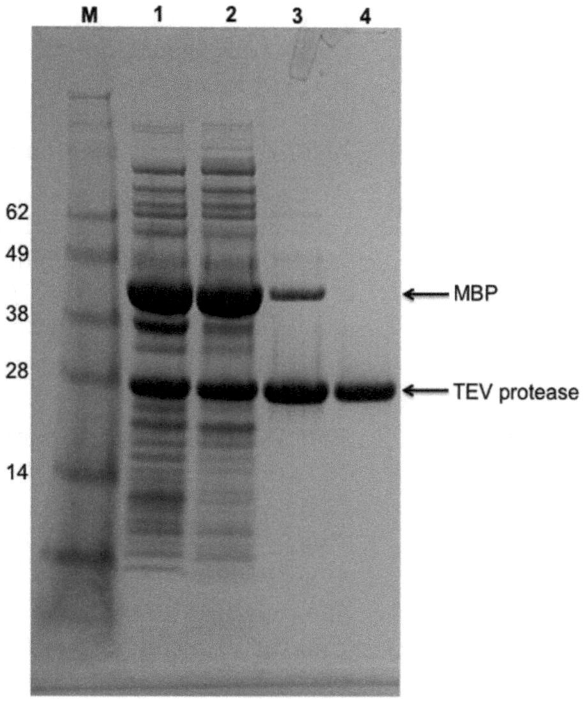

Fig. 2 Purification of His7-TEV (L56V, S135G, S219V) protease monitored by SDS-PAGE (NuPage 4–12% gradient MES gel). *M* molecular weight standards (kDa). *Lane 1*: total intracellular protein after induction. *Lane 2*: soluble cell extract. *Lane 3*: pooled peak fractions after IMAC. *Lane 4*: pooled peak fractions after gel filtration and concentration

3. Shake the flasks at 250 rpm and 37 °C until the cells reach mid-log phase (OD600nm ~ 0.5), approximately 2 h.
4. Shift the temperature to 30 °C and induce the culture(s) with IPTG at a final concentration of 1 mM (5 mL of 200 mM IPTG stock solution per liter of culture). Continue shaking at 250 rpm for 4–5 h.
5. Recover the cells by centrifugation at 5000 × *g* for 10 min at 4 °C, and store at −80 °C. A 6 L preparation typically yields 30–40 g of cell paste.

3.2 His7-TEV Protease Purification

His7-TEV (L56V, S135G, S219V) protease can be purified to homogeneity in two steps: immobilized metal affinity chromatography (IMAC) using Ni-NTA Superflow resin followed by size exclusion chromatography. An example of a purification monitored by SDS-PAGE is shown in Fig. 2 (*see* **Note 3**).

1. All procedures are performed at 4–8 °C. Thaw the cell paste from 6 L of culture on ice and suspend in ice-cold cell lysis/IMAC equilibration buffer (10 mL/g cell paste).

2. Lyse the cell suspension (*see* **Note 2**) and measure the volume using a graduated cylinder. Add polyethylenimine to a final concentration of 0.1% (1:50 dilution of the 5% stock solution at pH 8.0) and mix gently by inversion. Immediately centrifuge at 15,000 × *g* for 30 min.
3. Apply the supernatant to a 25 mL Ni-NTA superflow column equilibrated in cell lysis/IMAC equilibration buffer (*see* **Note 4**). Wash the column with equilibration buffer until a stable baseline is reached (approximately seven column volumes) and then elute the bound His7-TEV (L56V, S135G, S219V) with a linear gradient to 100% elution buffer over ten column volumes.
4. Pool the peak fractions containing the protease and measure the volume. Add EDTA to a final concentration of 2 mM (1:250 dilution of the 0.5 M EDTA, pH 8.0 stock solution), and mix well. Add DTT to a final concentration of 5 mM (1:200 dilution of the 1 M DTT stock solution), and mix well.
5. Concentrate the sample approximately tenfold using an Amicon stirred ultrafiltration cell fitted with a YM10 membrane. Remove any precipitate by centrifugation at 5000 × *g* for 10 min. Estimate the concentration of the partially pure protein solution spectrophotometrically at 280 nm using a molar extinction coefficient of 32,290 M^{-1} cm^{-1}. The desired concentration is between 5 and 10 mg/mL.
6. Apply 5 mL of the concentrated sample onto a HiPrep 26/60 Sephacryl S-100 HR column equilibrated with gel filtration buffer. The volume of sample loaded should be no more than 2% of the column volume and contain no more than 50 mg of protein.
7. Pool the peak fractions from the gel filtration column(s) of pure His7-TEV (L56V, S135G, S219V) protease and concentrate to 1–5 mg/mL (*see* Subheading 3.2, **step 5**). Filter through a 0.2 μm syringe filter, aliquot and flash freeze with liquid nitrogen. Store at −80 °C. The final yield of the purified TEV protease is approximately 7.0 mg per gram of wet *E. coli* cell weight (~250–300 mg from 6 L of cells).

3.3 Cleaving a Fusion Protein Substrate (MBP-NusG-His6) with TEV Protease

The standard reaction buffer for TEV protease is 50 mM Tris–HCl (pH 8.0), 0.5 mM EDTA and 1 mM DTT, but the enzyme has a relatively flat activity profile at pH values between 4 and 9 and will tolerate a range of buffers, including phosphate, MES, and acetate. TEV protease activity is not adversely affected by the addition of glycerol or sorbitol up to at least 40% (w/v). The enzyme is also compatible with some detergents [11]. TEV protease activity is not inhibited by PMSF and AEBSF (1 mM), TLCK (1 mM), Bestatin (1 mg/mL), pepstatin A (1 mM), EDTA (1 mM), E-64 (3 mg/mL), or "complete" protease inhibitor cocktail (Roche).

However, zinc will inhibit the activity of the enzyme at concentrations of 5 mM or greater, and reagents that react with cysteine (e.g., iodoacetamide) are potent inhibitors of TEV protease. The duration of the cleavage reaction is typically overnight. A good rule of thumb is to use 1 OD280 of TEV protease per 100 OD280 of fusion protein for an overnight digest. TEV protease is maximally active at 34 °C [12], but we recommend performing the digest at 4 °C.

1. 100 μg of a partially pure fusion protein (MBP-NusG-His6) with a canonical TEV protease recognition site (ENLYFQG) in the linker region [10] is incubated overnight at 4–8 °C in 50 μL of standard reaction buffer (*see* Subheading 3.3) in the absence or presence of 5.0 μg His7-TEV protease. The reaction products are separated by SDS-PAGE and visualized by staining with Coomassie Brilliant Blue.
2. 10 μg of pure His7-TEV protease is added to 100 μL of soluble crude extract containing the MBP-NusG-His6 fusion protein and incubated at room temperature for 1 h. The reaction products are separated by SDS-PAGE and visualized by staining with Coomassie Brilliant Blue (*see* **Note 5**).

The results of a typical TEV protease digest of a fusion protein substrate (MBP-NusG-His6) are shown in Fig. 3. Panels A and B represent the digestion of a partially purified sample of MBP-NusG-His6 fusion protein and the digestion of the same fusion protein performed in a crude cell lysate, respectively. The MBP-NusG-His6 fusion protein was affinity purified using an IMAC column [7].

3.4 Troubleshooting

Some fusion proteins are intrinsically poor substrates for TEV protease. This may be due to steric occlusion when the protease cleavage site is too close to ordered structure in the passenger protein, or when the fusion protein exists in the form of soluble aggregates. Sometimes this problem can be mitigated by using a large amount of TEV protease (we have occasionally used up to 1 OD280 of TEV protease per 5 OD280 of fusion protein) and/or performing the reaction at higher temperature (e.g., room temperature). Failing that, the addition of extra residues between the TEV protease cleavage site and the N-terminus of the target protein is advised. We have used polyglycine, polyhistidine, and a FLAG-tag epitope with good results.

4 Notes

1. *E. coli* BL21(DE3) CodonPlus-RIL cells containing pDZ2087 can be obtained from our laboratory or from the nonprofit distributor of biological reagents AddGene, Inc., Cambridge,

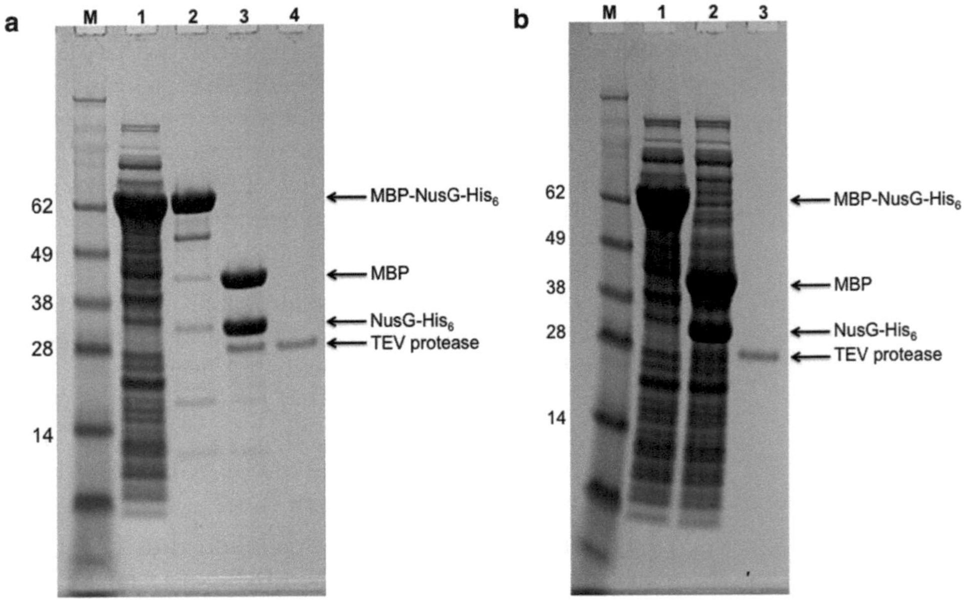

Fig. 3 Digestion of a fusion protein substrate by His7-TEV (L56V, S135G, S219V) protease. (**a**) 100 μg of the pure substrate, a fusion between *E. coli* maltose-binding protein (MBP) and *Aquifex aeolicus* NusG with a canonical TEV protease recognition site (ENLYFQG) in the linker region [10] was incubated overnight at 4 °C in 50 μl of standard reaction buffer (*see* Subheading 3.3) in the absence (*Lane 2*) or presence (*Lane 3*) of 5.0 μg His7-TEV protease. The reaction products were separated by SDS-PAGE (NuPage 4–12% gradient MES gel) and visualized by staining with Coomassie Brilliant Blue. *Lane 4* contains an equivalent amount of pure His7-TEV protease. *Lane 1* is crude MBP-NusG-His6 (soluble protein) before IMAC purification. (**b**) *Lane 1* is MBP-NusG-His6 soluble protein (crude sample); *Lane 2* is TEV protease digest of *lane 1* sample, soluble protein and *Lane 3* contains an equivalent amount of pure His7-TEV protease used in the cleavage reaction (~ 0.1 mg/mL). *M* molecular weight standards (kDa)

MA, USA (http://www.addgene.org) for a nominal shipping and handling fee. The pRIL plasmid is a derivative of the p15A replicon that carries the *E. coli argU*, *ileY*, and *leuW* genes, which encode the cognate tRNAs for AGG/AGA, AUA, and CUA codons, respectively. pRIL is selected for by resistance to chloramphenicol. Due to the presence of several AGG and AGA codons in the TEV protease coding sequence, the presence of pRIL dramatically increases the yield of TEV protease.

2. We routinely break cells using a APV-1000 homogenizer (Invensys, Roholmsvej, Germany) at 10–11,000 psi for 2–3 rounds. Other homogenization techniques such as French press, sonication, or manual shearing should yield comparable results. Centrifugation of the disrupted cell suspension for at least 30 min at 30,000 × *g* is recommended. Filtration through a 0.45 μm polyethersulfone (or cellulose acetate) membrane

is helpful to remove residual particulates and fines prior to chromatography.

3. We find it convenient to use precast gels for SDS-PAGE gels (e.g., 1.0 mm × ten well, 10–20% Tris–glycine gradient).
4. We use an ÄKTA Explorer chromatography system and Ni-NTA Superflow resin. A properly poured 25 mL Ni-NTA Superflow column (in an Amersham Biosciences XK26/20 column) can be run at 4–6 mL/min (backpressure less than 0.4 MPa) and will bind up to 200 mg of His7-TEV (L56V, S135G, S219V) protease. If a chromatography system is not available, the IMAC can be performed using a peristaltic pump or manually by gravity. If the latter is used, Ni-NTA agarose should be substituted for Superflow and the elution performed with step increases of imidazole in 25 mM increments. Binding and elution profiles can be monitored spectrophotometrically at 280 nm and by SDS-PAGE. Care must be taken to properly zero the spectrophotometer because imidazole has significant absorption in the UV range.
5. Digestion of a fusion protein by adding TEV protease to a crude cell lysate is a useful way to gauge the efficiency of processing and determine whether or not the passenger protein will remain soluble after it is cleaved from the affinity tag before any chromatography steps are performed. Samples of the TEV digest are compared by SDS-PAGE before and after centrifugation.

Acknowledgments

This research was supported by the Intramural Research Program of the NIH, National Cancer Institute, Center for Cancer Research.

References

1. Waugh DS (2005) Making the most of affinity tags. Trends Biotechnol 23:316–320
2. Goel A, Beresford GW, Colcher D et al (2000) Relative position of the hexahistidine tag effects binding properties of a tumor-associated single-chain Fv construct. Biochim Biophys Acta 1523:13–20
3. Bucher MH, Evdokimov AG, Waugh DS (2002) Differential effects of short affinity tags on the crystallization of *Pyrococcus furiosus* maltodextrin-binding protein. Acta Crystallogr D Biol Crystallogr 58:392–397
4. Fonda I, Kenig M, Gaberc-Porekar V et al (2002) Attachment of histidine tags to recombinant tumor necrosis factor-alpha drastically changes its properties. Scientific World J 2: 1312–1325
5. Chant A, Kraemer-Pecore CM, Watkin R et al (2005) Attachment of histidine tag to gene regulatory protein AreA causes a conformational change at the DNA-binding site. Protein Expr Purif 39:152–159
6. Parks TD, Leuther KK, Howard ED et al (1994) Release of proteins and peptides from fusion proteins using a recombinant plant virus protease. Anal Biochem 216:413–417
7. Kapust RB, Tozser J, Copeland TD et al (2002) The P1′ specificity of tobacco etch virus protease. Biochem Biophys Res Commun 294: 949–955

8. Kapust RB, Tozser J, Fox JD et al (2001) Tobacco etch virus protease: mechanism of autolysis and rational design of stable mutants with wild-type catalytic efficiency. Protein Eng 14:993–1000
9. Cabrita LD, Gilis D, Robertson AL et al (2007) Enhancing the stability and solubility of TEV protease using in silico design. Protein Sci 16:2360–2367
10. Tropea JE, Cherry S, Nallamsetty S et al (2007) A generic method for the production of recombinant proteins in *Escherichia coli* using a dual hexahistidine-maltose-binding protein affinity tag. Methods Mol Biol 363:1–19
11. Mohanty AK, Simmons CR, Wiener MC (2003) Inhibition of tobacco etch virus protease activity by detergents. Protein Expr Purif 27:109–114
12. Nallamsetty S, Kapust RB, Tozser J et al (2004) Efficient site-specific cleavage of fusion proteins by tobacco vein mottling virus protease in vivo and in vitro. Protein Expr Purif 38:108–115

Part III

Case Studies to Produce Challenging Proteins and Specific Protein Families

Chapter 15

Generation of Recombinant *N*-Linked Glycoproteins in *E. coli*

Benjamin Strutton, Stephen R.P. Jaffé, Jagroop Pandhal, and Phillip C. Wright

Abstract

The production of *N*-linked recombinant glycoproteins is possible in a variety of biotechnology host cells, and more recently in the bacterial workhorse, *Escherichia coli*. This methods chapter will outline the components and procedures needed to produce *N*-linked glycoproteins in *E. coli*, utilizing *Campylobacter jejuni* glycosylation machinery, although other related genes can be used with minimal tweaks to this methodology. To ensure a successful outcome, various methods will be highlighted that can confirm glycoprotein production to a high degree of confidence, including the gold standard of mass spectrometry analysis.

Key words Glycosylation, *N*-Linked glycoproteins, Posttranslational modifications, *E. coli*, Glycoprotein validation

1 Introduction

The process of glycosylation, a posttranslational modification estimated to be present on half of all human proteins [1, 2], can affect a wide variety of protein properties including its function, localization, and half-life [3–5]. Due to these properties, 40% of the drugs approved in the current recombinant therapeutics market are glycosylated, with this figure predicted to increase [6]. This sizeable market share means that glycoproteins contribute significantly to the biopharma industry, which in 2013 was calculated to be worth $140 billion USD [7].

Of the approved glycotherapeutics, approximately 70% are currently being produced in the eukaryotic cell line, Chinese hamster ovary (CHO) cells [8], which, through the utilization of their inherent glycosylation machinery, can mimic human type glycans, yielding therapeutic proteins with the desired glycosylation profile that will not induce an immunogenic response in humans [9].

Nicola A. Burgess-Brown (ed.), *Heterologous Gene Expression in E. coli: Methods and Protocols*, Methods in Molecular Biology, vol. 1586, DOI 10.1007/978-1-4939-6887-9_15, © Springer Science+Business Media LLC 2017

In the non-glycosylated therapeutic protein market, the main host cell factory is *E. coli,* which generates around 30% of the 151 recombinant therapeutics approved by either the EMA or the FDA [10]. With recent advancements in the synthetic biology toolkit for *E. coli,* this host now has the ability to express full-length monoclonal antibodies [11, 12], and generate recombinant *O*- and *N*-linked glycoproteins [13–17]. These new capabilities present the opportunity to generate recombinant glycoproteins within *E. coli.*

The initial discovery of bacterial *N*-linked glycosylation in *C. jejuni* and its successful transfer into *E. coli* [16, 18] opened up this area of research. As the field has advanced and accumulated knowledge on the bacterial glycosylation system, variations on the initial system containing the defined pgl pathway allowed the first steps in creating a model cell chassis. This has provided the opportunity for researchers to experiment with different genetic components within the system, replacing them with alternative parts from a variety of organisms, including eukaryotes [14, 19]. This led to the creation of a wide range of different glycan structures that could be attached [14, 16, 19], with a commercially successful example of this coming from the generation of glyco vaccines by GlycoVaxyn [20, 21].

With an ever-expanding number of glycan structures still being discovered in the archaeal and prokaryotic domains of life, the potential to transfer them to target proteins for currently unexplored uses is a promising prospect for the field [22, 23].

Although the option to engineer this system is apparent, in order to be able to carry out bacterial *N*-linked glycosylation, there are a few core components required. First, the genes required for sugar biosynthesis within the cytoplasm must be functional [24]. Secondly, the glycosyltransferases that transfer these monosaccharides and sequentially build the glycan on the lipid anchor must be present [25]. An enzyme to flip the anchored glycan across to the periplasm needs to be expressed [26, 27], and finally, an oligosaccharyl transferase (OSTase) that transfers the glycans to the target protein must also be functional within the system [28]. The structure and composition of the glycan, how efficiently it is flipped across the membrane and ultimately transferred to the target protein, are dependent on the genes introduced [14, 29, 30]. Here, we will outline the required components for producing glycoproteins in *E. coli* based upon the original *C. jejuni* system.

1.1 Oligosaccharyl Transferase and the Consensus Sequence That It Can Recognize Within the Target Protein

In eukaryotes, *N*-linked glycosylation occurs at the consensus site of N-X-S/T, where X can be any amino acid except proline. The recognition of this site and subsequent transfer of glycan onto the asparagine residue is dependent on a multi-subunit (OSTase) complex with a core functional unit known as STT3 [31]. In bacteria, this OST is a large single protein, with the most commonly utilized transferase being a periplasm located, membrane-bound protein called pglB [32–34]. The native form of this protein recognizes a stricter glycosylation sequon with the requirement of a negatively charged amino acid at the −2 position, giving the consensus sequence, D/E-X-N-X-S/T, again with X being any amino acid except proline. When thinking about utilizing this system for glycoprotein production, it is vital that this sequence is present, unless the machinery has been modified to include an OSTase that recognizes the eukaryotic glycosylation sequence [30]. Even so, this may not be sufficient for glycosylation to occur. Within eukaryotes, the process takes place in multiple steps as the protein is folded through two eukaryotic organelles, the ER and golgi [35, 36], making it a cotranslational process [37]. Within these organelles, the glycan structure can be built up and subsequently trimmed down by processing enzymes before final glycans are added to produce the mature glycan structure [38]. In bacteria, glycosylation typically takes place in the periplasm of the cell [22], and is seen as a single step block transfer of the final glycan to the target protein [39], which occurs posttranslationally. Due to this occurring on a fully folded substrate, it requires the consensus sequence to be situated in a flexible region of the protein that is accessible to the OSTase [39].

1.2 Periplasmic Localization

As mentioned in the previous section, attachment of the glycan onto the protein within bacteria occurs in the periplasm. Therefore, the target protein, glycan, and OSTase need to be localized to this compartment of the cell. PglB, as highlighted previously, is a membrane-bound protein located in the periplasm, so is already present. As for the target protein, there are multiple pathways available that will direct the target protein to this part of the cell, including the TAT export system [40], SRP pathway [41], and sec transport system [42]. The most utilized methodology is the sec transport system, which requires the addition of a 22 amino acid leader sequence at the N-terminus of the polypeptide chain. This can be engineered through molecular cloning of the expression plasmid, but it is recommended that protein expression and translocation rates be attenuated prior to glycoprotein production. Finally, the glycan of interest must be present in the periplasm. Within the bacterial system, the glycans are sequentially built upon an undecaprenyl-pyrophosphate (UND-pp) lipid anchor by various glycosyltransferases. Depending on the glycan structure, certain sugar biosynthesis genes must also be expressed that utilize molecules from the central carbon metabolism, modifying them to generate any unnatural glycan precursors such as bacillosamine, as well as the common monosaccharides like UDP-GlcNAc [43]. Once these have been generated, the glycosyltransferases transfer them to the glycan being built upon the lipid anchor. Once the glycans are attached, the UND-pp. linked glycan molecule can be flipped across the inner membrane by a flippase, such as the

C. jejuni pglK [34], or the native *E. coli* protein wzX [44] so the glycan now faces the periplasm [19]. With wzX and waaL, *E. coli* has a native system that utilizes periplasmic UND-pp. linked glycans for attachment to a lipid A core, for subsequent presentation on the cell surface [15]. It is therefore recommended that for recombinant *N*-linked glycoprotein expression, the competing protein that transfers the glycans to the lipid A core, waaL, is removed as seen in the W3110 mutant, CLM24 [15]. The presence of waaL may be utilized to detect the presentation of glycans on the surface as a means of checking whether they are being expressed (or not) [32, 45].

1.3 Conclusion of Requirements

The target recombinant protein must contain an asparagine residue situated in a consensus sequence, located in a flexible region of the protein, so that the OSTase of choice is able to recognize and bind. Once expressed, it must also be directed to the periplasm for this type of glycosylation to occur, which can be achieved by utilizing a number of export systems. Alongside expression of the target protein, glycosylation machinery must be incorporated into the cell, containing any genes needed for specific sugar biosynthesis, the required glycosyltransferases to build the glycan, a flippase that can recognize the glycan as a substrate, and an OSTase that will also recognize the glycan. This should all preferably take place in a bacterial strain where the waaL pathway is inhibited, unless glycan production is being checked by cell surface display.

1.4 General Analysis

In order to determine if the target protein has been successfully glycosylated, it is necessary to perform a periplasmic protein extraction, to release proteins that have the potential to be glycosylated. In order to determine whether *N*-glycosylation of the target protein is successful, a number of methodologies can be applied with varying degrees of speed and accuracy. These include western blots, where a mass shift for the addition of a glycan is observed [32], lectin peroxidase screen whereby a lectin that binds to a target glycan is bound [32] or the gold standard is the use of tandem mass spectrometry, which can provide both protein sequence and glycan structure information [33]. By following the methodology and workflow stated (*see* Fig. 1), the user should be able to generate *N*-linked glycoproteins in *E. coli* and validate this production using a variety of techniques. For more details on bacterial glycosylation, please refer to this extensive review paper [46].

2 Materials

All solutions should be prepared using either nuclease-free water or HPLC grade water along with analytical grade reagents.

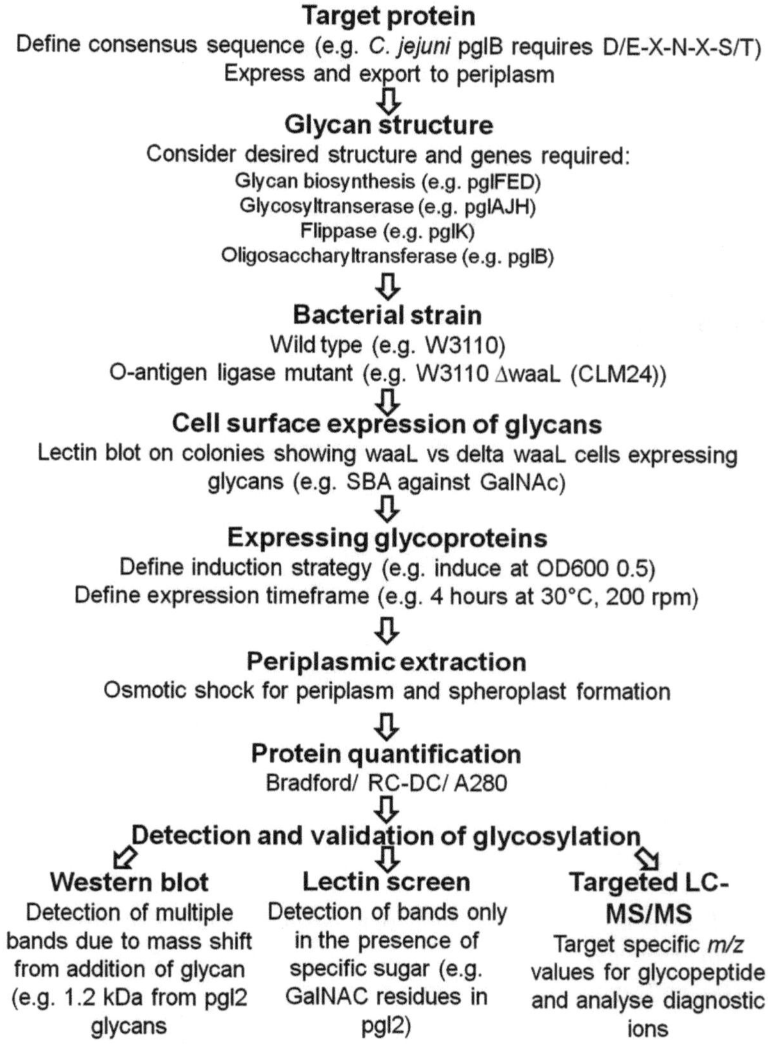

Fig. 1 A workflow showing all the key points required in designing, generating, and validating the expression of recombinant *N*-linked glycoproteins in *E. coli*

All solutions should be made up at room temperature unless otherwise stated. All examples and methodologies listed are for the generation of IFNα2b with an *N*-linked glycan structure of $GalNAc_5GlcNAC$ within *E. coli*, utilizing the plasmids pJExpress*IFNα2b* and pACYC*pgl2*.

2.1 General and Experiment Specific Reagents, Strains, and Plasmids

2.1.1 Antibiotic Stock Solutions

1. 50 mg/mL Kanamycin: Prepare in water and filter sterilize.
2. 30 mg/mL Chloramphenicol in 100% ethanol.

2.1.2 Bacterial Strains

1. *E. coli* W3110.
2. *E. coli* CLM24.

2.1.3 Glycosylation Machinery

1. pACYC*pgl2*.

2.1.4 Target Protein

1. pJexpress*IFNα2b*.

2.2 Cell Surface Expression of Glycans

1. Luria Broth: 10 g/L tryptone, 10 g/L NaCl 5 g/L yeast extract.
2. Petri dishes.
3. Protran Nitrocellulose membranes (Fisher Scientific).
4. PBS containing 2% (v/v) TWEEN® 20.
5. PBS with 0.05% (v/v) TWEEN® 20, 1 mM $CaCl_2$, 1 mM $MnCl_2$, 1 mM $MgCl_2$, and 3 μg of soybean agglutinin lectin peroxidase.
6. Immobilon™ chemiluminescent HRP (Fisher Scientific).
7. ImageQuant™ RT ECL (GE Healthcare) fitted with temperature-cooled 16-bit CCD Camera.
8. 1-Step TMB Ultra blotting solution (Thermo Fisher Scientific).

2.3 Target Protein Expression and Localization

1. Luria Broth: 10 g/L tryptone, 10 g/L NaCl 5 g/L yeast extract.
2. 1 mM IPTG.

2.4 Periplasmic Protein Extraction and Quantification

2.4.1 Extraction

1. Periplasmic lysis buffer composed of 20% (w/v) sucrose, 30 mM Tris–HCl, pH 8.0, and 1× HALT (Pierce) in HPLC grade water.
2. Centrifuge capable of spinning 50 mL falcon tubes at 3000 × *g*, 4 °C.

2.4.2 Quantification

1. Bradford reagent (Sigma Aldrich).
2. RC DC protein assay kit II (Bio-Rad).

2.5 SDS-PAGE Gels and Analysis

2.5.1 SDS-PAGE

1. NuPAGE Novex 4–12% Bis-Tris protein gels, 1.0 mm, 12 well (Life Technologies).
2. 20× Novex MOPS SDS running buffer (Life Technologies).
3. Novex Sharp Prestained protein standard (Life Technologies).
4. 10× NuPAGE® Sample Reducing Agent (Thermo Fisher Scientific).
5. Hoefer SE300 miniVE integrated vertical electrophoresis unit (Hoefer Inc).
6. Centrifuge capable of spinning 1.5 mL microcentrifuge tubes at 13,000 × *g*.

2.5.2 Western Blotting

1. iBlot™ Gel Transfer Device (Thermo Fisher Scientific).
2. iBlot™ Gel Transfer Stacks (Thermo Fisher Scientific).
3. Blocking Buffer consisting of: 5% (w/v) blocking powder (Bio-Rad) in TBS 0.05% (v/v) Tween® 20.
4. 6× Histidine HRP linked antibody (Abcam).
5. 1-Step TMB Ultra blotting solution (Thermo Fisher scientific).

2.5.3 Lectin Screen

1. iBlot™ Gel Transfer Device (Thermo Fisher Scientific).
2. iBlot™ Gel Transfer Stacks (Thermo Fisher Scientific).
3. Protran Nitrocellulose membranes from (Thermo Fisher Scientific).
4. PBS containing 2% (v/v) TWEEN® 20 (Sigma Aldrich).
5. PBS with 0.05% (v/v) TWEEN® 20, 1 mM $CaCl_2$, 1 mM $MnCl_2$, 1 mM $MgCl_2$, and 3 μg of soybean agglutinin lectin peroxidase.
6. Immobilon™ chemiluminescent HRP (Fisher Scientific).
7. ImageQuant™ RT ECL (GE Healthcare) fitted with temperature cooled 16-bit CCD Camera.
8. 1-Step TMB Ultra blotting solution (Thermo Fisher scientific).

2.6 Mass Spectrometry Analysis

2.6.1 In-Solution Digest

1. 100 mM ammonium bicarbonate made in HPLC grade water.
2. Protein Lobind tubes (Eppendorf).
3. 50 mM DTT stock solution.
4. 100 mM Iodoacetamide stock solution.
5. Trypsin solution made using lyophilized trypsin protease, MS grade (Pierce).
6. 94.5% (v/v) HPLC grade H_2O, 5% (v/v) acetonitrile, 0.5% (v/v) TFA.
7. Water bath.
8. Benchtop microcentrifuge for 1.5 mL microcentrifuge tubes.

2.6.2 C18 Cleanup

1. Pierce™ C18 Spin Columns (Thermo Fisher Scientific).
2. Protein Lobind tubes (Eppendorf).
3. Centrifuge capable of spinning down 1.5 mL microcentrifuge tubes at 13,000 × *g*.
4. Vacuum concentrator centrifuge (Eppendorf).

2.6.3 LC-MS

1. Sonicating water bath.
2. Vortex mixer.
3. Centrifuge capable of spinning down 1.5 mL microcentrifuge tubes at 13,000 × *g*.

4. Automated LC vials.
5. Automated LC vial caps.
6. maXis Q-ToF mass spectrometer (Bruker Daltonics).
7. Ultimate 3000 HPLC system (Dionex).
8. HPLC buffer A consisting of: 96.9% HPLC grade water, 3% Acetonitrile, 0.1% Formic acid.
9. HPLC buffer B consisting of: 3% HPLC grade water, 96.9% Acetonitrile, 0.1% Formic acid.

3 Methods

All examples and methodologies listed are for the generation of IFNα2b with an *N*-linked glycan structure of $GalNAc_5GlcNAC$ within *E. coli*, utilizing the plasmids pJExpress*IFNα2b* and pACYC*pgl2*.

3.1 Cell Surface Expression of Glycans

If checking glycan production, use bacterial strain W3110. If checking waaL deletion, use your modified strain or in our case CLM24 (*see* Figs. 2 and 3).

1. Transform the desired *E. coli* strain with pACYC*pgl2*.
2. Inoculate 1 mL of LB, with the appropriate antibiotics, in a sterile 1.5 mL centrifuge tube with a colony of the bacterial strain containing pACYC*pgl2*.
3. Incubate the tube at 37 °C for 16 h in a shaker at 180 rpm.
4. Measure the O.D. 600 and normalize the O.D. down to 0.6 using sterile LB.

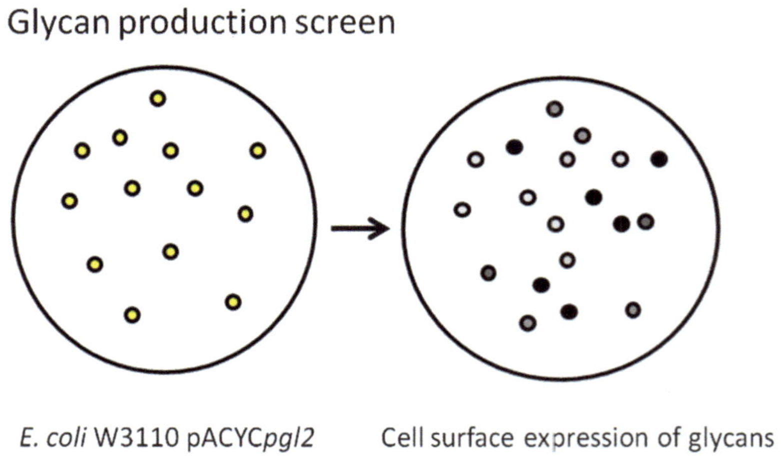

Fig. 2 Diagrammatic workflow outlining the strains for the glycan production screen

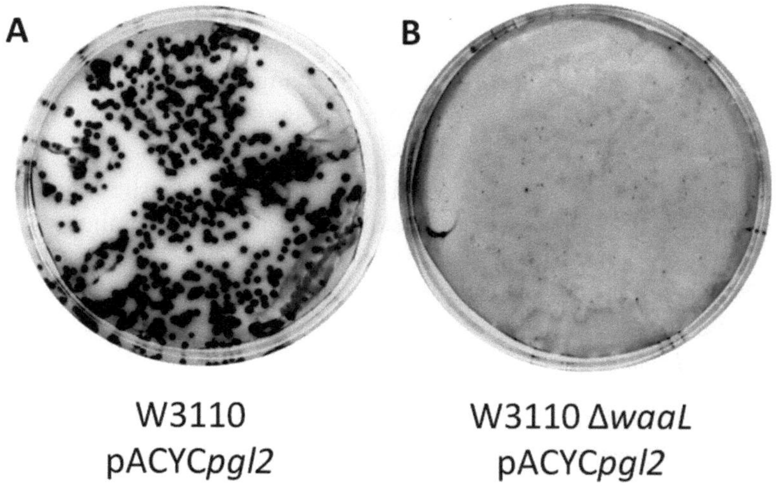

Fig. 3 Cell surface expression of glycans and confirmation of the production of a *waaL* knockout strain. (**a**) Cell surface screening with a functional *waaL* pathway showing that functional glycan machinery is being expressed. (**b**) Cell surface screening to show that glycans produced are not being exported, confirming the inhibition of the *waaL* pathway

5. Dilute the cells by a factor of 1 in 75,000 to a final volume of 1 mL.
6. Take 100 μL of the diluted culture and streak out onto an LB agar plate with correct antibiotics and leave to incubate for 16 hours at 37 °C (*see* **Notes 1** and **2**).
7. Soak a piece of nitrocellulose paper, cut to fit inside a petri dish, for 5 min in the required antibiotics for the maintenance of the plasmid in the strain.
8. Place the nitrocellulose paper in a flow hood for 5 min before placing the paper over the bacterial colonies using a clean pair of laboratory tweezers, and incubating for 3 h at 37 °C (*see* **Note 3**).
9. Remove the nitrocellulose paper from the agar plate and block the membrane in PBS containing 2% (v/v) TWEEN® 20 for 2 min at 20 °C.
10. Wash the membrane twice (10 min each) in PBS.
11. Incubate the membrane in PBS with 0.05% (v/v) TWEEN® 20, 1 mM $CaCl_2$, 1 mM $MnCl_2$, 1 mM $MgCl_2$, and 3 μg of soybean agglutinin lectin peroxidase (specific for GalNAc) for 16 h at 20 °C (*see* **Note 4**).
12. Wash the membrane for 2 × 10 min in PBS.
13. Detect colonies using Immobilon™ chemiluminescent HRP substrate with ImageQuant™ RT ECL (GE Healthcare), fitted with a cooled 16-bit CCD camera.

14. If this type of camera is not available, the screen can be developed using approximately 10 mL TMB-Ultra Blotting solution and leaving the membrane to develop for between 5 and 30 min depending on the desired intensity and the level of background development.
15. Wash with 2 × 5 min washes with HPLC grade water to stop development of the membrane (*see* **Note 5**).

3.2 Target Protein Expression and Localization

Required to check protein of interest is being expressed and translocated to the periplasm (*see* Fig. 4).

3.2.1 Bacterial Growth and Protein Expression

1. Inoculate 10 mL of LB containing the appropriate antibiotics with a colony of CLM24 pJexpress*IFNα2b* and grow for 16 h at 37 °C with shaking at 180 rpm.
2. Using the starter culture, inoculate 100 mL of LB containing the appropriate antibiotics and leave to grow at 37 °C with shaking at 180 rpm.
3. When the O.D. 600 of the culture reaches 0.5, induce target protein expression with 1 mM IPTG.
4. Leave the bacteria to express the protein of interest for 4 h at 30 °C with shaking at 180 rpm.
5. Measure the final O.D. 600 of the culture, collecting 40 O.D.'s worth of the culture by spinning at 3000 × *g* for 10 min at 4 °C.
6. Discard supernatant and continue to periplasmic protein extraction (*see* **Note 6**).

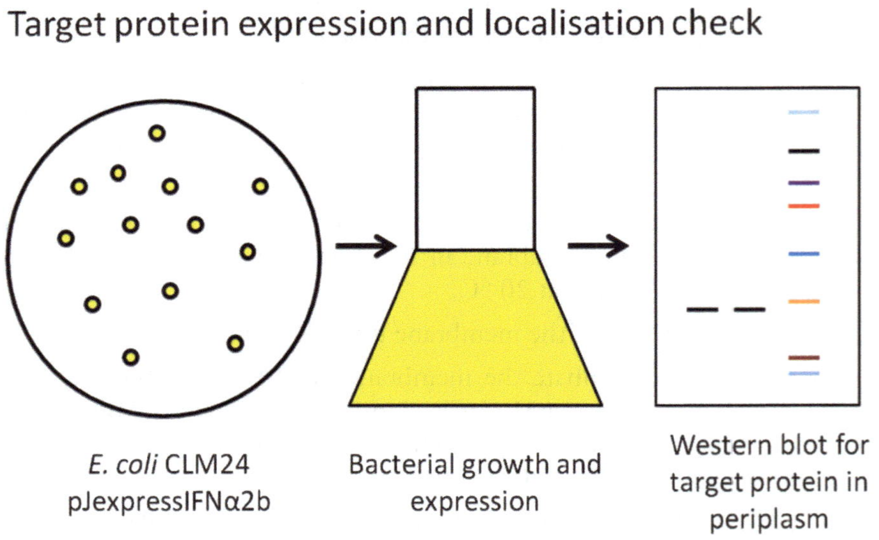

Fig. 4 Diagrammatic workflow outlining the process to check the target protein expression and localization

3.2.2 Periplasmic Protein Extraction

1. Gently resuspend the bacterial pellet in 1 mL of ice-cold periplasmic lysis buffer, keeping the suspension in a 50 mL Falcon tube.
2. Leave the pellet to gently roll on ice for 1.5–2 h.
3. Spin down the cell debris at 3000 × *g* for 10 min at 4 °C, collecting the supernatant as the periplasmic extract (*see* **Note 7**).
4. Pellet can be retained and the rest of the soluble fraction extracted to analyze the quantity of the target protein that has not been exported to the periplasm but has still been expressed.

3.2.3 Quantification of the Periplasmic Extract

Many common laboratory methods can be used here including a Bradford assay, RC/DC assay, or nanodrop. Due to the frequent nature of these protocols, we advise that the researcher follows the detailed protocols that are widely available for these quantification techniques.

3.2.4 SDS-PAGE

1. Precast protein gels are used to run the gel-based analysis.
2. In a single LoBind 1.5 mL centrifuge tube, add the required volume of the periplasmic extract to place 5 μg of protein into the tube. Along with the sample add, 5 μL of 4× LDS sample buffer, as well as 2 μL of 10× reducing agent. Using distilled water, make the total volume up to 24 μL.
3. Aliquot 10 μL of prestained protein standard into a fresh LoBind tube.
4. Boil the samples and protein ladder for 10 min at 70 °C.
5. Leave the samples to cool before spinning all tubes down at 13,000 × *g* for 1 min.
6. Load the gel into the gel tank and fill the internal chamber with fresh 1× MOPS SDS running buffer (*see* **Note 8**). Fill the rest of the tank with 1× MOPS SDS running buffer. Remove the plastic comb from the top of the gel and discard (*see* **Note 9**).
7. Load protein ladder and samples onto the protein gel (*see* **Note 10**).
8. Place the lid on the top of the gel tank and run the gel at a constant 200 V for 50 min (*see* **Note 11**).
9. Following the running of the gel, remove the gel tank lid, remove the plastic encased gel, and carefully open along the seam at the edge of the casing. Remove the gel and place into a clean plastic container (*see* **Note 12**). Discard the used plastic casing.

3.2.5 Western Blotting

1. Western blotting is conducted using the Invitrogen iBlot system and compatible transfer stacks.
2. Briefly wash the acrylamide gel with 2× MilliQ grade water to remove any excess SDS

3. Following the manufacturer's protocol place the protein gel onto the membrane and complete the stack setup (*see* **Note 13**).
4. Fasten lid and transfer the proteins onto the nitrocellulose membrane using the P3 transfer program for a run time of 7 min.
5. Once run, discard all of the excess stack, including the gel, and using tweezers, carefully remove the membrane, and submerge it in sufficient blocking buffer and leave rocking in this solution for 1 h at room temperature (*see* **Note 14**).
6. Wash the membrane with 3× washes in TBS-T (0.05% (v/v) Tween 20) for 10 min each at room temperature, discarding the wash solution after each repeat.
7. During the second wash, run a 1 in 10,000 dilution of the 6× Histidine residue antibody in 20 mL blocking buffer and place on the rocker during the rest of the wash steps.
8. Remove the final wash solution and cover the membrane in the antibody solution.
9. Leave this to incubate at 4 °C for 16 h, preferably with gentle shaking.
10. Discard the antibody solution and conduct 5 × 5 min wash steps with TBS-T (0.05% (v/v) Tween 20).
11. Post washing, cover the membrane with approximately 10 mL of TMB-Ultra Blotting solution, leaving the membrane to develop for between 5 and 30 min, depending on the desired intensity and the level of background development.
12. Wash with 2 × 5 min washes with HPLC grade water to stop development of the membrane.

3.3 Expression of Glycoprotein

See Subheading 3.2.1 as the protocol for this is the same needed for glycoprotein expression, except the starting bacterial strain must contain the plasmids required for both the target protein expression and the glycosylation machinery. Due to the presence of two plasmids, an extra antibiotic will be required for plasmid maintenance. If a mass spectrometry approach is going to be utilized for glycoprotein validation, it is important that when extracting the periplasmic proteins (Subheading 3.2.2), no protease inhibitor cocktail is present in the lysis buffer mentioned earlier and that the sample is kept cold at all times (*see* **Note 15**).

3.4 Validation of Glycoprotein Production Using Gel-Based Methods

3.4.1 Western Blot Analysis

Expecting to see multiple bands compared to the control due to the mass shift that occurs with the attached glycan (*see* Figs. 5 and 6). Run subsequent quantification steps (Subheading 3.2.3), SDS-PAGE (Subheading 3.2.4), and Western blot (Subheading 3.2.5) analysis as outlined above as well as a lectin screen specific to the sugars incorporated in the glycan.

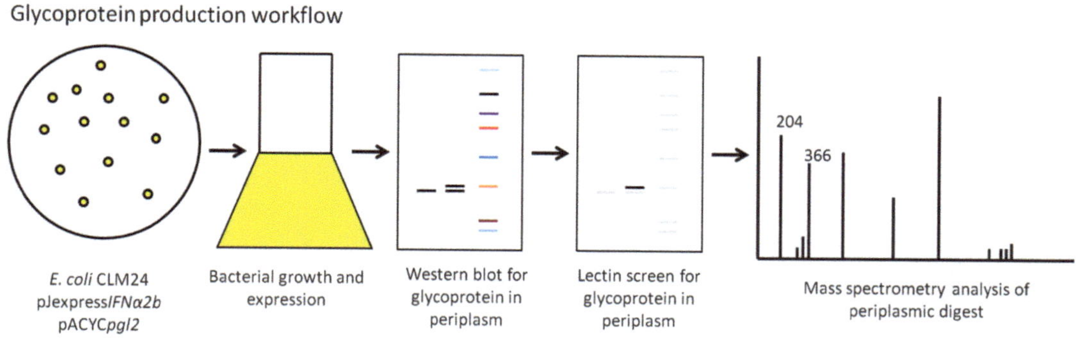

Fig. 5 Diagrammatic workflow outlining the process of glycoprotein production and validation

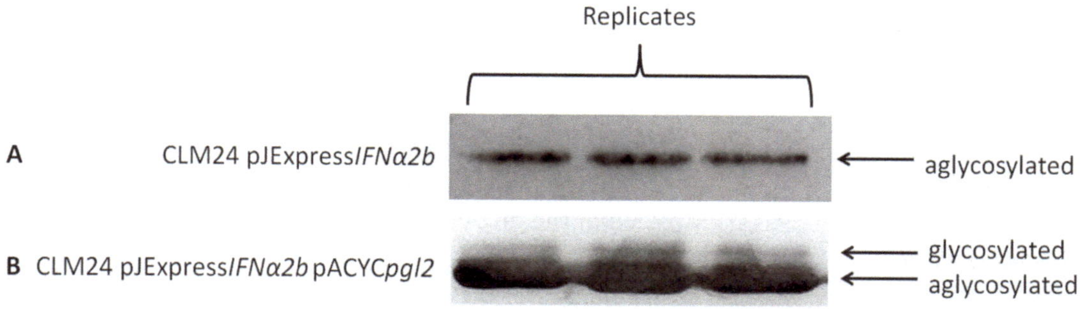

Fig. 6 Validation of glycoprotein production using Western Blotting. (**a**) Expression of the target protein without the presence of the glycosylation machinery. (**b**) Expression of the target protein with the glycosylation machinery producing the multiple glycoforms

3.4.2 Lectin Screen

Sugar-specific lectin should bind to the glycans if present.

1. Follow the same protocol as the western blotting up to **step 5** with dismemberment of the blotting stack but instead of placing the membrane in the western blocking buffer, the membrane must be washed with PBS containing 2% (v/v) Tween 20 for 2 min at room temperature.
2. Post blocking, wash the membrane twice with PBS for 10 min at room temperature.
3. Incubate the membrane for 16 h at room temperature on a shaker in PBS with 0.05% (v/v) TWEEN® 20, 1 mM $CaCl_2$, 1 mM $MnCl_2$, 1 mM $MgCl_2$, and 3 μg of soybean agglutinin lectin peroxidase (specific for GalNAc).
4. Repeat the wash steps conducted in **step 2** before covering the membrane with approximately 10 mL of TMB-Ultra Blotting solution, leaving the membrane to develop for between 5 and 30 min depending on the desired intensity and the level of background development.
5. Wash with 2 × 5 min washes with HPLC grade water to stop development of the membrane.

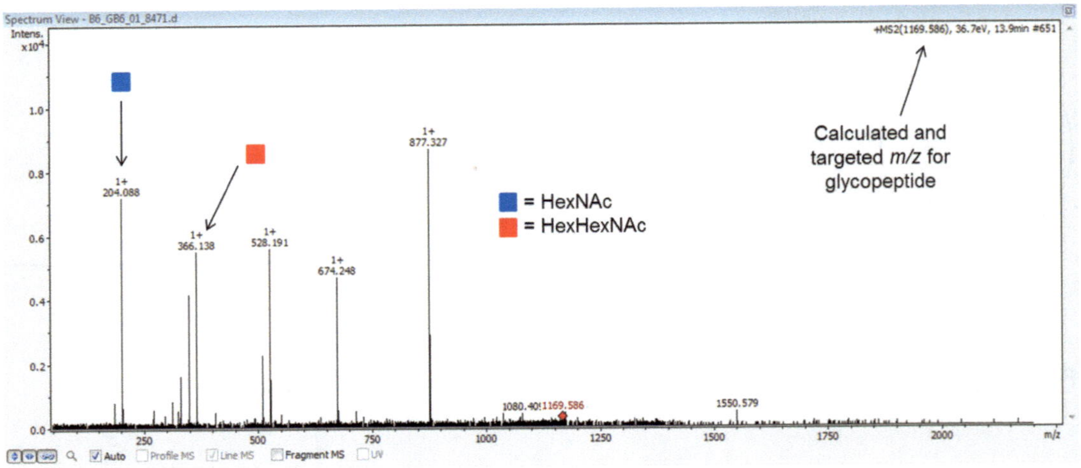

Fig. 7 Mass spectrometry validation of glycoprotein production. Spectra of a tryptically digested Interferon α2b glycopeptide with the diagnostic oxonium ions of 204 and 366 highlighted

3.5 Validation of Glycoprotein Production Using Mass Spectrometry Analysis

Expecting to see the presence of diagnostic ions correlating to the sugars involved in the glycan (*see* Figs. 5 and 7).

3.5.1 In-Solution Digest

1. In a single LoBind 1.5 mL centrifuge tube, add the required volume of the periplasmic extract to place 5 μg of protein into the tube.
2. Make up the total volume to 40 μL using 100 mM ammonium bicarbonate.
3. Add a sufficient quantity of 50 mM DTT to achieve a final DTT concentration of 4 mM in the final solution.
4. Incubate for 60 min in a 56 °C water bath.
5. Gently spin down at 1000 × *g* for 1 min.
6. Add a sufficient quantity of 100 mM Iodoacetamide to achieve a final Iodoacetamide concentration of 8 mM in the final solution.
7. Incubate in the dark at room temperature for 30 min.
8. Gently spin down at 1000 × *g* for 1 min.
9. Add the protease at a ratio of 1:25 (protease:protein).
10. Leave digest for 18 h at 37 °C.
11. Dry down the samples in a vacuum concentrator centrifuge (*see* **Note 16**).

12. Resuspend samples in 94.5% (v/v) HPLC grade water, 5% (v/v) ACN, and 0.5% (v/v) TFA.

3.5.2 C18 Cleanup

Prior to mass spectrometry analysis, the sample should undergo a C18 clean-up procedure. Due to the frequency of this technique and the wide range of columns available, it is recommended that the researcher follows the protocol that is supplied by the manufacturer. A recommended column is given in Subheading 2.6.2.

3.5.3 LC-MS

1. Add 10–20 μL of 94.5% (v/v) HPLC grade water, 5% (v/v) ACN, and 0.5% (v/v) TFA to the dried peptides post C18 cleanup.
2. Sonicate and vortex the samples for a minute each and then centrifuge the tubes at 13,000 × *g* for 30 s to pool the liquid at the bottom of the LoBind centrifuge tube.
3. Transfer liquid into a vial compatible with the automated LC system (*see* **Note 17**).

Using a maXis Q-TOF mass spectrometer (Bruker), perform high selectivity pseudo-selective reaction monitoring (pSRM) of the targeted glycopeptides.

3.5.4 Pseudo-Selective Reaction Monitoring (pSRM)

1. Select the correct *m/z* values for the desired glycopeptides (*see* **Note 18**) and operate the Q-TOF in MRM mode with an *m/z* window of 2 *m/z*.
2. Run the mass spectrometer in positive mode with an *m/z* window of 3.
3. Implement a rolling collision energy of 100–150% with an MRM scan mode.
4. Start online HPLC with 5% of HPLC buffer B and 95% of HPLC buffer A and run for 5 min.
5. Run a 30 min gradient program, increasing the concentration of HPLC buffer B up to 90%.
6. Once the gradient is at 90% of HPLC buffer B, maintain that level for 10 min.
7. End the program by returning the HPLC buffer B concentration to 5%.

3.5.5 Data Analysis

1. Select pSRM scans obtained from the analysis in subheading (Subheading 3.5.4) and analyze using DataAnalysis v.4.1.
2. Identify diagnostic ions present for the sugars generated in the expected glycan structure (204.08 for HexNAc, 366.14 for HexHexNAc) in MRM scans that add up to the expected *m/z* value for the glycopeptide.

4 Notes

1. This dilution factor should be significant enough to achieve single, non-touching colonies on the plate. If this is not the case, alterations may need to be made.
2. If very small colonies are seen, leave them to grow for another few hours to become more established.
3. Make sure tweezers are ethanol wiped before use and try to remove any air bubbles that can interfere with transfer.
4. Solution containing soybean agglutinin lectin peroxidase can be cloudy when all components are mixed together.
5. Try to take an image of the developed gel as soon as possible after development, if left the blot may dry out or continue to develop if not washed properly.
6. Try to remove as much residual LB as possible.
7. Supernatant may need to be centrifuged a second time to ensure no cell debris is present in the periplasmic extract as it can late interfere with quantification.
8. Ensure no leaks are present when filling the internal volume of the SDS-PAGE gel. This is an indication that the gel is not sealed in properly.
9. Carefully remove the plastic comb to expose the wells in the SDS-PAGE gel. Wells are very delicate and easily damaged.
10. Protein-loading tips (Bio-Rad) are recommended for loading SDS-PAGE gels.
11. If the target protein has a large MW protein, the gel can be run for longer to separate out bands of interest from one another more distinctly. Pgl2 glycans ($GalNAc_5GlcNAC$) are roughly 1.2 kDa.
12. Handle SDS-PAGE gel carefully as it is extremely delicate and partial to breaking (ripping). Wetting heavily with distilled water is advised to prevent this from occurring.
13. Ensure that when building up the stacks there are no air bubbles between layers as they can interfere with the transfer.
14. Colored protein ladder should now be visible on the membrane and no longer present in the acrylamide gel.
15. No protease inhibitor cocktail should be used as it may interfere with the protease used for protein digestion prior to mass spectrometry analysis.
16. At this point, the dried peptides can be stored at −20 °C and be analyzed at a later date.
17. Ensure there are no bubbles present in the LC vials once the sample has been transferred into it.

18. Be very careful when calculating MRM values. Make use of an in silico tool to calculate the m/z values for different peptides, taking into account the protease used, chemical modifications, and the glycan structure used.

Acknowledgments

The authors acknowledge and are deeply thankful for funding from the UK's Biotechnology and Biological Sciences Research Council (BBSRC) through the Bioprocess Research Industry Club (BRIC) programme (BB/K011200/1 and BB/F004842/1).

References

1. Apweiler R, Hermjakob H, Sharon N (1999) On the frequency of protein glycosylation, as deduced from analysis of the SWISS-PROT database. BBA-Gen Subjects 1473:4–8
2. Rich JR, Withers SG (2008) Emerging methods for the production of homogeneous human glycoproteins. Nat Chem Biol 5:206–215
3. Elbein AD (1991) The role of *N*-linked oligosaccharides in glycoprotein function. Trends Biotechnol 9:346–352
4. Lusch A, Kaup M, Marx U et al (2013) Development and analysis of alpha 1-antitrypsin neoglycoproteins: the impact of additional *N*-glycosylation sites on serum half-life. Mol Pharm 10:2616–2629
5. Moore KL, Stults NL, Diaz S et al (1992) Identification of a specific glycoprotein ligand for P-selectin (CD62) on myeloid cells. J Cell Biol 118:445–456
6. Walsh G (2010) Biopharmaceutical benchmarks 2010. Nat Biotechnol 28:917–924
7. Walsh G (2014) Biopharmaceutical benchmarks 2014. Nat Biotechnol 32:992–1000
8. Higgins E (2010) Carbohydrate analysis throughout the development of a protein therapeutic. Glycoconj J 27:211–225
9. Yoo EM, Yu LJ, Wims LA et al (2010) Differences in *N*-glycan structures found on recombinant IgA1 and IgA2 produced in murine myeloma and CHO cell lines. MAbs 2:320–334
10. Ferrer-Miralles N, Domingo-Espín J, Corchero JL et al (2009) Microbial factories for recombinant pharmaceuticals. Microb Cell Fact 8:1–8
11. Mazor Y, Van Blarcom T, Mabry R et al (2007) Isolation of engineered, full-length antibodies from libraries expressed in *Escherichia coli*. Nat Biotechnol 25:563–565
12. Chan CEZ, Lim APC, Chan AHY et al (2010) Optimized expression of full-length IgG1 antibody in a common *E. coli* strain. PLoS One 5:e10261
13. Faridmoayer A, Fentabil MA, Mills DC et al (2007) Functional characterization of bacterial oligosaccharyltransferases involved in *O*-linked protein glycosylation. J Bacteriol 189:8088–8098
14. Srichaisupakit A, Ohashi T, Misaki R et al (2015) Production of initial-stage eukaryotic *N*-glycan and its protein glycosylation in *Escherichia coli*. J Biosci Bioeng 119:399–405
15. Feldman MF, Wacker M, Hernandez M et al (2005) Engineering *N*-linked protein glycosylation with diverse *O*-antigen lipopolysaccharide structures in *Escherichia coli*. Proc Natl Acad Sci U S A 102:3016–3021
16. Wacker M, Linton D, Hitchen PG et al (2002) *N*-linked glycosylation in *Campylobacter jejuni* and its functional transfer into *E. coli*. Science 298:1790–1793
17. Strutton B, Pandhal J, Wright PC (2013) Expanding the bioprocessing toolbox of *Escherichia coli* through metabolic engineering and synthetic biology: an emerging glycosylation chassis. Pharm Bioprocess 1: 221–224
18. Szymanski CM, Yao R, Ewing CP et al (1999) Evidence for a system of general protein glycosylation in *Campylobacter jejuni*. Mol Microbiol 32:1022–1030
19. Schwarz F, Huang W, Li C et al (2010) A combined method for producing homogeneous glycoproteins with eukaryotic *N*-glycosylation. Nat Chem Biol 6:264–266
20. Wacker M, Wang L, Kowarik M et al (2013) Prevention of *Staphylococcus aureus* infections

by glycoprotein vaccines synthesized in *Escherichia coli*. J Infect Dis 209:1551–1561
21. Hatz CF, Bally B, Rohrer S et al (2015) Safety and immunogenicity of a candidate bioconjugate vaccine against *Shigella dysenteriae* type 1 administered to healthy adults: a single blind, partially randomized phase I study. Vaccine 33:4594–4601
22. Nothaft H, Szymanski CM (2010) Protein glycosylation in bacteria: sweeter than ever. Nat Rev Microbiol 8:765–778
23. Nothaft H, Szymanski CM (2013) Bacterial protein *N*-glycosylation: new perspectives and applications. J Biol Chem 288:6912–6920
24. Guerry P, Ewing CP, Schoenhofen IC et al (2007) Protein glycosylation in *Campylobacter jejuni*: partial suppression of pglF by mutation of pseC. J Bacteriol 189:6731–6733
25. Glover KJ, Weerapana E, Imperiali B (2005) In vitro assembly of the undecaprenylpyrophosphate-linked heptasaccharide for prokaryotic *N*-linked glycosylation. Proc Natl Acad Sci U S A 102:14255–14259
26. Lehrman MA (2015) Flipping a Lipid-Linked Oligosaccharide? You Must Whip It! Trends Biochem Sci 40:715–717
27. Alaimo C, Catrein I, Morf L et al (2006) Two distinct but interchangeable mechanisms for flipping of lipid-linked oligosaccharides. EMBO J 25:967–976
28. Ihssen J, Haas J, Kowarik M et al (2015) Increased efficiency of *Campylobacter jejuni* *N*-oligosaccharyltransferase PglB by structure-guided engineering. Open Biol 5:1–10
29. Ihssen J, Kowarik M, Dilettoso S et al (2010) Production of glycoprotein vaccines in *Escherichia coli*. Microb Cell Fact 9:1–13
30. Ollis AA, Zhang S, Fisher AC et al (2014) Engineered oligosaccharyltransferases with greatly relaxed acceptor-site specificity. Nat Chem Biol 10:816–822
31. Kelleher DJ, Gilmore R (2006) An evolving view of the eukaryotic oligosaccharyltransferase. Glycobiology 16:47R–62R
32. Pandhal J, Woodruff L, Jaffe S et al (2013) Inverse metabolic engineering to improve *Escherichia coli* as an N-glycosylation host. Biotechnol Bioeng 110:2482–2493
33. Pandhal J, Ow SY, Noirel J et al (2011) Improving *N*-glycosylation efficiency in *Escherichia coli* using shotgun proteomics, metabolic network analysis, and selective reaction monitoring. Biotechnol Bioeng 108:902–912
34. Pandhal J, Desai P, Walpole C et al (2012) Systematic metabolic engineering for improvement of glycosylation efficiency in *Escherichia coli*. Biochem Biophys Res Commun 419: 472–476
35. Aebi M (2013) *N*-linked protein glycosylation in the ER. Biochim Biophys Acta 1833: 2430–2437
36. D'Angelo G, Uemura T, Chuang C-C et al (2013) Vesicular and non-vesicular transport feed distinct glycosylation pathways in the Golgi. Nature 501:116–120
37. Ruiz-Canada C, Kelleher DJ, Gilmore R (2009) Cotranslational and posttranslational *N*-glycosylation of polypeptides by distinct mammalian OST isoforms. Cell 136:272–283
38. Bieberich E (2014) Synthesis, processing, and function of *N*-glycans in *N*-glycoproteins. Adv Neurobiol 9:47–70
39. Kowarik M, Numao S, Feldman MF et al (2006) *N*-linked glycosylation of folded proteins by the bacterial oligosaccharyltransferase. Science 314:1148–1150
40. Alanen HI, Walker KL, Suberbie MLV et al (2015) Efficient export of human growth hormone, interferon α2b and antibody fragments to the periplasm by the *Escherichia coli* Tat pathway in the absence of prior disulfide bond formation. Biochim Biophys Acta 1853:756–763
41. Lee YJ, Jeong KJ (2013) Enhanced production of antibody fragment via SRP pathway engineering in *Escherichia coli*. Bioprocess Biosyst Eng 18:751–758
42. Mergulhao F, Summers D, Monteiro G (2005) Recombinant protein secretion in *Escherichia coli*. Biotechnol Adv 23:177–202
43. Samuel G, Reeves P (2003) Biosynthesis of *O*-antigens: genes and pathways involved in nucleotide sugar precursor synthesis and *O*-antigen assembly. Carbohydr Res 338: 2503–2519
44. Valderrama-Rincon JD, Fisher AC, Merritt JH et al (2012) An engineered eukaryotic protein glycosylation pathway in *Escherichia coli*. Nat Chem Biol 8:434–436
45. Jaffé SR, Strutton B, Pandhal J et al (2015) Inverse metabolic engineering for enhanced glycoprotein production in *Escherichia coli*. Methods Mol Biol 1321:17–35
46. Jaffe SR, Strutton B, Levarski Z et al (2014) *Escherichia coli* as a glycoprotein production host: recent developments and challenges. Curr Opin Biotechnol 30:205–210

Chapter 16

Production of Protein Kinases in *E. coli*

Charlotte A. Dodson

Abstract

Recombinant protein expression is widely used to generate milligram quantities of protein kinases for crystallographic, enzymatic, or other biophysical assays in vitro. Expression in *E. coli* is fast, cheap, and reliable. Here I present a detailed protocol for the production of human Aurora-A kinase. I begin with transformation of a suitable plasmid into an expression strain of *E. coli*, followed by growth and harvesting of bacterial cell cultures. Finally, I describe the purification of Aurora-A to homogeneity using immobilized metal affinity and size exclusion chromatographies.

Key words Bacterial protein expression, Protein purification, Kinase, Aurora-A

1 Introduction

Protein kinases are a class of enzymes which catalyze the transfer of the γ-phosphate of ATP to the hydroxyl group of a serine, threonine or tyrosine residue of a protein substrate. Their catalytic action is central to many cellular processes including cell division, cell differentiation, and DNA damage response. Misregulation or aberrant activity of protein kinases occurs in a number of human diseases and protein kinases are major therapeutic targets in oncology and inflammatory disease [1–3]. Achieving inhibitor selectivity between human and parasitic kinases has also been proposed as a strategy for therapeutic intervention in parasitic diseases such as malaria [4–6].

A reliable source of recombinant protein is important for many studies on protein kinases—e.g., understanding the kinase's biological role, for biophysical study, for structure-based drug design or for inhibitor activity assays. Producing recombinant protein in *E. coli* is quick, easy, scalable, reliable, and cheap, and often achieves the large yields necessary for X-ray crystallography. However, not all proteins are amenable to soluble expression in *E. coli* systems and final expression levels need to be determined empirically. In particular, expression of some kinases which are

Nicola A. Burgess-Brown (ed.), *Heterologous Gene Expression in E. coli: Methods and Protocols*, Methods in Molecular Biology, vol. 1586, DOI 10.1007/978-1-4939-6887-9_16, © Springer Science+Business Media LLC 2017

expected to autophosphorylate during expression (e.g., Src and Abl), is improved by coexpressing the kinase with a suitable phosphatase [7, 8].

Protein kinases consist of a conserved kinase domain which carries out the phosphotransfer reaction together with an N- and/or C-terminal extension (e.g., Fig. 1 of [9] and the "Domains and repeats" section of the Uniprot entry for each kinase; www.uniprot.org). The N- and C-terminal sequences can be relatively short (as in the hydrophobic motif of PKA and related AGC kinases [9]) or may be long and consist of multiple domains (e.g., SH2 and SH3 domains found in the Src family of kinases, the C-terminal repeats and coiled coil region of Nek9 and the transmembrane domain of receptor tyrosine kinases). Depending on the assay to be performed, the protein construct expressed may consist of the kinase domain alone, or may include additional domains or regulatory segments. Additional domains may alter the ease with which a protein kinase can be expressed and much time can be spent optimizing the domain boundaries chosen (beyond the scope of this chapter). *E. coli* has been used to express both full-length proteins and shorter constructs of protein kinases (particularly the kinase domain alone). Fortunately for structure-based drug design, the kinase domain alone often gives sufficient information to design tight-binding selective inhibitors: structures of this domain will detail the binding mode of each inhibitor in the ATP binding site. The kinase domain alone is also often sufficient to determine the relative activity of different compounds in enzyme activity assays.

In this chapter, I give a detailed protocol for the expression of the kinase domain of the human mitotic serine/threonine kinase Aurora-A. Aurora-A has been the target of drug discovery programs in both academia and industry, including studies using structure-based drug design. In addition to activity assays used in these programs, kinetic studies of Aurora-A have enabled comparison of the roles of phosphorylation and protein partner binding in Aurora-A activation [10], quantification of the effectiveness of phospho-threonine mimics in the Aurora-A activation loop [11], and determination of the mechanism of Aurora-A autoactivation [12].

The construct of human Aurora-A I use consists of an N-terminal His-tag and the kinase domain (residues 122–403) in a pET30-based vector (kanamycin resistance; Novagen). Variations of this construct have been used to express numerous point mutants of the kinase and also dephosphorylated protein by coexpression with the serine/threonine phosphatase from bacteriophage λ (often known as λ phosphatase) [12]. When expressing unphosphorylated Aurora-A, I use a construct of λ phosphatase consisting of residues 2–221 (Uniprot P03772) in pCDF (NcoI/HindIII sites; Novagen). pCDF contains the antibiotic resistance gene for spectinomycin and has a CloDF13 replication origin, meaning that it coexpresses well with expression plasmids using other replication origins (e.g., pBR322 in pET or pGEX vectors; pGEX from GE Healthcare).

The protocol begins with transforming a plasmid encoding Aurora-A into an expression strain of *E. coli.* Small-scale pre-cultures derived from a single bacterial colony are used to inoculate large-scale expression cultures and protein expression is induced with isopropyl β-D-thiogalactoside (IPTG). Bacteria are grown overnight and are pelleted and frozen for storage. To purify the protein, defrosted pellets are resuspended and lysed by sonication. Soluble, overexpressed Aurora-A protein is purified to homogeneity using immobilized metal affinity (IMAC) and size-exclusion (gel filtration) chromatographies. The whole process (from transformation to pure protein) takes a minimum of 4 days (most comfortable in six) and yields around 4 mg of protein per liter of culture medium.

2 Materials

2.1 Solutions for Protein Expression

1. 35 mg/mL kanamycin stock solution: prepared in water and syringe filtered through a sterile 0.2 μm filter. Keep the filtrate frozen at −20 °C in 1 mL aliquots. Aliquots can be defrosted/refrozen as required.
2. 35 mg/mL chloramphenicol stock solution: prepared in ethanol (chloramphenicol is not soluble in water). Syringe filter the stock through a sterile 0.2 μm filter and keep the filtrate in the −20 °C freezer in a 50 mL plastic tube. Ethanol will not freeze at this temperature and the solution can be taken directly from the plastic tube for use as necessary.
3. 1 M IPTG solution: prepared in water and syringe filtered through a sterile 0.2 μm filter. Keep this solution frozen at −20 °C in 15 mL aliquots and defrost/refreeze as necessary.
4. LB medium: 1% (w/v) tryptone (10 g in 1 L), 0.5% (w/v) yeast extract (5 g in 1 L), 170 mM NaCl. For final expression, divide total culture volume into 800 mL per 2 L conical flask. Seal top of the flask with foil and autoclave (*see* **Notes 1** and **2**). For overnight pre-culture, culture volume depends on volume of final expression (Table 1). Unused, autoclaved media can be stored at room temperature.
5. SOC medium: 2% (w/v) tryptone (20 g in 1 L), 0.5% (w/v) yeast extract (5 g in 1 L), 10 mM NaCl, 2.5 mM KCl, 10 mM $MgCl_2$ or $MgSO_4$, 20 mM glucose. 0.5 mL of autoclaved medium required per transformation.

2.2 Other Reagents and Equipment for Protein Expression

1. pET30-based vector containing an N-terminal His-tag and Aurora-A kinase domain residues 122–403 in the MCS (*see* **Notes 3** and **4**).

2. BL21(DE3) RIL *E. coli* cells: 50 μL per transformation (*see* **Notes 5** and **6**).
3. 1.5% LB agar plates (*see* **Note 7**).
4. Shaking incubator at 37 °C.
5. Water bath at 42 °C.
6. Oven at 37 °C.
7. 1.5 mL Eppendorf tubes.
8. 50 mL plastic tubes.
9. Sterile plastic spreader.
10. Ice.
11. Virkon (or other laboratory disinfectant).

2.3 Solutions for Protein Purification (See Note 8)

Prepare all solutions using ultrapure water with a resistivity of 18 MΩ cm at room temperature and filter through a 0.2 μm filter prior to storage. This aids the long-term storage of buffers, and also reduces the chances of blocking narrow tubes on ÄKTA (or similar) fast protein liquid chromatography (FPLC) systems.

1. Lysis Buffer: 50 mM Tris–HCl pH 7.5, 200 mM NaCl, 10 mM imidazole, 10% (v/v) glycerol (*see* **Note 9**).
2. Elution Buffer: 50 mM Tris–HCl pH 7.5, 200 mM NaCl, 250 mM imidazole, 10% (v/v) glycerol (*see* **Note 9**). After making this buffer, place a drop on pH paper to check the pH. In my hands the pH is always as expected, but imidazole is basic and it may be necessary to adjust the pH of the solution to bring it back to pH 7.5.
3. Gel filtration Buffer: 50 mM Tris–HCl pH 7.5, 200 mM NaCl, 5 mM $MgCl_2$, 10% (v/v) glycerol, 1 mM DTT (*see* **Note 9**).

2.4 Other Reagents and Equipment for Protein Purification

1. Pierce Protease inhibitor tablets.
2. Sonicator (or equivalent).

Table 1
Volume of overnight pre-culture required

Final expression culture (number of 2 L flasks)	Overnight pre-culture
1	10 mL of autoclaved medium in 50 mL sterile plastic tube
2–4	50 mL of medium in 250 or 500 mL conical flask[a]
>4	10 mL of medium per 2 L flask + 10 mL in 250 mL or 500 mL conical flask[a]

[a]Conical flasks should be sealed with foil and autoclaved

3. 0.2 and 0.45 μm sterile syringe filters.
4. 10 mL plastic syringes (sterile).
5. 50 mL plastic tubes.
6. ÄKTA chromatography system (or equivalent) with superloop and 2 mL sample loading loop.
7. 5 mL His trap column (or equivalent).
8. 120 mL Superdex 200 gel filtration column.
9. Thin-walled 200 μL tubes (e.g., PCR tubes).
10. Liquid nitrogen for freezing protein.
11. Centrifuges and centrifuge tubes.
12. Ice.
13. 4–12% SDS-PAGE protein gel (or equivalent) with sample loading buffer, running tank, running buffer and power pack (optional).

3 Methods

3.1 Transformation and Initial Growth of E. coli

1. Warm SOC medium by placing in a water bath or incubator (*see* **Note 10**).
2. Place agar plate upside down in a 37 °C oven. This step enables cultures to be plated out onto a pre-warmed plate and increases the efficiency of transformation [13].
3. Incubate 2 μL of plasmid (*see* **Note 11**) with 30–50 μL of chemically competent *E. coli* BL21(DE3) RIL cells in a 1.5 mL Eppendorf tube on ice for 15 min. Do not mix by pipetting up and down as this will damage the cells. Instead, allow solutions to mix passively.
4. Using a foam floater to keep the top of the tubes above water level, place Eppendorf tubes in a water bath at 42 °C for 45 s. Remove and place on ice for 2 min.
5. Add 0.5 mL of preheated SOC to each Eppendorf, and place in a 37 °C shaking incubator for 30–60 min (1–2 replication cycles). No antibiotic selection is used at this stage.
6. Remove agar plate from oven and culture from shaking incubator. Pour the contents of the Eppendorf tube onto the surface of agar, and spread out with a sterile plastic spreader, rotating the plate with the other hand to ensure an even distribution of material. Dispose of Eppendorf tube and spreader as biohazard waste.
7. Replace lid of plate, turn plate over, and label with a marker pen. Place upside down in the 37 °C oven overnight (~16 h).
8. The next morning, remove plate from oven and place at 4 °C in fridge to store.

3.2 Overnight Culture and Growth of Large-Scale Cultures (See Note 2)

1. Using a "typical" single colony from the bacterial plate, inoculate the overnight pre-culture medium prepared in Sect. 2.1, **item 4** (*see* **Note 12**). Add kanamycin and chloramphenicol to 35 μg/mL final concentration (1 in 1000 dilution of each stock solution) and place pre-culture in a shaking incubator at 37 °C overnight (~16 h growth).
2. The next day, add 10 mL of pre-culture to each 2 L flask of LB expression medium. Add 1 mL of each of kanamycin and chloramphenicol stock solutions (i.e., 44 μg/mL final concentration).
3. Place 2 L flasks in a shaking incubator at 37 °C and grow until A_{600} is 0.8–1.2 AU (1 cm path length). This typically takes 3–4 h.
4. Add 400 μL of 1 M IPTG solution to each flask (i.e., 500 μM final concentration) and grow at 21 °C overnight (*see* **Note 13**).
5. Harvest the cells by pouring the bacterial culture into 1 L centrifuge bottles and spin at ~9000 × *g* for 15 mins at 4 °C.
6. Pour the supernatant back into the flasks (or other large container) and decontaminate with 1% (w/v) Virkon or other laboratory bactericide before discarding.
7. Transfer the pellet to a 50 mL plastic tube (with for example a spatula) and store at −80 °C until required.
8. Decontaminate centrifuge bottles with 1% (w/v) Virkon or other laboratory bactericide.

3.3 Lysing Pellet and Extracting Soluble Fraction

For maximum time efficiency, begin Sect. 3.5, **step 1** and Sect. 3.4, **step 1** now, before moving on to lysing the pellet and extracting the soluble fraction.

1. Remove tube containing pellet from −80 °C freezer and place on ice. Add approx. 40 mL of lysis buffer (the exact quantity is not important, but this volume is convenient for future centrifugation).
2. Add 1 protease inhibitor tablet and place tube on a rotator at room temperature to mix while the pellet defrosts and becomes resuspended in the buffer. Vortex every 5–10 min to aid pellet resuspension.
3. When the pellet is resuspended, place the tube on ice for 10 min to ensure that it is thoroughly chilled.
4. Lyse the pellet by sonicating for 2 min total on ice using a 130 W sonicator with a 6 mm tip and a pulse cycle of 1 s on, 2 s off (*see* **Note 14**). After 1 min, pause the sonication, remove the sample and mix by inverting the tube before returning for the final minute of sonication.
5. Transfer the lysed solution into a suitable centrifuge tube and spin at ~34,500 x *g* at 4 °C for at least 1 h.

6. After sonication, transfer the supernatant to a 50 mL plastic tube on ice and discard the pellet as biohazardous waste. Disinfect the centrifuge tube with a suitable laboratory disinfectant such as 1% (w/v) Virkon.

3.4 Affinity Chromatography

All chromatography steps should be carried out at 4 °C. This can be achieved by keeping the chromatography system in a cold cabinet or in the cold room.

1. Wash the flow path of a chromatography system (e.g., ÄKTA Prime) with lysis buffer (buffer A inlet) and elution buffer (buffer B inlet). Attach the His trap column to the system and equilibrate in buffer A. This step can be carried out after Sect. 3.5, **step 1** and while carrying out the final stages of lysing the pellet and extracting the soluble fraction (Sect. 3.3).
2. Using a 10 mL syringe and disposable sterile syringe filters, filter the supernatant from Sect. 3.3, **step 6**, through a 0.45 μm filter into a 50 mL plastic tube. After ~10 mL of solution the filter becomes blocked and it is necessary to use a new filter. Repeat filtration with a 0.2 μm filter, changing filters as required. This step removes any membrane fragments or unlysed bacteria, and ensures that the narrow tubing in the chromatography system does not become blocked.
3. Load the filtered protein sample into a sample loop on the chromatography system. A 50 mL glass superloop is convenient for this as it can be adjusted to load any volume of sample up to the maximum.
4. Equilibrate the column, load sample, wash and elute with a step gradient to buffer B, collecting fractions for the elution. My standard method for an ÄKTA Purifier 10 running Unicorn 5 software is given in Appendix 1 and a sample run is shown in Fig. 1a.
5. Select the fractions which contain the eluted protein (or the majority of the eluted protein) and place on ice.
6. Take a 10 μL sample from each fraction to run on an SDS-PAGE gel (optional). Mix with SDS-PAGE loading buffer and store until running the gel.

3.5 Size Exclusion Chromatography (See Note 15)

1. Wash the flow path of a chromatography system with gel filtration buffer (buffer A inlet only) and equilibrate a 120 mL Superdex 200 gel filtration column. This step will take around 2 h, and for maximum time efficiency can be started before lysing the pellet (Sect. 3.3).
2. Take 2 mL from the affinity chromatography fraction that has the highest protein concentration. Load this into a 2 mL sample loop and begin the first size exclusion chromatography run, collecting 2 mL fractions. This first run serves the dual

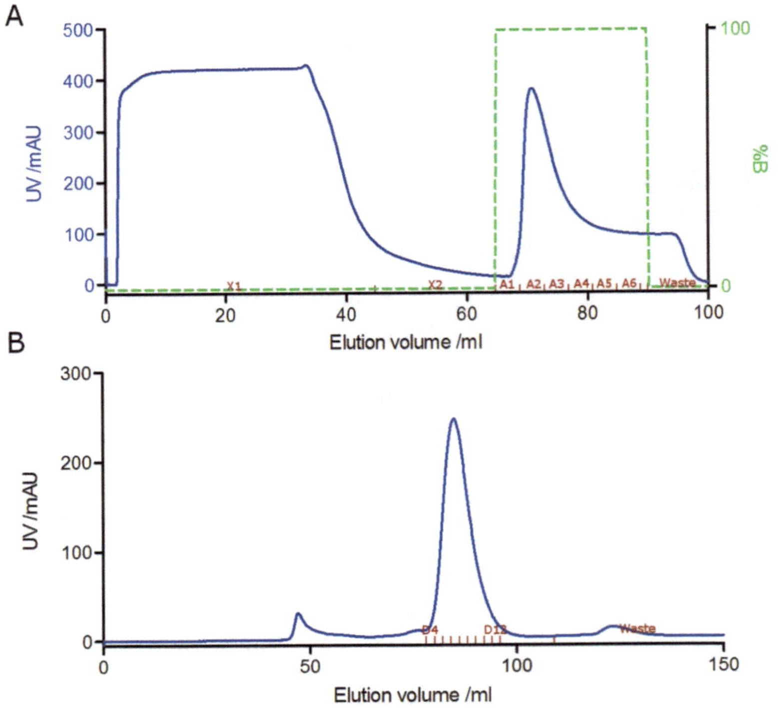

Fig. 1 (**a**) IMAC elution profile for Aurora-A. The pellet from 800 mL of bacterial cell culture was loaded onto a 5 mL His-trap *Ni*$^{2+}$ column. *Blue*—absorbance at 280 nm, *green*—proportion of buffer B (elution buffer), *red*—fraction numbers (*X* denotes 45 mL fractions, *A* denotes 4 mL fractions). Increasing the volume of bacterial cell culture to 1600 mL broadens the elution peak, but elution is still complete by fraction A6. The expected capacity of this column is 40 mg protein/mL resin. Note that the UV signal does not fall to zero in the presence of 100% elution buffer because imidazole absorbs at 280 nm. (**b**) Size exclusion elution profile for Aurora-A. 1 mL of concentrated IMAC eluent was loaded onto a 120 mL Superdex 200 column. Fractions (2 mL) were collected throughout the run until 109 mL, but only fractions containing monomeric protein are labelled. Increasing loading volume to 2 mL broadens the peak slightly (by 2–3 fractions)

purpose of preparing pure protein, and running a sample which confirms the aggregation/monomeric state of the protein immediately after affinity chromatography. My standard method for an ÄKTA Purifier 10 running Unicorn 5 software is given in Appendix 2 and a sample run is shown in Fig. 1b.

3. Add a small amount of buffer to a 25 mL spin concentrator (10 kDa membrane) and spin at 2600 × *g* for ~5 min to wash the membrane for protein concentration.
4. Mix remaining protein-containing fractions from affinity chromatography.
5. Empty out the buffer from the washed spin concentrator and replace with the mixed fractions from **step 4**. Spin at 4 °C until

the volume is around 4 mL. The time taken to achieve this will vary on the protein prep and the total concentration of protein, but the most important thing is to ensure that the protein concentration remains below 10 mg/mL (*see* **Note 16** for determination of protein concentration). Every 30–40 min, stop the centrifuge and mix the protein in the concentrator by briefly inverting or by pipetting up and down near the membrane. This step helps ensure that a high concentration of protein is not allowed to build up at the concentrator membrane.

6. When the concentration of the protein reaches 10 mg/mL, or when the volume reaches 4 mL (whichever happens first), remove the protein from the spin concentrator and place in a 15 mL plastic tube or other suitable container. Store on ice (or at 4 °C overnight) until needed (*see* **Note 17**).
7. After size exclusion, collect those fractions containing monomeric protein and take a 10 μL sample from each fraction to run on an SDS-PAGE gel (optional). Mix each sample with SDS-PAGE loading buffer and store until running the gel. A typical SDS-PAGE gel of Aurora-A purification is shown in Fig. 2.

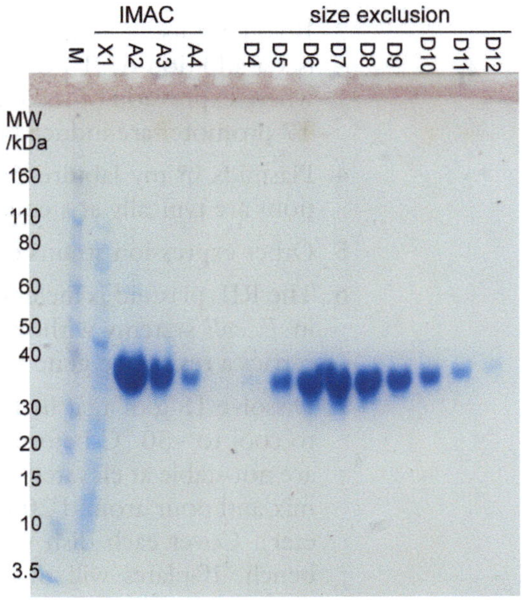

Fig. 2 Coomassie-stained SDS-PAGE gel of fractions from chromatographies outlined in Fig. 1. Most contaminants are removed during IMAC, and highly pure Aurora-A is obtained after size exclusion chromatography. Fractions X1 (wash) and A2–A4 (eluted protein) are from IMAC, and fractions D4–D12 from size exclusion chromatography. *M* protein standard; the position of the molecular mass markers are indicated on the *left* (in kDa)

8. Mix all fractions containing monomeric protein (*see* **Note 18**) and concentrate back to ~4 mL using a spin concentrator (pre-washing the concentrator as in **step 3**).
9. Measure protein concentration using a UV–Vis spectrometer, and using the flow-through in the spin concentrator as a blank (*see* **Note 16**). This way, contributions to A_{280} from ageing solutions of DTT are removed from the final calculation.
10. Aliquot into thin-walled 200 μL tubes (typically 5, 10, 100 or 200 μL aliquots depending on expected final purpose) and flash freeze in liquid nitrogen. Store at −80 °C.

4 Notes

1. Flasks can also be sealed with foam stoppers, or both stoppers and foil. The exact volume of medium is not important, as long as there is plenty of space above the liquid to enable aeration of the medium. I use 800 mL per flask as this volume fits easily into a 1 L or 2 × 500 mL centrifuge bottles (**step 5**, Sect. 3.2) without leakage.
2. I typically grow four flasks (3.2 L culture volume) of each construct and freeze pellets from two flasks (1.6 L) per 50 mL plastic tube.
3. If you are using a different plasmid, check whether induction is via addition of IPTG or uses another mechanism. Plasmids where expression of recombinant protein is under control of a T7 promoter are induced using IPTG in BL21(DE3) cells.
4. Plasmids in my lab are prepared by mini-prepping and solutions are typically at a concentration of 50–100 ng/μL.
5. Other expression strains of *E. coli* can be used (e.g., DH5α).
6. The RIL plasmid is included to aid expression of human genes in *E. coli* systems without codon optimization. This plasmid carries a resistance gene for chloramphenicol.
7. Dissolve 15 g of agar in 1 L of LB broth. Autoclave and allow to cool to ~50 °C before adding antibiotics (many antibiotics are not stable at elevated temperatures). Swirl and/or invert to mix and pour around 20 mL into each petri dish (9 cm diameter). Cover each dish with lid, and leave agar to set on the bench. If plates will not be used immediately, store upside down in the fridge or cold room at 4 °C. Wrapping the edge of the plates with parafilm reduces the rate at which the plate dries out and increases its shelf life.
8. I keep filtered stock solutions of the following common buffer ingredients at room temperature and mix these to make lysis

buffer, elution buffer and gel filtration buffer: 1 M Tris–HCl pH 7.5; 1 M imidazole; 1 M $MgCl_2$ and 5 M NaCl. I also keep a stock of 1 M dithiothreitol (DTT) in water in frozen aliquots of 1 mL which I defrost as needed for gel filtration buffer.

9. Including 10% (v/v) glycerol in buffers increases the stability of Aurora-A and improves the recovery of monomeric protein on thawing of frozen samples.
10. I place my bottle of SOC straight from the fridge into a 42 °C water bath and remove it after the heat shock in **step 4**, but it could also be placed in a 37 °C incubator.
11. Experiments in my lab have indicated that increasing the volume of DNA-containing solution beyond 2 μL decreases the efficiency of transformation.
12. The colony picked should be typical of the colonies on the plate (e.g., in size and density of other colonies nearby). If a colony is unusually large, or has an unusual amount of clear agar around it, this may indicate a spontaneous mutation within the colony.
13. Moving the flasks between two different incubators enables a step change in temperatures, but is often impractical. At the time of induction, I typically leave the lid of the incubator open to aid fast cooling and/or leave the flasks on the bench at room temperature until the temperature in the incubator has dropped. Alternatively, shortly before the expected induction time I lower the temperature of the incubator.
14. The exact power and diameter of the sonication tip are not important, although small diameter sonication tips (designed for small volumes) are unlikely to give complete lysis of the solution. Avoid long periods of continuous sonication as this may heat the sample and reduce the yield of soluble monomeric protein. Alternative lysis methods such as French press are also suitable.
15. The maximum capacity of commonly available size exclusion columns is smaller than that of an affinity column, so for Aurora-A and other proteins which express well, multiple size exclusion chromatography runs will be necessary to ensure that all protein is purified.
16. Protein concentration can be determined by measuring the absorbance of the protein solution at 280 nm, and using the Beer-Lambert Law: $A = \varepsilon \times c \times l$, where A is measured absorbance, ε is the extinction coefficient of the protein, c is concentration, and l is path length. The extinction coefficient at 280 nm for any protein can be easily calculated from its amino acid sequence using the ProtParam tool on the ExPASy website: http://web.expasy.org/protparam/.

17. In practice, I ensure that a second gel filtration run is started before the end of the first day. If this means that protein is not fully concentrated, a sample is taken for gel filtration, the remainder stored overnight in the fridge (not in the spin concentrator), and concentration continues on the next morning. I have found that running two size exclusion columns on the same day as lysis helps ensure that monomeric sample is prepared for less stable proteins. I often program an ÄKTA method with two sequential gel filtration runs, and refill the sample loop with 2 mL of concentrated protein at some point during the first run.
18. The approximate molecular mass of eluted protein can be determined from its elution volume (fractional elution volume is inversely proportional to log(molecular mass)). Monomeric, dimeric, and higher oligomeric species are resolved into separate peaks and so monomeric protein can be identified by comparing the peak elution volume with that of calibration standards of known molecular mass.

Acknowledgments

This work is supported by an Imperial College Junior Research Fellowship and grant BB/M021149/1 from the BBSRC.

5 Appendix 1

```
¤ (Main)
  0.00 Base CV 5.027 {mL} HisTrap_FF_5_mL
  0.00 Alarm_Pressure Enabled 1.5 {MPa} 0.00 {MPa}
  0.00 Flow 3 {mL/min}
¤ 0.00 Block Wash_column
  (Wash_column)
  0.00 Base SameAsMain
  3 End_Block
¤ 0.00 Block Load_sample
  (Load_sample)
  0.00 Base Volume
  0.00 Fractionation 30 mm 45 {mL} FirstTube Volume
  0.00 OutletValve F2
  0.00 InjectionMark
  0.00 InjectionValve Inject
```

```
  0.00  AutozeroUV
  60  InjectionValve Load
  65.00  Hold_Until UV Less_Than 20 {mAU} 25 {base}
  65.00  End_Block
¤ 0.00  Block Elute_sample
  (Elute_sample)
  0.00  Base SameAsMain
  0.00  Gradient 100 {%B} 0.00 {base}
  0.00  Fractionation 18 mm 4 {mL} FirstTube Volume
  5  End_Block
¤ 0.00  Block Back_to_A
  (Back_to_A)
  0.00  Base SameAsMain
  0.00  Gradient 0.0 {%B} 0.00 {base}
  0.00  FractionationStop
  2  End_Block
```

6 Appendix 2

```
¤ (Main)
  0.00  Base Volume
  0.00  Alarm_Pressure Enabled 1 {MPa} 0.00 {MPa}
  0.00  Flow 1 {mL/min}
  10  InjectionValve Inject
  10  OutletValve F2
  10  Fractionation 18 mm 2 {mL} FirstTube Volume
  10  AutozeroUV
  15  InjectionValve Load
  109  FractionationStop
  150.00  End_Method
```

References

1. Rask-Andersen M, Zhang J, Fabbro D et al (2014) Advances in kinase targeting: Current clinical use and clinical trials. Trends Pharmacol Sci 35:604–620
2. Zhang J, Yang PL, Gray NS (2009) Targeting cancer with small molecule kinase inhibitors. Nat Rev Cancer 9:28–39
3. Barnes PJ (2013) New anti-inflammatory targets for chronic obstructive pulmonary disease. Nat Rev Drug Discov 12:543–559
4. Lucet IS, Tobin A, Drewry D et al (2012) Plasmodium kinases as targets for new-generation antimalarials. Future Med Chem 4:2295–2310

5. Le Manach C, Paquet T, Gonzalez Cabrera D et al (2014) Medicinal chemistry optimization of antiplasmodial imidazopyridazine hits from high throughput screening of a softfocus kinase library: Part 2. J Med Chem 57:8839–8848
6. Crowther GJ, Hillesland HK, Keyloun KR et al (2016) Biochemical screening of five protein kinases from *Plasmodium falciparum* against 14,000 cell-active compounds. PLoS One 11:e0149996
7. Seeliger MA, Young M, Henderson MN et al (2005) High yield bacterial expression of active c-Abl and c-Src tyrosine kinases. Protein Sci 14:3135–3139
8. Wang W, Marimuthu A, Tsai J et al (2006) Structural characterization of autoinhibited c-Met kinase produced by coexpression in bacteria with phosphatase. Proc Natl Acad Sci U S A 103:3563–3568
9. Pearce LR, Komander D, Alessi DR (2010) The nuts and bolts of AGC protein kinases. Nat Rev Mol Cell Biol 11:9–22
10. Dodson CA, Bayliss R (2012) Activation of Aurora-A kinase by protein partner binding and phosphorylation are independent and synergistic. J Biol Chem 287:1150–1157
11. Rowan FC, Richards M, Bibby RA et al (2013) Insights into Aurora-A kinase activation using unnatural amino acids incorporated by chemical modification. ACS Chem Biol 8:2184–2191
12. Dodson CA, Yeoh S, Haq T et al. (2013) A kinetic test characterizes kinase intramolecular and intermolecular autophosphorylation mechanisms. Sci Signal 6:ra54
13. Pope B, Kent HM (1996) High efficiency 5 min transformation of *Escherichia coli*. Nucleic Acids Res 24:536–537

Chapter 17

Expression of Prokaryotic Integral Membrane Proteins in *E. coli*

James D. Love

Abstract

Production of prokaryotic membrane proteins for structural and functional studies in *E. coli* can be parallelized and miniaturized. All stages from cloning, expression, purification to detergent selection can be investigated using high-throughput techniques to rapidly and economically find tractable targets.

Key words High throughput, Membrane proteins, Detergent assay, Multiwell

1 Introduction

Prokaryotic overexpression of prokaryotic integral membrane proteins (IMPs) is now well established as a fruitful path to the production of these targets that have led many successful structural and functional studies [1–5]. High throughput (HT) approaches for protein production and structural determination have also matured largely since their inception in the late 1990s [6]. While these HT approaches, as pioneered by international structural genomics efforts, often required a large capital outlay to be exercised at full tilt, the technologies and methodologies that were developed by these efforts are applicable to a "normal" lab setting and should be implemented as they have a two-fold advantage: (1) the methods that have been developed have the distinct advantage that they have been optimized to be rigorous in their usefulness, and do not suffer from the standard anecdotal approaches that are encountered in many research labs [6, 7]. HT approaches have been successfully applied to the production and purification of tens of thousands of proteins, including membrane proteins and eukaryotic expression systems; (2) performing experiments in a HT-like manner, even without the expensive capital outlay for robotics, is much less reagent, consumable, and labor intensive [8–10]. A single researcher can miniaturize and parallelize experimental

Nicola A. Burgess-Brown (ed.), *Heterologous Gene Expression in E. coli: Methods and Protocols*, Methods in Molecular Biology, vol. 1586, DOI 10.1007/978-1-4939-6887-9_17, © Springer Science+Business Media LLC 2017

approaches so that they may conduct a vast number of experiments as compared with what can be done by more traditional approaches. Production and structural studies on membrane proteins still counts as a heroic endeavor—but the chance of success is of course proportional to the number of trials made—especially if a systematically valid rational is being applied [11]. Therefore, this chapter sets out a HT approach to the production of membrane proteins in *E. coli* that will enable investigators to increase their throughput with minimal capital expenditure.

2 Materials

2.1 Target Selection, Construct Design, and Expression Vector Construction

1. Integra Viaflo and Voyager II adjustable span pipettes.
2. 96 PCR plate (Eppendorf).
3. CloneAmp™ HiFi PCR Premix (Clontech).
4. Aluminum sealing films (VWR).
5. Brayer (VWR).
6. Eppendorf MixMate.
7. Centrifuge with plate rotor.
8. 96-well thermal cycler.
9. NucleoSpin 96 plasmid kit (or similar).
10. E-Gel 96 Gels with SYBR stain (Thermo Fisher).
11. Mother E-Base Device (Thermo Fisher).
12. UV visualization system.
13. In-Fusion® HD Cloning Plus.
14. DH5alpha (competent, in 96-well plates)—SOC medium is included.
15. Airpore breathable films (Qiagen).
16. Eppendorf Thermomix C or Eppendorf MixMate placed in 37 °C incubator.
17. 24-well deep well blocks.
18. Antibiotic selective agar growth medium, Kanamycin (Thermo Fisher).
19. EZ-spread Plating beads (Genlantis).
20. Sterile toothpicks.
21. LB medium (Teknova USA).
22. 96-deep well blocks (DWB) .
23. Machery Nagel 96 miniprep kit.
24. Plate reader for quantification of DNA.

2.2 Small Scale Expression and Purification

1. BL21(DE3) pLysS (Sigma-Aldrich or New England Biolabs)—recovery medium supplied.
2. 96-well thermal cycler.
3. Eppendorf Thermomix C.
4. Airpore breathable films.
5. 2× TY medium (Teknova USA).
6. 100 mM IPTG stock solution: Prepare in water and store at −20 °C.
7. 96-deep well blocks (DWB).
8. Aluminum sealing films (VWR).
9. 1 M AEBSF (Sigma-Aldrich): Prepare in water and store at −20 °C.
10. 1 M $MgSO_4$: Prepare in water.
11. Lysozyme: Add powder directly.
12. Benzonase Nuclease (Sigma-Aldrich).
13. Protease Complete tablet, EDTA-free (Roche).
14. Aluminum sealing films (VWR).
15. Eppendorf MixMate.
16. Sonicator—96-well plate horn (Qsonica) or single microprobe.
17. 10% (w/v) DDM (Anatrace) stock solution: Prepare fresh in water by gentle mixing and store at −20 °C.
18. Eppendorf Thermomix C.
19. Centrifuge with plate rotor.
20. 96-well filter plate (Thomson).
21. 96-well filter plate bottom seals (Thompson).
22. His60 Ni Superflow resin equilibrated into Lysis buffer.
23. Wide bore tips (Axygen Scientific) .
24. 96-well plate vacuum manifold (Sigma).
25. Lysis buffer: 50 mM Hepes pH 8, 250 mM NaCl, 10% (v/v) glycerol. Filter sterilize before use.
26. Wash buffer: 50 mM Hepes pH 8, 500 mM NaCl, 10% (v/v) glycerol, 10 mM imidazole, 0.02% (w/v) DDM.
27. Elution buffer: 50 mM Hepes pH 8, 250 mM NaCl, 10% (w/v) glycerol, 250 mM imidazole, 0.02% (w/v) DDM.
28. 96 PCR plate.
29. 4× Laemmli sample buffer (Bio-Rad). 100 mM DTT is freshly added before use.
30. Criterion Dodeca SDS-PAGE gels (4–20% Bio-Rad).
31. Criterion Cell PAGE apparatus and power supply (Bio-Rad).

32. 10× Tris/Glycine/SDS running buffer (Bio-Rad).
33. SimplyBlue safe stain (Thermo Fisher).
34. SeeBlue Plus2 Pre-stained protein standards (Thermo Fisher).

2.3 Medium Scale Expression and Purification

1. BL21(DE3) pLysS—recovery medium supplied.
2. 24-well deep well blocks.
3. Antibiotic selective agar growth medium, Kanamycin.
4. EZ-spread Plating beads (Genlantis).
5. 10 mL sterile plastic culture tube.
6. Shaker incubator, 1″ throw.
7. 2× TY medium.
8. Ultra Yield flasks and Airtop seals (Thompson).
9. 100 mM IPTG stock solution: Prepare in water and store at −20 °C.
10. 1 M AEBSF (Sigma-Aldrich): Prepare in water and store at −20 °C.
11. 1 M $MgSO_4$: Prepare in water.
12. Lysozyme: Add powder directly.
13. Benzonase Nuclease.
14. Protease Complete tablet, EDTA-free .
15. Sonicator single microprobe (Qsonica).
16. 10% (w/v) DDM stock solution: Prepare fresh in water by gentle mixing and store at −20 °C.
17. His60 Ni Superflow resin.
18. Bio-Rad Econo-Pac chromatography column.
19. Lysis buffer: 50 mM Hepes pH 8, 250 mM NaCl, 10% (v/v) glycerol. Filter sterilize before use.
20. Wash buffer: 50 mM Hepes pH 8, 500 mM NaCl, 10% (v/v) glycerol, 10 mM imidazole, 0.02% (w/v) DDM.
21. Elution buffer: 50 mM Hepes pH 8, 250 mM NaCl, 10% Glycerol, 250 mM imidazole, 0.02% (w/v) DDM).
22. 4× Laemmli sample buffer. 100 mM DTT is freshly added before use.
23. Criterion Dodeca SDS-PAGE gels.
24. Criterion Cell PAGE apparatus and power supply.
25. 10× Tris/Glycine/SDS running buffer.
26. SimplyBlue safe stain.
27. SeeBlue Plus2 Pre-stained protein standards.

Table 1
Suggested detergents and concentrations

Detergent	Catalog number (Anatrace)	% (w/v) for 2× CMC, approximate)
DDM	D310	0.02
DM	D322	0.2
OG	O311	1
NG	N324	0.4
OGNG	NG311	0.12
LMNG	NG310	0.002
LDAO	D360	0.04

2.4 Detergent Selection

1. Gel filtration buffer: 50 mM Hepes pH 8, 250 mM NaCl, 5% (v/v) glycerol, plus detergent of choice (Table 1).
2. HPLC system (Agilent 1200) or FPLC (AKTA prime, GE Life Sciences).
3. Small bore column (Sepax Technologies) or Superdex S200 Increase (GE Life Sciences).

3 Methods

3.1 Target Selection, Construct Design, and Expression Vector Construction

1. BLAST search for homologs that are available in your lab or from suppliers of genomic DNA, such as ATCC (http://www.atcc.org) or DSMZ. Optimize settings so that homology is over the maximized length of the target protein (*see* **Note 1**). Full details are given in Punta et al. [3]. Codon optimization may also be beneficial if synthetic genes are to be ordered [12].
2. It is necessary to design several constructs, starting with the full length (FL) protein and making a number of N, C or N and C terminal truncations to maximize the chances of protein expression [13]. It may be necessary to design at least 5–10 different constructs per target, but this is dependent on resources.
3. There are many available expression vectors for prokaryotic overexpression of membrane proteins and it may be useful to try several as they can express your target protein to different levels [14]. Pick an expression vector with a cleavable N or C terminal His tag from commercial suppliers or the DNASU plasmid repository (https://dnasu.org/DNASU/GetAllVectors.do?PSI=false&tab=2).

4. Use Primer prim'er (http://www-nmr.cabm.rutgers.edu/bioinformatics/Primer_Primer/) [15], Protein CCD (https://xtal.nki.nl/ccd/Welcome.html), or XtalPred (http://ffas.burnham.org/XtalPred-cgi/xtal.pl) to aid in construct design and design of PCR primers following the recommendations for length and annealing temperature—use the online tool supplied for In-Fusion: http://www.clontech.com/US/Products/Cloning_and_Competent_Cells/Cloning_Resources/Online_In-Fusion_Tools?sitex=10020:22372:US.
5. Order PCR primers from your supplier, requesting wet shipment, forward and reversed primer mixed, at 5 μM concentration.
6. Set up PCR reactions as per manufacturers' instructions (CloneAmp HiFi PCR Premix), scaled for 25 μL reactions in a 96-well format. Create a master mix components without template or primers, multiplying all values by 1.2 to ensure excess reagents to allow for minor losses on pipetting (*see* **Note 2**).
7. Aliquot master mix to 12 wells of a 96-well plate, and then using a 12-channel electronic pipette, use the multi-dispense program to pipette to each well of a plate (Fig. 1).
8. Add correct template to each well by hand, or by robot (if available). The ipad program (http://ipipet.teamerlich.org) may be of use in this step [16]. Typically, about 50–100 ng of genomic DNA template is required per reaction.
9. Add primers to each well using the multichannel pipette.
10. Seal plate with a metal foil, mix using the Eppendorf MultiMate (Fig. 1), 30 s, 1000 rpm, transfer to a centrifuge (30s, 1000 × *g*) to collect in the bottom of the wells.

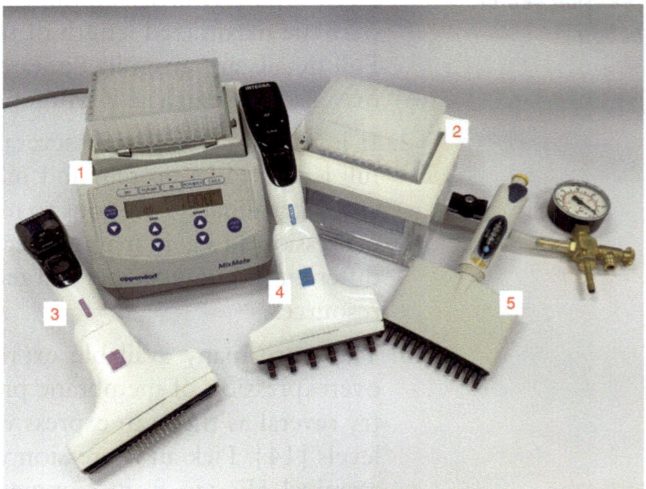

Fig. 1 Entry level high throughput liquid handling and processing equipment. (*1*) Small orbit (1.5–3 mm) high speed (~1000 rpm) shaker with 96-deep well block. (*2*) Vacuum manifold for 96-well filter plates. (*3*) Integra Viaflo variable span (4.5–9 mm) 12-channel pipettor. (*4*) Electronic fixed channel pipettor. (*5*) Manual fixed channel pipettor

11. PCR amplify the target gene using cycling parameters as suggested: 98 °C for 10 s, 55 °C for 10 s, 72 °C, 5 s per kb. If a variety of genes are being amplified with different base pair sizes, it may be necessary to split reactions by extension time, but in general this will not be necessary.
12. Remove 2 μL of the reaction mix and run out on a 96-well E-gel to test for amplification success. If the products are poor, optimize PCR conditions or clean up products of the correct length (if possible) by agarose gel electrophoresis and gel extraction using 96-well cleanup kits.
13. Prepare cut expression vector by cutting with appropriate restriction enzymes following the manufacturers' instructions . As a standard, prepare 1.5× excess expression vector as there are often losses. Scale the amount of expression vector cut for a 5 μL final reaction (*see* **Note 3**).
14. In 96-well plates, scaling as suggested in **step 6**, mix ~100 ng of purified PCR product with 50 ng of linearized vector, 2 μL of 5× In-fusion HD premix, and molecular biology grade water to 10 μL. Use the MixMate to mix well and spin down briefly to collect contents of the well.
15. Incubate for 15 min at 50 °C and then place on ice.
16. Mix on ice, 2 μL of the reaction mix with super competent DH5alpha cells in a 96-well format (*see* **Note 4**).
17. During dispensing of the expression vector, the tips of the multichannel pipette should be used to gently mix the cells/DNA. Pipetting up and down is not recommended.
18. Incubate the cells on ice for 30 min.
19. Heat shock for 45 s at 42 °C using a heated thermocycler.
20. Place the cells in the 96-well plate back on ice for 2 min.
21. Add 90 μL of pre-warmed SOC medium using a multi-dispense protocol of a 12-channel pipettor with 1250 μL tips. Take care to use a slow dispense speed so as not to shoot the contents of the well out of the plate or cross contaminate wells by touching the ends of the tips in the wells.
22. Seal the plate with an Air pore breathable film and incubate in the Thermomix R, at 37 °C, 900 rpm for 1 h.
23. Transfer 100 μL of transformation mix to a 24 DWB prefilled with 2 mL of LB/Agar/selective antibiotic: choice of antibiotic depends on the vector chosen in **step 3**.
24. Drop in 2–4 sterile glass balls and agitate until liquid has entered the agar. Discard balls by inversion.
25. Invert and grow overnight at 37 °C.

26. The next day, pick individual colonies using sterile toothpicks into 1 mL of pre-warmed LB medium with appropriate selective antibiotics in a 96 DWB.
27. Seal plate with an airpore strip and incubate in the Thermomix R overnight at 37 °C, with shaking at 900 rpm.
28. The next day, centrifuge the cell pellet at 4000 × *g* for 10 min and discard medium.
29. Extract the DNA using a NucleoSpin 96 plasmid kit (or similar) following the manufacturers' recommendations. Elute in the minimum volume of elution buffer.
30. Quantify by OD260nm the plasmid DNA recovered for a few samples to ballpark the amount of DNA recovered.

3.2 Small Scale Expression

1. Mix 1 μL (10–50 ng of vector) of sequence verified expression clone on ice with 20 μL of chemically competent BL21(DE3) pLysS T1R cells in a 96-well format.
2. During dispensing of the expression vector, the tips of the multichannel pipette should be used to gently mix the cells/DNA. Pipetting up and down is not recommended.
3. Incubate the cells on ice for 30 min.
4. Heat shock for 45 s at 42 °C using a heated thermocycler.
5. Place the cells in the 96-well plate back on ice for 2 min.
6. Add 90 μL of pre-warmed SOC medium using a multi-dispense protocol of a 12-channel pipettor with 1250 μL tips. Take care to use a slow dispense speed so as not to shoot the contents of the well out of the plate or cross contaminate wells by touching the ends of the tips in the wells.
7. Seal the plate with an airpore strip and incubate in the Thermomix R at 37 °C, 900 rpm for 1 h.
8. Transfer transformed cells to 900 μL of pre-warmed 2× TY medium with appropriate selective antibiotics in a 96 DWB.
9. Seal plate with an airpore strip and incubate in the Thermomix R overnight at 37 °C, with shaking at 900 rpm.
10. The next day, dilute the overnight culture 1:20 into 800 μL of fresh pre-warmed 2× TY medium plus antibiotics in a 96 DWB sealed with an airpore strip as in **step 9** (*see* **Note 5**).
11. When the OD 600 nm reaches ~0.6 (assayed by sampling a variety of wells), add 1 mM IPTG by addition of 8 μL of a 100 mM stock solution using the 12-channel pipette.
12. Grow the cells for a further 4–6 h.
13. Harvest by centrifugation at 4000 × *g* for 10 min.

14. Discard the spent growth medium by inverting the block, blot any drops with paper towels and seal the block with an adhesive metal foil.
15. Freeze the block at −80 °C until it is ready for use (*see* **Note 6**).

3.3 Small Scale Purification

1. Defrost the 96 DWB from the small scale expression.
2. Add 1 mL of Lysis buffer plus 0.5 mM AEBSF, 1 mM MgSO4, 1 U benzonase, 1 mg/mL lysozyme, and one protease cOmplete EDTA free table per 50 mL buffer.
3. Seal the plate with a metal foil and mix thoroughly by vortexing to ensure pellet is resuspended . Incubate for 30 min on ice.
4. Sonicate to lyse cells using a robot sonicator, or sonicator horn with a plate attachment (*see* **Note 7**) [10]. If these are not available, sonication can be performed using a sonicator with a microtip or use freeze/thaw cycles, however, detergent addition in **step 5** can alleviate this need. Keep samples cold during sonication.
5. Add DDM to ~1.5% final concentration by addition of 0.15 mL of 10% (w/v) stock DDM solution (DDM dissolved in water).
6. Shake the plate at 800 rpm on the Thermomix R/C for 2 h with temperature set at 20 °C.
7. Samples should have substantially clarified. Centrifuge for 30 min at 4000 × *g* to remove debris.
8. Seal the Thompson filter plate with the bottom seal (*see* **Note 8**).
9. Add 50 μL of a 50/50 slurry of His60 Ni Superflow resin to each well using a multichannel pipette with wide bore tips (*see* **Note 9**). Take care that resin is dispensed into the liquid.
10. Seal the block with a top seal, and incubate with shaking for 2 h to overnight on a Thermomix R at 800 rpm, 20 °C.
11. Unseal the block and remove lysate by vacuum or by centrifugation (1000 × *g*, 3 min) (Fig. 1).
12. Wash resin captured in filter plate with 1 mL of Wash buffer, allowing to drip through under gravity. Repeat twice.
13. Repeat **step 11** to partially dry the resin.
14. Seal the block with the bottom seal and add 30 μL of Elution buffer directly to the center of the well. The resin should look wet. Shake at 800 rpm for 30 min.
15. Unseal the block and place the filter plate on top of a 96-well PCR plate, then centrifuge at 1000 × *g* for 5 min to collect the eluate.
16. Transfer 20 μL of eluted protein to a fresh 96-well PCR plate and add 5 μL of 4× Laemmli SDS-PAGE sample buffer with freshly added 100 mM DTT. Seal with a metal foil.

17. Mix by brief vortexing on the MixMate (96-well program) and centrifuge samples to collect in the bottom of the wells (1000 × *g*, 1 min).
18. Incubate on ice for 20 min (*see* **Note 10**).
19. Prepare four, 26 lane Bio-Rad criterion gels per 96 sample by making fresh SDS-PAGE Tris/Glycine/SDS running buffer. Remove bottom seals and top comb as per the manufacturers' instructions and gently wash out un-polymerized reagents with running buffer using a 12-channel pipette.
20. Load 10 μL of pre-stained molecular weight markers in the first and last well.
21. Load gels with 12.5 μL of sample using the Integra Viaflo variable span 12-channel pipette so that all samples from adjacent wells are loaded on the gel next to each other (e.g., A1, A2, A3). If this pipette is not available, then 12-channel fixed pipettors can be used and sequential rows will now be interleaved on the gel due to the tip spacing (A1, B1, A2, B1, etc.).
22. Run the gel according to the manufacturers' specifications.
23. Stain and destain using SimpleBlue safe stain according to the specifications. Other Coomassie stains will work equally well. Example of the results expected are given in [10], Fig. 2, which were prepared via the method outlined above with some modifications. (*see* **Note 10**).

3.4 Medium Scale Purification

1. Clones showing positive expression in the small scale assay are expressed at a larger scale for the detergent selection analysis.
2. Prepare fresh transformations of the expression positive clones into BL21(DE3) pLysS T1R cells. Transform and recover as **steps 1–6** from Subheading 3.2.
3. Plate out transformed cells onto selective agar plates and incubate overnight at 37 °C.
4. The next day, pick single colonies and inoculate into 2 mL of 2× TY with appropriate antibiotics in a 10 mL culture tube.
5. Incubate overnight with shaking at 37 °C making sure the cap is loose for good aeration.
6. The next day, inoculate 50 mL of 2× TY (pre-warmed to 37 °C) with 1 mL of the overnight culture containing appropriate antibiotics in an Ultra Yield flask with a breathable seal.
7. Shake in an incubator shaker at 250 rpm (1″ throw) at 37 °C until OD600 nm ~1.
8. Add 1 mM IPTG and grow for a further 4–6 h.
9. Transfer culture to a 50 mL centrifuge tube and harvest the cell pellet by centrifugation at 4000 × *g* for 10 min.

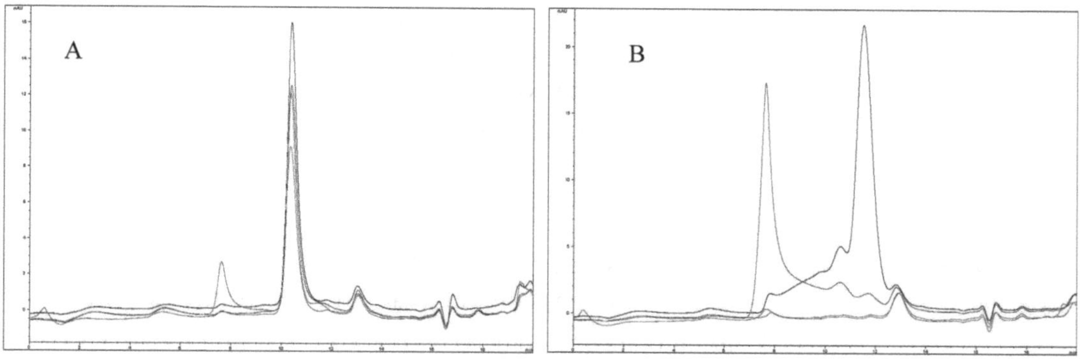

Fig. 2 Gel filtration elution profiles from size exclusion chromatography runs utilizing four detergents at 2× CMC. The membrane protein in panel **a** is well behaved in all four detergents with a sharp, symmetrical elution profile, suggesting monodisperse sample. Material in the void volume would elute at ~7.5 min. The membrane protein in panel **b** is less well behaved, showing elution in the void fraction in the detergent LDAO (*red trace*) and no elution profile from the C8E4 (*green trace*) or ßOG (*pink trace*) detergent solutions. The DDM elution (*blue*) shows some multimerization/aggregation, but the majority of sample in the included volume. Overall, the protein assayed in panel A would be more favored for further studies than that in panel **b**. Figure reproduced with permission [5]

10. Discard the spent medium, weigh the pellet and freeze at −80 °C until ready for use.
11. Defrost the pellet on ice and resuspend in 10 mL of Lysis buffer plus 0.5 mM AEBSF, 1 mM MgSO4, 1 U benzonase/ml, 1 mg/mL lysozyme, and one protease cOmplete EDTA free table per 50 mL of buffer.
12. Sonicate on ice for 1 min until the pellet is fully disrupted. Avoid significant heating of the sample.
13. Add DDM detergent at approximately 0.1–0.2 g per g of cell pellet.
14. Cap the tube and mix end over end for 2 h; the lysate should clear substantially.
15. Clarify further by centrifugation for 30 min at maximum g (15,000 × *g*).
16. Transfer to fresh tube and add 100 μL of a 50:50 slurry of His60 Ni Superflow resin. Incubate with gentle mixing at 4 °C for 1 h to overnight.
17. Pour the lysate/resin into a fresh Econo-Pac column and allow lysate to flow through.
18. Wash captured resin with 10 mL of Wash buffer and repeat twice.

19. Allow all the Wash buffer to flow through. Cap the column and add 75 µL of Elution buffer. Allow to sit for 5 min before de-capping and collecting the elution.
20. Check the protein by SDS-PAGE using 5 µL of sample and following **steps 19–23** in Subheading 3.3. Clear bands of the target protein should be visible.

3.5 Detergent Selection

1. Attach a Sepax SRT-C SEC-500 or Superdex 200 Increase (3.2 × 300 mm) column to an HPLC or FPLC system.
2. Equilibrate the column with a least 4 column volumes of gel filtration column buffer contain the detergent to be assayed at 2× CMC; typically starting with DDM at 0.02% (w/v). Follow the manufacturers' recommendations for flow rate and back pressure.
3. Inject sample from the medium scale purification (10 µL) onto the column and monitor the elution profile by recording the absorbance at 280 nm (*see* **Note 11**).
4. After the run has completed, make all further protein injections.
5. Equilibrate the column in the new detergent containing buffer and repeat **steps 2–4** for all other detergents to be assayed (see Table 1 for suggested detergents and concentrations).
6. From the elution profiles, identify proteins that do not aggregate (i.e., little material in the void volume) and have symmetrical, mono disperse elution profiles. Proteins that have these properties in multiple detergents, particularly short chain detergents such as OG and LDAO, can be confidently perused for further study and scale up, in particular crystallization trials [10, 17].

4 Notes

1. As single amino acid changes or indeed single codons can radically alter expression or stability of the final protein product, it is worth maximizing diversity of target templates to maximize successes. Most high throughput structural genomics type approaches operate on a "funnel" type method—where many targets enter the start of the pipeline with only a few successes exiting the end; so it is necessary to have many targets initially [11].
2. If the same template is being used in all reactions (i.e., creating a series of truncation constructs) then this DNA can be included in the master mix.

3. The cut vector must be checked for very low number of background colonies. Do not proceed with further experiments until this has been minimized. Vectors with negative selection elements in the cloning region are useful, such as those containing SacB or ccdB [18].
4. It is vital to pretest the success of the ligation on a few samples, in conjunction with the batch of competent cells that are to be used, to ensure that an adequate number of colonies will appear on the agar plate. Do not attempt 96 transformations until the system has been 'ranged in'.
5. Other growth media, such as TB or auto-induction media may be investigated as they can give different expression patterns. Auto-induction media is particularly useful as it is labor saving due to no OD600nm measurments to assess IPTG induction density is necessary [19].
6. Adhesive foils may become dislodged in the −80 °C freezer due to distortion of the plasticware or failure of the adhesive. Heat sealable adhesives are more effective for long term storage.
7. Sonic disruption in 96-well plates is not standard for most labs. Multi-probe horns and plate horns are available, but may suffer from uneven energy distribution so care must be taken in their usage.
8. The bottom seal for the filter plate must be properly applied and firmly pressed down. Do not attempt to seal filter plates with adhesive films etc., as they will not adhere. If bottom seals are not available, then the incubation would be better performed in a deep well block before transfer to a filter plate.
9. It can be problematic to dispense small volumes of resin slurries due to settling. Frequent mixing of the resin slurry is necessary and the user should observe the aspiration of the slurry to ensure equal amounts are dispensed.
10. Membrane protein samples must NOT be boiled as the may not enter the gel matrix using electrophoresis. Membrane proteins often migrate more rapidly than soluble proteins on SDS-PAGE gels (~10%) and may maintain oligomeric states (dimers, trimers, etc.) in the gel matrix [20]. Silver staining or western blotting could also be utilized if very low expression levels are generated, however, proteins that express very poorly may not be tractable for further study.
11. The amount of sample to be loaded onto the column may need to be empirically determined. Other absorption wavelengths (e.g., 220 nm) could also be monitored as they may give greater signal. If available, multi-angle light scattering coupled with refractive index measurements can be used accurately to

determine the aggregation status and molecular weight of the protein/detergent complex and the amount of free detergent. These values can be useful in optimizing protein samples in crystallization trials [21, 22].

References

1. Kloppmann E, Punta M, Rost B (2012) Structural genomics plucks high-hanging membrane proteins. Curr Opin Struct Biol 22:326–332
2. Willis MS, Koth CM (2008) Structural proteomics of membrane proteins: a survey of published techniques and design of a rational high throughput strategy. Methods Mol Biol 426:277–295
3. Punta M, Love J, Handelman S et al (2009) Structural genomics target selection for the New York consortium on membrane protein structure. J Struct Funct Genomics 10: 255–268
4. Eshaghi S (2009) High-throughput expression and detergent screening of integral membrane proteins. Methods Mol Biol 498: 265–271
5. Love J, Mancia F, Shapiro L et al (2010) The New York Consortium on Membrane Protein Structure (NYCOMPS): a high-throughput platform for structural genomics of integral membrane proteins. J Struct Funct Genomics 11:191–199
6. Graslund S, Nordlund P, Weigelt J et al (2008) Protein production and purification. Nat Methods 5:135–146
7. Xiao R, Anderson S, Aramini J et al (2010) The high-throughput protein sample production platform of the northeast structural genomics consortium. J Struct Biol 172:21–33
8. Chaudhary S, Pak JE, Pedersen BP et al (2011) Efficient expression screening of human membrane proteins in transiently transfected human embryonic kidney 293S cells. Methods 55: 273–280
9. Li M, Hays FA, Roe-Zurz Z et al (2009) Selecting optimum eukaryotic integral membrane proteins for structure determination by rapid expression and solubilization screening. J Mol Biol 385:820–830
10. Mancia F, Love J (2010) High-throughput expression and purification of membrane proteins. J Struct Biol 172:85–93
11. Lewinson O, Lee AT, Rees DC (2008) The funnel approach to the pre-crystallization production of membrane proteins. J Mol Biol 377:62–73
12. Gustafsson C, Minshull J, Govindarajan S et al (2012) Engineering genes for predictable protein expression. Protein Expr Purif 83:37–46
13. Savitsky P, Bray J, Cooper CDO et al (2010) High-throughput production of human proteins for crystallization: the SGC experience. J Struct Biol 172:3–13
14. Eschenfeldt WH, Lucy S, Millard CS et al (2009) A family of LIC vectors for high-throughput cloning and purification of proteins. Methods Mol Biol 498:105–115
15. Everett JK, Acton TB, Montelione GT (2004) Primer Prim'er: a web based server for automated primer design. J Struct Funct Genomics 5:13–21
16. Zielinski D, Gordon A, Zaks BL et al (2014) iPipet: sample handling using a tablet. Nat Methods 11:784–785
17. Newby ZE, O'Connell JD, Gruswitz F et al (2009) A general protocol for the crystallization of membrane proteins for X-ray structural investigation. Nat Protoc 4:619–637
18. Bernard P, Couturier M (1992) Cell killing by the F plasmid CcdB protein involves poisoning of DNA-topoisomerase II complexes. J Mol Biol 226:735–745
19. Studier FW (2005) Protein production by auto-induction in high density shaking cultures. Protein Expr Purif 41:207–234
20. Rath A, Glibowicka M, Nadeau VG et al (2009) Detergent binding explains anomalous SDS-PAGE migration of membrane proteins. Proc Natl Acad Sci U S A 106:1760–1765
21. Slotboom DJ, Duurkens RH, Olieman K et al (2008) Static light scattering to characterize membrane proteins in detergent solution. Methods 46:73–82
22. Miercke LJ, Robbins RA, Stroud RM (2014) Tetra detector analysis of membrane proteins. Curr Protoc Protein Sci 77:1–30

Chapter 18

Multiprotein Complex Production in *E. coli*: The SecYEG-SecDFYajC-YidC Holotranslocon

Imre Berger, Quiyang Jiang, Ryan J. Schulze, Ian Collinson, and Christiane Schaffitzel

Abstract

A modular approach for balanced overexpression of recombinant multiprotein complexes in *E. coli* is described, with the prokaryotic protein secretase/insertase complex, the SecYEG-SecDFYajC-YidC holotranslocon (HTL), used as an example. This procedure has been implemented here in the ACEMBL system. The protocol details the design principles of the monocistronic or polycistronic DNA constructs, the expression and purification of functional HTL and its association with translating ribosome nascent chain (RNC) complexes into a RNC-HTL supercomplex.

Key words Multiprotein complexes, Membrane proteins, Protein insertion and secretion, Holotranslocon HTL, ACEMBL system, Cre recombinase, Donor–acceptor fusion, Subunit stoichiometry, Ribosome nascent chain complex RNC

1 Introduction

About a third of the proteome in living cells is integrated into or transported across membranes, catalyzed by the ubiquitous Sec machinery which is conserved in all kingdoms of life. In prokaryotes, the Sec machinery contains the core translocon, SecYEG. Regulatory subunits including SecDF, YajC and YidC can associate with the core translocon giving rise to the SecYEG-SecDFYajC-YidC holotranslocon complex [1–4]. All subunits in this heptameric holotranslocon are transmembrane proteins [3, 5].

Heterologous expression has been instrumental to advance protein research in the life sciences, and *E. coli* has largely dominated the field of recombinant expression, a trend which remains unbroken to date, although mammalian and insect cell based eukaryotic expression systems have been making their impressive mark more recently [6–9]. *E. coli* remains unsurpassed as a simple and cost-effective expression host for protein production, and a very large number of plasmids and host strains are conveniently

Nicola A. Burgess-Brown (ed.), *Heterologous Gene Expression in E. coli: Methods and Protocols*, Methods in Molecular Biology, vol. 1586, DOI 10.1007/978-1-4939-6887-9_18, © Springer Science+Business Media LLC 2017

available to the research community for recombinant expression in this prokaryotic system.

In living cells, the catalysts of biological activity are typically not single, isolated proteins, but protein complexes which can contain many different subunits [10, 11]. Many complexes are characterized by low abundance and heterogeneity in their native hosts, which is refractory to their extraction from native source material, and necessitates recombinant overexpression to purify them in sufficient amounts for detailed mechanistic studies. The SecYEG-SecDFYajC-YidC holotranslocon transmembrane protein complex, HTL, has been discovered in *E. coli* membranes, where it catalyzes the transport of membrane-bound or periplasmic proteins as they emerge from the actively translating ribosome [3–5, 12–14]. A particular challenge in expressing functional complexes such as HTL can be encountered when many heterologous subunits are co-produced from individual expression cassettes, resulting in imbalanced expression levels of the individual proteins compromising proper complex assembly. Certain subunits can be expressed either more weakly or strongly, and, occasionally, one subunit may dominate the recombinant overexpression experiment to a degree that it becomes detrimental to overall yield. As a consequence, a complex containing all subunits at their physiological stoichiometric ratios cannot be obtained. Moreover, the HTL consists of assembly blocks which by themselves can form stable entities (SecYEG, SecDF, and YidC), further complicating HTL production and purification as a homogeneous, functional assembly, pre-necessitating elaborate promoter and expression cassette design, tag placement and systematic trial-and-error to achieve success [13–15]. Originally, only six subunits (SecYEG, SecDF, and YidC) were thought to be present in HTL and thus included in the overexpression experiments [15]. Later, a seventh subunit, YajC, was identified and added to the overexpression setup to yield complete, functional and stable heptameric HTL [12–14].

In order to meet these challenges and to produce and purify HTL in the quality and quantity required for detailed study, we developed ACEMBL, a modular overexpression system for protein complex expression in *E. coli* (Fig. 1) [15–17]. ACEMBL affords the means to assemble monocistronic or polycistronic expression cassettes to optimally balance expression levels and achieve properly assembled complexes with correct subunit stoichiometry. ACEMBL consists of small, custom designed expression plasmids containing all elements required for expressing recombinant proteins in *E. coli* as an expression host (Fig. 1). Two families of plasmids exist, Acceptors and Donors (Fig. 1). Acceptors and Donors can each contain one or several genes of interest, arranged in (1) single gene expression cassettes consisting of a promoter, a gene, and a transcriptional terminator, or (2) in the form of polycistrons where several genes are transcribed from the same promoter, or a

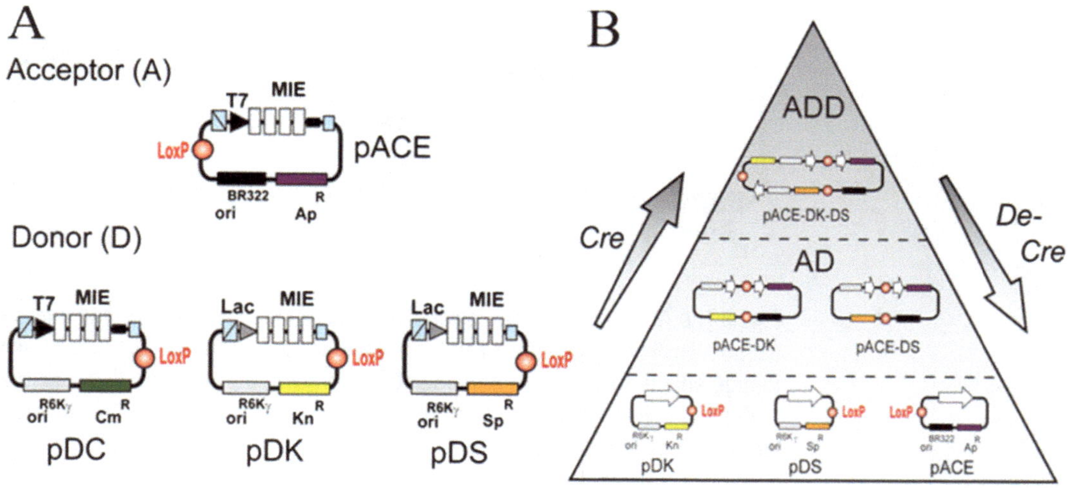

Fig. 1 ACEMBL system components. (**a**) ACEMBL consists of Donor (pDC, pDK, pDS) and Acceptor (pACE) plasmids into each of which one or several genes, optionally as polycistrons, can be inserted by a variety of cloning routines. *Ap* ampicillin, *Cm* chloramphenicol, *Kn* kanamycin, *Sp* spectinomycin. R6Kψ, phage-derived conditional origin of replication; BR322, regular *E. coli* replicon. Transcriptional terminators are indicated as *boxes* filled in *black*. A multiplication modules is shown as *boxes* filled in *blue*. LoxP sequences are indicated as *circles* filled in *red*. T7 and Lac promoters are shown as *triangles* filled in *black* or *grey*, respectively. Promoters and terminators can be exchanged to different promoters (ara, trc, others) if needed. MIE stands for multi-integration element and indicates a polylinker that facilitates polycistron assembly [15]. (**b**) Cre-LoxP mediated fusion of one or several Donors with an Acceptor, each carrying one or several expression cassettes of interest, results in multigene expression constructs for protein complex production (diagram adapted from [16]). The Cre reaction is an equilibrium reaction with the excision reaction being preferred over fusion. When equilibrium is reached, Donor and Acceptor plasmids coexist with higher order multigene fusion constructs (AD, ADD, …). All DNA entities, educts and products, present in the Cre reaction are quasi bar-coded by their resistance marker combinations and can be identified by transformation and subsequent challenge with antibiotics [15]

combination of both [15]. Donors and Acceptors, each containing one or several genes of interest, are conveniently fused by Cre recombinase to yield Donor–Acceptor fusions which contain all genes that need to be co-produced. Promoters, genes and terminators are in BioBrick design [18] and exchangeable in the individual Donors and Acceptors, a feature which enabled us to test the best combinations for producing functional HTL in parallel by trial and error within a reasonable timeframe until a suitable HTL expression construct was identified (Fig. 2). ACEMBL can be implemented in a robotics environment [15, 16], however, for most applications in our laboratory (including HTL) production, the manual approach, optionally with a multichannel pipette, is sufficient to achieve success. We have described earlier the conceptual use of ACEMBL to generate HTL expression constructs [13–15]. In the present contribution, we describe the principle of HTL multigene expression construct assembly which can be used as a blueprint for other transmembrane or soluble complexes. We highlight important considerations for the production and purification

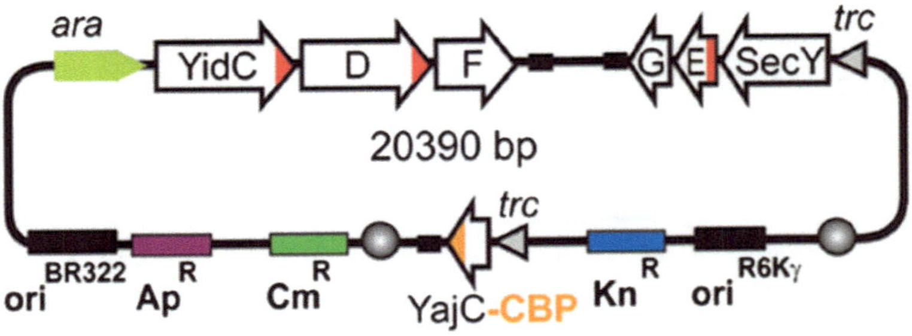

Fig. 2 The ACEMBL HTL expression construct. The hexa-histidine tags and the calmodulin-binding peptide (CBP) tag are marked in *red* and *orange* respectively (image adapted from [13]). *ara* arabinose promoter, *trc* modified, tight lac promoter. LoxP sites are depicted as *circles* filled in *grey*, transcriptional terminators are shown as *boxes* filled in *black*

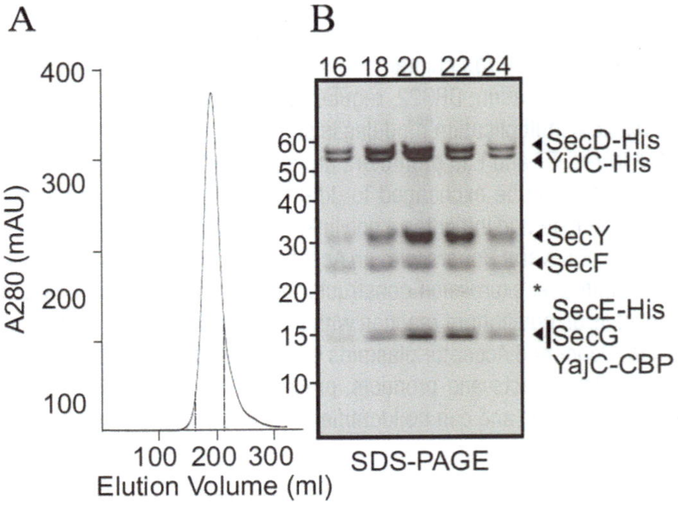

Fig. 3 Purification of the *E. coli* HTL. (**a**) Representative gel filtration/ion exchange elution profile of the HTL evidencing a single peak, demonstrating the co-purification of all HTL subunits in a single complex. (**b**) SDS-PAGE analysis of peak fractions 16–24 (corresponding to elution volumes of the profile displayed in (**a**) marked by *dotted lines*). The *asterisk* represents a proteolytic break-down fragment of SecY

of functional HTL (Fig. 3), and describe the assembly of HTL with actively transcribing ribosome nascent chain complexes (RNCs) into a RNC-HTL supercomplex (Fig. 4).

2 Materials

We strongly recommend carrying out the design of all expression constructs in silico using a DNA cloning software of choice (i.e., VectorNTI, ApE, others). Gene synthesis has become very affordable,

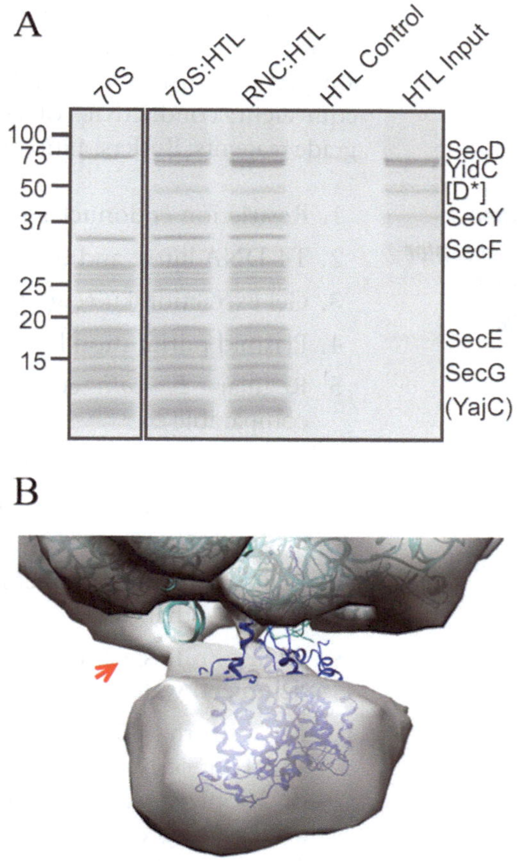

Fig. 4 HTL-RNC supercomplex. (**a**) Interaction of the HTL^{DDM} with ribosome nascent chain complexes (RNC) analysed by sedimentation experiments and SDS-PAGE. Molecular weight marker is shown on the left. Coomassie stained protein bands corresponding to HTL components are marked on the right. (**b**) Atomic model of the ribosome-SecYEG complex fitted into a preliminary cryo-EM structure of the RNC^{FtsQ}-HTL^{DDM} complex. The density of the HTL is larger than SecYEG alone, and an additional strong connection to the ribosome was identified (indicated by the *red arrow*)

and we prefer synthetic genes for generating the individual genes of interest, in which internal restriction sites for subcloning into the ACEMBL Donor and Acceptor plasmids are eliminated by design. We also recommend codon optimization for expression of the genes of interest, which is a service provided at no cost from synthetic DNA suppliers, and typically includes removal of potentially problematic RNA secondary structure elements in the transcripts. If synthetic genes are used, we further recommend to additionally eliminate any restriction sites that are part of the so-called multiplication modules in the plasmids [15, 16]. This allows for maximum flexibility of gene assembly for co-expression (e.g., if purification tag placement needs to be revised in later iterations).

The modular concept of ACEMBL furthermore allows transferring expression cassettes between various plasmids [15]. All reagents are prepared using ultrapure water (Millipore Milli-Q system or equivalent; conductivity of 18.2 MΩ cm at 25 °C) and analytical grade reagents. Buffers, antibiotics, and enzymes are stored at −20 °C.

2.1 Generation of Donor and Acceptor Plasmids for HTL Subassemblies

1. Restriction endonucleases and reaction buffers.
2. T4 DNA ligase and buffer.
3. Gel extraction kit (e.g., Qiagen).
4. Plasmid purification kit (e.g., Qiagen).
5. Regular *E. coli* competent cells (TOP10, HB101, or comparable).
6. *E. coli* competent cells containing *pir* gene (for Donor plasmids).
7. 100 mg/mL ampicillin (1000×): Dissolve in distilled or deionized water.
8. 25 mg/mL chloramphenicol stock solution (1000×): Dissolve in 100% ethanol.
9. 50 mg/mL spectinomycin (1000×): Dissolve in distilled or deionized water.
10. Luria Broth (LB): Weigh 10 g of tryptone, 5 g of yeast extract, and 10 g of NaCl. Suspend the solids in ~800 mL of distilled or deionized water. Add further distilled or deionized water in a measuring cylinder to ensure accuracy, to make a total of 1 L. Autoclave at 120 °C for 20 min.
11. LB agar: Weigh 10 g of tryptone, 5 g of yeast extract, 10 g of NaCl, and 15 g of agar. Suspend the solids in ~800 mL of distilled or deionized water. Add further distilled or deionized water in a measuring cylinder to ensure accuracy, to make a total of 1 L. Autoclave at 120 °C for 20 min.

2.2 Generation of Multigene Donor–Acceptor Fusions Expressing Complete HTL

1. *E. coli* competent cells (TOP10, HB101, or comparable).
2. Cre recombinase enzyme (e.g., New England Biolabs).
3. Antibiotics listed in Subheading 2.1.
4. Media (LB and agar) listed in Subheading 2.1.
5. DNA gel extraction kits (e.g., from Qiagen).
6. UV spectrophotometer (e.g., Thermo Scientific NanoDrop 2000).
7. 1.5 mL Eppendorf tubes.

2.3 Production of HTL

1. *E. coli* competent cells C43(DE3).
2. ACEMBL:holo-translocon (HTL) multigene expression plasmid encoding for YidC, SecYEG, SecDF, and YajC fitted with appropriate affinity purification tags [13, 14].

3. 2× TY broth with antibiotics: Measure ~900 mL of distilled H_2O. Add 16 g of Bacto tryptone, 10 g of Bacto yeast extract, and 5 g of NaCl. Adjust pH to 7.0 with 5 N NaOH and make up to 1 L with distilled H_2O. Sterilize by autoclaving. After cooling, add ampicillin, chloramphenicol and spectinomycin (stock concentrations listed in Subheading 2.1).
4. 100 mM isopropyl β-D-1-thiogalactopyranoside IPTG (100×).
5. L-(+)-Arabinose (>99%).
6. Cell disruptor (e.g., Constant Systems, Ltd.).
7. TSG130 buffer: 20 mM Tris–HCl (pH 8.0), 130 mM NaCl, 10% (v/v) glycerol.
8. *n*-Dodecyl-β-D-maltoside (DDM) (e.g., Sigma Aldrich).
9. Chelating Ni^{2+}-Sepharose Fast Flow column (GE Healthcare).
10. Superdex 200, 26/60 gel filtration column (GE Healthcare).
11. Q-Sepharose ion exchange column (GE Healthcare).
12. 50-kDa molecular weight cutoff centrifugation filter (Amicon).

2.4 Preparing HTL-Ribosome Nascent Chain Supercomplex

1. Purified ribosome nascent chain (RNC) complexes comprising a 108-amino acid long FtsQ nascent chain with a signal sequence (for detailed preparation protocol *see* [22]).
2. DDM solubilized purified HTL from Subheading 2.3.

3 Methods

The genes encoding for the individual subunits are designed in silico, and then inserted into the Donor and Acceptor plasmid of choice. Once designed, monocistronic or polycistronic expression cassettes can be created by a variety of means including DNA synthesis, restriction/ligation cloning, ligation independent cloning (LIC) or sequence and ligation independent cloning (SLIC) or other methods [19, 20], according to individual user preference. We recommend custom DNA synthesis to facilitate expression cassette construction, in particular if polycistronic expression cassettes are used as in the case of HTL.

3.1 In Silico Design

1. Group genes into functional units based on a set of criteria (known interaction partners, physiological (sub)assemblies, here YidC, SecYEG, SecDF, and YajC). GenBank identifiers for HTL subunits are: SecY, WP_001118868.1; SecE, WP_001275702.1; SecG, AAN82372.1; SecD, ODA87210.1; SecF, AAN78997.1; YidC, WP_000378250.1; YajC, WP_000007629.1.
2. Decide on the number of expression cassettes to be co-expressed and the Donor and Acceptor plasmids you plan to use.

3. Decide on placement of tags and, optionally, on proteolytic sites to remove them (we recommend Tobacco etch virus (TEV) NIa protease and/or precision protease).
4. Decide on the use of polycistrons or individual expression cassettes, and the promoters to be tested. These can be the promoters provided by the standard ACEMBL plasmids (T7, lac), or include different promoters, as in the case of HTL, for example arabinose and trc promoters.
5. Generate the DNA sequence. Decide on DNA assembly strategy (SLIC, restriction/ligation, PCR assembly, others). Add custom restriction sites to 5′ and 3′ ends which are compatible with the polylinkers in the Donor and Acceptor plasmids in case you use restriction/ligation-based cloning.
6. Create all DNA sequences in silico and validate by simulating the reading frame in the in silico plasmid sequences.

3.2 Preparation of Donor and Acceptor Plasmids

1. Choose from ACEMBL system components pACE1, pACE2, pDK, pDC, pDS to generate the expressing construct of choice. If promoters other than those already present on the original plasmids (T7, lac) are to be used, we recommend to create the "empty" expression cassette first in the format promoter-polylinker-terminator, substitute the original expression cassette, and keep the resulting new Donor or Acceptor plasmid, fitted with the preferred promoter, in stock for future expressions of different complexes. All ACEMBL plasmids contain a LoxP site for generating Donor–Acceptor fusions by Cre recombinase. In the case of the HTL expression, an arabinose promoter (controlling a YidC-SecDF polycistron) and two trc promoters (controlling the SecYEG polycistron or the YajC gene, respectively), resulted in the construct yielding properly assembled complex [13, 14].
2. Digest several micrograms of Donor or Acceptor plasmid by the restriction enzymes selected by in silico design according to manufacturers' recommendation.
3. Analyze the digestions by agarose gel electrophoresis to confirm that the digestions are complete.
4. Purify digested plasmid by using commercial gel extraction kits. Elute the extracted DNA in the minimal volume defined by the manufacturer. Determine the concentration of the extracted DNA by UV measurement with spectrophotometer. Store in frozen aliquots in Eppendorf tubes.

3.3 Inserting Genes into Digested Donor or Acceptor

1. Digest several micrograms of the DNA (generated by DNA synthesis, SLIC, PCR assembly or other method of choice) encoding for the desired individual gene or polycistron (the "insert") with the enzymes identified by in silico design,

according to the manufacturers' recommendation. For polycistron generation, follow the guidelines and protocols [15].

2. Purify digested insert DNA by using a commercial gel extraction kit. Elute the extracted DNA in the minimal volume defined by the manufacturer. Determine the concentration of the extracted DNA spectrophotometrically.
3. Set up ligation reactions by mixing purified insert and digested and gel extracted Donor or Acceptor, respectively (*see* Subheading 3.2) in a 10–20 μL reaction volume with T4 DNA ligase and perform ligation reactions at 25 °C overnight. Optionally, analyze the ligation reaction by agarose gel electrophoresis to evaluate the ligation efficiency.
4. Transform *E. coli* competent cells (TOP10 or HB101 for Acceptors, *pir* gene containing cells for Donors; *see* **Note 1**) with ligation reaction mixture. Incubate the transformation reaction in a 37 °C shaker for several hours and plate on agar plates in a dilution series to ensure optimal colony separation.
5. Pick colonies, grow minicultures and purify plasmids using standard procedures.
6. Identify positive clones by restriction digestion and DNA sequencing of the insert.

3.4 Cre-LoxP Fusion of Donors and Acceptor

1. Prepare a 20 μL reaction mixture for Cre reaction by combining 100 ng each of Donor and Acceptor plasmids. Add Cre enzyme and Cre buffer according to the recommendations of the supplier. Incubate at 30 °C.
2. Transform regular *E. coli* competent cells (TOP10, HB101, or comparable) with Cre reaction mixture following standard transformation protocols.
3. Incubate the transformation reaction in a 37 °C shaker for one or several hours.
4. Plate the transformation reaction on agar plates containing the proper antibiotics combination (for HTL: ampicillin, cloramphenicol, spectinomycin) and incubate at 37 °C overnight.
5. Pick colonies and inoculate 25 mL aliquots of LB medium supplemented with corresponding antibiotics (*see* Subheading 2.1).
6. After overnight incubation, prepare Acceptor–Donor fusion plasmid using standard kits (e.g., Qiagen).
7. Predict Acceptor–Donor fusion plasmid sequence by using web-based Cre-ACEMBLER software [21].
8. Check plasmids by restriction digestion using appropriate enzymes identified by restriction pattern prediction (*see* **Note 2**).

3.5 Producing Recombinant HTL

1. Grow freshly transformed *E. coli* C43(DE3) with ACEMBL plasmid encoding for HTL (Fig. 2) in 2× YT broth with antibiotics to $OD_{600} = 0.8$.
2. Induce for 3 h by adding 1 mM IPTG and 0.2% (w/v) arabinose.
3. Pellet cells by centrifugation (4000 × *g*) and break cells at 25 kpsi with cell disruptor in TSG130 buffer.
4. Collect membrane fraction containing HTL and solubilize by rotation in TSG130 containing 2% (w/v) DDM for 1 h at 4 °C.
5. Clarify DDM-soluble fraction by centrifugation and purify by using Ni^{2+} metal affinity chromatography (column pre-equilibrated in TSG130 containing 0.1% DDM).
6. Wash column thoroughly (ten column volumes) with buffer containing 30 mM imidazole.
7. Elute bound HTL with Elution buffer containing 500 mM imidazole.
8. Pool peak fractions and purify immediately using a S200 size exclusion column, placed in-line with a Q-Sepharose column, both columns equilibrated in TSG130 + 0.05% DDM.
9. Pool peak fractions and concentrate (50 kDa cutoff Amicon filter).
10. Concentrate to 5–10 mg/mL (molar extinction coefficient: $\varepsilon_{HTL} = 497{,}000\ M^{-1}\ cm^{-1}$) [13].

3.6 Assembling RNC-HTL Supercomplex

1. Mix DDM-solubilized HTL and purified RNC^{FtsQ} in a 25:1 molar ratio.
2. Isolate resulting RNC^{FtsQ}-HTL supercomplex by sedimentation centrifugation through a sucrose cushion to separate unbound HTL from ribosome associated HTL as described earlier for the core translocon, SecYEG, bound to RNC [21]. The sucrose cushion separates particles according to their densities. Unbound HTL will float while RNC-HTL supercomplexes will permeate the cushion.

4 Notes

1. Donors and their derivatives can only be propagated in cells that express the *pir* gene (such as BW23473, BW23474, or PIR1 and PIR2, Invitrogen) due to the conditional origin present on these plasmids [23]. In contrast, Acceptors and their derivatives contain regular ColE1 origin of replication and can be propagated in regular *E. coli* strains (TOP10, HB101, or comparable).

2. Cre-ACEMBLER application software can be downloaded from https://github.com/christianbecke/Cre-ACEMBLER/downloads/.

Accompanying information can be downloaded from www.embl.fr/multibac/multiexpression_technologies/cre-acembler/.

Acknowledgments

We thank all members of the Berger, Schaffitzel, and Collinson laboratories for helpful discussions. The authors thank Sir John Walker for the *E. coli* C43 expression strain. J.K. was supported by a doctoral training grant from the BBSRC. C.S. is supported by a European Research Council ERC Starting Grant Award. I.C. acknowledges support by the BBSRC (Project Grants BB/M003604/1 and BB/I008675/1) and is recipient of a Wellcome Trust Senior Investigator Award. I.B. acknowledges support from the European Commission Framework Programme 7 ComplexINC project (contract no. 279039).

References

1. Arkowitz RA, Wickner W (1994) SecD and SecF are required for the proton electrochemical gradient stimulation of preprotein translocation. EMBO J 13:954–963
2. Beck K, Eisner G, Trescher D et al (2001) YidC, an assembly site for polytopic *Escherichia coli* membrane proteins located in immediate proximity to the SecYE translocon and lipids. EMBO Rep 2:709–714
3. Duong F, Wickner W (1997) The SecDFyajC domain of preprotein translocase controls preprotein movement by regulating SecA membrane cycling. EMBO J 16:4871–4879
4. Scotti PA, Urbanus ML, Brunner J et al (2000) YidC, the *Escherichia coli* homologue of mitochondrial Oxa1p, is a component of the Sec translocase. EMBO J 19:542–549
5. Duong F, Wickner W (1997) Distinct catalytic roles of the SecYE, SecG and SecDFyajC subunits of preprotein translocase holoenzyme. EMBO J 16:2756–2768
6. Nettleship JE, Assenberg R, Diprose JM et al (2010) Recent advances in the production of proteins in insect and mammalian cells for structural biology. J Struct Biol 172:55–65
7. Bieniossek C, Imasaki T, Takagi Y et al (2012) MultiBac: expanding the research toolbox for multiprotein complexes. Trends Biochem Sci 37:49–57
8. Nettleship JE, Watson PJ, Rahman-Huq N et al (2015) Transient expression in HEK 293 cells: an alternative to *E. coli* for the production of secreted and intracellular mammalian proteins. Methods Mol Biol 1258:209–222
9. Van Oers MM, Pijlman GP, Vlak JM (2015) Thirty years of baculovirus-insect cell protein expression: from dark horse to mainstream technology. J Gen Virol 96:6–23
10. Kodadek T (2008) Rise of the machines: Bruce Alberts and the biochemistry of multi-protein complexes. Mol Biosyst 4:1043–1045
11. Nie Y, Viola C, Bieniossek C et al (2009) Getting a grip on complexes. Curr Genomics 10:558–572
12. Komar J, Alvira S, Schulze R et al. (2016) Membrane protein insertion and assembly by the bacterial holo-translocon SecYEG-SecDF-YajC-YidC. Biochem J, July 19. pii: BCJ20160545
13. Schulze RJ, Komar J, Botte M et al (2014) Membrane protein insertion and proton-motive-force-dependent secretion through the bacterial holo-translocon SecYEG-SecDF-

YajC-YidC. Proc Natl Acad Sci U S A 111: 4844–4849

14. Komar J, Botte M, Collinson I, Schaffitzel C et al (2015) ACEMBLing a multiprotein transmembrane complex: the functional SecYEG-SecDF-YajC-YidC Holotranslocon protein secretase/insertase. Methods Enzymol 556:23–49
15. Bieniossek C, Nie Y, Frey D et al (2009) Automated unrestricted multigene recombineering for multiprotein complex production. Nat Methods 6:447–450
16. Vijayachandran LS, Viola C, Garzoni F et al (2011) Robots, pipelines, polyproteins: enabling multiprotein expression in prokaryotic and eukaryotic cells. J Struct Biol 175:198–208
17. Haffke M, Marek M, Pelosse M et al (2015) Characterization and production of protein complexes by co-expression in *Escherichia coli*. Methods Mol Biol 1261:63–89
18. Shetty RP, Endy D, Knight TF Jr (2008) Engineering BioBrick vectors from BioBrick parts. J Biol Eng 2:5
19. Celie PH, Parret AH, Perrakis A (2016) Recombinant cloning strategies for protein expression. Curr Opin Struct Biol 38: 145–154
20. Benoit RM, Ostermeier C, Geiser M et al (2016) Seamless insert-plasmid assembly at high efficiency and low cost. PLoS One 11:e0153158
21. Nie Y, Chaillet M, Becke C et al (2016) ACEMBL tool-kits for high-throughput multigene delivery and expression in prokaryotic and eukaryotic hosts. Adv Exp Med Biol 896:27–42
22. Schaffitzel C, Ban N (2007) Generation of ribosome nascent chain complexes for structural and functional studies. J Struct Biol 158:463–471
23. Metcalf WW, Jiang W, Wanner BL (1994) Use of the rep technique for allele replacement to construct new *Escherichia coli* hosts for maintenance of R6Kλ origin plasmids at different copy numbers. Gene 138:1–7

Chapter 19

Membrane Protein Production in *E. coli* Lysates in Presence of Preassembled Nanodiscs

Ralf-Bernhardt Rues, Alexander Gräwe, Erik Henrich, and Frank Bernhard

Abstract

Cell-free expression allows to synthesize membrane proteins in completely new formats that can relatively easily be customized for particular applications. Amphiphilic superstructures such as micelles, lipomicelles, or nanodiscs can be provided as nano-devices for the solubilization of membrane proteins. Defined empty bilayers in the form of nanodiscs offer native like environments for membrane proteins, supporting functional folding, proper oligomeric assembly as well as stability. Even very difficult and detergent-sensitive membrane proteins can be addressed by the combination of nanodisc technology with efficient cell-free expression systems as the direct co-translational insertion of nascent membrane proteins into supplied preassembled nanodiscs is possible. This chapter provides updated protocols for the synthesis of membrane proteins in presence of preassembled nanodiscs suitable for emerging applications such as screening of lipid effects on membrane protein function and the modulation of oligomeric complex formation.

Key words G-protein coupled receptors, Nanodiscs, Synthetic biology, Membranes, Oligomerization, Lipid screening

1 Introduction

Cell-free production of proteins is a core technology in the rapidly emerging field of synthetic biology. The absence of cell boundaries, the elimination of physiological requirements and the general reduction in biological system complexity has opened new avenues for protein biosynthesis. Cell lysates can be combined with synthetic environments in order to customize protein expression processes [1]. Specific requirements for protein folding and stability can thus be addressed and fine-tuned for optimal sample quality. Cell-free synthetic biology has in particular completely redesigned production pathways for membrane proteins. The co-translational insertion of membrane proteins into supplied nanodisc membranes enables functional or structural studies in natural lipid environments of defined composition. Synthesis of membrane protein/nanodisc

Nicola A. Burgess-Brown (ed.), *Heterologous Gene Expression in E. coli: Methods and Protocols*, Methods in Molecular Biology, vol. 1586, DOI 10.1007/978-1-4939-6887-9_19, © Springer Science+Business Media LLC 2017

complexes can be completed within 1 day and the complexes are subsequently purified out of the reaction mixtures by conventional affinity chromatography using small purification tags either attached to the target protein or to the nanodiscs. The complete production and purification process can be performed without any contacts to detergents or critical processing steps such as membrane extraction or exchange of hydrophobic environments.

Both cell-free synthetic biology and the nanodisc technology are rapidly evolving platforms and new protocols, modifications or refinements are continuously emerging. Nanodiscs can be assembled in different sizes, they are very stable, and they can be combined with a fast growing diversity of lipids and lipid mixtures [2]. The planar nanodisc membranes are accessible from both sides and provide ideal environments for the analysis of membrane integrated enzymes or ligand binding proteins such as G-protein coupled receptors. The efficient production of such membrane proteins by cell-free synthetic biology is an excellent synergy giving fast access to these formerly very difficult to obtain targets [3]. Cell-free systems are highly tolerant for a large variety of additives and the open accessibility of the expression reactions allows the supply of preformed nanodiscs even in combination with ligands or other stabilizers [4, 5].

The combination of the two technical platforms enables the straightforward evaluation of lipid effects on stability, functional folding, or conformational features of membrane proteins. We exemplify a strong lipid dependent variation in the functional folding of the membrane integrated lipid I forming enzyme MraY and of a G-protein coupled receptor [6, 7]. Moreover, by systematic titration of synthesized membrane proteins with supplied nanodiscs, the formation of oligomeric complexes within a nanodisc membrane can be modulated. In this chapter, we give a current update on the protocol development for the co-translational insertion of membrane proteins into preformed nanodiscs.

2 Materials

All stock solutions should be prepared with ultrapure water and stored at −20 °C, if not stated otherwise.

2.1 General Materials

1. Fermenter for bacterial cultures, e.g., 5–10 L volume.
2. French Press or similar mechanic cell disruptor.
3. Fluorescence spectrophotometer for green fluorescent protein (GFP) measurement.
4. Standard centrifuges and set of rotors.
5. Thermoshaker for incubation in between 20 and 37 °C.

6. Chromatographic system (e.g., ÄKTA purifier, GE Healthcare).
7. Immobilized Metal Affinity Chromatography (IMAC) material or column (Cube Biotech).
8. Centriprep filter devices, 10 kDa MWCO (Millipore).
9. Columns for size exclusion chromatography (SEC): Superdex 200 3.2/30, Superdex 200 10/300 GL (GE Healthcare).

2.2 *E. coli* Lysate Preparation

Commercial *E. coli* A19 S30 lysates supplemented with T7 RNA polymerase (Cube Biotech) may be used as controls.

1. *E. coli* strains A19, BL21, or C43.
2. 1× TPG medium: 10 g/L yeast extract, 16 g/L tryptone, 5 g/L NaCl, 100 mM glucose, 22 mM KH_2PO_4, 40 mM K_2HPO_4.
3. Antifoam (e.g., Sigma).
4. 40× LY-A/B buffer: 400 mM Tris-acetate pH 8.2, 560 mM $Mg(OAc)_2$, 2.4 M KCl.
5. 1× LY-A buffer (washing buffer) diluted from the 40× LY-A/B stock, supplemented with 6 mM ß-mercaptoethanol.
6. 1× LY-B buffer (lysis buffer) diluted from the 40× LY-A/B stock, supplemented with 1 mM DTT and 1 mM phenylmethanesulfonyl fluoride (PMSF).
7. 40× LY-C buffer: 400 mM Tris-acetate pH 8.2, 560 mM $Mg(OAc)_2$, 2.4 M KOAc.
8. 1× LY-C buffer (dialysis buffer): diluted from the 40× LY-C stock, supplemented with 0.5 mM DTT.
9. 5 M NaCl.

2.3 Two-Compartment Cell-Free Expression Reactions

1. MD100 dialysis cartridges as reaction mix containers (Scienova).
2. 96-deep-well microplates as feeding mix containers (Ritter riplate PP, 2 mL).
3. Dialysis tubes, 12–14 kDa MWCO (Spectrum).
4. Stock solutions required for expression reactions are listed in Table 1.

2.4 Membrane Scaffold Protein Preparation and Nanodisc Formation

1. Nanodisc preparation is exemplified by using the membrane scaffold protein derivative MSP1E3D1. Commercial preformed nanodiscs (Cube Biotech) may be used as controls.
2. pET-28-MSP1E3D1 vector.
3. BL21(DE3) Star cells.
4. LB-medium: 10 g/L peptone, 5 g/L yeast extract, 5 g/L NaCl.

Table 1
Pipetting scheme for two-compartment cell-free screening reactions

(A) Master mixtures				
Compound	**Stock concentration**	**Final concentration**	**Amount/example (µL)**	
MM-F:				
Complete protease inhibitor cocktail	50×	1×	212.3	
PEG 8000	40%	2%	530.8	
20 amino acid mix	25 mM	1 mM	424.6	
Acetyl phosphate (Li^+, K^+), pH 7.0	1 M	20 mM	212.3	
Phospho(enol)pyruvic acid (K^+), pH 7.0	1 M	20 mM	212.3	
75× NTP mix, pH 7.0	75×	1×	141.5	
HEPES/KOH, pH 8.0	2.73 M	100 mM	389.2	
DTT	500 mM	2 mM	42.5	
Folinic acid	10 mg/mL	0.1 mg/mL	106.2	
$Mg(OAc)_2$	1 M	7.1 mM	75.4	
KOAc	10 M	130 mM	138	
NaN_3	10%	0.05%	53.1	
			2538	
			−171 (= MM-F1)[b]	= 2367
Ly-C				3465
20 amino acid mix	25 mM	0.55		396
			Total	6228
MM-R:				
MM-F1	–	–	171	
DNA template	0.75 mg/mL	0.015 mg/mL	14.3	
t-RNA *(E. coli)*	40 mg/mL	0.5 mg/mL	8.94	
Pyruvate kinase	10 mg/mL	0.04 mg/mL	2.86	
RiboLock	40 U/µL	0.3 U/µL	5.36	
			Total 202.46	

(continued)

Table 1
(continued)

(B) Screening matrix for Mg^{2+} ions[c]

RM/Mg^{2+} (mM)	12[d]	14	16	18	20
MM-R	36.8	36.8	36.8	36.8	36.8
E. coli lysate + T7RNAP	45.5	45.5	45.5	45.5	45.5
100 mM $Mg(OAc)_2$	0	2.6	5.2	7.8	10.4
H_2O	47.7	45.1	42.5	39.9	37.3
Final volume:	130	130	130	130	130

FM/Mg^{2+} (mM)	12	14	16	18	20
MM-F	1132.4	1132.4	1132.4	1132.4	1132.4
100 mM $Mg(OAc)_2$	0	36	72	108	144
H_2O	667.6	631.6	595.6	559.6	523.6
Final volume:	1800	1800	1800	1800	1800

[a] Volumes are calculated for a total of 715 μL RM and 9900 μL FM in order to include some excess volume
For the final reactions, 650 μL RM and 9000 μL FM are used
[b] This volume has to be removed for MM-R preparation before MM-F is completed
[c] The basic Mg2+ ion concentration in the reaction is already 12 mM (4.9 mM^{2+} Mg are added with the S30 lysate and 7.1 mM^{2+} Mg are added in the MM-F)

5. 10% (w/v) glucose stock solution.
6. 1 M IPTG stock solution.
7. Complete EDTA-free protease inhibitor (Roche).
8. 10% (v/v) Triton X-100 stock solution in H_2O.
9. MSP-A buffer: 40 mM Tris–HCl, pH 8.0, 300 mM NaCl, 1% (v/v) Triton X-100.
10. MSP-B buffer: 40 mM Tris–HCl, pH 8.9, 300 mM NaCl, 50 mM cholic acid.
11. MSP-C buffer: 40 mM Tris–HCl, pH 8.0, 300 mM NaCl.
12. MSP-D buffer: 40 mM Tris–HCl, pH 8.0, 300 mM NaCl, 50 mM imidazole.
13. MSP-E buffer: 40 mM Tris–HCl, pH 8.0, 300 mM NaCl, 300 mM imidazole.

Table 2
Recommended lipid-to-MSP ratios for the in vitro assembly of nanodiscs

Lipid	Ratio to MSP1	Ratio to MSP1E3D1
Aso-PC	40	50
BPL	45	70
EPL	40	60
ETL	40	50
HTL	–	50
DEPG	–	85
DMPA	–	110
DMPC	80	115
DMPG	70	110
DOPA	–	90
DOPE	30	80
DOPC	30	80
DOPG	30	80
DOPS	–	90
POPC	55	85
POPG	–	90
POPS	–	90
SOPG	–	80

14. MSP-F (dialysis) buffer: 40 mM Tris–HCl, pH 8.0, 300 mM NaCl, 10% (v/v) glycerol.
15. Lipid-cholate stock solutions: 50 mM lipid (Table 2), 100–500 mM sodium cholate (for complete solubilization ultrasonic water bath may be required). Suitable lipids are listed in Fig. 2 and obtained from Avanti Polar Lipids.
16. 10% dodecylphosphocholine (DPC) stock solution in H_2O.
17. ND-A buffer: 10 mM Tris–HCl, pH 8.0, 100 mM NaCl.
18. Ni-NTA resin (Cube Biotech).

3 Methods

3.1 Lysate Preparation from E. coli Strains

E. coli is by far the most analyzed bacterial organism and cell-free synthetic biology profits from the large variety of engineered strain derivatives available and from the expertise accumulated during the

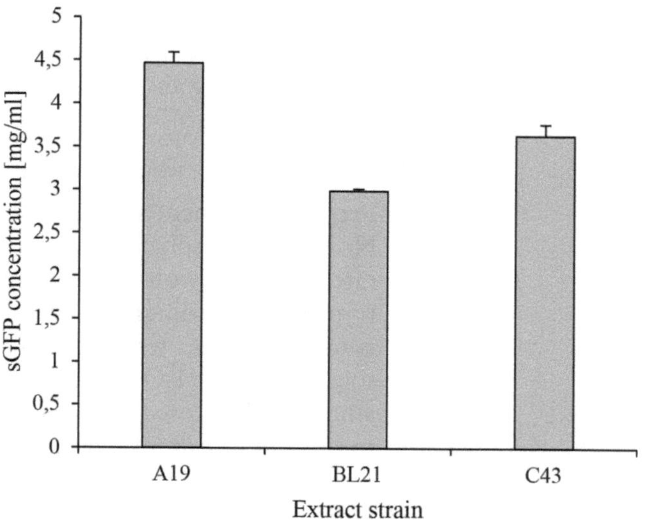

Fig. 1 Protein expression efficiency with S30 lysates prepared from different *E. coli* strains. The genotypes are A19 (rna19, gdhA2, his95, relA1, spoT1, metB1), BL21 (*E. coli* B F– ompT gal dcm lon hsdS B (r B – m B –) [malB +] K-12 (λS)), C43 (DE3) F– ompT gal dcm hsdSB (rB – mB –). Cells were fermented at identical conditions and S30 lysates were prepared with identical protocols. The production of shifted GFP was analyzed in two-compartment cell-free expression reactions

past decades. In addition, much of the knowledge on optimizing protein production in *E. coli* such as fusion technologies or purification strategies could become useful for cell-free expression in *E. coli* lysates as well. Potentially any strain of *E. coli* could be used for the preparation of cell-free lysates by using a protocol similar to that described below. For particular applications, lysate production out of specific constructs containing mutations for example in amino acid scrambling enzymes, proteases, or chaperones might be considered. We routinely prepare lysates out of strains such as A19, BL21, or C43 (DE3). However, variations in the basic recombinant protein production efficiencies are notable (Fig. 1). In our hands, lysates from strain A19 have among the highest protein production capacities and yields in between 4 and 5 mg of fluorescent GFP per 1 mL of reaction mixture can be obtained by using the protocol described below.

The protocol exemplifies the preparation of *E. coli* lysate from strain A19 in a 10 L fermenter with vigorous stirring and providing good aeration throughout the fermentation process (*see* **Note 1**). Up- or down-scaling to other volumes is certainly possible. The production protocol offers numerous options for tuning the resulting lysate properties according to specific applications. An essential step is to harvest the cells at mid-log phase in order to obtain highly active translation machineries. For each new combination of strain, media, growth condition or fermentation vessel, an initial pilot study to define the particular growth curve and optimal time point for harvesting should thus be determined.

1. Prepare preculture of the selected strain in 100 mL TBG medium and incubate overnight at 37 °C (*see* **Note 2**).
2. Prepare a fermenter with 10 L of sterile TPG medium (*see* **Note 3**). Phosphate and glucose should be sterilized separately. 5 L of twofold TPG medium containing yeast extract, tryptone, and sodium chloride supplemented with 3 L H_2O is autoclaved. The tenfold concentrated phosphate buffer dissolved in 1 L H_2O is autoclaved separately. The appropriate amount of glucose dissolved in 1 L H_2O is sterile filtered. The sterile TPG medium, phosphate buffer, and glucose are finally combined in the fermenter. Add an appropriate aliquot of antifoam (*see* **Note 4**) and inoculate in a ratio of 1:100 with the fresh preculture.
3. Incubate the culture at vigorous stirring (approximately 400–600 rpm) and good aeration at 37 °C (*see* **Note 5**). Monitor the growth curve by recording the optical density at 600 nm.
4. Chill the fermenter broth at mid-log growth phase (in our hands OD600 nm = 4–4.5) to approximately 20 °C. Continue stirring and aeration during cooling. Cooling should be completed in between 20 and 30 min (*see* **Note 6**).
5. Harvest the cells by centrifugation at 8000 × *g* for 15 min.
6. Completely suspend the pellet in 300 mL LY-A buffer, centrifuge at 8000 × *g* for 10 min, and discard the supernatant. Repeat this step three times.
7. Combine and weigh the pellets. Completely suspend the pellet in 110% (w/v) LY-B buffer.
8. Disrupt the cells via French press at 1000 psi. Similar devices could be used as well. The lysate will become viscous.
9. Centrifuge the lysate at 30,000 × *g* for 30 min. Transfer the supernatant into a fresh tube and repeat centrifugation step.
10. Carefully transfer the upper ¾ of the supernatant into a fresh tube. The solution is then adjusted to a final concentration of 400 mM NaCl. This will cause dissociation of the ribosomes from endogenous mRNA templates.
11. Incubate the solution at 42 °C for 45 min (*see* **Note 7**). A significant amount of proteins and other compounds will precipitate and the solution will become turbid.
12. Transfer the turbid solution into a dialysis tube with 12–14 kDa cutoff and dialyze against 100 times volume of LY-C buffer. The LY-C buffer should be exchanged twice.

13. Centrifuge the solution at the desired *g*-force for 30 min. Recommended *g*-forces are 15,000 × *g* (S15 lysate), 30,000 × *g* (S30 lysate), 60,000 × *g* (S60 lysate), and 100,000 × *g* (S100 lysate). The S15 lysate is still relatively crude, while the S30 lysate is the most common standard lysate. However, it still contains approximately one third of the total *E. coli* proteome as background. In addition, some 100 μg of lipids per 1 mL of lysate are still present as small liposomes and contain significant amounts of residual membrane proteins such as OmpF. This background could cause problems in subsequent highly sensitive assays, e.g., by analyzing recombinant channels by electrophysiology, and it is gradually decreased in S60 or S100 lysates. The overall protein synthesis efficiency is similar in S15 and S30 lysates, but might be might be decreased for some 20% in S60 and S100 lysates.
14. Transfer the supernatant into a fresh tube and repeat centrifugation step at the desired *g*-force. Remove supernatant in a fresh tube and mix carefully. The final total protein concentration of the lysate should be in between 30 to 50 mg/mL.
15. The lysate might now be supplemented with T7 RNA polymerase. The enzyme can either be obtained from commercial sources (e.g., Sigma, New England BioLabs) or purified by standard fermentation in *E. coli* cells as described [8, 9]. Alternatively, the T7 RNA polymerase can later be added to the expression reactions (*see* **Note 8**).
16. Immediately aliquot the lysate into suitable volumes, shock-freeze in liquid nitrogen and store at −80 °C.

3.2 Preparation of Preformed Empty Nanodiscs

Preformed empty nanodiscs are used as matrix for the co-translational insertion of membrane proteins. The nanodiscs are formed by combining purified derivatives of the apolipoprotein ApoA1 with detergent solubilized lipids (Fig. 2). The ApoA1 derivatives act as membrane scaffold proteins and two copies of a derivative assemble with a certain number of lipids to the final nanodisc. The assembly is directed by removal of the detergent either with biobeads or by dialysis. Choice of the scaffold proteins determines the final diameter of the nanodisc and commonly used are MSP1D1ΔH5 (8-9 nm), MSP1 (10 nm), and MSP1E3D1 (12 nm). The suitable size of the nanodisc should be selected according to both the number of transmembrane segments of the analyzed membrane proteins and its assumed oligomeric state formation. The scaffold proteins are synthesized and isolated out of BL21(DE3) Star cells upon standard fermentation as described below. Alternatively, purified scaffold proteins are also commercially available (e.g., Cube Biotech).

Nanodiscs can be filled with a large variety of different lipids or lipid mixtures and are thus an ideal tool to screen lipid dependent effects on membrane protein activity and stability. The lipids might

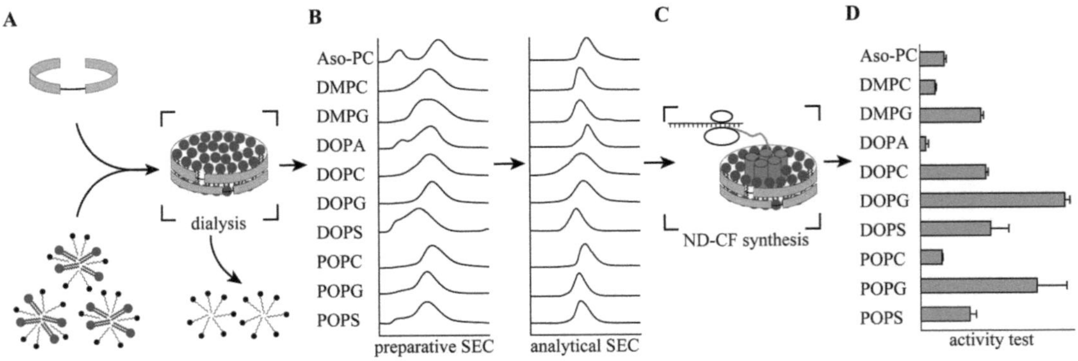

Fig. 2 Detergent-free lipid screen of cell-free synthesized membrane proteins. (**A**) Purified membrane scaffold protein and detergent-solubilized lipids are mixed in a suitable ratio and detergent is removed by dialysis. (**B**) Assembled nanodiscs are purified by preparative SEC and final concentrated stocks are controlled for homogeneity by analytical SEC. (**C**) Cell-free synthesized membrane proteins are co-translationally inserted in supplied nanodiscs. (**D**) Synthesized membrane protein/nanodisc complexes are quality controlled, e.g., by analyzing specific activity

differ in chain length, flexibility or charge and each individual characteristic could have a significant effect on membrane protein function (Fig. 3). In addition, certain amounts of structurally different lipids such as cholesterol or cardiolipin could be supplied as additives. For reproducible nanodisc quality, it is important to keep a certain ratio of scaffold protein to lipid. Examples of most commonly used combinations are given in Table 2. For other lipids or lipid combinations it is recommended to determine first the MSP to lipid ratios giving the most homogenous peaks upon size exclusion profiling (Fig. 2).

3.2.1 Production and Purification of Membrane Scaffold Proteins

1. MSP derivatives are synthesized in BL21(DE3) Star cells by conventional fermentation in baffled Erlenmeyer flasks with LB medium supplemented with 0.5% (w/v) glucose and the appropriate antibiotics (usually kanamycin as most available MSP derivatives are cloned into the vector pET28). Expected yields of finally purified protein per 1 L of culture are in between 10 mg for MSP1E3D1 and 30 mg for MSP1D1ΔH5. The flasks are inoculated 1:12 with a fresh pre-culture and incubated at 37 °C with vigorous shaking (approx. 200 rpm).
2. Expression of the MSP protein is induced at log phase (OD600 of 1.0) by addition of IPTG at 1 mM final concentration. Fermentation is then continued for 1 h at 37 °C.
3. After 1 h, reduce the fermentation temperature to 28 °C and continue incubation for further 4 h.
4. Harvest the cells by centrifugation (6000 × *g* for 15 min at 4 °C). Discard supernatant and wash the pellets once with MSP-C buffer. The washed pellets can be stored at −80 °C as thin plates wrapped in aluminum foil.

Lipid name	Synonyms	Structure
1-stearoyl-2-oleoyl-*sn*-glycero-3-phospho-(1'-rac-glycerol)	18:0-18:1 PG SOPG	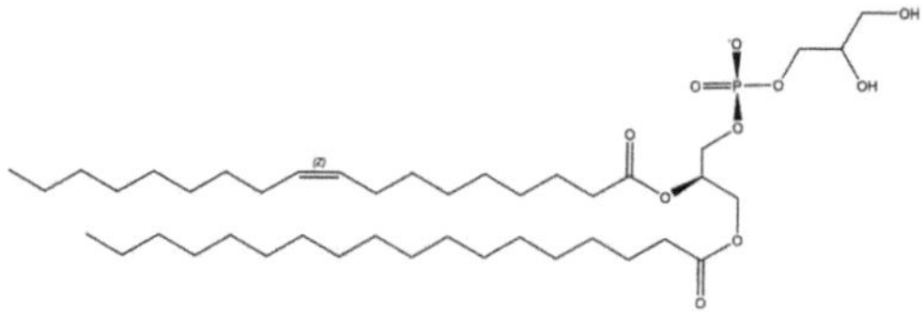
1-palmitoyl-2-oleoyl-*sn*-glycero-3-phospho-(1'-rac-glycerol)	16:0-18:1 PG POPG	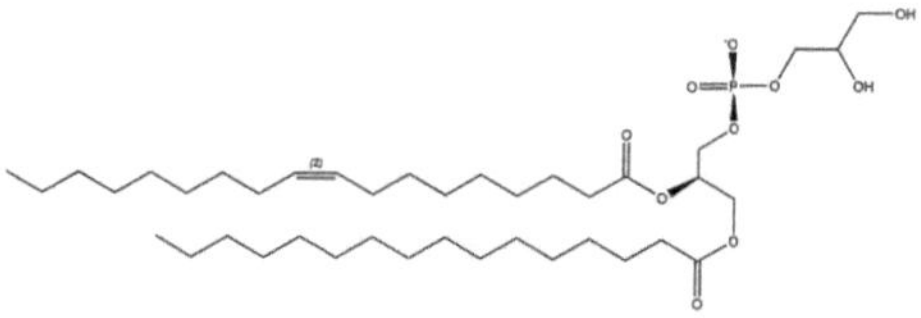
1,2-dioleoyl-*sn*-glycero-3-phosphocholine	18:1 (Δ9-Cis) PC DOPC	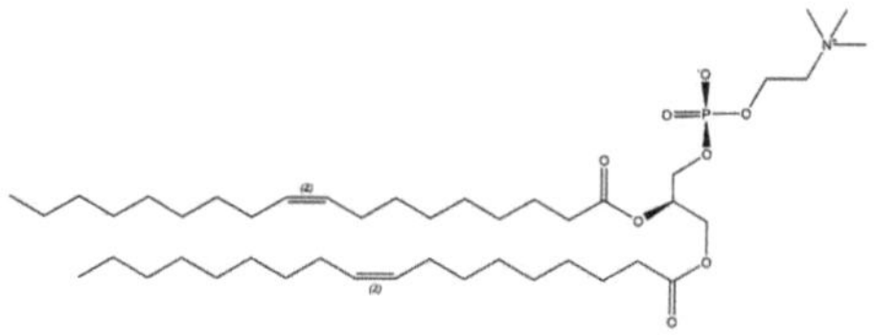
1,2-dimyristoyl-*sn*-glycero-3-phosphocholine	14:0 PC DMPC	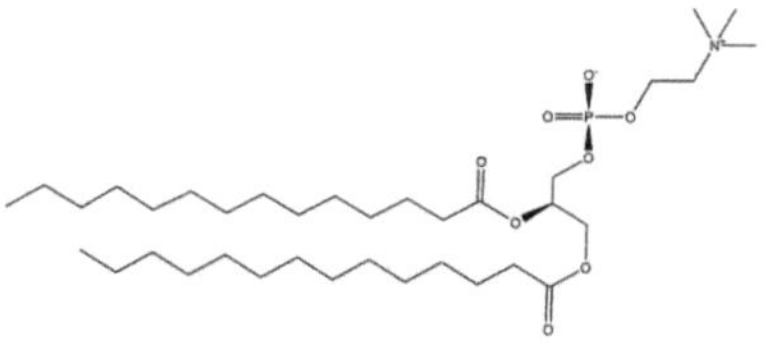
1,2-dimyristoyl-*sn*-glycero-3-phospho-(1'-rac-glycerol)	14:0 PG DMPG	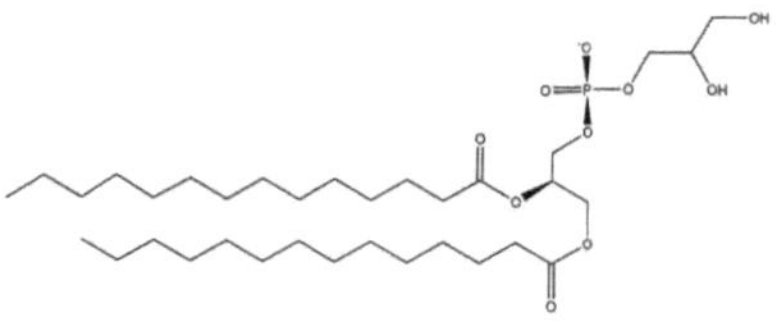
1,2-dioleoyl-*sn*-glycero-3-phospho-(1'-rac-glycerol)	18:1 (Δ9-Cis) PG DOPG	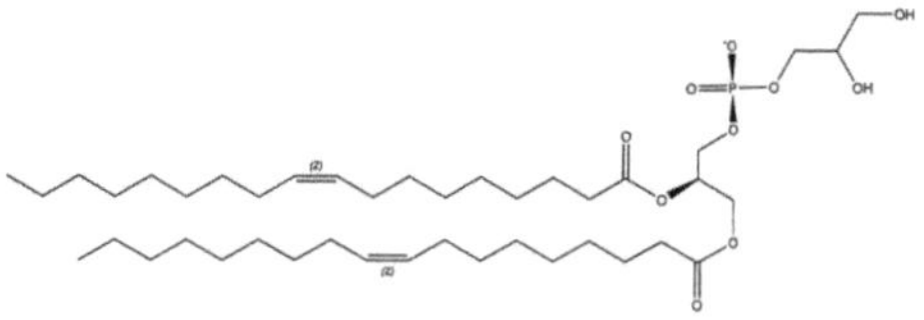
1,2-dielaidoyl-*sn*-glycero-3-phospho-(1'-rac-glycerol)	18:1 (Δ9-Trans) PG DEPG	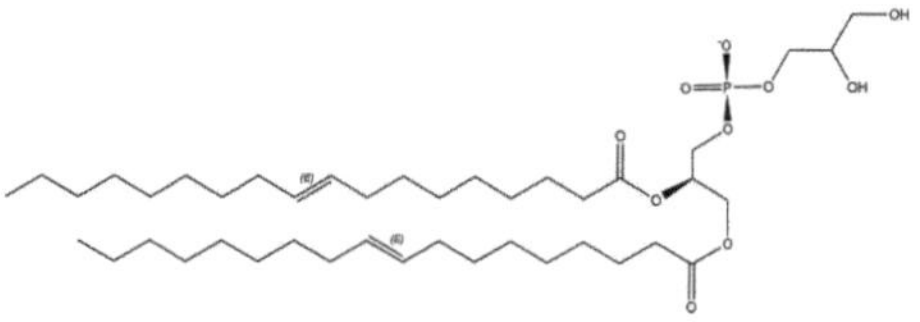

Fig. 3 Selected lipids suitable for the assembly of nanodiscs

5. Suspend 15-20 g of pellet in 50 mL MSP-C buffer supplemented with one dissolved tablet of Complete protease inhibitor.
6. Adjust to a final concentration of 1% (v/v) Triton X-100 by using a 10% Triton X-100 stock in H_2O (*see* **Note 9**).
7. Disrupt cells by sonication (3 × 60 s and 3 × 45 s), with a rest period of at least 60 s on ice between each cycle. Gently mix after each sonication cycle.

8. Centrifuge the suspension for 30 min at 30,000 × *g* at 4 °C and filter supernatant through a 0.45 μm filter (*see* **Note 10**).
9. Equilibrate a Ni^{2+} loaded IMAC column (15 mL bed volume) with 5 CV MSP-A buffer and load filtered supernatant on the column. Flow rate should be 1–2 mL/min.
10. Wash the column with each 5 CV MSP-A, -B, -C, and -D buffer and with a flow rate of 2-3 mL/min.
11. Elute protein in 1 mL fractions with MSP-E buffer and pool MSP1E3D1 containing fractions. Adjust to 10% (v/v) glycerol.
12. Dialyze for 3 h at 4 °C against 5 L MSP-F buffer, change to fresh 5 L MSP-F buffer and continue dialysis overnight.
13. Transfer solution into a centrifugation vial and remove potential precipitates by centrifugation at 18,000 × *g* for 30 min at 4 °C. The protein concentration in the supernatant should be in between 80 and 100 μM. Aliquot the supernatant, shock-freeze in liquid nitrogen and store frozen at −80 °C.

3.2.2 Assembly of Empty Nanodiscs

1. Select a lipid and prepare 1–2 mL of a 50 mM lipid stock in sodium cholate in ND-A buffer.
2. Combine the selected purified MSP in the appropriate ratio (*see* Table 2) with a corresponding volume of the lipid stock and the DPC stock. The final DPC concentration should be at 0.1% (*see* **Note 11**). The solution is completed up to the final volume with buffer ND-A. Mix the solution by gently inverting for 1.5 h on a rotary shaker at room temperature. The final volume of the mix should be in between 11 and 12 mL. An example for preparing MSP1E3D1 (DEPG) nanodiscs is given in Table 3.
3. Fill the mixture in a 12 mL Slide-A-lyzer (MWCO 10 kDa) and dialyze extensively for several h against 5 L of ND-A buffer to remove the detergent. The ND-A buffer is exchanged two times and dialysis is completed after approximately 48 h. The nanodiscs are formed upon the continuous removal of the detergent (*see* **Note 12**).

Table 3
Pipetting scheme for nanodisc assembly

Compound	Stock concentration	Volume	Final concentration
MSP1E3D1	80 μM	9 mL	65.45 μM
DEPG/Na^+-cholate/ND-A	50 mM	1.224 mL	5.564 mM
DPC	10%	110 μL	0.1%
ND-A		666 μL	

4. Transfer the solution from the Slide-A-Lyzer into a fresh tube and centrifuge at 30,000 × *g* for 30 min in order to completely remove potential precipitates. Depending on the selected lipid type, some notable precipitate might be present.
5. Transfer supernatant into a fresh tube and concentrate to a final volume of 0.5 mL in a Centriprep concentrating unit (MWCO 10 kDa) equilibrated with ND-A buffer at 2000 × *g*.
6. Separate the sample on a preparative scale Superdex 200 10/300 GL column equilibrated with ND-A buffer at a flow rate of 0.5 mL/min. With most lipid types, the elution profile shows a relatively homogenous peak eluting in between 11 and 13 mL depending on the nanodisc size (Fig. 1). Some aggregates are eluting in the void volume. Elution profiles of nanodiscs assembled with rather complex lipid mixtures such as commercial brain or heart lipid extracts can be of higher heterogeneity.
7. The main peak from the size exclusion column is collected and concentrated in a Centriprep concentrating unit (MWCO 10 kDa) equilibrated with ND-A buffer by centrifugation at 2000 × *g* until the final MSP concentration is approximately 1–2 mM, corresponding to 0.5–1 mM nanodisc concentration. The homogeneity of the concentrated nanodisc samples should finally be checked by analytical scale size-exclusion chromatography using a Superdex 200 3.2/30 column (*see* **Note 13**).
8. Freeze appropriate aliquots of nanodisc samples in liquid nitrogen and store the aliquots at −80 °C until further use. Multiple refreezing of the nanodisc samples should be avoided as some precipitation might occur.

3.3 Two-Compartment Cell-Free Expression Reactions

Cell-free expression systems based on *E. coli* lysates are rapidly evolving and numerous variations of reaction setup, lysate preparation, reaction composition, or transcriptional regulation are possible. Each modification can affect system efficiency. We exemplify the protein production in a two-compartment cell-free system based on A19 S30 lysates and with T7-RNA polymerase controlled transcription. Other systems or system modifications may be used with similar results.

Efficient basic reaction conditions are prerequisite for subsequent lipid screening or other approaches of sample quality tuning. Quality of the cell-free expression system should be routinely controlled via production of a convenient monitor such as GFP. The production efficiency of the system could then be quantified directly in the reaction mixture by fluorescence measurement (excitation wavelength 485 nm and emission wavelength 510 nm). The system is highly sensitive to several common compounds such as Mg^{2+}, K^{+} or other ions. For Mg^{2+} ions, the critical working range

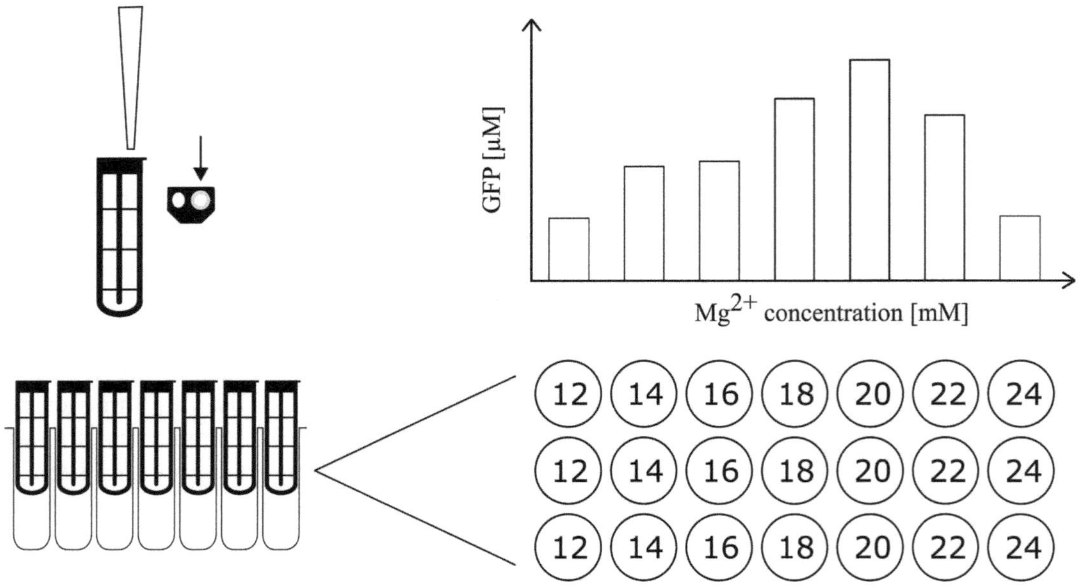

Fig. 4 Screening for optimal Mg^{2+} ion concentration in the two-compartment cell-free expression system. Scheme of a typical Mg^{2+} ion screen with the synthesis of GFP as monitor. The reactions are performed in triplicates within a range of 12–24 mM Mg^{2+} ion concentration in MD100 cartridges and 96-deep-well plates. The inlet for the reaction mixture of the MD100 cartridges is indicated

is very narrow (Fig. 4). Corresponding routine adjustment of each new lysate batch or compound stock is thus recommended.

In the two-compartment system, the reaction mixture is separated from a feeding compartment of 13–20 times higher volume to the RM by a semipermeable membrane with a molecular weight cut-off (MWCO) of 12–14 kDa. It should be considered that variation of the ratio of reaction mixture to feeding mixture would affect the final product yield. With the described system, routine synthesis efficiencies of correctly folded GFP in between 4 and 5 mg per mL of reaction mixture are obtained. This efficiency should be taken as benchmark for the quality control of the system. Expression efficiencies of membrane proteins in presence of nanodiscs are usually lower and yields in between 0.5 and 1 mg per mL of reaction mixture can be expected.

If necessary, routine optimization of template design should be performed as described earlier [8]. We generally recommend to construct at least two different derivatives of a target, (1) a fusion protein with a C-terminally attached superfolderGFP-tag for fast protocol development in analytical scale reactions, and (2) a nonfused derivative with only a small C-terminal purification tag such as a StrepII-tag for subsequent preparative scale expression and functional or structural analysis. The superfolderGFP monitor will significantly accelerate the evaluation of target solubilization in lipid screens. However, the folding of the superfolderGFP moiety should not be taken as indication of the functional folding of the attached membrane protein. Commercial dialysis cartridges (Xpress Micro Dialyzer MD100, 12–14 kDa MWCO, Scienova) are appropriate as convenient reaction containers for analytical or semi-preparative scale reactions (*see* **Note 14**). The cartridges hold the reaction mixture volume and are placed into cavities of standard 96-deep-well plates (Ritter) holding appropriate volumes of feeding mixture (Fig. 4). Analytical scale expression reactions necessary for protocol optimization or compound screening are best performed in reaction mixture volumes of 25–75 μL. Reactions can be scaled up in a linear ratio to volumes of 30 mL by using commercial Slide-A-Lyzer devices (Thermo Scientific). The Slide-A-Lyzers should be placed into suitable containers holding appropriate volumes of feeding mixture such as plastic or glass trays or custom-made Plexiglas containers [9]. The following protocol exemplifies the preparation of ten analytical scale standard reactions (e.g., for screening of five different Mg^{2+} ion concentrations in duplicates) with 65 μL of reaction mixture and 900 μL of feeding mixture each (*see* **Note 15**). The concentrations of the individual stocks and the resulting required volumes for this example are given in Table 1.

1. The calculated total volume for ten reactions plus some excess volume (*see* **Note 16**) is 715 μL for the reaction mixture and 9.9 mL for the feeding mixture. The pipetting workflow starts with preparing a master mixture MM-F containing all common compounds of reaction mixture and feeding mixture (*see* Table 1A). Then an aliquot of MM-F is removed to prepare the master mixture MM-R for the reaction mixtures. The MM-F and MM-R mixtures are then completed by addition of the compounds specific for reaction and feeding mixtures. The master mixtures are aliquoted, supplemented with the screening compound Mg^{2+} and the lysate according to a screening matrix, and completed with water to give the appropriate final volumes (*see* Table 1B).
2. Prepare MM-F containing the common compounds of reaction mixture and feeding mixture according to Table 1A and mix. The total volume should be 2.538 mL.
3. Transfer 171 μL of this still incomplete MM-F into a fresh tube for preparation of MM-R.
4. Complete MM-F by addition of 3.465 mL of Ly-C buffer and 396 μL of AA-Mix to give a final volume of 6.228 mL.
5. Complete MM-R with high molecular weight compounds. According to Table 1A, this would be 14.3 μL of template DNA, 5.36 μL of Ribolock RNAse inhibitor, 8.94 μL of tRNA, and 2.86 μL of pyruvate kinase giving a final volume of 202.46 μL.

6. Set up a screening matrix for different Mg^{2+} ion concentrations (*see* **Note 17**). An example matrix for screening final Mg^{2+} concentrations from 12 to 20 mM is given in Table 1B. The *E. coli* lysate is added as last compound and the reaction and feeding mixtures are finally completed by addition of the indicated volumes of H_2O.
7. Transfer 900 μL of feeding mixture into cavities of 96-deep-well plates.
8. Transfer 55 μL of reaction mixture into MD100 cartridges and place the cartridges into the corresponding cavity of the 96-deep-well plate (*see* **Note 15**).
9. Incubate reaction for 6–9 h or overnight at 25–30 °C. The reaction should be slightly agitated in order to promote reagent exchange through the membrane.
10. After incubation, mix the reaction mixture thoroughly in order to suspend potentially formed precipitates. Then remove the reaction mixture from the container with a pipette and analyze protein expression, e.g., by superfolderGFP fluorescence.

3.4 Co-translational Solubilization of Membrane Proteins with Nanodiscs

The co-translational insertion of membrane proteins into preformed empty membranes of defined compositions is a new strategy and will continuously be refined. Insertion mechanisms within this artificial system are still poorly understood. Natural insertion or translocon machineries are incomplete, low abundant or even completely absent in the cell lysates. Membranes of nanodiscs are furthermore most likely quite different from that of liposomes in view of lateral pressure, curvature and other topological features. They are accessible from both sides and the dynamic interface in between membrane and scaffold protein might serve as a preferred entry site for membrane proteins. Insertion efficiencies of membrane proteins into nanodisc membranes or liposomes of identical composition could therefore be very different. Important determinants of membrane protein insertion efficiency are (1) nanodisc size, (2) nanodisc concentration, and (3) membrane composition. All three parameters might be subject of individual screens. The larger MSP1E3D1 nanodiscs usually provide a good starting point, in particular if larger membrane proteins or oligomeric complexes should be analyzed that might not insert properly into smaller nanodiscs. Suitable solubilization conditions could thus first be identified with MSP1E3D1 nanodiscs and then be used as background for the evaluation of the smaller MSP1 or MSP1D1ΔH5 nanodiscs.

The quantitative solubilization of a cell-free synthesized membrane protein is correlated to its expression efficiency and to the supplied nanodisc concentration. A systematic titration of membrane protein solubilization with increasing nanodisc concentrations should thus be performed first (Fig. 5). We recommend to

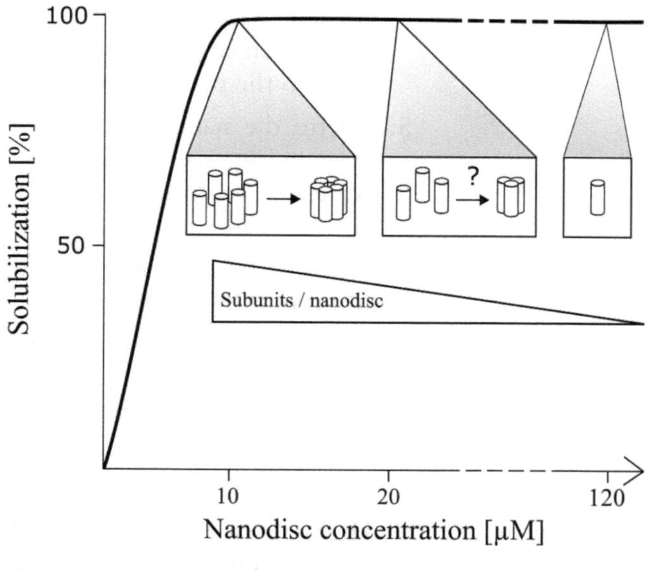

Fig. 5 Adjusting nanodisc concentrations for the co-translational solubilization of membrane proteins. The example illustrates the solubilization screen of proteorhodopsin in two-compartment cell-free expression reactions supplemented with increasing concentrations of MSP1E3D1 nanodiscs containing DMPC membranes [5]. Solubilization was quantified by absorbance at 520 nm due to cofactor retinal incorporation of folded proteorhodopsin. The expression efficiency of proteorhodopsin is approximately 60 μM at these conditions. Complete solubilization is already achieved at 10–15 μM final nanodisc concentrations, indicating the formation of higher oligomeric complexes up to hexamers within one nanodisc. Increased nanodisc to proteorhodopsin ratios will gradually result into the formation of lower oligomeric complexes

use MSP1E3D1 nanodiscs assembled with anionic lipids such as DMPG, DOPG, or POPG for these initial screens. Anionic lipids may be more likely to promote the membrane insertion of membrane proteins.

1. Prepare a stock of MSP1E3D1 nanodiscs assembled with either DMPG or DOPG or POPG lipids. The concentration of the nanodisc stock should be 0.5–1 mM (*see* **Note 18**).
2. Determine the expression efficiency of the target protein without any supplemented hydrophobic compounds in order to get a rough estimate of its concentration in the reaction mixture. The membrane protein will precipitate and the pellet could be analyzed by immunodetection or Coomassie-staining after SDS-PAGE and by using appropriate markers.
3. Set up a screening matrix for the reaction mixtures according to Table 1B for different nanodisc concentrations. The nanodiscs will not be added to the feeding mixtures. The screening range could be up to 120 μM final nanodisc concentration in the reaction mixture.

4. Use a target protein-superfolderGFP fusion for the concentration screen and supply the appropriate amount of template DNA to the master mixture MM-R (Table 1A).
5. Pipette the nanodisc concentration screen according to the screening matrix and assemble the two-compartment cell-free expression reactions according to Subheading 3.3 in MD100 cartridges and deep-well plates.
6. Incubate the reactions overnight at 30 °C with shaking at 150–200 rpm (GFL 3005 orbital shaker).
7. Transfer the reaction mixtures from the MD100 cartridges into fresh tubes and centrifuge in a tabletop centrifuge at 18,000 × *g* for 10 min.
8. Transfer the supernatants into a fresh tube and determine the superfolderGFP fluorescence.
9. The molarity of the solubilized target protein-superfolderGFP fusion should be determined from the measured fluorescence according to a GFP calibration curve and plotted against the nanodisc concentrations (Fig. 5). The data can already give a rough estimate of the number of membrane protein monomers within one nanodisc. Membrane insertion of individual monomers may occur rather stochastically and the monomer number per nanodisc could thus be modulated by the supplied nanodisc concentration.

3.5 Tuning Membrane Protein Quality with Different Lipid Environments

Besides nanodisc properties, individual characteristics of the membrane protein targets themselves are further crucial determinants for membrane insertion and functional folding. Requirements for particular lipid compositions or other membrane features might be very specific and could significantly differ with even closely related homologues (Fig. 6a). The screening of a variety of different membrane compositions is therefore recommended in order to identify suitable lipid environments for a membrane protein. The fine tuning of lipid environments by analyzing lipid mixtures might be considered as well in subsequent steps. Target protein-superfolderGFP fusions can be used as initial monitor for lipid screens. The GFP fluorescence will enable the fast evaluation of solubilization efficiency of the fusion construct. However, association or only partial integration of the membrane protein into nanodiscs will most likely also result into folding and fluorescence of the GFP moiety. The membrane protein-superfolderGFP fusion will therefore help to identify a subset of promising membrane compositions, but the functional folding of the membrane protein has to be specifically analyzed in subsequent assays.

1. Prepare a set of nanodisc stocks with the selected scaffold protein and assembled with a variety of different lipids according to Table 2. Membranes could be composed out of synthetic

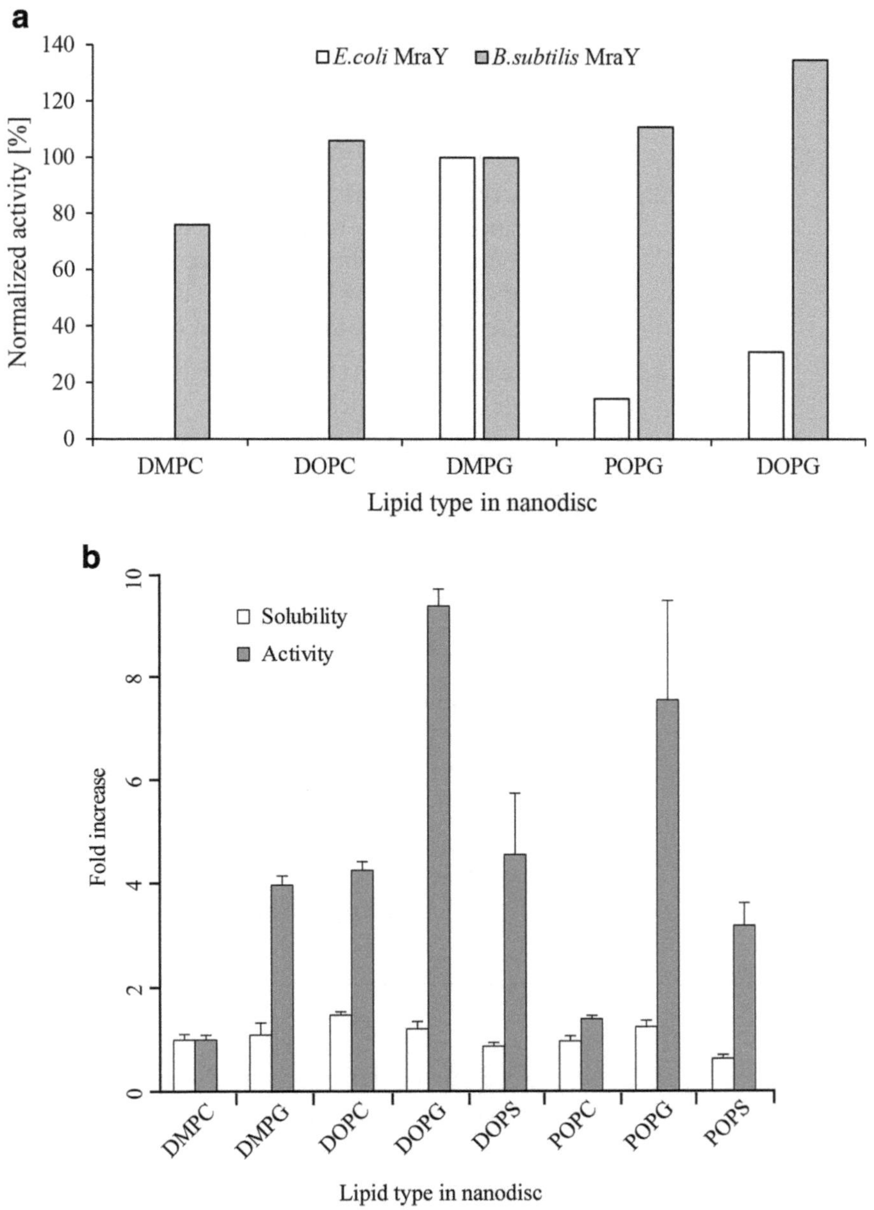

Fig. 6 Lipid dependent solubilization and folding of membrane proteins. Membrane proteins are synthesized in two-compartment cell-free expression reactions in presence of preformed MSP1E3D1 nanodiscs containing different membrane compositions. The specific activity of the membrane proteins in the different lipid environments was analyzed in subsequent assays. (**a**) Lipid I formation of the MraY homologues of *E. coli* and *Bacillus subtilis* inserted in different nanodiscs [6]. Ec-MraY shows only activity in negatively charged lipids, whereas Bs-MraY activity is rather lipid nonspecific. (**b**) Co-translational solubilization and alprenolol binding activity of a thermostabilized turkey ß1-adrenergic receptor derivative fused to superfolderGFP [7]. Solubilization was determined by GFP fluorescence. While solubilization efficiency is similar with the analyzed lipids, major variations in ligand binding activity are detectable

lipids, complex natural lipid extracts or of defined lipid combinations (*see* **Note 13**). Initial preference may be given to lipids or lipid mixtures similar to those of the natural environment of the target protein. Nanodisc stocks should be concentrated up to 0.5–1 mM by ultrafiltration.

2. Prepare a screening matrix according to the intended number of different nanodisc types. The final concentration of the nanodiscs in the reaction mixture should be in accordance to the previously determined optimum (*see* Subheading 3.4). The different nanodisc types are supplied at identical final concentrations.
3. Prepare the reaction mixtures and the feeding mixtures according to Table 1 (*see* Subheading 3.3).
4. Transfer the reaction mixtures into MD100 cartridges and the feeding mixtures into corresponding cavities of 96-deep-well plates (*see* Subheading 3.3). Incubate overnight at 30 °C with shaking at 150–200 rpm (GFL 3005 orbital shaker).
5. Transfer the reaction mixtures from the MD100 cartridges into fresh tubes and centrifuge in a tabletop centrifuge at 18,000 × *g* for 10 min.
6. Transfer the supernatants into a fresh tube and determine the superfolderGFP fluorescence (Fig. 6b). The fluorescence can provide a first estimate on lipid compatibility for membrane protein insertion.
7. The membrane protein folding must be analyzed in subsequent specific assays, e.g., ligand binding assays in case of G-protein coupled receptors (Fig. 6b).

4 Notes

1. Growing cells in baffled flasks might work as well, although with some reduced efficiency of the resulting lysate.
2. A fresh grown culture should be taken for inoculation. Do not inoculate directly from glycerol stocks.
3. Other fermentation media might work as well. It is important to first record a growth curve of the cells in modified medium to determine the optimal harvesting point.
4. Depending on the medium and fermentation condition, repeated addition of antifoam might be necessary.
5. Temperature might be modified during the fermentation process, e.g., to increase chaperone production.
6. If cooling is not possible in the fermenter, chilling of the culture broth in an ice bath or by addition of frozen medium might be possible. It would be beneficial if fermentation (i.e.,

stirring and aeration) is going on during cooling, otherwise the cells might enter stationary phase and inactivate ribosomes. A final temperature in between 20 and 25 °C will be fine.

7. In our hands, this step improves lysate quality and efficiency. However, variation of temperature and incubation time might be considered.
8. We recommend to purify T7 RNA polymerase by standard overexpression in *E. coli* cells. The protein is highly expressed and the isolation procedure is reliable and efficient. Once a source of the enzyme has been selected, a concentration screen for the cell-free synthesis of GFP should be performed. T7 RNA polymerase is essential if the target gene is transcribed under control of the T7 promoter (e.g., from conventional pET or pIVEX vectors). Alternatively, transcription can be operated with promoters (e.g., tac) recognized from the endogenous *E. coli* RNA polymerase, which is still present in the S30 lysates.
9. Due to its partial hydrophobicity, the scaffold proteins stay attached to the membrane fraction. Addition of detergent is therefore essential to release and to solubilize the scaffold proteins into the supernatant.
10. The supernatant may still be very viscous, thus rapidly clocking the filter. Short sonication for 30 s or DNAse treatment will significantly reduce viscosity.
11. Sodium cholate has a high critical micellar concentration of 9–15 mM. The addition of DPC shall therefore prevent precipitations or liposome formation by keeping the lipids detergent solubilized.
12. Removal of detergent by Biobeads may be considered as well. In our hands, the assembly of nanodiscs by dialysis resulted into more homogenous samples.
13. The final homogeneity of the preformed nanodiscs is somehow variable while they still could give good results upon membrane protein solubilization. In our hands, nanodiscs assembled with complex lipid mixtures such as total *E. coli* lipids, brain or heart lipid extracts are most heterogeneous.
14. Commercial pre-autoclaved cartridges may give some 10–15% better efficiencies in protein synthesis.
15. The calculation of 65 μL for the reaction mixture includes excess volume and only 55 μL are finally pipetted in order to compensate volume loss.
16. The calculation of a 10–15% excess volume of the individual compounds is recommended in order to compensate for the loss of volume effect upon mixing of the individual reagent volumes.

17. The basic Mg^{2+} concentration is already 12 mM ± 2 mM (7.1 mM are added into MM-F and 4.9 mM are contributed from the lysate).
18. The final concentration of the nanodisc stock may depend on the lipid type.

Acknowledgments

This work was funded by the Collaborative Research Center (SFB) 807 of the German Research Foundation (DFG). Support was further obtained by Instruct, part of the European Strategy Forum on Research Infrastructures (ESFRI).

References

1. Hein C, Henrich E, Orbán E et al (2014) Hydrophobic supplements in cell-free systems: designing artificial environments for membrane proteins. Eng Life Sci 14:365–379
2. Schuler MA, Denisov IG, Sligar SG (2013) Nanodiscs as a new tool to examine lipid-protein interactions. Methods Mol Biol 974: 415–433
3. Henrich E, Hein C, Dötsch V et al (2015) Membrane protein production in *Escherichia coli* cell-free lysates. FEBS Lett 589: 1713–1722
4. Katzen F, Fletcher JE, Yang JP et al (2008) Insertion of membrane proteins into discoidal membranes using a cell-free protein expression approach. J Proteome Res 7:3535–3542
5. Roos C, Zocher M, Mueller D et al (2012) Characterization of co-translationally formed nanodisc complexes with small multidrug transporters, proteorhodopsin and with the *E. coli* MraY translocase. Biochim Biophys Acta 1818:3098–3106
6. Henrich E, Ma Y, Engels I et al (2016) Lipid requirements for the enzymatic activity of MraY translocases and in vitro reconstitution of the lipid II synthesis pathway. J Biol Chem 291:2535–2546
7. Rues RB, Dötsch V, Bernhard F (2016) Co-translational formation and pharmacologiocal characterization of beta1-adrenergic receptor/nanodisc complexes with different lipid environments. Biochim Biophys Acta 1858: 1306–1316
8. Orbán E, Proverbio D, Haberstock S et al (2015) Cell-free expression of G-protein-coupled receptors. Methods Mol Biol 1261: 171–195
9. Reckel S, Sobhanifar S, Durst F et al (2010) Strategies for the cell-free expression of membrane proteins. Methods Mol Biol 607:187–212

Chapter 20

Not Limited to *E. coli*: Versatile Expression Vectors for Mammalian Protein Expression

Katharina Karste*, Maren Bleckmann*, and Joop van den Heuvel

Abstract

Recombinant protein expression is not limited to *E. coli* or other prokaryotic systems. It is inevitable to use eukaryotic systems in order to express challenging mammalian proteins. Eukaryotic systems are able to perform complex posttranslational modifications like protein processing, phosphorylation, glycosylation, which are essential for stability and functionality of many proteins. Different eukaryotic protein expression systems employing yeast, insect, or mammalian cell lines are established with each having its own advantages and disadvantages. Often it is quite difficult to decide which will be the most optimal expression system as this depends highly on the protein itself. Expression in stable cell lines requires substantial screening of expressible constructs prior to developing a stable expression cell line. To achieve fast screening by transient expression in multiple hosts, versatile vectors can be applied. In this chapter, we present an overview of the most common multi-host vectors, which allow for fast expression analysis without tedious (re)cloning of the gene of interest in several different protein production systems. The protocols in this chapter describe the latest methods for fast transient expression in insect and mammalian cell lines.

Key words Versatile vectors, Transient expression, HEK293-6E, Hi5, Expression vector

1 Introduction

The availability of large amounts of desired target proteins in high quality is a prerequisite for functional and structural analysis, drug-target screening strategies or production of complex biological therapeutics [1]. To reach optimal recombinant protein production, expressible constructs are often tested in more than one host expression system for various reasons. Bacterial systems may be used, e.g., for initial studies of proteins with a size below 100 kDa or for single protein domains in the range of 10–30 kDa to investigate their solubility, activity, and ability to produce large amounts for structural and functional studies. However, many viral and mammalian proteins either require specific posttranslational

*Both the Authors Contributed Equally.

Nicola A. Burgess-Brown (ed.), *Heterologous Gene Expression in E. coli: Methods and Protocols*, Methods in Molecular Biology, vol. 1586, DOI 10.1007/978-1-4939-6887-9_20, © Springer Science+Business Media LLC 2017

modification for proper folding and full biological activity [2] or have to be co-expressed in the presence of their binding partners to assemble in a multiprotein complex for full function and stability [3]. In order to obtain full-length active proteins with correct and functional eukaryotic posttranslational modifications, more complex eukaryotic cell systems have to be employed for expression. Current available eukaryotic expression systems are yeast, plant, insect, and mammalian cell lines with each having its own advantages and disadvantages. Depending on the required amount and properties of the target protein, a single host cell line might be the optimal expression system while others are not suitable for expression of the same target protein at all. Thus, it is undoubtedly important to test different eukaryotic expression hosts. Such a protein expression screen in eukaryotic systems is not straightforward as each host cell line requires different expression cassettes containing optimal promoter and transcriptional termination sequences, secretion signal sequences, tags for increasing solubility or protein purification as well as additional elements to control the level of expression.

Commercial suppliers like Novagen, Life Technologies (Invitrogen), Stratagene, Clontech, GE, and other companies offer diverse sets of bacterial, yeast, baculoviral, and mammalian expression vectors. These catalogues of expression vectors are extended by the extensive variety of expression vector series developed by academic research institutes, which often favour their own specialized expression vector variants. However, most of the available expression vectors require substantial (re)cloning of the desired target gene into individual vectors, each specific for a particular host system. The cloning process even with efficient strategies like "Golden Gate" [4], "SLIC-Fusion" [5] has an essential limiting time and cost factor, hampering fast analysis of expressible construct in different expression hosts. Thus, the overall process can be clearly accelerated using a single versatile expression vector to determine the best expression vector–host combination for the target protein. Simultaneously, the achievable expression level with a versatile vector is only moderately reduced compared to expression in an optimized expression vector for a single expression system [1]. Especially, while performing high-throughput expression analysis such a fast versatile vector approach is beneficial [6, 7].

Below, we give an overview of the different available versatile expression vectors with its individual advantages and disadvantages. Furthermore, we describe their application in HEK293-6E cells and Hi5 insect cells as the expression in such eukaryotic systems are suitable for producing complex and/or difficult-to-express mammalian target proteins. In addition, the HEK293-6E and Hi5 cell lines are especially qualified for fast, inexpensive, and simple screening as they are compatible with the plasmid-based expression using polyethylenimine (PEI) [8, 9]. In conclusion, the use of versatile vectors combined with plasmid-based expression simplifies eukaryotic expression to a point where it is hardly more effort than using prokaryotic systems.

2 Materials

In the first part, we present an overview of most of the available versatile vectors, to enable the reader to decide for the optimal vector system, which is compatible to the individual required application. In the second and third part, we describe the necessary materials needed for the fast plasmid-based screening techniques presented in the methods section.

2.1 Selection of Expression Vector

Versatile multi-host vectors can be acquired from commercial suppliers or from specialized academic laboratories. In Table 1 an overview of the most common multi-host-vectors is presented. All of them allow protein production in different hosts almost comparable to an optimized single host expression vector. Most of the versatile vectors described here harbor a backbone or promoter cassette derived from pTriEx, which is the most commonly used vector commercially available from Novagen. It combines the proven features of the pET, BacVector™, and Bac-to-Bac™ Systems and provides all of the benefits of each system in only one vector backbone. The hereto appropriated promoter and enhancer elements for each host system were placed head to tail in tandem orientation without hindering the expression characteristics in the other hosts. The pTriEx vector harbors a mammalian promoter (CMV enhancer fused to the Chicken β-actin or CMV promoter), baculoviral promoter (p10 derived from *Autographa californica* nucleopolyhedrovirus (AcMNPV)), and *E. coli* promoter (T7*lac*) which is IPTG inducible. Translation initiation signals include a ribosome binding site (RBS) for bacterial expression [10] followed by an optimal Kozak consensus sequence [11] for mammalian cell expression. The promoter and cloning regions are flanked on each side by segments of baculovirus genomic DNA that facilitate the generation of baculovirus recombinants at the viral *polh* locus. The pTriEx vector family contains a set of different N-terminal fusion tags, comprehensive cloning regions, signal sequences, protease cleavage sites, and optional C-terminal fusion tags (Table 2). Additional, pTriEx-1.1, pTriEx-2, pTriEx-3, and pTriEx-4 are available as stable expression vectors for mammalian systems. Here, a downstream selection marker gene (hygromycin (Hyg) or neomycin (Neo)) simplifies the isolation of the stably integrated cell clones expressing the target genes.

An alternative to pTriex is the pDual® Expression Vector from Agilent Technologies Inc. It offers high-level expression of heterologous genes in both mammalian and prokaryotic systems [12, 13]

Table 1
Overview of multi-host vectors and their application

Vector	Origin	Promoters	Hosts	Applications	Special features
pTriEx™	pET BacVector™ BacMam™	T7*lac*, p10, CAG or CMV	*E. coli*, CHO, insect cells (BEVS)	Protein Expression in three different hosts	Several N-and C-terminal tags available
PDual	pCI	T7*lac*, CMV	*E. coli*, mammalian cell lines	Protein Expression in two different hosts	
pEXPRESS	pRSV-CAT PET-8c	T7*lac*, RSV-LTR	*E. coli*, Hela, in vitro	Screening of cDNA libraries	
pFlpBtM-II	pFastBac pTT5 pTriEx-7™	CMV, p10	CHO, HEK293-6E, insect cell (BEVS)	Production of structural relevant proteins	Tn7-sites FRT sites for RMCE, OriP
phr5-OpIE2-p10	pIEx/Bac-5 pIZT/V5-His	OpIE2, p10	Insect cells (BEVS and plasmid-based)	Versatile vector in insect cells—fast plasmid-based screening and production	hr5 enhancer for higher expression
pBVboostFG	pFastBac pTriEx-1™	T7*lac*, polH, p10, CAG	Mammalian, bacterial, insect cells (BEVS), in vivo	Functional genomics; analysis of cDNA libraries	Tn7-sites
pOPIN	pTriEx-2™ pTT3	T7*lac*, p10, CAG	*E. coli*, HEK293-6E, insect cells (BEVS)	HTP cloning test run for scale-up experiments	OriP

Table 2
Overview of available pTriEx-expression vectors

Vector	Promoters	Fusion tags		Protease cleavage sites	Optional for stable expression
		N-terminal	C-terminal		
pTriEx-1.1	T7 *lac*, p10, CAG	–	HSV-Tag His-Tag	–	Hyg or neo
pTriEx-2	T7 *lac*, p10, CAG	His-Tag S Tag	HSV-Tag His-Tag	Thrombin enterokinase	Hyg or neo
pTriEx-3	T7 *lac*, p10, CMV	–	HSV-Tag His-Tag	–	Hyg or neo
pTriEx-4	T7 *lac*, p10, CMV	His-Tag S Tag	HSV-Tag His-Tag	Thrombin enterokinase	Hyg or neo

but not in insect cells. For constitutive expression in mammalian cells, the pDual expression vector contains a mutagenized version of promoter/enhancer of the human CMV gene. Inducible gene expression in prokaryotes is directed from the hybrid T7/lacO promoter. In addition the vector carries the lac repressor gene (LacI) for inducible expression using IPTG and a tandemly arranged bacterial Shine–Dalgarno [10] and mammalian Kozak [11] consensus sequence for efficient translation of mRNA in both host systems. cDNA inserts encoding proteins are inserted into the vector using the unique seamless cloning method [14]. A special feature of the pDual vector is an uncommon calmodulin-binding peptide (CBP) tag which can be used for affinity purification and identification by Western blot. In addition, the vector harbors a c-Myc epitope and 6× His purification tag. Selection of stable cell clones in both mammalian and bacterial cells is possible due to a neomycin phosphotransferase gen (neo) under the control of the SV40 promoter (mammalian) and the β-lactamase promoter (bacterial). Next to pDual® offered from Agilent Technologies Inc. there is another vector available which has an equal name but is written in capital letters, the pDUAL. This so called multipurpose vector series enables both the introduction of multiple copies of genes with episomal maintenance and also a single copy chromosomal integration into the fission yeast, *Schizosaccharomyces pombe* [15].

The versatile expression vector, pEXPRESS, was developed by Forman and Samuels in 1991 and serves as shuttle expression plasmid in Hela cells as well as in *E. coli* and in vitro transcription–translation systems [16]. For functionality in all these systems, it combines the mammalian RSV-LTR (Rous sarcoma virus-long terminal repeat) promoter with the *E. coli* T7 promoter ϕ10 system, as well as a Kozak and a Shine–Dalgarno sequence. The promoter region is followed by a multiple cloning site and a SV40 Poly(A) Signal but lacks a T7 terminator. This results in long transcripts in *E. coli* which are more stable than transcripts with a SV40 Poly(A) Signal according to reference [16]. pEXPRESS is as efficient as the standard eukaryotic vectors in Hela cells and is also suitable for high expression in *E. coli*. As a special application, pEXPRESS can also be employed for direct synthesis of active proteins by coupled in vitro transcription–translation. Thereby, pEXPRESS allows fast screening of cDNA libraries without cloning of separate vectors for eukaryotic, prokaryotic, or in vitro systems. pEXPRESS has been successfully applied for production of various thyroid hormone receptor mutants and for the synthesis of neurofilament proteins [16].

The pFlp-Bac-to-Mam II (pFlpBtM II) vector was developed in the Helmholtz Protein Sample Production Facility for multi-host expression and is specially designed to produce proteins for structural analysis [1]. It contains all elements required for the Baculovirus Expression Vector System (BEVS), transient plasmid-based expression in HEK293-6E cells and stable expression in CHO Lec.3.2.8.1 based on the recombinase mediated cassette exchange (RMCE) approach. To enable BEVS expression, Tn7-transposition sites flank the gene of interest, and for selection of the recombinant baculovirus, a gentamicin resistance gene is present as well. The expression in BEVS is derived from the strong late phase baculoviral p10 promoter. The promoter region also includes the strong CMV (Cytomegalovirus) promoter variant with intron for the transient expression in HEK293-6E cells [17]. Additionally, the backbone of pFlpBtM-II contains the Epstein–Barr virus oriP, which ensures increased nuclear transport and episomal replication of the plasmid in EBNA-1 expressing cells (e.g., HEK293-6E). This offers semi-stable protein expression which leads to high protein yields [18]. Moreover, pFlpBtM-II can be directly applied to generate stable CHO Lec3.2.8.1 cell lines via RMCE [19]. CHO Lec3.2.8.1. RMCE master cell line harbors a "tagging cassette" in its chromosomal DNA, flanked by two heterospecific *Fl*ipase *R*ecombination *T*arget sites (FRT) [20], which can be easily exchanged by co-transfection of a "targeting" vector carrying the gene of interest and the Flipase coding vector. The pFlpBtMII vector comprises all features necessary for successful RMCE, like the compatible FRT sites, a PGK promoter with an ATG start codon for activation of the selection trap (Δ*neo* or Δ*puro*) after the cassette exchange.

A further versatile vector developed in the Helmholtz Protein Sample Production Facility is the phr5-OpIE2-p10 vector [21]. This vector is only usable in insect cell lines like Sf21 or Hi5 cells, however it can be applied for expression in insect cell lines using the BEVS or without baculovirus. It contains the baculoviral p10 promoter and the expression cassette is flanked by segments of baculovirus genomic DNA that facilitate the generation of baculovirus recombinants at the viral *polh* locus. Additionally to the BEVS based expression, it also facilitates direct virus-free, plasmid-based expression by the strong OpIE2 promoter [22] in Hi5 cells. Due to the combination of the hr5 enhancer element and the p10 promoter, a protein production by transactivation is likewise possible [21]. The main application of this vector is fast and efficient plasmid-based screening, e.g., by using the SplitGFP method [23] before finally starting production of expressible constructs in BEVS.

The pBVboostFG comprises a tetra-promoter vector which enables screening of gen/cDNA libraries for functional genomic studies [6]. It thereby facilitates gene expression in bacteria, insect and mammalian cells and furthermore expression in vivo. For that purpose, the plasmid contains a promoter cassette composed of T7*lac*, pPolh and p10 and CAG (CMVie enhancer + chicken beta-actin promoter). The insertion of mini Tn7 transposition sites into the vector allows fast and easy production of recombinant baculoviruses [24]. The pBVboostFG extends the advantages of this versatile host vector with an efficient and flexible cloning of desired DNA fragments into the vector by the site-specific recombination cassette of the bacteriophage lambda containing attR1/2 sites. Thus, the production of baculoviral libraries using the pBVboostFG vector is fast and compatible with high-throughput technology as it produces a high diversity of clones in one single step without background. The derived baculovirus can be used directly for the successful transduction of CHO cells. Additional, also a direct plasmid-based transfection of CHO cells is possible. As an example, the viability genes or cDNAs of 18 different proteins were cloned into pBVboostFG and their expression was tested in different hosts. The system was also used for in vivo eGFP expression in rat by baculoviral infection [25].

The Oxford Protein Production Facility (OPPF) is a structural proteomics facility funded to produce high quality structural data for proteins from several host organisms, including viruses, bacterial human pathogens and human proteins associated with the etiology of human diseases such as cancers. In order to investigate these proteins either in large numbers or to investigate, in parallel, many multiple domains of smaller numbers of these proteins, a highly efficient cloning and expression screening strategy was required. For that purpose Berrow et al. [26] developed a vector system expressible in the three available hosts (*E. coli*, insect and mammalian cells) named as pOPINE, pOPINF, pOPINJ or pOPINM based on pTriEx2 (Novagen) which can be easily constructed in HTP mode by utilizing the In-Fusion™ technology of Clontech. In combination with a 15 bp homology region, this technique allows ligation-independent cloning of PCR products.

In only 1 year, the OPPF has constructed a total of 661 vectors from 703 PCR products, with an overall cloning efficiency of 94% [26]. Thus, similar to the pBVboostFG vector family, the pOPIN vectors combine a fast and efficient cloning strategy with a versatile vector system to be most efficient in high-throughput mode. The pOPIN vectors are based on several different plasmid backbones, including the three-promoter vector pTriEx2 (T7*lac*, p10, CAG) for multi-host expression and pTT3 for enhanced eukaryotic expression. They offer various N- and C-terminal fusion tags for affinity purification for improved downstream processing, which can be removed by protease cleavage sites for crystallization and 3D structural analysis.

2.2 Cell Lines for Protein Production

1. The HEK293-6E cell line was licensed from National Research Council (NRC), Biotechnological Research Institute (BRI), Montreal, Canada.
2. The Hi5 insect cell line (officially called BTI-Tn-5B1-4) were isolated by the Boyce Thompson Institute for Plant Research, Ithaca, USA.

2.3 Cell Culture Conditions

1. 125 mL up to 1 L polycarbonate shake flasks (e.g., Corning).
2. Minitron™ CO_2 orbital shaker with 25 mm orbital (e.g., Infors).
3. Complete Cultivation Medium for HEK293-6E: FreeStyle F17 (Gibco, Life Technologies/Thermo Scientific) supplemented with 7.5 mM-glutamine, 0.1% Pluronic-F68 and 25 μg/mL G418. Prepare by adding components as follows to 1 L of fresh F17 medium: 40 mL of 200 mM stock solution of L-glutamine, 0.5 mL of 50 mg/mL G418-solution, and 10 mL of 10% (w/v) Pluronic-F68 (20 g in 200 mL of water and filter through 0.2 μm). Mix well and store at 4 °C.
4. Complete Cultivation Medium for Hi5 insect cells: ExCell405.

2.4 Transfection Reagents and Additional Chemicals for Protein Production

Prepare all solutions using ultrapure water and analytical grade reagents. All reagents are sterile filtered and stored at 4 °C (unless otherwise mentioned).

1. 1 mg/mL polyethylenimine (PEI), linear: Dissolve 0.05 g of PEI in 50 mL of water and heat until it is completely dissolved. Store aliquots at −70 °C.
2. Use 15 mL polystyrene tubes for preparing DNA–PEI complexes.
3. 20% (w/v) tryptone N1: Weigh 100 g of tryptone N1, dissolve and make up to 500 mL with water.
4. 300 g/L glucose: Weigh 150 g of glucose, dissolve and make up to 500 mL with water.

5. 75 mM valproic acid (20×): Dissolve 1.2 mL and make up to 100 mL with F17 medium (supplemented).
6. Guava EasyCyte™ Mini (Merck, Millipore).

3 Methods

Carry out all procedures at room temperature unless otherwise specified.

3.1 Transient Protein Expression in HEK293-6E Cells

The protocol describes a 50 mL expression in HEK293-6E cells. The employed vectors should contain a suitable expression cassette and the OriP to ensure optimal expression levels.

1. **[Day 3, e.g., Friday]** Prepare a 40 mL culture containing 0.3 × 10^6 HEK293-6E c/mL. Incubate the culture for 72 h at 37 °C, 110 rpm and 5% CO_2.
2. **[Day 0, e.g., Monday]** Count the cells of the preparatory culture and prepare a 25 mL culture containing 1.5–2 × 10^6 HEK293-6E c/mL by centrifuging the required volume of the cell suspension at 180 × *g* for 4 min. Discard the supernatant carefully and resolve the cell pellet in 25 mL of fresh F17. Place the cells back into the incubator (37 °C, 90 rpm and 5% CO_2) until use (*see* **Note 1**).
3. Prepare the DNA solution in 1.25 mL of F17 in a 15 mL tube (*see* **Note 2**). Hereto, mix 23.75 µg of the plasmid containing your gene of interest with 1.25 µg control plasmid coding, e.g., for eGFP. Mix by vortexing briefly.
4. Prepare the PEI solution in 1.25 mL of F17 in a 15 mL Polystyrene tube (*see* **Note 2**). Hereto, pipette 62.5 µL PEI of the 1 mg/mL PEI stock solution into the. Mix by vortexing briefly.
5. Mix the PEI solution with the DNA solution directly (*see* **Note 3**).
6. Incubate the PEI–DNA solution for 15 min up to 1 h, at room temperature to preform DNA–PEI complexes.
7. Pipette the DNA–PEI complexes (2.5 mL) to the cells and mix gently. Incubate the cells until harvest at 37 °C, 90 rpm and 5% CO_2.
8. **[Day 1–7, e.g., Tuesday–Monday]** Take samples daily and determine transfection efficiency in the flow cytometer (Guava EasyCyte) and/or determine target protein expression by a suitable technique (SDS-PAGE, slot blot or western blot).
9. **[Day 2, e.g., Wednesday]** Add 25 mL of fresh F17 to the cells as well as 1.25 mL of the TN1-Stock (20%) = 0.5%.

10. **[Day 3, e.g., Thursday]** Add 0.75 mL of the glucose stock (300 g/L) = 4.5 g/L.
11. **[Day 4, e.g., Friday]** Add 2.5 mL of the valproic acid stock (75 mM) = 3.75 mmol/L.
12. **[Day 7, e.g., Monday]** Harvest secreted target proteins by centrifuging the cell suspension 15 min at 1000 × *g* and store the supernatant at 4 °C until purification. Depending on the protein, it might be required to add protease inhibitor. For intracellular proteins, snap freeze the cell pellet and store at −80 °C until cell lysis and purification.

3.2 Transient Protein Expression in Hi5 Cells

The described protocol is for a 100 mL expression in Hi5 cells. The employed vectors should contain a suitable expression cassette with a strong promoter element (*see* **Note 4**).

1. **[Day 1, e.g., Monday]** Prepare an 80 mL culture containing 0.5 × 10^6 Hi5 c/mL. Incubate the culture for 24 h at 27 °C and 90 rpm (*see* **Note 5**).
2. **[Day 0, e.g., Tuesday]** Count the cells of the preparatory culture and prepare a 20 mL culture containing 5 × 10^6 Hi5 c/mL by centrifuging the required volume of the cell suspension at 180 × *g* for 4 min. Discard the supernatant and resolve the cell pellet in 25 mL fresh ExCell405.
3. Incubate the culture at 27 °C and 90 rpm for 1 h.
4. Mix 95 μg of the expression plasmid and 5 μg of a control plasmid e.g., coding for eGFP (*see* **Note 6**).
5. Pipette the DNA mix directly to the prepared cells and mix gently (*see* **Note** 7).
6. Immediately pipette 400 μL PEI of the 1 mg/mL PEI stock solution to the cells and mix gently.
7. Incubate the culture at 27 °C and 90 rpm for 3 up to 5 h.
8. Add 80 mL of fresh ExCell405 (*see* **Note 8**).
9. **[Day 1–3, e.g., Wednesday–Friday]** Take samples daily, count the cells and determine transfection efficiency in the cytometer or/and determine target protein expression by a suitable technique (SDS-PAGE, slot blot or western blot). If cells reach a density above 3 × 10^6 c/mL, adjust the concentration back to 2 × 10^6 c/mL by adding fresh culture medium.
10. **[Day 3, e.g., Friday]** Harvest secreted target proteins by centrifuging the cell suspension 15 min at 1000 × *g* and store the supernatant at 4 °C until purification. For intracellular proteins, freeze the cell pellet at −20 °C until cell lysis and purification.

4 Notes

1. Replacing used culture medium for fresh F17 (without antibiotic) ensures a sufficient amount of nutrients to keep the cells growing until addition of fresh culture medium at day 2.
2. Using polystyrene tubes instead of polypropylene tubes is important to ensure that the PEI–DNA complexes do not stick to the tube which would lead to a lower transfection rate due to a lower total amount of available complex [27].
3. Directly mix the DNA with PEI to form stable DNA–PEI complexes.
4. The optimal combination in our facility comprises the OpIE2 promoter, the IE1 terminator and a FlashBac compatible backbone [21].
5. This step ensures that the cells are in the exponential growth phase the next day which will increase the transfection efficiency and protein expression vastly.
6. Using higher amounts of DNA (up to 3 μg per 1×10^6 cells) leads to higher yields if the PEI amount is increased accordingly and the PEI–DNA ratio of 4:1 is kept stable. Further increasing the amount of DNA will decrease cell viability due to the required higher PEI concentration, as PEI is known to be cytotoxic in high concentrations. This will decrease overall productivity.
7. Direct transfection without preforming the DNA–PEI complexes led to the highest transfection rates and protein levels in our lab. In contrast, Shen et al. observed no differences between direct transfection or transfection after preforming the DNA–PEI complexes [9]. The difference might be caused by different performance of other batches of Hi5 cells. However, a direct transfection is easier and faster. Thus, the described direct transfection is the optimal method in our hands.
8. Diluting with fresh media is important to keep the cells in the exponential growth phase until the time of harvest.

Acknowledgments

Katharina Karste and Maren Bleckmann are equally contributing authors. We thank Nadine Konisch, Anke Samuels and Daniela Gebauer for their excellent technical support during the experiments. This work was supported by the Helmholtz Protein Sample Production Facility and by Instruct, part of the European Strategy Forum on Research Infrastructures (ESFRI).

References

1. Meyer S, Lorenz C, Baser B et al (2013) Multi-host expression system for recombinant production of challenging proteins. PLoS One 8:e68674
2. Hochkoeppler A (2013) Expanding the landscape of recombinant protein production in *Escherichia coli*. Biotechnol Lett 35:1971–1981
3. Willett RA, Kudlyk TA, Lupashin VV (2015) Expression of functional Myc-tagged conserved oligomeric Golgi (COG) subcomplexes in mammalian cells. Methods Mol Biol 1270:167–177
4. Engler C, Marillonnet S (2014) Golden Gate cloning. Methods Mol Biol 1116:119–131
5. Li MZ, Elledge SJ (2012) SLIC: a method for sequence- and ligation-independent cloning. Methods Mol Biol 852:51–59
6. Laitinen OH, Airenne KJ, Hytönen VP et al (2005) A multipurpose vector system for the screening of libraries in bacteria, insect and mammalian cells and expression in vivo. Nucleic Acids Res 33:e42
7. Braun P, LaBaer J (2003) High throughput protein production for functional proteomics. Trends Biotechnol 21:383–388
8. Swiech K, Kamen A, Ansorge S et al (2011) Transient transfection of serum-free suspension HEK 293 cell culture for efficient production of human rFVIII. BMC Biotechnol 11:114
9. Shen X, Pitol AK, Bachmann V et al (2015) A simple plasmid-based transient gene expression method using High Five cells. J Biotechnol 216:67–75
10. Shine J, Dalgarno L (1974) The 3′-terminal sequence of *Escherichia coli* 16S ribosomal RNA: complementarity to nonsense triplets and ribosome binding sites. Proc Natl Acad Sci U S A 71:1342–1346
11. Kozak M (1986) Point mutations define a sequence flanking the AUG initiator codon that modulates translation by eukaryotic ribosomes. Cell 44:283–292
12. Mullinax RL, Wong DT, Davis HA et al (2000) Expression in both *E. coli* and mammalian cells. Strategies 13:41–43
13. Davis HA, Wong DT, Padgett KA et al (2000) High-level dual mammalian and bacterial protein expression vector | biocompare: the buyer's guide for life scientists [Internet]. Available: http://www.biocompare.com/Application-Notes/42069-High-Level-Dual-Mammalian-And-Bacterial-Protein-Expression-Vector/
14. Padgett KA, Sorge JA (1996) Creating seamless junctions independent of restriction sites in PCR cloning. Gene 168:31–35
15. Matsuyama A, Shirai A, Yashiroda Y et al (2004) pDUAL, a multipurpose, multicopy vector capable of chromosomal integration in fission yeast. Yeast 21:1289–1305
16. Forman BM, Samuels HH (1991) pEXPRESS: a family of expression vectors containing a single transcription unit active in prokaryotes, eukaryotes and in vitro. Gene 105:9–15
17. Durocher Y, Butler M (2009) Expression systems for therapeutic glycoprotein production. Curr Opin Biotechnol 20:700–707
18. Jäger V, Büssow K, Wagner A et al (2013) High level transient production of recombinant antibodies and antibody fusion proteins in HEK293 cells. BMC Biotechnol 13:52
19. Schlake T, Bode J (1994) Use of mutated FLP recognition target (FRT) sites for the exchange of expression cassettes at defined chromosomal loci. Biochemistry (Mosc) 33:12746–12751
20. Wilke S, Groebe L, Maffenbeier V et al (2011) Streamlining homogeneous glycoprotein production for biophysical and structural applications by targeted cell line development. PLoS One 6:e27829
21. Bleckmann M, Schürig M, Chen F-F et al (2016) Identification of essential genetic baculoviral elements for recombinant protein expression by transactivation in Sf21 insect cells. PLoS One 11:e0149424
22. Bleckmann M, Fritz MH-Y, Bhuju S et al (2015) Genomic analysis and isolation of RNA polymerase II dependent promoters from *Spodoptera frugiperda*. PLoS One 10:e0132898
23. Bleckmann M, Schmelz S, Schinkowski C et al (2016) Fast plasmid based protein expression analysis in insect cells using an automated SplitGFP screen. Biotechnol Bioeng 9999:1
24. Airenne KJ (2003) Improved generation of recombinant baculovirus genomes in *Escherichia coli*. Nucleic Acids Res 31:101e–101
25. Lehtolainen P, Tyynelä K, Kannasto J et al (2002) Baculoviruses exhibit restricted cell type specificity in rat brain: a comparison of baculovirus- and adenovirus-mediated intracerebral gene transfer in vivo. Gene Ther 9:1693–1699
26. Berrow NS, Alderton D, Sainsbury S et al (2007) A versatile ligation-independent cloning method suitable for high-throughput expression screening applications. Nucleic Acids Res 35:e45
27. González-Sánchez C, González-Quesada M, de la Orden MU et al (2008) Comparison of the effects of polyethylenimine and maleated polypropylene coupling agents on the properties of cellulose-reinforced polypropylene composites. J Appl Polym Sci 110:2555–2562

Chapter 21

A Generic Protocol for Intracellular Expression of Recombinant Proteins in *Bacillus subtilis*

Trang Phan, Phuong Huynh, Tuom Truong, and Hoang Nguyen

Abstract

Bacillus subtilis (*B. subtilis*) is a potential and attractive host for the production of recombinant proteins. Different expression systems for *B. subtilis* have been developed recently, and various target proteins have been recombinantly synthesized and purified using this host. In this chapter, we introduce a generic protocol to express a recombinant protein in *B. subtilis*. It includes protocols for (1) using our typical expression vector (plasmid pHT254) to introduce a target gene, (2) transformation of the target vector into *B. subtilis*, and (3) evaluation of the actual expression of a recombinant protein.

Key words *Bacillus subtilis*, pHT01 vector, P*grac* promoter, pHT254, P*grac*100 promoter

1 Introduction

The demand for recombinant proteins in biotechnology has become higher with every passing day. For producing a successful native-like protein, the most important consideration is the host system. This choice will initiate the outline of the whole process from selecting respective expression vectors to taking up the technology along with a variety of molecular tools, equipment, and reagents [1, 2]. Hosts extensively used these days include bacteria, yeast, filamentous fungi, unicellular algae, and a few animal cells with their particular strengths and weaknesses. Among them, bacteria are preferred choices if posttranslational modifications are not needed [3].

The utilization of bacteria for producing recombinant proteins has advanced greatly. The gram-positive soil bacterium *B. subtilis* is emerging as an attractive host for the production of various proteins such as proteases, amylases, and lipases. Homologous proteins can be produced to several grams per culturing liter [4, 5]. Unlike *E. coli*, *B. subtilis* is classified as a "generally recognized as safe" (GRAS) organism and is endotoxin-free. *B. subtilis* are also highly appreciated as they can secrete proteins directly into the medium via

Nicola A. Burgess-Brown (ed.), *Heterologous Gene Expression in E. coli: Methods and Protocols*, Methods in Molecular Biology, vol. 1586, DOI 10.1007/978-1-4939-6887-9_21, © Springer Science+Business Media LLC 2017

the Sec and Tat pathways [6], and are easy for genetic manipulation and experiment handling. Additionally, the growing time for this bacterium to reach a specific density is only within hours, and thus can be applied in required large-scale industrial processes with ease, for example with the production of laundry enzymes and riboflavin [4].

Our current expression system is initially constructed using a plasmid that replicates via theta-mode of replication and confers structural stability [7]. Besides this backbone, we also generate a strong synthetic promoter, P*grac* that consists of wild-type *groESL* promoter from *B. subtilis* in combination with the *lac* operator from *E. coli*, and strong *gsiB* ribosome binding site [8]. Based on these discoveries, we generated the first pHT expression vector, pHT01, that can produce protein up to 16% of the cellular proteins of *B. subtilis* cells [9]. From this success, we created a P*grac* family, containing modifications of the P*grac* promoter in combination with modified *lacO* and mRNA controllable stabilizing elements (CoSE). These promoters could be controlled by using IPTG as an inducer allowing massive recombinant protein production, up to 40% of the total cellular proteins of *B. subtilis* cells [10, 11].

In this chapter, we describe a routine protocol for protein expression in *B. subtilis* using one of a representative vector in our laboratory. Plasmid pHT254 carries strong promoter, P*grac*100 that is proved to produce high protein expression levels in fusion with His-tag in *B. subtilis* [12]. A reporter β-galactosidase (BgaB) [13] is used as a model to demonstrate all the steps of the protocols.

2 Materials

2.1 Bacterial Strains and Vectors

1. *E. coli* OmniMAX ordered from Life Technologies (*see* **Note 1**).
2. *B. subtilis* 1012 wild-type (Genotype: *leu*A8 *met*B5 *trp*C2 *hsd*RM1) [14].
3. Plasmids, pHT01-*bgaB* (P*grac*-BgaB), pHT254 (P*grac*100-MCS-His) (Fig. 1), pHT1170 (P*grac*100-GFP-His), and pHT1179 (P*grac*100-BgaB-His) constructed by our laboratory [9, 12]. Plasmid pHT254 can be obtained from MoBiTec.

2.2 Medium for Bacteria Growth and Kits for Cloning

1. LB medium: Dissolve 5 g of yeast extract, 10 g of tryptone, and 5 g of NaCl in 1 L of dH_2O and autoclave.
2. LB agar plates: Dissolve 5 g of yeast extract, 10 g of tryptone, 5 g of NaCl, and 15 g of agar in 1 L of dH_2O and autoclave. Cool down the medium to 50 °C and add 100 μg/mL ampicillin and 40 μg/mL X-gal or 10 μg/mL chloramphenicol and 40 μg/mL X-gal before pouring to Petri dishes.
3. 10 mg/mL chloramphenicol stock: Prepared in 100% ethanol.

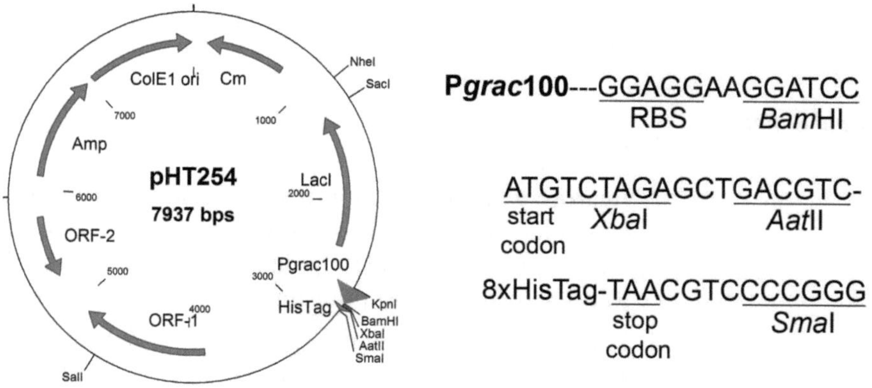

Fig. 1 Map of plasmid pHT254 and the typical sequence of pHT vectors from the ribosome binding site (RBS) to the end of the multicloning site [12]

4. 100 mg/mL ampicillin stock: Prepared in Dimethylformamide (DMF).
5. 40 mg/mL X-gal stock: Prepared in Dimethylformamide (DMF) water.
6. QIAquick PCR Purification Kit (Qiagen).
7. QIAprep Spin Miniprep Kit (Qiagen).

2.3 Buffer and Media for Transformation

1. 10× S-Base: Dissolve 20 g of $(NH_4)_2SO_4$, 140 g of K_2HPO_4, 60 g of KH_2PO_4, 10 g of Na_3Citrate, and 8 g of $MgSO_4$ in 1 L of dH_2O. Add a sterile stock solution of 120 g/L $MgSO_4$ to complete the 10× S-Base before use.
2. HS medium: 5 g of glucose, 0.05 g of l-tryptophan, 0.2 g of casein acid hydrolysate (protein hydrolysate amicase), 5 g of yeast extract, 8 g of arginine, 0.4 g of histidine, and 1× S-Base. Prepare separate filter sterile stock solutions of 40% (w/v) glucose, 0.1% (w/v) l-tryptophan, 2% (w/v) casein acid hydrolysate, 10% (w/v) yeast extract, 8% (w/v) arginine, 0.4% (w/v) histidine, and 10× S-Base and mix them at desired amounts to complete the medium.
3. LS medium: 5 g of glucose, 0.005 g of l-tryptophan, 0.1 g of casein acid hydrolysate, 1 g of yeast extract, 2.5 mM $MgCl_2$, 0.5 mM $CaCl_2$, and 1× S-Base. Prepare separate filter sterile stock solutions of 40% (w/v) glucose, 0.1% (w/v) l-tryptophan, 2% (w/v) casein acid hydrolysate, 10% (w/v) yeast extract, 1 M $MgCl_2$, 1 M $CaCl_2$, and 10× S-Base and mix them at desired amounts to complete the medium.
4. 0.1 M EGTA (ethylene glycol tetraacetic acid).
5. Pure glycerol.

2.4 Enzymes

1. *Pfu* DNA polymerase (2.5 U/μL), *Taq* DNA polymerase (2.5 U/μL), T4 DNA ligase (5 U/μL), *Bam*HI (10 U/μL), and *Aat*II (10 U/μL) from Life Technologies.

2. 10 mg/mL lysozyme solution: Prepared in 10 mM Tris–HCl, pH 8.0.

2.5 Gels and Buffers

1. 1× TAE buffer: 40 mM Tris base, 20 mM acetic acid, 1 mM EDTA pH 8.0.
2. Agarose gel: Weigh 1.5 g of agarose and transfer to a 250 mL container, add 100 mL of 1× TAE buffer and boil the mixture until agarose is completely dissolved. Add 10 μL of 10,000× SYBR Green, stir and pour into the mold (*see* **Note 2**).
3. Polyacrylamide stacking gel: 4% (v/v) acrylamide–bis-acrylamide 29:1, 125 mM Tris–HCl pH 6.8, 0.1% (w/v) SDS, 0.07% (w/v) APS, and 0.1% (v/v) TEMED.
4. Polyacrylamide separating gel: 12% (v/v) acrylamide–bis-acrylamide 29:1, 375 mM Tris–HCl pH 8.8, 0.1% (w/v) SDS, 0.07% (w/v) APS, and 0.1% (v/v) TEMED (*see* **Note 3**).
5. 5× SDS-PAGE sample buffer: 250 mM Tris–HCl pH 6.8, 10% (w/v) SDS, 30% (v/v) glycerol, 500 mM DTT, 0.02% (w/v) bromophenol blue.
6. 1× SDS-PAGE running buffer: 25 mM Tris base, 192 mM glycine, 0.1% (w/v) SDS (*see* **Note 4**).
7. SDS-PAGE Stain solution: Prepare 0.625 g of Coomassie Brilliant Blue, 25 mL of glacial acetic acid, and 112.5 mL of absolute ethanol in 1 L of dH_2O (*see* **Note 5**).
8. SDS-PAGE Destain solution: Prepare 100 mL of absolute ethanol and 75 mL of absolute acetic acid in 1 L of dH_2O.
9. BgaB lysis buffer: 57.7 mM Na_2HPO_4, 42.4 mM NaH_2PO_4, 10 mM $MgCl_2$, 200 μg/mL Lysozyme, 1 μg/mL DNase.
10. BgaB reaction buffer: 57.7 mM Na_2HPO_4, 42.4 mM NaH_2PO_4, 10 mM $MgCl_2$, 10 mM mercaptoethanol.
11. BgaB stop buffer: 1 M Na_2CO_3.
12. 4 mg/mL ONPG (ortho-nitrophenyl-β-galactoside).

2.6 Plastics, Glassware, and Equipment

1. 1.5 and 2 mL Eppendorf tubes.
2. Erlenmeyer flask for cultures.
3. Petri dish for LB plates.
4. Hockey stick for plating.
5. Glass tubes (ϕ 18 mm).
6. Pipettes.
7. Horizontal electrophoresis system for agarose gel running (e.g., Bio-Rad).
8. Plate reader for reading color development in BgaB assay.
9. PCR machine for gene amplification and colony PCR.

10. Dry incubator for 42, 65, and 95 °C.
11. Vertical electrophoresis system for SDS-PAGE analysis (e.g., Bio-Rad).
12. Shaking incubator.
13. Spectrophotometer.
14. Centrifuge.

3 Methods

3.1 Using Plasmid pHT254 to Construct pHT1179 Containing bgaB Gene

Plasmid pHT1179 is constructed by inserting the *bgaB* gene into plasmid pHT254 using standard molecular cloning techniques as described below.

3.1.1 Preparation of bgaB Gene

1. Amplify the *bgaB* gene encoding for a heat-stable β-galactosidase by PCR reaction using primers ON941 (AAAGGAGGAA GGATCCATGAATGTGTTATC) and ON1250 (CTGCCCC GGGGACGTCAACCTTCCC GGCTTCATCATGC) with template pHT01-*bgaB*.
2. Prepare a 100 μL reaction mixture containing 1× PCR buffer for *Pfu* with $MgSO_4$, 200 μM of each dNTP, 50 ng of template, 0.25 μM of each primer, and 2.5 U *Pfu* DNA polymerase (*see* **Note 6**).
3. After an initial incubation at 95 °C for 5 min, perform 30 three-step cycles as follows: 30 s at 95 °C, 30 s at 55 °C, 2 min at 72 °C, and a final extension at 72 °C for 10 min (*see* **Notes** 7 and **8**).
4. Examine the PCR results by agarose gel electrophoresis running in TAE 1× buffer. The predicted amplicon size of *bgaB* is 2048 bps.
5. Use the QIAquick PCR Purification Kit to purify the PCR product before cutting with restriction enzymes, *Bam*HI and *Aat*II (*see* **Note 9**).
6. Prepare a 100 μL cutting reaction mixture containing 1× Tango buffer, 20 U of each restriction enzyme and 5 μg of *bgaB* gene (*see* **Note 10**).
7. Incubate the reactions at 37 °C for 2 h and clean up the DNA fragments using QIAquick PCR Purification Kit before the ligation steps.

3.1.2 Preparation of Linearized Plasmid pHT254

1. Cleave the plasmid pHT254 with *Bam*HI and *Aat*II to make sticky-ends (*see* **Note 9**). Other restriction enzymes could be selected for cloning (*see* Fig. 1).
2. The 100 μL reaction mixture contains 1× Tango buffer, 20 U of each restriction enzyme, and 5 μg of pHT254 (*see* **Note 10**).

3. Incubate the reactions at 37 °C for 2 h and clean up the DNA fragments using QIAquick PCR Purification Kit before the ligation steps.

3.1.3 Formation of the Recombinant Vector pHT1179

1. The *bgaB* gene is introduced into pHT254 at downstream region of the P*grac*100 promoter resulting in the pHT1179 (*see* **Note 11**).
2. Prepare a 20 μL ligation mixture containing 1× *T4* DNA ligation buffer, 150 ng of linearized plasmid, 150 ng of the gene fragment, and 5 U of *T4* DNA ligase (*see* **Note 12**).
3. Incubate the reaction at 25 °C or room temperature for 60 min and inactivate the ligase at 65 °C for 10 min.

3.1.4 Selection of Accurate Target Vector

1. Use 10 μL of the ligation mixture for the transformation into *E. coli* as described elsewhere [15], then plate on an LB plate containing ampicillin and X-Gal (*see* **Note 13**).
2. Choose the blue colonies for colony PCR using the primer pair of ON1249 (5′-CGTTTCCACCGGAATTAGCTTG-3′) which anneals to the plasmid and ON1384 (5′-CGGTTCGATCTT GCTCCAACTG-3′) which anneals to the inserted *bgaB*.
3. Choose the colonies that show a correct DNA band size (302 bp) on the agarose electrophoretic gels after colony-PCR for sequencing. Use forward primer ON1249 and reversed primer ON314 (5′-GTTTCAACCATTTGTTCCAGGTA AG-3′) for sequencing.
4. Extract the plasmid using QIAprep Spin Miniprep Kit.
5. Analyze the sequences using ON1249, ON314 and store the fully verified and correct plasmid, pHT1179, at −20 °C for future use.

3.2 Transformation of pHT1179 into B. subtilis 1012

3.2.1 Preparation of B. subtilis 1012 Competent Cells

1. Transfer a fresh colony of *B. subtilis* 1012 on an LB plate into 5 mL of HS medium and shake vigorously at 37 °C for 12 h.
2. Inoculate a particular volume of the pre-culture solution into 50 mL of fresh HS medium so that the OD_{600} of the new culture is about 0.05 and continue shaking vigorously at 37 °C.
3. Measure OD_{600} of the *B. subtilis* culturing solution every 1 h (*see* **Note 14**) and follow the growth curve. When the culture reaches OD_{600} at 0.5, follow every 30 min.
4. Once the cells reach the stationary phase, transfer 10 mL of cell culture into an ice-cold tube containing 1 mL of cold 100% glycerol.
5. Divide the mixture into aliquots of 1 mL each (*see* **Note 15**).
6. Continue to collect the cell culture every 15 min until the end, similarly as **steps 4** and **5**. Store all the aliquots at −80 °C.
7. Test the transformation efficiency of each collection (*see* **Note 16**).

3.2.2 Transformation of Plasmid pHT1179 into B. subtilis *1012*

1. Inoculate 1 mL of an aliquot of competent cells into 10 mL of LS medium and gently shake at 50 rpm at 30 °C for 2 h.
2. Add 100 μL of 0.1 M EGTA to the competent cell solution and continue to incubate at 30 °C but without shaking for 10 min.
3. Transfer 1 mL of competent cell solution to 2 mL Eppendorf tubes that contain 10 μg of pHT1179 and vigorously shake at 37 °C for another 2 h.
4. Centrifuge the Eppendorf tubes at 3000 × *g* and keep the cell pellets and about 100 μL of medium. Resuspend the cells and spread onto an LB plate containing 10 μg/mL chloramphenicol and 40 μg/mL X-gal.
5. Incubate the plates at 37 °C overnight.

3.3 Examination of the Expression of Recombinant BgaB in B. subtilis

3.3.1 Analysis of Protein Expression by Activity

1. Transfer the blue colonies from the X-gal plate into the 25 mL of LB liquid medium containing 10 μg/mL chloramphenicol and shake at 250 rpm with incubation at 37 °C.
2. Measure the culture during the time. When OD_{600} reaches 0.4–0.8 (log phase), add IPTG to the final concentration of 0.1 mM (*see* **Note 17**). Collect one sample before the induction for analysis.
3. Collect the cells after 2 and 4 h induction. The amount of the cells collected should be equivalent to an OD_{600} of 1.2.
4. Add 500 μL of BgaB lysis buffer and incubate at 37 °C for 30 min.
5. Centrifuge at 13,000 × *g* for 2 min at 4 °C and collect the supernatant.
6. Add 50 μL of supernatant into the tubes, then supplement with 37 μL of BgaB reaction buffer and incubate at 55 °C for 10 min.
7. Finally, add 13 μL of 4 mg/mL ONPG and keep incubating at 55 °C until the solution changes to yellow color.
8. Add 100 μL of 1 M Na_2CO_3 to stop the reactions and measure OD_{420} of the solution.
9. The higher value of OD_{420} indicates the higher of BgaB activity and thus the higher level of BgaB expression.

3.3.2 Analysis of Protein Expression by SDS-PAGE

1. Culture the cells as in Subheading 3.3.1 and collect the cells before and after addition of IPTG. The amount of the cells collected should be equivalent to an OD_{600} of 2.4.
2. Resuspend in 100 μL of water (*see* **Note 18**).
3. Add 3 μL of 50 mg/mL lysozyme and incubate samples at 37 °C for 5 min.
4. Add 25 μL of 5× SDS-PAGE sample buffer and heat at 95 °C for 5 min, then centrifuge at 13,000 × *g* for 5 min.

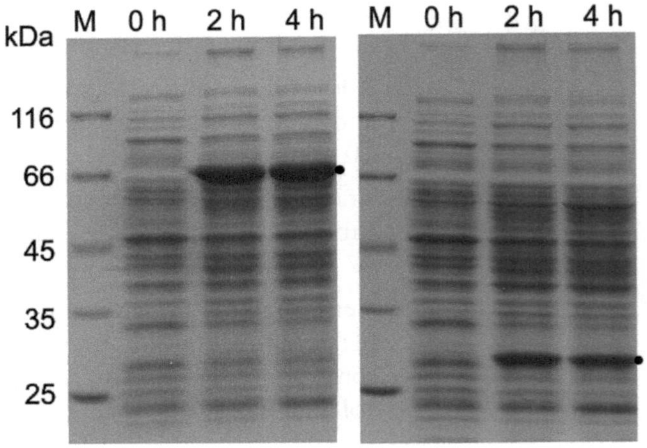

Fig. 2 SDS-PAGE analysis of the expression of the recombinant proteins. *B. subtilis* 1012/pHT1179 (pHT254-bgaB) and/pHT1170 (pHT254-gfp) were grown in LB medium to mid-log, and production of the recombinant proteins was induced by addition of 0.1 mM IPTG. Aliquots were taken before addition of IPTG and 2 and 4 h later. Lane 0 h, before induction; lanes 2 and 4 h, 2 and 4 h after induction. *Black dots* indicate the positions of BgaB (*left panel*) or GFP (*right panel*), respectively [12]

5. Load the supernatants into the wells. Load protein ladder and control samples on two separate wells.
6. Connect the electrodes to the electricity for electrophoresis running at the current of 25 mA for one mini gel, maximum voltage. The time for running depends on the size of the target protein.
7. After finishing electrophoresis, take out the gels and remove the stacking gels. Stain by soaking the gels in SDS-PAGE Stain solution for 1 h. Transfer the gel to the Destain solution and change several times.
8. Analyze the position and intensity of the right protein band on the gels (*see* Fig. 2).

4 Notes

1. In principle, any *E. coli* that harbors *lac*I^{q} can be used as a cloning strain to construct plasmids.
2. Wait for the temperature of the melted agarose come down to approximately 60 °C before adding SYBR Green.
3. TEMED is added into the gels immediately before pouring into the mold.
4. SDS-PAGE running buffer should be prepared as stock (usually 5×) and diluted immediately before use.

5. Gently heat the gels with SDS-PAGE Stain solution for more efficient staining.
6. *Pfu* or other high-fidelity DNA polymerases should be used instead of *Taq* which is notorious for its highly inaccurate DNA amplification.
7. It is typically lower than the primers' Tm by 5 °C.
8. The time of the elongation steps depend on the length of the final PCR products and the synthesis speed of the DNA polymerase. For *Pfu*, this speed is approximately 2000 bp per minute.
9. Some restriction enzymes can effectively cut the DNA within the buffer in PCR reactions. If it is the case, the PCR product purification steps could be skipped.
10. Beware of star activity of the restriction enzymes, which likely affect the outcome of the cutting reactions.
11. Be sure to insert the coding genes into the plasmids at the right open reading frames (ORFs).
12. 10× T4 DNA ligase buffer is supplied together with ATP that is degraded through constant freezing–thawing cycles. Therefore, this buffer should be stored in small aliquots at −20 °C and used completely after being thawed.
13. Other screening methods for other indicator proteins could be used appropriately.
14. Failing determination of the time, when the cell enters stationary phase to the collect sample, might influence the quality of competent cells.
15. An aliquot of 1 mL of *B. subtilis* competent cells is used for 10–20 transformation reactions. For fewer transformation reactions, a smaller aliquot could be used.
16. An identical plasmid with a simple indicator such as antibiotic resistance or fluorescent protein should be used for the testing of transformation efficiency for all competent cell fractions.
17. The concentration of an inducer, such as IPTG could be different depending on the purpose of the experiment and expression levels.
18. Lysis buffer (20 mM Tris–HCl, pH 7.2 and 15% (w/v) sucrose) can be used instead of water to obtain sharper protein bands on the gels.

Acknowledgments

This research was funded by the Vietnam National Foundation for Science and Technology Development (Nafosted) under grant number 106-NN.02-2015.24. The research was also supported by TWAS under research grants 14-201 RG/BIO/AS_G.

References

1. Adrio JL, Demain AL (2010) Recombinant organisms for production of industrial products. Bioeng Bugs 1:116–131
2. Demain AL, Vaishnav P (2009) Production of recombinant proteins by microbes and higher organisms. Biotechnol Adv 27:297–306
3. Sahdev S, Khattar SK, Saini KS (2008) Production of active eukaryotic proteins through bacterial expression systems: a review of the existing biotechnology strategies. Mol Cell Biochem 307:249–264
4. Schallmey M, Singh A, Ward OP (2004) Developments in the use of *Bacillus* species for industrial production. Can J Microbiol 50: 1–17
5. Schumann W (2007) Production of recombinant proteins in *Bacillus subtilis*. Adv Appl Microbiol 62:137–189
6. Tjalsma H, Antelmann H, Jongbloed JDH et al (2004) Proteomics of protein secretion by *Bacillus subtilis*: separating the "secrets" of the secretome. Microbiol Mol Biol Rev 68: 207–233
7. Nguyen HD, Nguyen QA, Ferreira RC et al (2005) Construction of plasmid-based expression vectors for *Bacillus subtilis* exhibiting full structural stability. Plasmid 54:241–248
8. Phan TTP, Nguyen HD, Schumann W (2006) Novel plasmid-based expression vectors for intra- and extracellular production of recombinant proteins in *Bacillus subtilis*. Protein Expr Purif 46:189–195
9. Nguyen HD, Phan TTP, Schumann W (2007) Expression vectors for the rapid purification of recombinant proteins in *Bacillus subtilis*. Curr Microbiol 55:89–93
10. Phan TTP, Nguyen HD, Schumann W (2012) Development of a strong intracellular expression system for *Bacillus subtilis* by optimizing promoter elements. J Biotechnol 157: 167–172
11. Phan TTP, Nguyen HD, Schumann W (2013) Construction of a 5′-controllable stabilizing element (CoSE) for over-production of heterologous proteins at high levels in *Bacillus subtilis*. J Biotechnol 168:32–39
12. Phan TTP, Tran LT, Schumann W et al (2015) Development of P*grac*100-based expression vectors allowing high protein production levels in *Bacillus subtilis* and relatively low basal expression in *Escherichia coli*. Microb Cell Fact 14:72
13. Hirata H, Fukazawa T, Negoro S et al (1986) Structure of a beta-galactosidase gene of *Bacillus stearothermophilus*. J Bacteriol 166: 722–727
14. Saito H, Shibata T, Ando T (1979) Mapping of genes determining nonpermissiveness and host-specific restriction to bacteriophages in *Bacillus subtilis* Marburg. Mol Gen Genet 170:117–122
15. Inoue H, Nojima H, Okayama H (1990) High efficiency transformation of *Escherichia coli* with plasmids. Gene 96:23–28

Part IV

Applications of *E. coli* Expression

Chapter 22

In Vivo Biotinylation of Antigens in *E. coli*

Susanne Gräslund, Pavel Savitsky, and Susanne Müller-Knapp

Abstract

Site-specific biotinylation of proteins is often the method of choice to enable efficient immobilization of a protein on a surface without interfering with protein folding. The tight interaction of biotin and streptavidin is frequently used to immobilize an antigen during phage display selections of binders. Here we describe a method of in vivo biotinylation of proteins during expression in *E. coli*, by tagging the protein with the short biotin acceptor peptide sequence, Avi tag, and co-expression of the *E. coli* biotin ligase (BirA) resulting in precise biotinylation of a specific lysine residue in the tag.

Key words Biotinylation, BirA, Avi-tag, Antigen capturing, Antigen immobilization, Streptavidin, IMAC, SEC

1 Introduction

Antibodies and other affinity reagents are highly valuable tools in investigating the function and localization of proteins. Ideally, high-quality renewable binders should be available for every human protein. However, the selection of recombinant antibodies requires high-quality, stable, and well-folded antigens. For in vitro selection methods like phage display or ribosome display, the antigen also needs to be appended with a tag to enable immobilization on a surface during selections. Several different tagging/immobilization systems can be used, but one of the best options for phage display has been shown to be the biotin/streptavidin system [1]. The biotinylated protein can be efficiently immobilized on streptavidin-coated surfaces or beads during selections. Biotin is a component of the vitamin B2 complex and has a very high affinity to the fungal streptavidin protein (i.e., dissociation constant of ~10^{-15} [2, 3]). Although unspecific biotinylation of a protein can be achieved in vitro, biotinylation in vivo is an attractive alternative route for achieving site-specific modification without the risk of interfering with protein folding or function of the antigen and reducing batch-to-batch differences due to the site-specificity of

Nicola A. Burgess-Brown (ed.), *Heterologous Gene Expression in E. coli: Methods and Protocols*, Methods in Molecular Biology, vol. 1586, DOI 10.1007/978-1-4939-6887-9_22, © Springer Science+Business Media LLC 2017

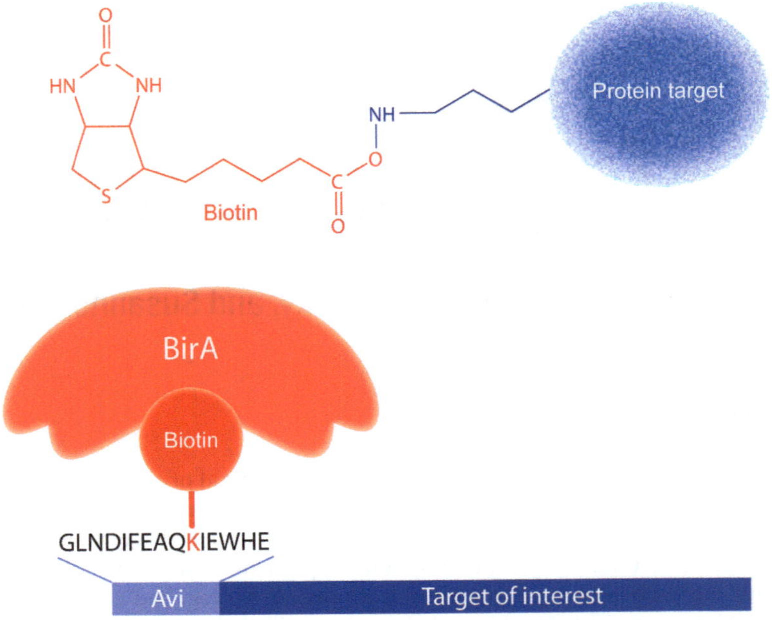

Fig. 1 Schematic representation of enzymatic biotinylation by BirA. A biotin moiety is added to the epsilon amino group of a lysine residue (*upper panel*). For site specific biotinylation the target of interest is cloned in frame with the 15 amino acid long Avi tag, which is biotinylated on a central lysine residue by the *E. coli* enzyme BirA (*lower panel*)

this reaction. The protein of interest is therefore cloned in frame with a 15 amino acid segment, known as the Avi tag, (GLNDIFEAQ(K^)IEWHE), which can be specifically biotinylated at the lysine residue by the *E. coli* biotin holoenzyme synthetase, BirA ligase (Uniprot ID P06709) (*see* Fig. 1). BirA ligase normally catalyses transfer of biotin to the epsilon amino group of a specific lysine residue of the biotin carboxyl carrier protein (BCCP) subunit of the acetyl-CoA carboxylase. The Avi tag was shown to be the minimal substrate sequence required for efficient BirA-catalyzed biotinylation [4]. The Avi tag can either be biotinylated in vivo through co-expression of the BirA ligase and addition of biotin [5], or in vitro biotinylated after purification of the Avi-tagged target protein by incubation with purified BirA ligase [6]. Proteins can be produced being 100% biotinylated and experience in our lab has shown that at least 90% of Avi-tagged, biotinylated proteins can be produced in milligram quantities. For proteins for which no prior construct knowledge is available, it is advisable to generate two protein fusion constructs with the Avi tag at either the N- or C-terminus, as sometimes one fusion is less soluble.

Since in vitro selection methods generally work better with a highly pure and mono-disperse protein, we recommend to use at least a two-step purification procedure including an affinity step, for example, immobilized metal affinity chromatography (IMAC) followed by size exclusion chromatography (SEC) which is the combination that will be described in Subheading 3. After purification, the approximate level of biotinylation can be analyzed by mass spectrometry where the biotinylated version of the protein has a molecular weight of an additional 226 Da [7].

2 Materials

2.1 Cloning and Transformation

1. Expression vector for subcloning of antigens adding an Avi tag and possibly also other tags enabling affinity purification, for example, pNIC-Bio3 (GenBank JN792439) or p28BIOH-LIC (GenBank KC164371) which both carry T7 promoters, hexahistidine tags and a kanamycin resistance marker (*see* Fig. 2 for suitable vectors) (*see* **Note 1**).
2. Expression vector for co-expression of BirA, for example pBirAcm (Avidity, Aurora, USA) or pCDF-BirA (GenBank JF914075). The enzyme could also be subcloned into another vector of choice that is compatible with the antigen expression vector (i.e., not with the same antibiotic resistance marker).
3. Competent cells of a suitable *E. coli* strain for gene expression from the expression vectors used, for example BL21(DE3).

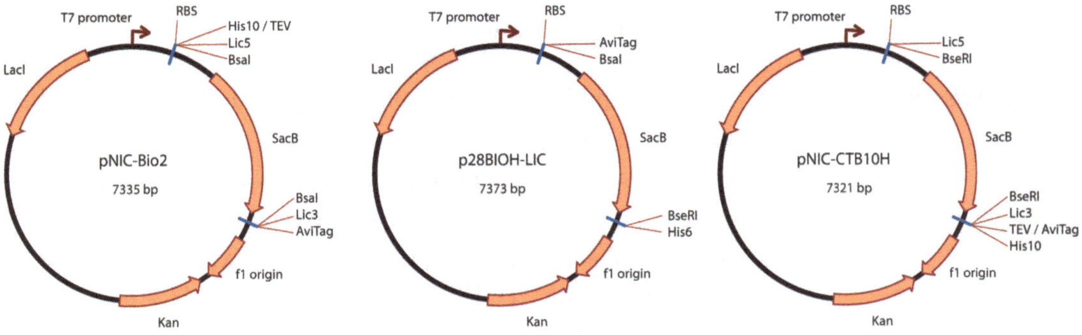

Vector	Genbank	N-tag	C-tag
pNIC-Bio1		MHHHHHHHHHHSSGVDLGTENLYFQ*SM	SKGGYGLNDIFEAQKIEWHE
pNIC-Bio2	JF912191	MHHHHHHHHHHDLGTENLYFQ*SM	SKGGYGLNDIFEAQKIEWHE
pNIC-Bio3	JN792439	MHHHHHHSSGVDLGTENLYFQ*SM	SKGGYGLNDIFEAQKIEWHE
pNIC-CTB10H	KX139199	none	ENLYFQ*SGGGLNDIFEAQKIEWHEHHHHHHHHHH
p28BIOH-LIC	KC164371	MSGLNDIFEAQKIEWHEGSAGGSG	GGSGHHHHHH

* indicates TEV cleavage site

Fig. 2 A selection of available Avi tag containing vectors. Vector maps of pNIC-Bio2, p28BIOH-LIC, and pNIC-CTB10H. Shown are GenBank accession numbers and the respective fusion tags at N- and C-termini of the target protein

4. Luria Bertani (LB) agar: Dissolve 15 g of agar, 10 g of tryptone, 5 g of yeast extract and 5 g of NaCl in 1 L of dH_2O and autoclave.
5. Sterile filtered stock solutions of relevant antibiotics, for example, 50 mg/mL kanamycin prepared in dH_2O, 34 mg/mL chloramphenicol prepared in 95% ethanol and 50 mg/mL spectinomycin prepared in dH_2O.
6. Sterile petri dishes.

2.2 Expression

1. Terrific broth (TB): Dissolve 12 g of Enzymatic Casein Digest, 24 g of yeast extract, 9.4 g of potassium phosphate (dibasic, anhydrous), 2.2 g of potassium phosphate (monobasic, anhydrous) and 8 mL of glycerol in 1 L of dH_2O and autoclave. Other growth media like Luria–Bertani broth can be used as well.
2. Sterile filtered stock solution of 1 M isopropyl-β-D-thiogalactopyranoside (IPTG) prepared in dH_2O.
3. Sterile filtered stock solution of 100 mM D-Biotin prepared in dH_2O with constant stirring and dropwise addition of 1 M NaOH until pH reaches 8.
4. 100 mL Erlenmeyer flask for starter culture.
5. Cultivation flask, for example 2.5 L TunAir bottle (Sigma).
6. Shake incubator.
7. High-speed centrifuge, rotors and centrifuge tubes for harvest and sample preparation.

2.3 Purification

1. Protease inhibitors, for example, Complete EDTA-free (Roche).
2. Benzonase (Sigma).
3. Lysis buffer: 100 mM HEPES, 500 mM NaCl, 5% (v/v) glycerol, 0.5 mM TCEP, 10 mM imidazole, pH 8.0.
4. Wash buffer 1: 50 mM HEPES, 500 mM NaCl, 5% (v/v) glycerol, 0.5 mM TCEP, 10 mM imidazole, pH 7.5.
5. Wash buffer 2: 50 mM HEPES, 500 mM NaCl, 5% (v/v) glycerol, 0.5 mM TCEP, 25 mM imidazole, 1 mM D-Biotin, pH 7.5.
6. Elution buffer: 50 mM HEPES, 500 mM NaCl, 5% (v/v) glycerol, 0.5 mM TCEP, 500 mM imidazole, pH 7.5.
7. Gel Filtration buffer: 20 mM HEPES, 300 mM NaCl, 5% (v/v) glycerol, 0.5 mM TCEP, pH 7.5.
8. Sonicator with tip suitable for 50–100 mL volume.
9. 100 mL glass beakers.
10. Filters with 0.45 μm pore size.

11. Purification system, for example, ÄKTA Xpress system (GE Healthcare).
12. Affinity chromatography column, for example HiTrap Chelating HP (GE Healthcare) charged with Ni^{2+} ions or Ni sepharose resin (GE Healthcare) and manual gravity column.
13. Size exclusion chromatography column, for example HiLoad XK16/60 Superdex 200 (GE Healthcare).
14. SDS-PAGE system including gels, loading dye, size marker, running buffer and gel staining dye.

3 Methods

The method will be described using pNIC-Bio3, pCDF-BirA, and BL21(DE3)-R3-pRARE2. This is a T1 phage resistant strain already harboring the pRARE2 plasmid, providing seven rare-codon tRNAs (produced in house at SGC).

3.1 Cloning and Transformation

1. Subclone the antigen coding DNA sequences into pNIC-Bio3 by Ligation-independent cloning [8].
2. Transform pCDF-BirA into BL21(DE3)-R3-pRARE2 and prepare competent cells of strain named BL21(DE3)-R3-pRARE2 [9].
3. Transform the antigen expression construct to the expression strain and plate on LB-Agar plates containing 50 μg/mL kanamycin, 34 μg/mL chloramphenicol, and 50 μg/mL spectinomycin.

3.2 Expression

1. Using a 100 mL Erlenmeyer flask, start an inoculation culture by picking several colonies and inoculate 10 mL of TB medium supplemented with 50 μg/mL kanamycin, 34 μg/mL chloramphenicol, and 50 μg/mL spectinomycin. Incubate overnight at 30 °C with shaking.
2. Use the inoculation culture to inoculate 750 mL of TB medium in a 2.5 L TunAir bottle. Add 50 μg/mL kanamycin, and 50 μg/mL spectinomycin and D-biotin to a final concentration of 50 μM. *Note*: there is no need to add chloramphenicol at this step. The pRARE2 plasmid will not be lost and this will ensure more predictable growth. (*see* **Note 2**).
3. Cultivate at 37 °C with shaking and monitor OD_{600} until mid-log phase is reached, approximately 1–2 in TB. Down-temper the cultivation to 18 °C for 30–60 min and add IPTG to a final concentration of 0.5 mM. Continue incubation overnight at 18 °C.
4. Harvest cells by centrifugation, 5000 × *g* for 10 min, and store cell pellets at −80 °C.

3.3 Purification

1. Resuspend cell pellets from 750 mL culture in 50 mL lysis buffer supplemented with protease inhibitors and benzonase (2000 U) and sonicate on ice (regular tip, 70% amplitude, 1 s on/1 s off, 3 min total time).
2. Clarify lysate by centrifugation at 49,000 × *g* for 30–60 min at 4 °C.
3. Decant soluble fraction into a beaker and filtrate through a 0.45 μm filter.
4. Load the lysate onto HiTrap Chelating HP 1 mL columns on an ÄKTA Xpress equilibrated with Lysis buffer. After washing with Wash buffer 1 until stable baseline has been reached and Wash buffer 2 (20 column volumes) to remove unbound fractions, elute with elution buffer (10 column volumes) and apply eluate to a HiLoad XK16/60 Superdex 200 column equilibrated with Gel Filtration buffer (*see* **Notes 3** and **4**).

3.4 Quality Control

1. Analyse relevant fractions by SDS-PAGE by loading 10 μL samples from each fraction in loading dye onto a SDS-PAGE gel.
2. Pool the fractions corresponding to a mono-disperse peak (*see* **Note 5**).
3. Take a sample to analyse protein identity and degree of biotinylation (Addition of a 226 Da biotin moiety) by mass spectrometry (LC-ESI-MS) (*see* **Note 6**).
4. Flash freeze pooled protein in liquid nitrogen and store at −80 °C.

4 Notes

1. Several different expression vectors can be used with different tag combinations and protease cleavage site. *See* Fig. 2 for different options made within the Structural Genomics Consortium.
2. Yields of Avi-tagged proteins can be reduced as compared to those obtained from corresponding constructs made with only the His_6 tag. If this is a problem it can in most cases be overcome by increasing the culture volume. Trying additional construct variants can be attempted. Another alternative might be to produce the protein in a eukaryotic expression system instead, for example *Sf9* insect cells, with the vector pFBD-BirA.
3. Biotin is a small hydrophobic molecule and biotinylation can sometimes make the protein slightly sticky, causing a small increase in peak tailing in size exclusion chromatography. Usually this is not a problem for most antigens.

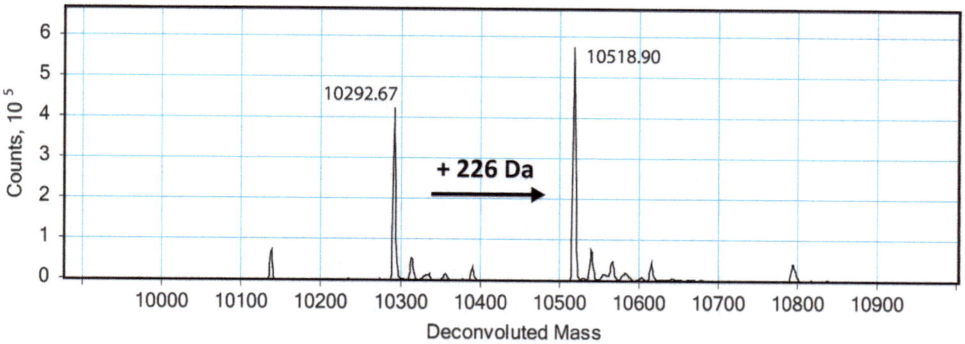

Fig. 3 Mass spectrum of biotinylated and non-biotinylated JARID1B. Mass spectrum of JARID1B showing peaks corresponding to both biotinylated and non-biotinylated protein with a mass difference of 226 Da. Mass spectrometry is not a quantitative method, but the presence of a significant peak for the non-biotinylated protein indicates non-complete biotinylation. Ideally, only a peak for the biotinylated version is detected

4. The protein purification can be carried out using a gravity flow column on the bench instead of using the automated procedure on the ÄKTA-Xpress following the same steps and volumes as described above. Many proteins will also benefit from being purified at 4 °C in a cold room or cold cabinet.
5. BirA ligase is sometimes co-purified with the biotinylated target protein. To avoid this, a small amount of biotin is added to the Wash 2 buffer, and the amount can be increased if needed.
6. Usually 50 μM D-biotin in the culture medium is sufficient to achieve complete biotinylation. However, if MS analysis shows peaks for both biotinylated and non-biotinylated target protein several alternative strategies can be employed to increase the level of biotinylation (*see* Fig. 3). First, the amount of biotin added could be increased up to 100 μm. Alternatively, Biotin can also be added at multiple occasions besides culture start, like at inoculation, at induction or at harvest after cell lysis. Using a less rich medium giving slower growth might also improve the level of biotinylation. Finally, biotinylation can also be performed in vitro after the protein has been purified through the use of purified BirA ligase, biotin and ATP. Alternatively for highly expressing proteins, expression levels can be reduced, e.g., by adapting IPTG concentration. Low level of biotinylation might also be due to the Avi tag not being displayed properly for the BirA ligase. In that case, trying to clone the tag at the other end of the protein might help.
7. Many vectors allow protease cleavage of the His tag, for example with TEV protease. If desired add His-tagged TEV protease to a 1:20 mg/mL ratio and store protein at 4 °C overnight for digestion. Apply protein on a Ni-column and collect flow-through which contains cleaved protein. Any uncleaved protein as well as the TEV protease itself will stick to the column.

Acknowledgments

The SGC is a registered charity (number 1097737) that receives funds from AbbVie, Bayer Pharma AG, Boehringer Ingelheim, Canada Foundation for Innovation, Eshelman Institute for Innovation, Genome Canada, Innovative Medicines Initiative (EU/EFPIA) [ULTRA-DD grant no. 115766], Janssen, Merck & Co., Novartis Pharma AG, Ontario Ministry of Economic Development and Innovation, Pfizer, São Paulo Research Foundation-FAPESP, Takeda, and Wellcome Trust [092809/Z/10/Z].

References

1. Zhong N, Loppnau P, Seitova A et al (2015) Optimizing production of antigens and Fabs in the context of generating recombinant antibodies to human proteins. PLoS One 10:e0139695
2. Wilchek M, Bayer EA, Livnah O (2006) Essentials of biorecognition: the (strept)avidin-biotin system as a model for protein-protein and protein-ligand interaction. Immunol Lett 103: 27–32
3. Laitinen OH, Nordlund HR, Hytonen VP et al (2007) Brave new (strept)avidins in biotechnology. Trends Biotechnol 25:269–277
4. Beckett D, Kovaleva E, Schatz PJ (1999) A minimal peptide substrate in biotin holoenzyme synthetase-catalyzed biotinylation. Protein Sci 8:921–929
5. Keates T, Cooper CD, Savitsky P et al (2012) Expressing the human proteome for affinity proteomics: optimising expression of soluble protein domains and in vivo biotinylation. N Biotechnol 29:515–525
6. Cull MG, Schatz PJ (2000) Biotinylation of proteins in vivo and in vitro using small peptide tags. Methods Enzymol 326:430–440
7. Sundqvist G, Stenvall M, Berglund H et al (2007) A general, robust method for the quality control of intact proteins using LC-ESI-MS. J Chromatogr B Analyt Technol Biomed Life Sci 852:188–194
8. Gileadi O, Burgess-Brown NA, Colebrook SM et al (2008) High throughput production of recombinant human proteins for crystallography. Methods Mol Biol 426:221–246
9. Hanahan D, Jessee J, Bloom FR (1991) Plasmid transformation of *Escherichia coli* and other bacteria. Methods Enzymol 204:63–113

Chapter 23

Cold-Shock Expression System in *E. coli* for Protein NMR Studies

Toshihiko Sugiki, Toshimichi Fujiwara, and Chojiro Kojima

Abstract

The cold-shock system using the pCold vector is one of the most effective *Escherichia coli* heterologous protein expression systems. It allows the improvement of the expression level of the protein of interest in a soluble fraction. In this chapter, we describe practical procedures for the overexpression of heterologous protein of interest by using the pCold vector or the single-protein production system. The latter is one of the most advanced pCold technologies for isotope labeling of the target protein and its NMR studies.

Key words Protein expression system, *Escherichia coli* (*E. coli*), pCold-GST vector, Isotope labeling, Single protein production (SPP) system, Nuclear magnetic resonance (NMR)

1 Introduction

Low-temperature cell cultivation is one of the most effective ways to improve the expression level of heterologous proteins in a soluble form in *Escherichia coli* (*E. coli*) host cells [1]. Protein overexpression at a low temperature makes it possible to suppress the production of undesirable background proteins in the host cells. In many cases, however, it also results in insufficient expression of the heterologous target protein.

The cold-shock expression system is an advanced method with strong potential to overexpress heterologous proteins in a soluble fraction, with minimal background protein production [2]. A rapid drop in temperature to below 15 °C causes a cold-shock response in the *E. coli* cells. The *cspA* promoter is strongly activated, leading to a specific and marked expression of a series of cold-shock proteins, such as CspA [2]. At the same time, *E. coli* cell growth is temporarily arrested, and thereby, de novo protein synthesis is significantly suppressed (except for CspA protein) via the low-temperature antibiotic effect of truncated CspA expression [3]. A soluble heterologous protein can be specifically overexpressed in the cold-shock expression system by placing its gene under the

Nicola A. Burgess-Brown (ed.), *Heterologous Gene Expression in E. coli: Methods and Protocols*, Methods in Molecular Biology, vol. 1586, DOI 10.1007/978-1-4939-6887-9_23, © Springer Science+Business Media LLC 2017

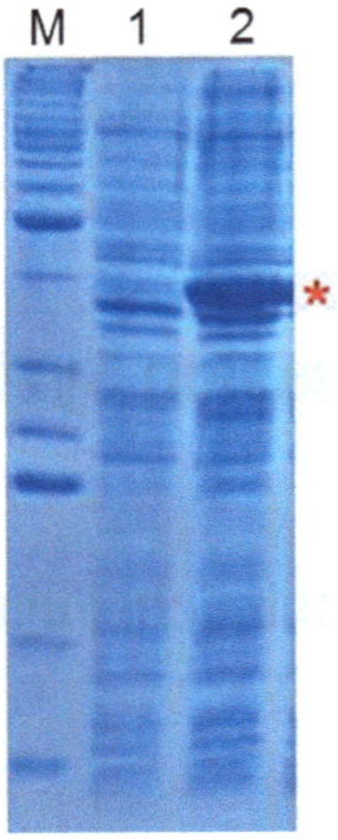

Fig. 1 Overexpression of heterologous proteins using the pCold-GST system, confirmed by SDS-PAGE. The gel was stained with Coomassie Brilliant Blue. The ***red asterisk*** indicates the position of the GST-fused target protein. M, molecular weight markers. ***Lanes 1*** and ***2***, before and after the induction of target protein expression, respectively. The amino acid sequence of the target protein was designed by the authors and their coworkers using computational methods to generate a novel functional protein (unpublished data)

control of the *cspA* promoter on the pCold vector and cultivating the *E. coli* transformants under low-temperature (<15 °C) conditions.

Various recently developed applications based on the pCold technology, offer successful overexpression of difficult-to-produce proteins and fulfill a wide variety of research purposes. For example, pCold-glutathione-S-transferase (GST) system, in which a heterologous protein is expressed with fused GST as a solubility enhancement tag (SET), promotes the overexpression of soluble target proteins more strongly than the pCold system (Fig. 1) [4, 5]. In the single-protein production (SPP) system, the undesired mRNAs coding for the background proteins are specifically digested without damaging the mRNA coding for the desired heterologous protein. In this system, the recombinant nuclease, MazF, is expressed in the host cells, ensuring the translation of just the target protein (Fig. 2) [6].

1.1 pCold-GST System

If the expression level of target proteins in the soluble fraction is insufficient in the pCold vector system, the combined use of the pCold vector and SETs can increase the yield of these proteins. Using the pCold-GST vector developed by Hayashi and Kojima, many heterologous proteins that were previously difficult to express or prone to aggregation, were successfully overexpressed in *E. coli* [4]. Hayashi and Kojima have tested a wide variety of SETs,

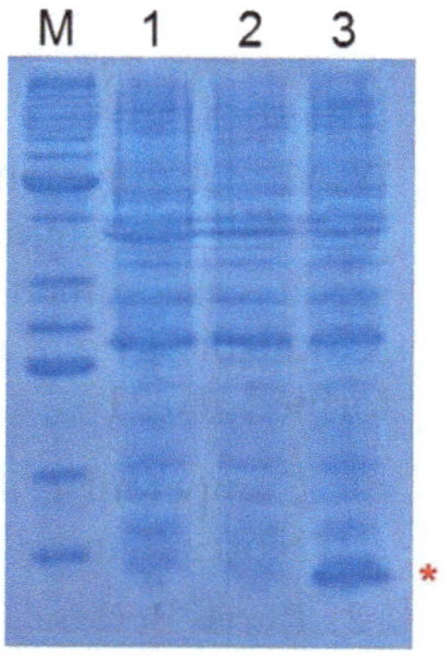

Fig. 2 Overexpression of a heterologous protein (ubiquitin) using the SPP system, confirmed by SDS-PAGE. The gel was stained with Coomassie Brilliant Blue. The ***red asterisk*** indicates the position of ubiquitin band. M, molecular weight markers. ***Lane 1***, before the induction of MazF and ubiquitin expressions; ***Lane 2***, after the induction of MazF expression without ubiquitin; ***Lane 3***, after the induction of ubiquitin expression

including MBP, GB1, and Trx to identify an optimal SET, which maximizes the power of the pCold system. They have found that the GST tag led to the most successful results [5] (*see* **Note 1**).

1.2 SPP System

E. coli possesses an intrinsic toxin–antitoxin system, which participates in the regulation of *E. coli* cell growth and programmed cell death in response to environmental changes. The MazF toxin has an mRNA interferase activity and selectively degrades the 5′ moiety of the ACA sequence in mRNA. Since most native mRNA sequences in *E. coli* include an ACA sequence, they can be digested by MazF, resulting in almost complete suppression of de novo protein synthesis. Even under these conditions, the basic and minimal vital activities of the *E. coli* cells, such as ATP synthesis, can be maintained since the molecules required for these processes already exist before the MazF-mediated mRNA clearance. However, the cell growth is arrested in a "quasi-dormant state" [7].

Using this approach, the heterologous protein will be almost exclusively expressed in the *E. coli* if any ACA sequences in its gene are substituted for other bases (without altering the amino acid sequence of the protein). This system has been developed by Inouye and his coworkers as the single protein production (SPP) method [6]. The SPP system has several advantages, especially in the area of $^{2}H/^{13}C/^{15}N$ isotope labeling. Isotope labels can be selectively incorporated into the protein of interest, and therefore, the efficiency of labeling is improved in comparison with the conventional labeling procedures [1]. This method minimizes the NMR signals from non-target protein and/or chemical groups, even if the sample contains some impurities [8–10].

2 Materials

Water mentioned in this chapter is deionized water prepared using the Milli-Q system (Merck Millipore, MA). Analytical or higher grade is recommended for all chemicals. The pCold plasmid DNA is available from TaKaRa, Japan. All autoclave sterilization is performed at 121 °C for 15–20 min.

2.1 *E. coli* Culture for pCold-GST

1. Competent cells of an appropriate strain of *E. coli* BL21 (*see* **Notes 2** and **3**).
2. Luria–Bertani (LB) broth medium containing ampicillin: LB broth solution (Dissolve 10 g of tryptone, 5 g of yeast extract and 5 g of NaCl in 1 L of water in a glass bottle) and sterilize by autoclaving. It can be stored at room temperature. For cell cultivation, the required amount of LB broth medium is transferred into a sterilized tube or a flask aseptically and a stock solution of ampicillin (50–100 mg/mL; sterilized by filtration and stored at −20 °C) is added to a final concentration of 50–100 μg/mL (*see* **Note 4**).
3. Agar plate of LB broth containing ampicillin: LB broth medium containing 18–20 g/L agar powder. Sterilize by autoclaving. When the temperature of the LB medium falls below 50 °C, add 50–100 μg/mL ampicillin as described above, and mix well. Pour the mixture into disposable dishes aseptically on a clean bench and allow to solidify. Further dry the agar by turning the dish upside down on a clean bench. The agar plates can be stored at 4 °C.
4. M9 minimal medium: Dissolve 7.0 g of Na_2HPO_4, 3.0 g of KH_2PO_4, 0.5 g of NaCl, and 0.5 g of NH_4Cl (*see* **Note 5**) in 1 L of water in a 2-L baffled Erlenmeyer glass flask (*see* **Note 6**).
5. 40% (w/v) D-glucose stock solution (400 mL): Dissolve 160 g of D-glucose (dextrose) in water and mix with a magnetic stirrer with heating (40–50 °C) as needed. Sterilize by aseptic filtration and aseptically transfer into a pre-autoclaved glass medium bottle (500-mL volume). Store the solution at room temperature (*see* **Note 7**).
6. 0.5 M isopropyl-1-thio-β-D-galactopyranoside (IPTG) stock solution: Dissolve IPTG powder in water at 0.5 M final concentration. Store at −20 °C.
7. 0.85% (w/v) NaCl solution: Dissolve 8.5 g of NaCl in 1 L of water in a glass medium bottle and sterilize by autoclaving. Store at 4 °C.
8. Centrifuge, rotor, and tubes: Centrifuge, Avanti HP-26XP (Beckman Coulter, CA); centrifugation rotor, JLA-8.1000; and 1-L centrifugation tubes, or their functional equivalents.

2.2 *E. coli* Culture for SPP

1. pCold vector containing cDNA coding heterologous protein of interest (*see* **Note 8**).
2. Plasmid vector encoding MazF (with G27K/H28R amino acid substitutions) (*see* **Note 9**).
3. Competent cells of *E. coli* BL21(DE3) (*his*) (*see* **Note 10**).
4. M9 (10×) minimal medium stock solution: Dissolve 132 g of $Na_2HPO_4{\cdot}7H_2O$, 30 g of KH_2PO_4, and 5 g of NaCl in 1 L of water in a glass medium bottle and sterilize by autoclaving (*see* **Note 11**). Store at room temperature.
5. Sterilized water (SW): Transfer 1 L of water into an appropriate glass medium bottle and sterilize by autoclaving. Store at room temperature.
6. $MgSO_4$ stock solution (0.81 M): Dissolve 100 g of $MgSO_4{\cdot}7H_2O$ in 500 mL of water in an appropriate glass bottle and sterilize by autoclaving. Stored at room temperature.
7. 40% (w/v) D-glucose stock solution: Prepare and store as described above.
8. 0.5 mg/mL vitamin B1 stock solution (VB1): Dissolve 25 mg of vitamin B1 powder in 50 mL of water and sterilize by aseptic filtration. Store at 4 °C.
9. 5 mg/mL L-tryptophan (Trp) stock solution: Dissolve 50 mg of Trp powder in 10 mL of water and sterilize by aseptic filtration. Store at 4 °C.
10. 10 mg/mL L-histidine (His) stock solution: Dissolve 100 mg of His powder in 10 mL of water and sterilize by aseptic filtration. Store at 4 °C.
11. 20% (w/v) casamino acids (CA) stock solution: Dissolve 20 g of CA powder in 100 mL of water in an appropriate glass medium bottle and sterilize by autoclaving (*see* **Note 12**). Store at room temperature.
12. 0.5 M IPTG stock solution: Prepare and store as described above.
13. Agar plate in M9 minimal medium: Mix 50 mL of 10× M9 minimal medium, 440 mL of SW, 0.5 g of NH_4Cl, and 7.5 g of agar powder in a 1-L Erlenmeyer flask and sterilize by autoclaving. When the temperature of the liquid drops to ca. <50 °C, aseptically add 630 μL of 0.81 M $MgSO_4$, 5 mL of 40% (w/v) D-glucose, 2 mL of VB1, 5 mL of CA, 50–100 μg/mL ampicillin, and 25 μg/mL chloramphenicol and mix well. Pour the mixture into a disposable dish, aseptically, on a clean bench. Further dry the agar by turning the dish upside down on a clean bench.
14. M9 minimal medium (for culture-1): Mix 10 mL of 10× M9 minimal medium, 87 mL of SW and 0.1 g of NH_4Cl in a 500-mL volume Erlenmeyer flask (no baffles) and sterilize by

autoclaving. When the medium is at room temperature, aseptically add 126 μL of 0.81 M $MgSO_4$, 1 mL of 40% (w/v) D-glucose, 400 μL of VB1, 1 mL of CA, 400 μL of Trp, 200 μL of His, ampicillin, and chloramphenicol as described above and mix well. The medium should be prepared on Day 1 and stored at room temperature.

15. M9 minimal medium (for culture-2): Mix 100 mL of 10× M9 minimal medium, 900 mL of SW, and 1.0 g of NH_4Cl in a 5-L volume Erlenmeyer flask (no baffles) and sterilize by autoclaving. When the temperature of the liquid decreases to room temperature, aseptically add 1.26 mL of 0.81 M $MgSO_4$, 10 mL of 40% (w/v) D-glucose, 4 mL of VB1, 10 mL of Trp, 5 mL of His, ampicillin, and chloramphenicol as described above and mix well. Prepare on Day 3 and store at room temperature.
16. M9 minimal medium (for cell washing): Aseptically mix 10 mL of 10× M9 minimal medium and 90 mL of SW in two sterile 50-mL Corning tubes. Prepare immediately before use.
17. M9 minimal medium (for culture-3): Mix 10 mL of 10× M9 minimal medium, 87 mL of SW, and 0.1 g of NH_4Cl in a 300-mL Erlenmeyer flask and sterilize by autoclaving. When the temperature of the liquid decreases to room temperature, aseptically add 126 μL of 0.81 M $MgSO_4$, 1 mL of 40% (w/v) D-glucose, 400 μL of VB1, 400 μL of Trp, 200 μL of 0.5 M IPTG, ampicillin, and chloramphenicol as described above and mixed well. Prepare on Day 3 and store at room temperature.
18. Sterilized empty 2-L Erlenmeyer flask: Sterilize an empty 2-L Erlenmeyer flask (no baffles) by autoclaving.
19. M9 minimal medium (for culture-4): Sterilize an empty 1-L Erlenmeyer flask (no baffles) by autoclaving. Aseptically mix 5 mL of 10× M9 minimal medium, 44 mL of SW, and 0.05 g of $^{15}NH_4Cl$ (*see* **Note 5**) in the sterilized flask. Then, aseptically add 63 μL of 0.81 M $MgSO_4$, 500 μL of 40% (w/v) D-glucose (*see* **Note 7**), 200 μL of VB1, 500 μL of Trp, 250 μL of His, 100 μL of 0.5 M IPTG, ampicillin, and chloramphenicol as described above and mix well. Prepare on Day 4 and store at room temperature.

3 Methods

3.1 E. coli Culture for Recombinant Protein Overexpression Using pCold-GST Vector

Experimental procedure described in this section is applicable, not only for pCold-GST vector but also for other pCold vectors, such as pCold-TF or pCold-ProS. The experimental procedure takes 4 days.

In addition to the materials and solutions described in Subheading 2, prepare 150 mL of LB broth in a 300-mL baffled Erlenmeyer flask.

(**Day 1**)

1. Transform *E. coli* competent cells with a pCold-GST plasmid encoding the protein of interest and select the transformants by growing the cells on an LB agar plate containing ampicillin. Incubate the LB agar plate at 37 °C overnight (12–20 h).

(**Day 2**)

2. Pick a single colony of the transformants and inoculate into 10 mL of fresh LB broth medium in a 50-mL Corning tube. Cultivate at 37 °C with reciprocal shaking at 180–200 rpm overnight (12–20 h).

(**Day 3**)

3. Aseptically add the cultured whole cell suspension to 150 mL of fresh LB broth medium, prepared in a 300-mL baffled Erlenmeyer flask. Cultivate at 37 °C with horizontal rotational shaking at 160–200 rpm for 4–5 h.
4. Harvest the *E. coli* cells by centrifugation (2000 × *g*, for 10 min at room temperature).
5. Resuspend the cell pellets in 1 L of fresh M9 minimal medium and cultivate at 37 °C with horizontal rotational shaking at 120–160 rpm until the turbidity (optical density at 600 nm wavelength, OD_{600} value) of the medium reaches 0.6–1.0 (*see* **Note 13**).
6. Decrease the temperature of the culture medium to less than 15 °C as promptly as possible by dipping the flask in ice water (*see* **Notes 14** and **15**). Incubate the medium for a further 30–60 min to chill it sufficiently (*see* **Note 16**).
7. Add 0.5–1.0 mM IPTG stock solution to the cell cultivation medium and continue cultivation at 15 °C with horizontal rotational shaking at 120–160 rpm overnight (20–24 h).

(**Day 4**)

8. Harvest whole *E. coli* cells by centrifugation (3000 × *g*, for 15 min at 4 °C).
9. After the removal of the supernatant, resuspend the precipitated cell pellets in 100 mL of 0.85% (w/v) NaCl solution and transfer into disposable plastic tubes, such as 50-mL Corning tubes.
10. Harvest whole *E. coli* cells by centrifugation (3000 × *g*, for 15 min at 4 °C).
11. Discard the supernatant and store the cell pellets at −80 °C until purification of the overexpressed heterologous protein of interest.

3.2 E. coli *Culture for Recombinant Protein Overexpression in SPP System*

In this system, the co-expression of MazF and target protein in host *E. coli* cells is important. First, only MazF should be expressed to eliminate the background mRNAs, except for the target protein mRNA, before expressing the protein of interest. Then, overexpression of the target protein is carried out by replacing the culture medium with fresh M9 medium containing the desired stable isotopes. To accomplish the stepwise overexpression of two different proteins, we suppress the expression of the target protein during the MazF expression step. We construct plasmids containing cDNA encoding the target protein without any ACA sequences in the reading frame. To achieve this, we fuse hexahistidine to the N-terminus of the target protein; the plasmid encoding MazF does not contain any histidine codons in its cDNA sequence.

The *E. coli* host cells, transformed by those two plasmids, are cultivated in medium without histidine. If protein overexpression is initiated by adding IPTG, the polypeptide synthesis and elongation of MazF will progress because the MazF does not have any histidine residues. However, synthesis of the N-terminal hexahistidine-fused target protein will be arrested in the absence of histidine. The strict regulation of protein synthesis can be further ensured by using a histidine-auxotrophic *E. coli* strain.

Cell death of the *E. coli* host cells can be suppressed by keeping the temperature below 15 °C. This will arrest the vital activity of the cells; the lack of histidine in the culture medium severely damages the growth and maintenance of the *E. coli* host cells in a general culture environment. After the first step of MazF expression, overexpression of isotopically labeled target proteins is carried out by changing the culture medium to fresh medium containing histidine and isotopes such as ^{13}C and/or ^{15}N.

This experimental procedure takes 6 days.

(**Day 1**)

1. Transform competent cells of *E. coli* BL21(DE3) (*his*) with plasmids encoding MazF (*see* **Note 17**), and select the transformants by growing *E. coli* on an M9 agar plate containing appropriate antibiotics (*see* **Note 18**). Incubate the M9 agar plate at 37 °C overnight (12–20 h). Pick the generated single colonies and store their glycerol stocks at −80 °C.

(**Day 2**)

2. Prepare competent cells of the MazF-transformants of *E. coli* BL21(DE3) (*his*) and store their glycerol stocks at −80 °C (*see* **Note 19**).
3. Transform competent cells of MazF-transformed *E. coli* BL21(DE3) (*his*) with pCold plasmids encoding the heterologous protein of interest (*see* **Note 17**). Select transformant cells by growing them on an M9 agar plate containing ampicillin and chloramphenicol. Incubate the M9 agar plate at 37 °C overnight (12–20 h).

(**Day 3**)

4. **Culture-1:** Pick a single colony of the MazF/pCold dual-transformants and inoculate into 100 mL of fresh M9 minimal medium in a 500-mL glass Erlenmeyer flask (no baffles). Cultivate at 37 °C with horizontal rotational shaking at 180 rpm overnight (12–20 h) (*see* **Note 20**).

(**Day 4**)

5. Harvest the whole *E. coli* cells by gentle centrifugation (2000 × *g*, for 12 min at 4 °C).
6. **Culture-2:** Gently resuspend the cell pellets in 1 L of fresh M9 minimal medium (*see* **Note 21**) in a 5-L glass Erlenmeyer flask (no baffles). Cultivate the cells at 37 °C with horizontal rotational shaking at 150 rpm until the OD_{600} value reaches 0.5 (*see* **Note 22**).
7. Decrease the temperature of the culture medium to below 15 °C by incubation in ice water box for 5 min, and continue incubation at 15 °C for 1 h. During this interval, the buffers and empty centrifugation tubes for the next step should be chilled to 4 °C (*see* **Note 23**).
8. Harvest the whole *E. coli* cells by gentle centrifugation (2000 × *g*, for 12 min at 4 °C).
9. Discard the supernatants. and wash the cells by adding 25 mL of chilled M9 medium and gently resuspending by vortex mixing. Transfer the cell suspension to chilled 50-mL Corning tubes. After rinsing the inner wall of the centrifugation tubes with 10 mL of fresh M9 medium, transfer the rinse liquid to the same 50-mL Corning tubes.
10. Harvest the whole *E. coli* cells by gentle centrifugation (2000 × *g*, for 12 min at 4 °C).
11. **Culture-3:** Discard the supernatants and resuspend the cell pellets in the chilled M9 medium. Add an aliquot of M9 medium (25 mL) to each Corning tube, and gently resuspend the cell pellet completely by vortex mixing, then transfer into a chilled empty 2-L Erlenmeyer flask (no baffles). Use the residual fresh M9 medium for washing the inner wall of the Corning tubes and transfer into the 2-L Erlenmeyer flask (total volume of the cell suspension becomes 100 mL).
12. Cultivate the cell suspension at 15 °C with horizontal rotational shaking at 120–160 rpm overnight (*see* **Note 24**).

(**Day 5**)

13. Harvest the whole *E. coli* cells by gentle centrifugation (2000 × *g*, for 12 min at 4 °C) and discard the supernatant.
14. **Culture-4:** Resuspend the cell pellets in chilled M9 medium prepared for the main culture-4, and transfer into a sterilized

and chilled empty 1-L Erlenmeyer flask (no baffles) as follows: Add the M9 medium, 15 mL per Corning tube, and completely resuspend the pellet by gentle vortexing. Transfer the suspension into the chilled empty 1-L Erlenmeyer flask. Use the residual fresh M9 medium for washing the inner wall of the Corning tubes and transfer. into the 1-L Erlenmeyer flask (final volume of the cell suspension is 50 mL).

15. Cultivate the cell suspension at 15 °C with horizontal rotational shaking at 120–160 rpm overnight (24 h).

(**Day 6**)

16. Harvest the whole *E. coli* cells by centrifugation (3000 × *g*, for 15 min at 4 °C).
17. Discard the supernatant and store the cell pellets at −80 °C until purification of the overexpressed heterologous protein of interest.

4 Notes

1. It might be necessary to try other SETs, depending on the characteristics of the target proteins. The pCold ProS2 vector, encoding protein S derived from *Myxococcus xanthus* (ProS2) as a SET on the pCold vector has been developed and is commercially available from TaKaRa [10].
2. Almost all the commercially available *E. coli* strains can be utilized for the pCold expression system because the induction of heterologous protein expression in the pCold system is mediated by the CspA promoter derived from *E. coli* host cells. Furthermore, the presence or absence of λDE3 gene within *E. coli* host cells does not affect heterologous protein expression in the pCold system because T7 RNA polymerase is not needed in this system. BL21 *E. coli* strain is widely used in the pCold expression system because it lacks several intracellular proteases.
3. The expression level of the target proteins and/or its yield in the soluble fraction could be improved by co-expressing a molecular chaperone. A set of plasmid DNA encoding several kinds of molecular chaperones (DnaK-DnaJ-GrpE, GroES-GroEL, tig) and *E. coli* BL21 competent cells that are already transformed by those chaperone-plasmids are available from TaKaRa. The competent cells of *E. coli* BL21 strain, transformed with plasmids encoding cold-chaperonins Cpn60 and Cpn10 from psychrophilic bacteria *Oleispira antarctica*, are commercially available from Stratagene. If the expression level of the target protein is insufficient in the pCold system, we can decrease the temperature of cultivation during protein expression to 10–12 °C.

4. Carbenicillin is a chemically stable alternative to ampicillin. It is recommended instead of ampicillin for stricter selection of *E. coli* transformants or for prevention of the loss of heterologous plasmids during cell cultivation. However, carbenicillin is more expensive than ampicillin.
5. $^{15}NH_4Cl$ (>98 atom % of ^{15}N) is used as a sole nitrogen source in uniform ^{15}N-labeling of the target protein.
6. Other appropriate shaking flasks are also available. Optimal aeration is important during cell cultivation to maximize the expression level of the target protein. This depends on the characteristics of the target protein; the most appropriate shaking flasks should be used to achieve the desired aeration efficiency.
7. D-[$^{13}C_6$]-glucose (>99 atom % of ^{13}C) is used as a sole carbon source for uniform ^{13}C-labeling of target proteins. It is not used as a stock solution; a required amount (2 g/L) is dissolved in M9 medium immediately before starting the cell culture.
8. All ACA sequences must be eliminated from the cDNA encoding the target protein by mutagenesis, without altering the amino acid sequence of the protein.
9. The plasmid vector contains cDNA-encoding MazF, without histidine residues. We use a pACYC vector with chloramphenicol resistance, co-expressing it with a pCold vector coding for the target protein.
10. It is a histidine-auxotrophic *E. coli* strain.
11. It does not contain any nitrogen sources such as NH_4Cl.
12. Casamino acids mixture, an acid hydrolysate of casein, contains few cysteine or tryptophan. Casein itself does not have cysteine residues and tryptophan can be lost during the acid hydrolysis reaction. Therefore, we add tryptophan to the casamino acids stock solution.
13. Optimization of the cell cultivation parameters is important; the induction of protein expression at the optimal OD_{600} can maximize the target protein yield. This OD_{600} value varies depending on the type and characteristics of the target protein.
14. In the pCold system, triggering the cold-shock response in the host cells by a rapid chilling of the medium (to below 15 °C) is critical for an effective increase in the expression level of the target proteins.
15. Several kinds of antibiotics, such as chloramphenicol or tetracycline, can induce a cold-shock response in the host cells. We can take advantage of this phenomenon to boost the efficiency of cold-shock response by adding such antibiotics (to

a low final concentration of ~1 ng/mL) to the cell culture media [1, 11].

16. During this interval, the temperature setting of the shaker should be changed from 37 to 15 °C to ensure that the cell cultivation starts at 15 °C in the next step.
17. Transformation of the auxotroph *E. coli* BL21 (DE3) is performed as follows: After heat shock at 42 °C and prompt chilling on ice for 90 s, the culture is incubated at 38 °C for 20 min without shaking.
18. The type of antibiotic depends on the properties of the plasmid encoding MazF. We are using a pACYC vector to express MazF, so the appropriate antibiotic is chloramphenicol (25 μg/mL).
19. Please refer to other general protocols for the preparation of *E. coli* competent cells. During the entire procedure, you should use buffers containing 25 μg/mL chloramphenicol to prevent the loss of MazF encoding plasmids.
20. If OD_{600} value of the medium does not reach 1.5–1.8 on the following day, further cultivation will probably fail; the experiments should be restarted, beginning with the transformation of *E. coli*.
21. At this stage, the OD_{600} of the cell suspension should be 0.15–0.18.
22. It would take 4–6 h.
23. In the procedures in **step** 7, ensure that the *E. coli* cells are maintained below 15 °C by using sufficiently prechilled flasks, tubes, and buffers. At temperatures above 15 °C, the cells are more vulnerable to MazF toxicity, which severely affects the final expression level of the target proteins.
24. The incubation time should be only as long as necessary; a period of 3 h–7 days is acceptable.

Acknowledgments

We are grateful to M. Yoneyama, K. Hayashi, S. Kataoka, I. Kawahara, K. Furuita, Y. Hattori, K. Harada, S. Shinya (Osaka University, Japan), L. Mao, Y. Ishida, M. Inouye (Rutgers University, NJ), and Y. Yamaguchi (Osaka City University, Japan) for many useful suggestions and comments. This work was supported in part by KAKENHI (26840050, 26291029, 26291013, 26650018, 24000017, 24657074, 23370053) for T.S., T.F., and C.K., the Platform Project for Supporting Drug Discovery and Life Science Research, and AMED-CREST from MEXT and AMED.

References

1. Sugiki T, Fujiwara T, Kojima C (2012) Latest approaches for efficient protein production in drug discovery. Expert Opin Drug Discovery 9:1189–1204
2. Qing G, Ma LC, Khorchid A et al (2004) Cold-shock induced high-yield protein production in *Escherichia coli*. Nat Biotechnol 22:877–882
3. Jiang W, Fang L, Inouye M (1996) Complete growth inhibition of *Escherichia coli* by ribosome trapping with truncated cspA mRNA at low temperature. Genes Cells 1:965–976
4. Hayashi K, Kojima C (2008) pCold-GST vector: a novel cold-shock vector containing GST tag for soluble protein production. Protein Expr Purif 62:120–127
5. Hayashi K, Kojima C (2010) Efficient protein production method for NMR using soluble protein tags with cold shock expression vector. J Biomol NMR 48:147–155
6. Suzuki M, Mao L, Inouye M (2007) Single protein production (SPP) system in *Escherichia coli*. Nat Protoc 2:1802–1810
7. Suzuki M, Roy R, Zheng H et al (2006) Bacterial bioreactors for high yield production of recombinant protein. J Biol Chem 281:37559–37565
8. Schneider WM, Inouye M, Montelione GT et al (2009) Independently inducible system of gene expression for condensed single protein production (cSPP) suitable for high efficiency isotope enrichment. J Struct Funct Genomics 10:219–225
9. Mao L, Vaiphei ST, Shimazu T et al (2010) The *E. coli* single protein production system for production and structural analysis of membrane proteins. J Struct Funct Genomics 11:81–84
10. Vaiphei ST, Tang Y, Montelione GT et al (2011) The use of the condensed single protein production system for isotope-labeled outer membrane proteins, OmpA and OmpX in *E. coli*. Mol Biotechnol 47:205–210
11. VanBogelen RA, Neidhardt FC (1990) Ribosomes as sensors of heat and cold shock in *Escherichia coli*. Proc Natl Acad Sci USA 87:5589–5593

Chapter 24

High-Throughput Production of Proteins in *E. coli* for Structural Studies

Charikleia Black, John J. Barker, Richard B. Hitchman, Hok Sau Kwong, Sam Festenstein, and Thomas B. Acton

Abstract

We have developed a standardized and efficient workflow for high-throughput (HT) protein expression in *E. coli* and parallel purification which can be tailored to the downstream application of the target proteins. It includes a one-step purification for the purposes of functional assays and a two-step protocol for crystallographic studies, with the option of on-column tag removal.

Key words Protein purification, Parallel purification, High-throughput, ÄKTA-Xpress, *E. coli* expression, Expression triage

1 Introduction

Proteins are involved in almost all aspects of cellular function and as such have an immeasurable contribution to disease biology. Nearly all pharmaceuticals act on proteins and increasingly proteins themselves are being used therapeutically [1]. Understanding the relationship between biological function and disease involves investigating all traits of proteins from enzymatic activity and interactions to their tertiary structure. Such studies require significant amounts of high quality and functional recombinant protein. The explosion of genetic information made available by sequencing the human genome, has resulted in an unprecedented demand for purified protein. This post-genomic era demand has propelled technological advances in protein purification methods and has led to the development of cost-effective HT approaches. Methods for screening large number of proteins have accelerated disease study and facilitated the identification of new disease targets.

A number of laboratories around the world have developed methodologies and instrumentation to fit HT purposes [2–4] most notably the Structural Genomics Consortium (SGC) [5, 6] with

Nicola A. Burgess-Brown (ed.), *Heterologous Gene Expression in E. coli: Methods and Protocols*, Methods in Molecular Biology, vol. 1586, DOI 10.1007/978-1-4939-6887-9_24,

innovative cloning solutions [7], the use of magnetic beads [8], and custom-designed equipment [9]. With automation and miniaturization [10] having been the catalyst to this plethora of HT research projects, the quantity of protein delivered from such efforts is typically not sufficient for extensive structural studies. We set out to develop a flexible workflow that can respond to the varying needs of different projects, can handle modest to large numbers of constructs in parallel and deliver protein in the quantity and purity levels required. As such, we have developed a fast and efficient workflow for medium to high-throughput protein purification which allows educated predictions on the success of each construct to be made early on in the process; repeated construct triage eliminates constructs unlikely to reach the amount and purity requirements. The workflow can be tailored to the downstream use of the target proteins and comprises a one-step purification for the purposes of functional assays and a two-step protocol for crystallographic studies, with the option of tag cleavage. The schematic in Fig. 1 describes the various scenarios and outcomes of such projects. This process can be run using standard protein purification laboratory equipment with no necessity for expensive or custom-made robotics. While the throughput of this experimental set up is not as high as with fully automated and miniaturized protocols, it still delivers a much higher number of pure proteins than the traditional "one protein/one person/one week" capacity [11]. Its modular approach also allows greater flexibility in its application to different systems.

Requirements for assay-ready protein tend to include tens of proteins at a time, whereas for crystallography material, the number of constructs is much lower. In many cases, an initial set of 6–10 constructs are initially investigated, with additional constructs designed on the basis of the experimental information derived from the first set. Crystallographic projects generally require lower number of constructs, allowing higher numbers of expression conditions to be tested, typically in the range of 6–8 per construct. Construct design and cloning will not be discussed in detail in this chapter except to note that an affinity tag is required for this workflow. Routinely, this is an N-terminal 6× His-tag but longer tags (e.g., 8× or 10× His) and larger solubility tags (e.g., MBP or GST) can be used with equal chance of success (*see* **Note 1**).

The protocol described below focuses on expression in *E. coli*; however, it has been used successfully with other expression hosts, including baculovirus infected insect cells. Expression levels are assessed using densitometry and comparison to a labeled protein of known amount (e.g., 500 ng His-tagged GFP). This method does not provide absolute quantitation of the expressed protein. It does provide a useful comparative tool for an initial construct triage. This has been found to be essential for eliminating poorly

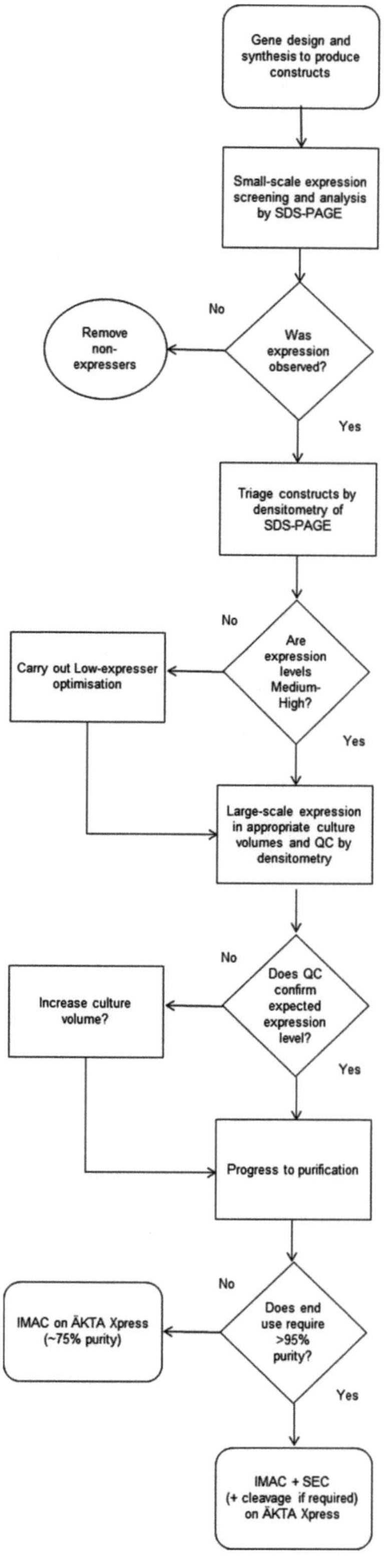

Fig. 1 A flexible workflow designed to identify "best expressers" at an early stage

expressing constructs at the earliest stage possible allowing considerable savings in downstream time, material, and overall cost (*see* **Note 2**). Using information gathered from a large number of constructs processed in the way described above, we have developed a model which can, at the scouting stage, predict the likelihood of success at purification. For this workflow, success is defined by final protein purity of ≥75% for assay-ready protein and ≥95% for crystallographic material with a final quantity of 5 mg.

Briefly, the workflow shown in Fig. 1 is typically initiated with 96 constructs entering expression scouting. Smaller sets of constructs can be used at a cost to the overall efficiency of the workflow. Whilst multiple conditions can be tested at this stage, we have standardized it to 2 that we have found to give the best results for the large numbers of constructs we have tested to date. Having used densitometry to categorize constructs into High, Medium, Low, and Non-expressers, then Non-expressers are rejected. Low Expressers may go through a second round of more extensive expression testing and optimization as part of a rescue strategy (*see* **Note 3**). Medium and High expressers are progressed directly to large-scale expression and volumes are adjusted to 2 and 1 L, respectively (*see* **Note 4**). Following analysis of the large-scale expression samples, the same densitometric decision gate is applied to account for potential scale-up variability. Constructs that pass this decision gate progress to purification in units of four constructs per ÄKTA-Xpress per day. In a HT process, it is absolutely essential that purification does not depend on time consuming optimization of conditions that exploit subtle differences in protein size, charge, or hydrophobicity [7]; we have standardized the buffers and purification conditions to a generic formulation that appears to be suitable for the majority of constructs tested to date (*see* Table 1 and **Note 5**).

Table 1
Standardized storage parameters

Aliquot size:	50 μL and 0.5 mL; all tubes within one box
Concentration:	1–5 mg/mL for assays; 10 mg/mL for crystallography
Tube/container:	0.5 mL Matrix vials
Tube labeling:	Protein ID, Date, Concentration
Matrix box:	1 protein per box
Freezing method:	Flash frozen
Storage:	−80 °C

The protocol below provides the experimental details for a workflow for medium to high-throughput protein purification that relies on parallelization and process standardization. This platform produces reliable and reproducible material and has been validated successfully for assay-ready through to crystal grade material. Due to the predictive element of the method, high success rates can be achieved at purification (>80%) with minimum cost.

2 Materials

2.1 Transformation and Expression Scouting

1. Expression plasmids (*see* **Note 6**).
2. 1.5 mL Eppendorf tubes.
3. Sterile petri dishes.
4. Chemically competent BL21(DE3) *E. coli* cells and derivatives, e.g., BL21(DE3) pLysS or Star, Rosetta(DE3), Rosetta 2(DE3), BL21-CodonPlus(DE3)-RIPL (Millipore, Stratagene).
5. Water bath set at 42 °C.
6. Super Optimal broth with Catabolite repression (SOC) medium.
7. Agar plates: Prepare LB agar by dissolving 30 g of LB agar powder in 1 L of ddH$_2$O. Autoclave in volumes of 500 mL or less and store at room temperature. To make agar plates, heat LB agar in the microwave until liquid and let to cool slowly to approximately 40 °C. Under sterile conditions, add appropriate antibiotic and pour ~4 mL into each petri dish. Leave to set and store at 4 °C in an inverted position.
8. LB medium: Dissolve 22 g of LB powder in ~950 mL ddH$_2$O. Autoclave and store at room temperature.
9. Antibiotics: 50 mg/mL ampicillin prepared in water, 50 mg/mL kanamycin prepared in water, and 34 mg/mL chloramphenicol prepared in 100% ethanol. Store stocks at 4 °C and use at a 1 in 1000 dilution.
10. 0.5 M stock solution isopropyl β-D-1-thiogalactopyranoside (IPTG). Store stocks at −20 °C and use at a 1 in 1000 dilution.
11. Spectrophotometer.
12. 24 deep-well blocks with capacity for 10 mL.
13. Shaking incubator set at 37 °C.
14. Bio-Rad 4–20% gradient gels.
15. Bio-Rad 2× SDS dye.
16. Precision Plus All Blue Bio-Rad Markers.

17. His-tagged GFP protein (GFP-His) mixed with 2× SDS loading dye at a concentration of 5 ng/μL.
18. Benchtop centrifuge with 25Ti rotor.
19. Sonicator Sonics Vibra-cell VCX 500 with eight element 3 mm and single element 13 mm probes.
20. PhyNexus MEA2 robot.
21. Ni^{+2}-NTA PhyTips (10 μL resin).
22. Bio-Rad Imager with ImageLab software.
23. NanoDrop with 16 position LVis-plate.
24. Eppendorf Varispan multipipette.

2.2 Large-Scale Expression

1. Materials from Subheading 2.1.
2. 4 L baffled Erlenmeyer flasks.

2.3 Purification

1. IKA homogenizer.
2. Beckman ultracentrifuge with 45Ti rotor.
3. GE Healthcare ÅKTA Xpress Twin.
4. Lysis buffer: 25 mM Tris-HCl pH 8.0, 300 mM NaCl, 1 mM TCEP, 5% (v/v) glycerol, benzonase 10 U/mL, EDTA-free protease inhibitor (PI) tablets 1/50 mL (Roche), 1 mM PMSF, 0.25% (w/v) CHAPS (Affymetrix).
5. Wash buffer: 25 mM Tris-HCl pH 8.0, 300 mM NaCl, 1 mM TCEP, 5% (v/v) glycerol, 20 mM imidazole.
6. Elution buffer: 25 mM Tris-HCl pH 8.0, 300 mM NaCl, 1 mM TCEP, 5% (v/v) glycerol, 500 mM imidazole.
7. Size exclusion chromatography (SEC) buffer: 25 mM Tris-HCl pH 8.0, 300 mM NaCl, 1 mM TCEP, 5% (v/v) glycerol.

3 Methods

3.1 Transformation

1. Transform BL21 cells by adding 50 ng of supercoiled plasmid DNA to 20 μL of cells in 1.5 mL Eppendorf tubes on ice.
2. After incubating for 30 min on ice, heat-shock the cells for 45 s by incubating them in a water bath at 42 °C.
3. Add 100 μL of SOC medium and incubate the cells at 37 °C with shaking for 1 h to allow the cells to recover.
4. Spread 50 μL of cells onto an LB agar plate containing an appropriate antibiotic and incubate at 37 °C for ~18 h.
5. Once colonies are visible on the plate, pick one colony and use to inoculate a 5 mL starter culture. Grow culture at 37 °C for ~18 h with rigorous shaking (~ 220 rpm).

3.2 Expression Scouting

1. Use 0.1 mL of the overnight starter culture to inoculate 5 mL of LB in 24-deep well blocks. Grow the cultures at 37 °C until the OD reaches 0.8, induce with 0.5 mM IPTG and incubate for 4 h at 37 °C or overnight at 18 °C.
2. At the end of expression, take an OD measurement and then harvest cells by centrifugation at 3500 × *g* for 15 min; lyse by sonication for 45 s, 5 s on 5 s off at 40% power in lysis buffer on ice.
3. Move lysates from the 24-deep well block to Eppendorf tubes using a multi-pipette and clarify by centrifugation at 14,000 × *g* for 30 min. Retain a 20 μL sample for SDS-PAGE analysis.
4. Apply the lysate onto a PhyTip and run a pre-set method where the tips are washed two times and the 50 μL elution is collected into a new plate.
5. Normalize samples based on the OD at harvest by adding more lysis buffer as appropriate.
6. Prepare samples for Coomassie-stained SDS-PAGE gel and anti-His western blot by mixing 15 μL of the eluted protein with 15 μL of 2× SDS dye. Load 10 μL each on two separate gels. Use one lane on each gel to load 500 ng of GFP-His which will serve as a quantitative control. Semi-quantify the amount of expressed protein recovered post-IMAC from a de-stained gel by densitometric comparison of each sample to the GFP-HisExpression:scouting.
7. Plot the densitometry value of each sample against their respective purity using TIBCO Spotfire or similar software and choose the constructs with estimated expression level above 2 mg/L and purity above 50% to progress to large-scale expression (Fig. 2).
8. Constructs with expression levels above 3.75 mg/L will be expressed at the 1 L scale and those with expression levels between 2 and 3.5 mg/L will be produced at the 2 L scale.
9. Low expresser constructs will be optimized by expression testing in different strains and conditions. Those constructs whose expression levels reach 2 mg/L will then progress to large-scale expression.

3.3 Large-Scale Expression

1. For the constructs selected to progress into large-scale expression, use a fresh plate to pick a colony and amplify cells by inoculating a 50 mL LB starter culture and growing overnight at 37 °C with shaking at 170 rpm.
2. Use 10 mL of starter culture to inoculate 1 L of medium (containing appropriate antibiotics) until A_{600} reaches 0.8.
3. Induce the culture with 0.5 mM IPTG and leave to express for the required length of time and temperature as established by the scouting experiments.

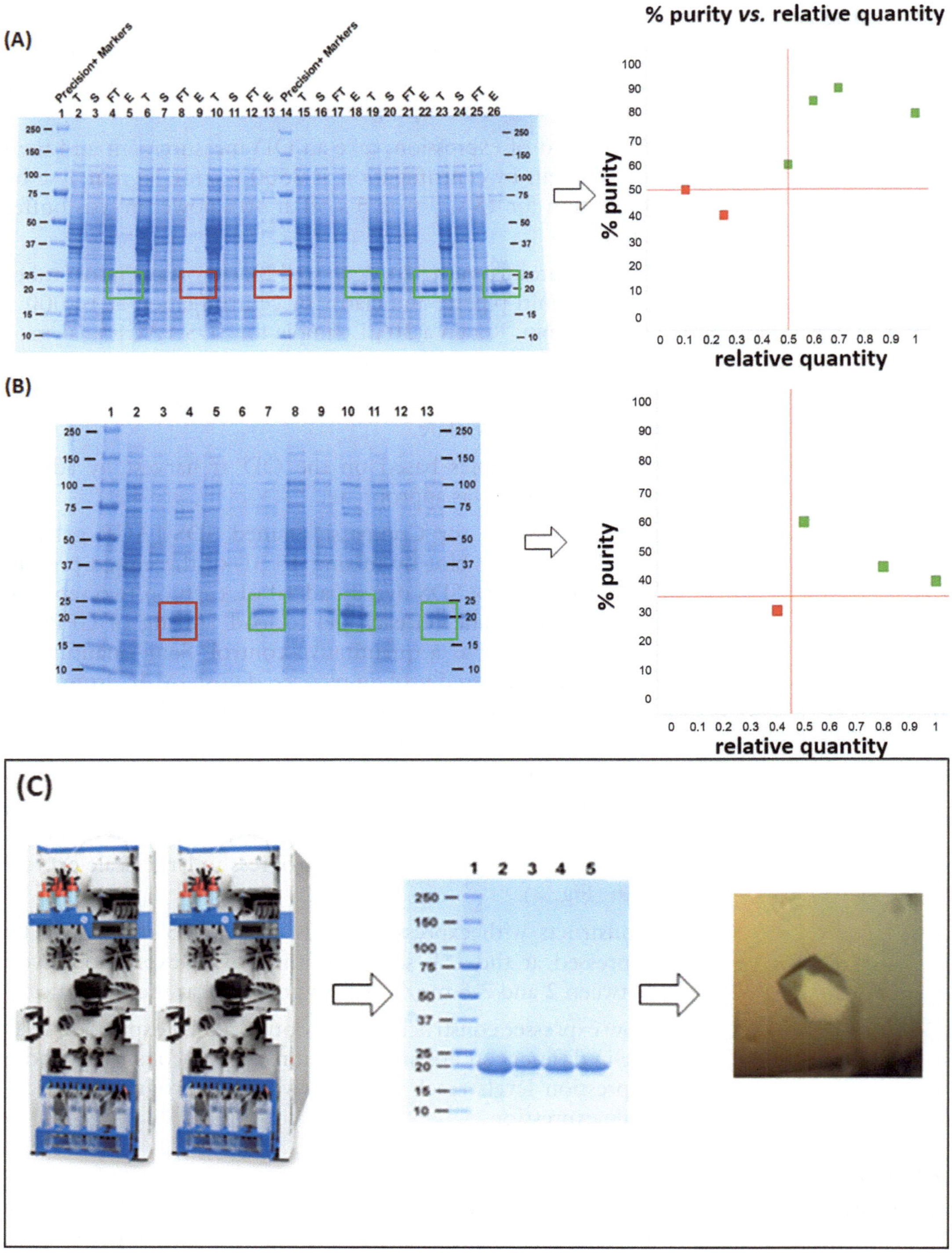

Fig. 2 Generally, for HT applications we process constructs in batches of 96. However, for illustration purposes, an example is shown of a smaller crystallography project. (**a**) Coomassie-stained SDS-PAGE showing expresser scouting results of six constructs (mutants of a wild-type protein). *Lanes 1* and *14* show molecular weight ladders. *Lanes 2*, *6*, *10*, *15*, *19*, and *23* show Total (T) fractions; *lanes 3*, *7*, *11*, *16*, *20* and *24* show soluble (S) fractions; *lanes 4*, *8*, *12*, *17*, *21*, and *25* show flow-through (FT) fractions; *lanes 5*, *9*, *13*, *18*, *22*, and *26* show

4. Take an OD measurement, keep a 5 mL sample and harvest cells by centrifugation at 3500 × *g* for 15 min at 4 °C. Freeze the pellets at −80 °C until ready to purify.
5. Normalize collected samples based on their OD at harvest by adding more lysis buffer as appropriate and prepare samples for Phynexus MEA2 Ni^{+2}-NTA PhyTip pulldown, Coomassie-stained SDS-PAGE gel, and anti-His western blot as described above. Analyze gels and plot densitometric values as described above for expression scouting. Using the same thresholds as previously, select the constructs to proceed to purification.

3.4 Purification

Perform all purifications in a cold room and handle all samples on ice. Prepare and pH buffers using cold ddH_2O and store at 4 °C for no more than 2 days. Add PI tablets and reducing agents just prior to use (*see* **Note 7**).

1. Thaw pellets for four constructs on ice and resuspend with lysis buffer using a ratio of 6 mL of buffer/g of cells (*see* **Note 8**). Homogenize cells using an IKA homogenizer or equivalent and lyse on ice by sonication for 5 min (30 s on/30 s off at 5% power).
2. Clarify cells by centrifugation at 150,000 × *g* for 45 min and collect supernatant.
3. Set up the ÄKTA-Xpress by placing 4× 1 mL HisTrap FF crude columns, and placing the inlets S1–S4 in lysis buffer, the outlet A1 in Wash Buffer and the outlet B1 in Elution Buffer. For crystal grade material, attach an S200 16/60 SEC column on port 5 and place outlet A4 in SEC buffer (*see* **Notes 9** and **10**).
4. Run a pre-set method corresponding to the intended workflow:
   (a) For a one-step purification, equilibrate column, load lysate, and wash the column with Wash buffer for ten column volumes (CV). Elute with a step elution 0–100% over ten CV while collecting 2 mL fractions.

Fig. 2 (continued) eluted (E) fractions. Densitometry was carried out using the protein loaded in the Eluted fraction lanes and plotted using Spotfire (% purity vs. relative density). Constructs showing % purity greater than 50% and relative density equal to or above 2 mg/L are highlighted in *green* (four constructs). Those failing purity and/or relative density thresholds are indicated in *red* (two constructs). Successful constructs were progressed to large-scale expression and their expression QC is shown in (**b**). Following densitometry three constructs exceeded the required thresholds for purification (indicated in *green*) and one construct failed (highlighted in *red*). (**c**) The three successful mutant constructs were purified in parallel on an ÄKTA Xpress alongside their corresponding wild-type construct (*lane 1* shows molecular weight ladder; *lane 2* shows wild-type construct; *lanes 3–5* show mutant constructs). Protein showed high purity and yield for all constructs and crystals were obtained for mutants

(b) For a two-step purification with no cleavage, attach and equilibrate a Superdex 16/60 S200 on port 5 and run a method for IMAC in tandem with SEC. If running four samples simultaneously, peak fractionation is necessary during SEC to ensure all eluted proteins are collected in a single 2 mL 96-well plate.

(c) For a two-step purification with cleavage, incorporate an 8 h on-column cleavage step in the above tandem method and inject 0.8 mL of ~2.5 mg/mL TEV protease when setting up the Xpress.

5. Pool fractions corresponding to the single peak. If multiple peaks are present in the chromatogram, analyze fractions by SDS-PAGE prior to pooling the fractions together (*see* **Note 11**).

3.5 Final Concentration and Purity Assessment

1. Measure OD_{280} and calculate protein concentration by the value of absorbance at 280 nm and the molecular weight and the extinction coefficient for each protein. Adjust concentration to required final concentration. For parallel measurements using NanoDrop, a 16 position LVis-plate can be used.
2. Prepare samples for analysis as above and load 5 μg of each sample on SDS-PAGE gel and anti-His western blot.
3. Determine purity by densitometry of reducing SDS-PAGE with 5 μg of total protein applied per lane (Coomassie-stained SDS-PAGE).

3.6 Quality Control

While high purity of the target protein is essential, chemical and biophysical homogeneity is also paramount for crystallization and further quality assessment is required in addition to concentration and purity.

1. Confirm the identity and quality of every protein batch by mass spectrometry with single Dalton accuracy.
2. Evaluate the sample homogeneity and suitability for crystallography by DLS using the calculated % of mass intensity and mass distribution for a protein model.
3. Analytical SEC can provide invaluable information on the oligomeric or aggregation state of a protein whilst consuming only trivial quantity of material. A high-throughput setup using an Agilent HPLC instrument and a Superdex 200 5/150 column offers a rapid and cost-effective way of gathering such information.
4. N-terminal protein sequencing can be used for confirmation of protein identity but can have limited expediency in projects where all proteins have been cloned with an N-terminal His-tag.
5. On a more restricted basis, Tandem MS of trypsin-digested proteins can been used for complete sequence identification of proteins from low level expression.

4 Notes

1. While usually a 6× His tag is used in the construct design, it is possible to use an 8× or 10× His tag for increased binding to the HisTrap column. Stronger binding can offer the advantage of higher tolerance during the imidazole wash with increased purity as a consequence but at the expense of a longer, less native and more elastic protein.
2. Sufficient expression levels are paramount for the success of this workflow and cannot be compensated for by improvements in purification. Generally, the higher the expression level, the higher the final purity.
3. A number of constructs are likely to show poor expression under the conditions tested. These constructs can be expressed in a small set of different conditions varying usually the cell strain and reducing the expression temperature. Following a large body of such expression optimization rounds, we have found that while some constructs improve, the improvement is usually modest and therefore expression optimization is performed only for constructs with expression levels ≥1 mg/L. Constructs with lower expression levels are unlikely to improve enough to reach the threshold of 2 mg/L required to progress to large-scale expression and purification.
4. A small number of 3 L expressions for expression levels below 2 mg/L can be processed with this workflow but the chance of success has been found to be substantially lower. Constructs expressing less than 1 mg/L are unlikely to produce protein of suitable quantity and quality and are not purified with this workflow.
5. Following expression scouting experiments, it is possible that a number of proteins show acceptable expression levels but only in the total cell fraction indicating that these proteins are not soluble in the buffer conditions used. In such cases, screening a panel of different buffer conditions can prove useful. The buffers tested depend on compatibility with the downstream applications and can include detergents of varying properties. A typical panel of buffer scouting conditions, is shown in Table 2.
6. For increased efficiency and impartiality, constructs are identified by a short prefix denoting the project name and a four-digit number.
7. Expressed proteins may be degraded and protease inhibitors need be added when necessary. Generally, the higher the MW of the protein, the higher the chance of degradation.
8. No more than 3 L of culture can be used for purification using this workflow as loading of the supernatant becomes a limiting

Table 2
Typical buffer scouting conditions

Buffer Matrix						
Buffer name	**Standard buffer**	**Buffer set 1**	**Buffer set 2**	**Buffer set 3**	**Buffer set 4**
Lysis buffer	25 mM Tris-HCl pH 8.0, 300 mM NaCl, 1 mM TCEP, 5% glycerol, benzonase 10 U/mL, PI tablets 1/50 mL, 1 mM PMSF, 0.25% CHAPS	SB + 0.2% Triton X-100	SD + 0.1% n-dodecylmaltoside (DDM)	SD + 0.1% NP-40	SB + 1% n-octylglucoside (OG)
IMAC buffer	25 mM Tris-HCl pH 8.0, 300 mM NaCl, 1 mM TCEP, 5% glycerol, 20 mM imidazole				
Elution buffer	25 mM Tris-HCl pH 8.0, 300 mM NaCl, 1 mM TCEP, 5% glycerol, 500 mM imidazole				

factor and the chances of clogging the column during overnight loading become high.

9. Extreme caution should be exercised when labeling tubes for this kind of higher throughput experiments. We have found that setting up samples always in the same order (e.g., in ascending order) can reduce the risk of errors.

10. Metalloproteins have been known to lose their metal ion when purified by IMAC. In such cases, purification can be performed with a column loaded with the same metal ion as the protein rather than Ni^{+2}. While the binding capacity of alternative metals may not be as strong as that of Ni^{+2}, it provides a satisfactory solution to the loss of metal from the protein.

11. Data management is an important part of HT projects and a standardized platform can increase the efficiency of the project. We have found that recording all information for each construct, from expression testing to purification, in one datasheet is a simple but effective method of cross-checking results and keeping track of variability of scale-up expressions. To make recording fast and less error-prone, auto-calculations are used where possible. Report preparation and result presentation have also been standardized and automated to a large extent with the use of a custom-designed scripts.

References

1. Braun P, LaBaer J (2003) High throughput protein production for functional proteomics. Trends Biotechnol 21:383–388
2. Quevillon-Cheruel S, Collinet B, Trésaugues L et al (2006) Cloning, production, and purification of proteins for a medium-scale structural genomics project. Methods Mol Biol 363: 21–38
3. Saez NJ, Vincentelli R (2014) High-throughput expression screening and purification of recombinant proteins in *Escherichia coli*. Methods Mol Biol 1091:33–53
4. Stevens RC (2000) Design of high-throughput of protein production for structural biology. Structure 8:R177–R185
5. Gräslund S, Nordlund P, Weigelt J et al (2008) Protein production and purification. Nat Methods 5:135–146
6. Savitsky P, Bray J, Cooper CD et al (2010) High-throughput production of human proteins for crystallization: the SGC experience. J Struct Biol 172:3–13
7. Shih YP, Kung WM, Chen JC et al (2002) High-throughput screening of soluble recombinant proteins. Protein Sci 11:1714–1719
8. Murthy TVS, Shi Z, Fernandez M et al (2005) Automated high-throughput protein purification with MagneHis™ particles. Promega Notes 91:2–5
9. Yoo D, Provchy J, Park C et al (2014) Automated high-throughput protein purification using an ÄKTApurifier and a CETAC autosampler. J Chromatogr A 1344:23–30
10. Tworak A, Podkowinski JMF (2011) High throughput protein production. J Biotechnol Comput Biol Bionanotechnol Perspect 92:119–131
11. Edwards A (2005) Purifying proteins for structural biology. In: Sundström M (ed) Structural genomics and high-throughput structural biology. Taylor and Francis, New York

Chapter 25

Mass Spectrometric Analysis of Proteins

Rod Chalk

Abstract

Mass spectrometry is a generic technique for the structural and functional analysis of purified proteins. Instrument capabilities and the possibilities of intact protein, peptide fragmentation and native analyses are discussed. Detailed experimental protocols are described for the most commonly applied techniques of protein identification, posttranslational modification (PTM) characterization, PTM mapping, native mass spectrometry, and analysis of membrane proteins using electrospray mass spectrometry.

Key words Mass spectrometry, Electrospray, Liquid chromatography–mass spectrometry, LC-MS, Tandem mass spectrometry, LC-MSMS, Protein, Posttranslational modification, PTM, Native mass spectrometry, Integral membrane proteins, IMPs, SDS-PAGE, Size exclusion chromatography, SEC, High pressure liquid chromatography, HPLC, Phosphorylation, Glycan

1 Introduction

Mass spectrometry (MS) is unmatched in its capability to generate information quickly on protein purity, identity, and structure. Depending upon the technique, femptomoles to picomoles of protein are required without the need for customized reagents. MS should not be viewed as a single technique, but rather a range of different methods in which a mass spectrometer generates the analytical output. For the non-specialist, it is important to understand which of these techniques are available, and which is appropriate for their particular experimental context. Protein mass spectrometry can yield both structural and functional data which can be both qualitative and quantitative. While mass spectrometers can analyze moderately complex protein mixtures, it should not be seen as a separation technique. Best results are always obtained from pure samples, and maximum sample information is generally gained when MS data is used alongside orthogonal and complimentary techniques such as SDS-PAGE or size exclusion chromatography (SEC).

Nicola A. Burgess-Brown (ed.), *Heterologous Gene Expression in E. coli: Methods and Protocols*, Methods in Molecular Biology, vol. 1586, DOI 10.1007/978-1-4939-6887-9_25, © Springer Science+Business Media LLC 2017

If the researcher is fortunate enough to have easy access to a mass spectrometer, the instrument type will to some extent dictate which techniques are possible. If they are able to choose between instruments, some knowledge of instrument capabilities is desirable. Two instrument types, matrix assisted laser desorption ionization MALDI [1] and electrospray [2], are used in protein analysis. MALDI is fast, robust, simple, and performs equally with electrospray for the analysis of peptides. These advantages are offset somewhat by the requirement for sample preparation prior to analysis [3]. Electrospray instruments are coupled directly to a high pressure liquid chromatography HPLC system which means that sample preparation is minimal, and critically, samples of greater complexity may be analyzed. In all respects, electrospray outperforms MALDI at high mass. While all instruments have the ability to derive mass, some have the capacity to generate fragments from a parent compound and in turn derive the mass of these fragments, a process known as tandem mass spectrometry or MSMS. Such instruments have greatly enhanced analytical capability.

Although similar analytical methods may be used, MS analysis of proteins is not synonymous with proteomics [4]. For the purposes of this chapter, mass spectrometric analysis of proteins is the detailed structural and functional characterization of individually purified proteins, where the sequence and mass are presumed to be known in advance, and the goal is their validation.

1.1 What Can Be Measured?

A wealth of structural and functional information can be obtained using mass spectrometry [5]. Which type of information is dependent upon the chosen method of sample preparation and the method of mass spectrometry employed. Sample preparation may involve chemical derivatization [6], heavy isotope labeling [7], or enzymatic pretreatment, purification, or enrichment [8]. Three distinct categories of protein mass spectrometry exist:

1. The protein being analyzed is subjected to endoproteolytic cleavage, generating peptides readily amenable to fragmentation within the mass spectrometer, the so-called bottom-up analysis.
2. The protein remains intact under denaturing conditions, making it amenable to accurate mass analysis, and in some cases also fragmentation.
3. The protein remains intact and under native conditions, allowing measurement of native conformation making it amenable to functional analysis.

Within these categories there are numerous sample preparation, MS acquisition, and data analysis options allowing in-depth structural and functional characterization, summarized in Fig. 1. It should not be forgotten that indirect analysis of enzyme function is also possible via quantitative MS analysis of the conversion of

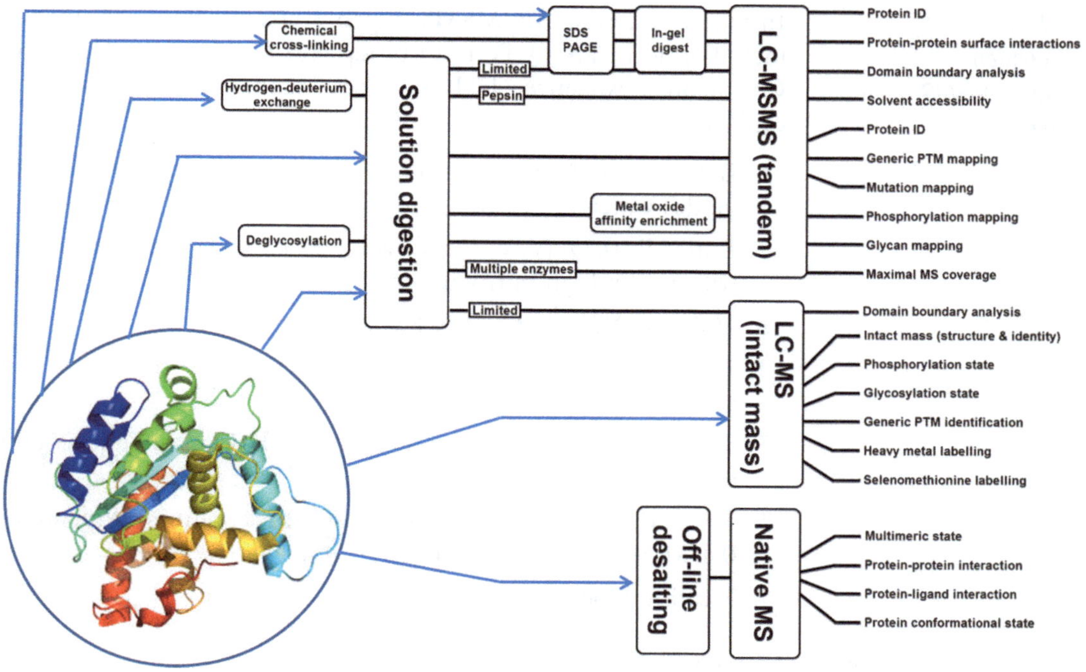

Fig. 1 Mass spectrometry-based strategies for structural and functional analysis of purified proteins

substrate to product [9], though this falls outside the scope of this chapter. Mass spectrometry is the most powerful analytical tool available to the protein scientist at the bench. Intact mass measurement and tryptic digest MSMS analyses greatly facilitate protein expression, scale-up, and purification when they are made fast and freely available. Posttranslational modification analysis is critical in understanding of protein function and requires both top-down and bottom-up analytical approaches. MS of membrane proteins and native MS can be routine even in a high-throughput environment. These two techniques in particular will become increasingly important in the post-proteomics future.

2 Materials

2.1 Intact Mass Analysis

Instrument types: All accurate mass capable electrospray instruments including TOF, Q-TOF, Orbitrap, and FTICR. This method uses a standard ESI source and an analytical HPLC.

1. Buffer A: 0.1% (v/v) formic acid in ultra-high purity (LC-MS grade) water.
2. Buffer B: 0.1% (v/v) formic acid in LC-MS grade methanol.
3. Buffer C: 0.1% (v/v) formic acid in LC-MS grade acetonitrile.
4. 2.1 mm × 12.5 mm Zorbax 5 μm 300SB-C3 guard column and column holder (Agilent).

2.2 Solution Digestion and LC-MSMS Analysis

Instrument types: All MSMS capable electrospray instruments including ion trap, Q-TOF, Orbitrap and Fourier Transform Ion Cyclotron Resonance, FTICR. Uses standard ESI source and capillary/nano HPLC.

1. Buffer A: 0.1% (v/v) formic acid in ultrahigh purity (LC-MS grade) water.
2. Buffer B: 0.1% (v/v) formic acid, 80% LC-MS grade acetonitrile in ultrahigh purity water.
3. 100 mM NH_4HCO_3 (pH 8.0).
4. 1 M dithiothreitol (DDT): Prepare in H_2O, pre-aliquot and store at −20 °C.
5. Saturated iodoacetamide: Prepare in H_2O, pre-aliquot and store at −20 °C.
6. 1 mg/mL sequencing grade trypsin stock solution: Prepare in 0.1 M HCl and store at 4 °C.
7. Heating block.

2.3 In-Gel Digestion and LC-MSMS Analysis

Materials as for solution digestion (Subheading 2.2) plus:

1. Gel cutting tips, 4 mm (GeneCatcher, Web Scientific).
2. 96-well PCR plate.
3. Microtiter plate seals with adhesive-free zone.
4. 10% (v/v) methanol.
5. 12-channel, 200 μL multichannel pipettor.
6. Plastic trough.
7. Thermal cycler.
8. Round-bottomed polypropylene microtiter plate (Agilent).

2.4 Phosphorylation Mapping

Materials as for solution digestion (Subheading 2.2) plus:

1. Titansphere TiO_2 10 μm medium (GL-Sciences, supplied via Hichrom).
2. 3 M Empore C18 High performance extraction (filter) disc 2215.
3. Blunt cannulation needles.
4. ZrO_2 powder, 6 μm sieve or greater.
5. Short length of 375 μm O.D. fused silica tubing.
6. Wash 1: Aqueous ammonia solution (25% (v/v) ammonium hydroxide stock, 75% (v/v) water).
7. Wash 2: Low DHB solution (80% (v/v) ACN, 2% (v/v) TFA, 3 mg/mL DHB).
8. Elution 1: High DHB solution (80% (v/v) ACN, 300 mg/mL DHB, 0.1% (v/v) TFA).

=(Neutral_mass*3+D2)/F2

	C	D	E	F	G	H	I	J	K	L
	Neutral mass	proton mass	total mass	z	Charge state	m/z	Charge state	m/z	Charge state	m/z
1	20000.000	1.007276	20001.007276	1	M1 z1	20001.007	M2 z1	40001.007	M3 z1	60001.007
2	40000.000	2.014552	40002.014552	2	M1 z2	10001.007	M2 z2	20001.007	M3 z2	30001.007
3	60000.000	3.021828	60003.021828	3	M1 z3	6667.674	M3 z3	13334.341	M3 z3	20001.007
4	80000.000	4.029104	80004.029104	4	M1 z4	5001.007	M2 z4	10001.007	M3 z4	15001.007
5	100000.000	5.036380	100005.036380	5	M1 z5	4001.007	M2 z5	8001.007	M3 z5	12001.007
6	120000.000	6.043656	120006.043656	6	M1 z6	3334.341	M2 z6	6667.674	M3 z6	10001.007
7	140000.000	7.050932	140007.050932	7	M1 z7	2858.150	M2 z7	5715.293	M3 z7	8572.436
8	160000.000	8.058208	160008.058208	8	M1 z8	2501.007	M2 z8	5001.007	M3 z8	7501.007
9	180000.000	9.065484	180009.065484	9	M1 z9	2223.229	M2 z9	4445.452	M3 z9	6667.674
10	200000.000	10.072760	200010.072760	10	M1 z10	2001.007	M2 z10	4001.007	M3 z10	6001.007
11	220000.000	11.080036	220011.080036	11	M1 z11	1819.189	M2 z11	3637.371	M3 z11	5455.553

Fig. 2 Excel ion table to calculate m/z values for all charge states and all multimeric states for a given neutral mass

9. Wash 3: 80% (v/v) ACN, 0.1% (v/v) TFA.
10. Wash 4: 80% (v/v) ACN, 0.1% (v/v) FA.
11. Elution 2: Phosphopeptide elution buffer (NH_3 water, 40% (v/v) ACN).
12. pH indicator paper strips.

2.5 Glycan Mapping

Materials as for solution digestion (Sect. 2.2) plus:

1. Peptide N-Glycosidase F (PNGaseF).

2.6 Native Mass Spectrometry

Protein or complexes less than 50 kDa can be analyzed using any ESI instrument. Those greater than 50 kDa require an ion trap or orbitrap while those greater than 100 kDa require a high m/z TOF or QTOF.

1. Syringe pump: Gastight syringe with blunt needle and 200 μL capacity.
2. 20 μm ID PEEK tubing connected to a finger-tight fitting and a fillport.
3. Micro Bio-Spin 6 columns (Bio-Rad).
4. 50 mM ammonium acetate pH 6.5.
5. Agilent TOF tune mix MS calibrants.
6. Ion table: Create an ion table in Excel as shown in Fig. 2 to calculate m/z values for all charge states and all multimeric states for a given neutral mass.

2.7 Membrane Proteins

Materials as in Subheading 2.1 plus:

1. RP-18e Chromolith 25 × 2 mm (Merck) (*see* **Note 1**).

3 Methods

3.1 Intact Mass Analysis of Soluble Proteins

1. Adjust the concentration of the protein to 1 mg/mL using buffer A. Depending upon the HPLC sample loop volume, further dilute the sample to achieve a final protein loading of 1 μg. Higher protein loading may be required for larger (>80 kDa) proteins, but avoid overloading the cartridge.
2. Program the LC-parameters according to Table 1 using a C-3 guard cartridge. These are optimized for high-throughput desalting and do not allow for chromatographic separation of complex protein mixtures.
3. Program the mass spectrometer according to Table 2. Acquisition parameters will vary somewhat from instrument to instrument. Those described in Table 2 are appropriate for an Agilent 6530 QTOF. Key parameters are source parameters to provide adequate desolvation at the relatively high flow rate of 1 mL/min.

Table 1
LC parameters for intact mass analysis

Time (min)	Buffer A (%)	Buffer B (%)	Flow (mL/min)	Max pressure (bar)
0.00	90	10	1.0	600
0.34	20	80	1.0	600
0.36	5	95	1.0	600
0.75	5	95	1.0	600
0.77	90	10	1.0	600
0.90	90	10	1.0	600

Table 2
MS acquisition parameters for intact mass

Parameter	Value
Ion mode	Positive
Nebulizer pressure	60 psi
Drying gas flow rate	12 L/min
Drying gas temperature	350 °C
Capillary voltage	4000 V
Fragmentor voltage	250 V
Skimmer voltage	60 V
Octopole RF voltage	250 V

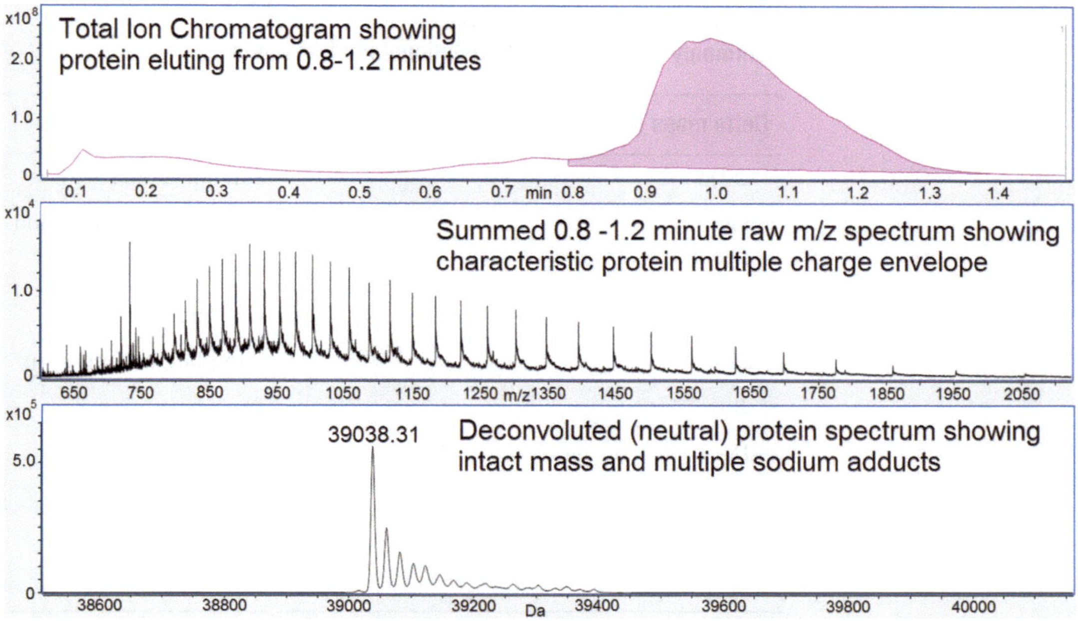

Fig. 3 Intact mass data for typical protein showing different stages of data analysis

4. Open the acquired data file using proprietary software, in this case Masshunter Qualitative Analysis version 7.0 (Agilent).
5. Sum spectra across the total ion chromatogram peak of interest, typically 0.8–1.2 min and apply maximum entropy deconvolution using the default parameters across an appropriate output mass range as illustrated in Fig. 3. For batch analyses, this can be automated and performed on-the-fly.
6. Perform data interpretation using the expected protein sequence. Also required is an understanding of which posttranslational modifications (PTMs) are (a) possible and (b) likely.
7. Subtract the observed from the expected mass to obtain the delta mass value, which should be less than 50 ppm for a match to the unmodified sequence.
8. Identify sequence truncations by mass match to a C-terminal or N-terminal fragment with the same 50 ppm mass accuracy.
9. Use Table 3 to identify commonly observed mass discrepancies.
10. Use the Unimod web server (http://www.unimod.org) to attempt to identify mass discrepancies not represented in Table 3, bearing in mind that most of the 1357 modifications listed are extremely rare or not possible.

Table 3
Commonly observed posttranslational modifications

Delta mass (Da)	Likely interpretation
(+22)n	Sodium adduct
−89	Met loss with acetylation
−131	Met loss
+178	Gluconolyation
+256	Phosphogluconoylation
(+80)n	Phosphorylation
−18	Pyroglutamic acid
+42	Acetylation
(+16)n	Oxidation
+14	Methylation

3.2 In-Gel Digestion (See Note 2)

1. Use gel cutting tips to excise Coomassie Blue stained bands from a polyacrylamide gel.
2. Expel the cut bands into a 96-well PCR plate, cover with 200 μL 10% (v/v) methanol, seal the plate with film and store at 4 °C until ready for analysis.
3. Using a 200 μL 12-channel pipettor, remove and discard 10% (v/v) methanol solution.
4. Shrink the gel band by addition of 200 μL acetonitrile and stand for 30 s.
5. Remove and discard the supernatant.
6. Freshly prepare an appropriate volume of reduction buffer using Table 4.
7. Add 200 μL of reduction buffer to each gel band, seal the plate, place in a thermal cycler and incubate at 56 °C for 40 min.
8. Remove the plate. Remove and discard the reduction buffer.
9. Repeat **steps 4** and **5**.
10. Freshly prepare an appropriate volume of alkylation buffer using Table 4.
11. Add 200 μL of alkylation buffer to each gel band, seal the plate and incubate at room temperature in the dark for 20 min.
12. Remove and discard the alkylation buffer.
13. Repeat **steps 4** and **5**.
14. Freshly prepare an appropriate volume of digestion buffer using Table 4.
15. Add 50 μL of digestion solution to each gel band.

Table 4
Appropriate reagent volumes for in-gel digestion of 12–96 samples

	Reduction		Alkylation		Digestion		
Number of samples	100 mM NH_4HCO_3	1 M DTT	100 mM NH_4HCO_3	Saturated iodoacetamide	H_20	100 mM NH_4HCO_3	Trypsin
12	2.5 mL	250 μL	2.5 mL	125 μL	496 μL	156 μL	2 μL
24	5 mL	500 μL	5 mL	250 μL	938 μL	312 μL	4 μL
36	7.5 mL	750 μL	7.5 mL	375 μL	1.4 mL	468 μL	6 μL
48	10 mL	1 mL	10 mL	500 μL	1.88 mL	625 μL	8 μL
60	12.5 mL	1.25 mL	12.5 mL	625 μL	2.38 mL	781 μL	10 μL
72	15 mL	1.5 mL	15 mL	750 μL	2.81 mL	937 μL	12 μL
84	17.5 mL	1.75 mL	17.5 mL	875 μL	3.3 mL	1.1 mL	14 μL
96	20 mL	2 mL	20 mL	1 mL	3.75 mL	1.25 mL	16 μL

16. Visually inspect the plate to ensure that each gel band is submerged.
17. Seal the plate and incubate at 37 °C overnight.

3.3 Digestion in Solution

1. Add 20 μl of purified protein at a concentration between 1 mg/mL and 5 mg/mL to 100 μL of 100 mM NH_4HCO_3 (pH 8.0).
2. Add 1 μL of 1 M DTT solution and incubate at 56 °C for 40 min.
3. Add 4 μL of saturated iodoacetamide solution and incubate at room temperature in the dark for 20 min.
4. Add a further 100 μL of ammonium bicarbonate buffer.
5. Add 1–5 μL of trypsin stock solution and incubate at 37 °C overnight (*see* **Note 3**).

3.4 LC-MSMS Analysis of Protein Digests

1. Using a multichannel pipettor, transfer 46 μL of the digest supernatant from the 96-well PCR plate to a round bottomed microtitre plate compatible with your LC autosampler. Retain the gel bands for further analysis.
2. Program the micro LC-parameters according to Table 5 using a 200 μm × 5 cm Pepswift PS-DVB monolithic column.
3. Program the mass spectrometer according to Table 6. Acquisition parameters will vary somewhat from instrument to instrument. Those described are appropriate for a Bruker HCT ion trap. Key factors are source parameters matched to the capillary flow rate and choice of data dependent fragmentation parameters to match the HPLC peak width.
4. Program the autosampler to inject 1 μL of digest solution.

Table 5
LC parameters for MSMS analysis

Time (min)	Buffer A (%)	Buffer B (%)	Flow (μL/min)	Max pressure (bar)
0.0	98	2	2.5	800
1.0	98	2	2.5	800
5.0	83	17	2.5	800
7.0	58	42	2.5	800
7.1	8	92	2.5	800
8.1	8	92	2.5	800
8.3	98	2	2.5	800
14.0	98	2	2.5	800

Table 6
MSMS acquisition parameters for protein identification

Parameter	Value
Ion mode	Positive
Nebulizer pressure	16 psi
Drying gas flow rate	5 L/min
Drying gas temperature	300 °C
Capillary voltage	−4000 V
Skimmer voltage	40 V
Capillary exit	145 V
Scan range	200–2000 m/z
Scan rate	26,000 m/z/s
Fragmentation mode	Data dependent
Precursors per MS1 scan	3
Active exclusion	20 s

5. Program the automated data analysis using proprietary software, in our case Data Analysis version 4.0 (Bruker).
6. Extract peptide fragment datasets using the function: find compounds using auto MS^n.
7. Deconvolute peptide data to zero charge state.
8. Generate a .mgf (Mascot generic format) file for each sample.
9. Either use proprietary batch control software to submit multiple .mgf file for an MSMS ion search on-the-fly (in our

case we use Biotools version 3.1 to submit to an in-house Mascot server) or submit files manually using the a web server (e.g., www.matrixscience.com).

10. Specify the database, enzyme, and number of missed cleavages. We use Uniprot, trypsin, and four missed cleavages.
11. Set carbamidomethyl cysteine as the sole global modification.
12. Restrict the number of variable modifications to only those expected (e.g., methionine oxidation) to minimize false positives.
13. Specify mass search tolerances and ion types. These will be dependent upon the instrument and the type of fragmentation used. In our case, these are MS tolerance ± 1.5 Da, MSMS tolerance ± 1.3 Da, number of ^{13}C = 1. An example of a high confidence Mascot MSMS peptide fragment match is shown in Fig. 4 (*see* **Note 4**).

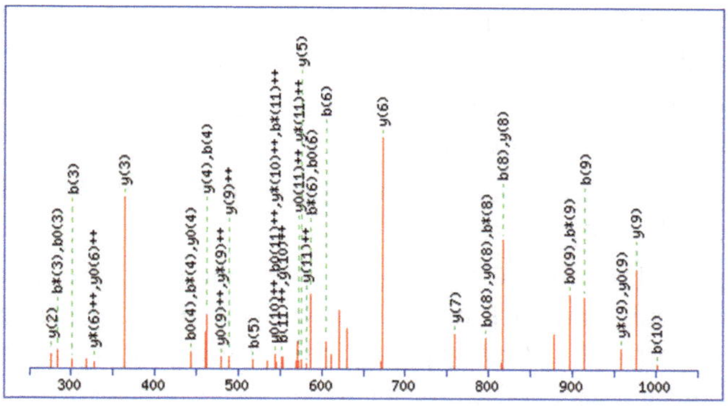

Monoisotopic mass of neutral peptide Mr(calc): 1275.5878
Fixed modifications: Carbamidomethyl (C) (apply to specified residues or termini only)
Ions Score: 70 **Expect:** 0.00041
Matches : 43/126 fragment ions using 38 most intense peaks (help)

#	b	b^{++}	b^*	b^{*++}	b^0	b^{0++}	Seq.	y	y^{++}	y^*	y^{*++}	y^0	y^{0++}	#
1	116.0342	58.5207			98.0237	49.5155	D							12
2	173.0557	87.0315			155.0451	78.0262	G	1161.5681	581.2877	1144.5415	572.7744	1143.5575	572.2824	11
3	**301.1143**	151.0608	284.0877	142.5475	283.1037	142.0555	Q	1104.5466	552.7769	1087.5201	544.2637	1086.5361	543.7717	10
4	**461.1449**	231.0761	444.1184	222.5628	443.1343	222.0708	C	***976.4880***	488.7477	959.4615	480.2344	958.4775	479.7424	9
5	**518.1664**	259.5868	501.1398	251.0736	500.1558	250.5815	G	***816.4574***	408.7323	799.4308	400.2191	798.4468	399.7271	8
6	**605.1984**	303.1028	588.1719	294.5896	587.1878	294.0976	S	***759.4359***	380.2216	742.4094	371.7083	741.4254	371.2163	7
7	702.2512	351.6292	685.2246	343.1159	684.2406	342.6239	P	***672.4039***	336.7056	655.3774	328.1923	654.3933	327.7003	6
8	**815.3352**	408.1713	798.3087	399.6580	797.3247	399.1660	L	***575.3511***	288.1792	558.3246	279.6659	557.3406	279.1739	5
9	**914.4036**	457.7055	897.3771	449.1922	896.3931	448.7002	V	***462.2671***	231.6372	445.2405	223.1239	444.2565	222.6319	4
10	**1001.4357**	501.2215	984.4091	492.7082	983.4251	492.2162	S	***363.1987***	182.1030	346.1721	173.5897	345.1881	173.0977	3
11	1102.4834	551.7453	1085.4568	543.2320	1084.4728	542.7400	T	***276.1666***	138.5870	259.1401	130.0737	258.1561	129.5817	2
12							R	175.1190	88.0631	158.0924	79.5498			1

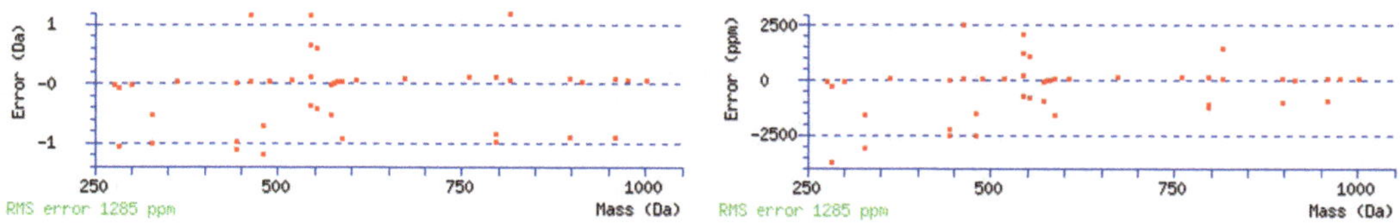

Fig. 4 Mascot ion fragmentation match for the TEV peptide DGQCGSPLVSTR showing a high confidence MOWSE score of 70

3.5 Phosphorylation Mapping

Phosphorylation mapping is a variation of PTM identification using the LC-MSMS methods described in Subheadings 2.3 and 2.4. It deserves special attention because of the central role of phosphorylation in protein activation and molecular signaling. Phosphopeptides often constitute 1% of a protein digest or less. Consequently, phosphopeptide enrichment is usually, though not always required. In our laboratory, protein phosphorylation state and occupancy is always determined by intact mass analysis in advance of phosphorylation mapping. This is critical for proper evaluation and validation of MSMS data. Phosphorylation mapping should be performed both on whole and on metal oxide affinity enriched protein digests.

1. Perform an intact mass analysis of the target protein using the techniques described in Subheading 3.1. Confirm the presence of one or more +80 Da mass shifts and determine the percentage occupancy for each phosphorylation state.
2. Perform a tryptic digestion and LC-MSMS analysis as described in Subheadings 3.3 and 3.4 with the following alterations.
3. In Subheading 3.4, **step 9**, set the variable modifications to phospho (ST) and phospho (Y).
4. Set the data dependent acquisition parameters to select MSMS on all peptide charge states.
5. Punch out a disc of the C18 medium using the 18 gauge cannulation needle and pack it into a P10 tip using a 29 gauge needle, to form a frit. Gently pack the frit using a short length of fused silica.
6. Use fine forceps to punch a hole of approximately 3 mm diameter in the lid of a 1.5 mL microcentrifuge tube and mount the tip through this hole so that it is held by the bevel half-way up the tip and the end of the tip is 5 mm from the bottom of the tube.
7. Using a 1.5 mL microcentrifuge tube, weigh out 2.5 mg of TiO_2 beads and 2.5 mg of ZrO_2 powder per tip.
8. Add 50 μL of water per tip.
9. Pipette up and down to form a suspension immediately prior to pipetting, then add 50 μL of the TiO_2, ZrO_2 suspension to the tip (*see* **Note 5**).
10. Pack the medium by centrifugation at 1100 × *g* for 1 min (*see* **Note 6**).
11. Wash the tips with 50 μL of Wash 1.
12. Wash the tips with 50 μL of Wash 2 and repeat. The packed medium will turn yellow.
13. Check that the pH of the Wash 2 eluant is acidic with indicator paper.

14. Label the tips (not the collection tubes).
15. Adjust the digest sample solution to 10% (v/v) formic acid.
16. Load the sample on to the spin column in 50 μL stages.
17. Use a Pasteur pipette to remove and discard flow-through after each spin.
18. Desalt the sample using 50 μL 2% (v/v) ACN, 0.1% FA (v/v) and repeat.
19. Elute non-phosphopeptides using 50 μL of Elution 1. Discard flow-through.
20. Wash the tip with 50 μL of Wash 3.
21. Wash the tip with 50 μL of Wash 4.
22. Transfer the tip to a new, labeled collection tube.
23. Elute phosphopeptides with 50 μL of Elution 2 and retain flow-through. The packed medium will turn grey.
24. Dry down the flow-through using a rotary evaporator at 60 °C for 45 min. The dried deposit will be dark red in color and should not smell of ammonia.
25. Resuspend the dried deposit in 5 μL of 2% (v/v) ACN, 0.1% (v/v) FA and transfer to a round-bottomed 96-well microtiter plate for LC-MSMS. Seal the plate using adhesive-free zone plate sealing film.
26. Perform LC-MSMS as described in the previous Subheading 3.4.
27. Manually evaluate all putative phosphorylation assignments. Do not rely on the search algorithm as false positives are common.
28. Is the MOWSE score statistically significant?
29. Are there neutral losses for phosphoserine or phosphothreonine?
30. Was the peptide captured by metal oxide affinity?
31. Is the difference between the top peptide MOWSE score and the next highest scoring peptide greater than 10?
32. "Walk" through the b and y ion series from low to high mass and check that they confirm phosphorylation at the site nominated by Mascot.
33. Check for ambiguity if there is more than one potential site. A comparison of MSMS data from enriched and unenriched phosphoprotein is shown in Table 7 (*see* **Note 7**).

3.6 Glycan Mapping

Glycan mapping is important to the protein crystallographer, who may wish to remove this modification. The lack of consensus sequence for *O*-linked glycans plus the lack of *O*-glycan specific deglycosylases makes mapping these more desirable still. Glycans are challenging to map in comparison with other PTMs because multiple glycoforms are usually present and their masses are unknown. Glycoforms may be readily characterized by intact mass

Table 7
Comparison of peptides identified from whole and enriched phosphoprotein (peptide MOWSE values shown in *brackets*)

Whole tryptic digest 65% coverage, MOWSE 1177	Enriched tryptic digest 44% coverage, MOWSE 829	PI	Neg charges
AEYEFDSPYWDDISDSAK (109)	AEYEFDSPYWDDISDSAK (94)	3.4	7
ETLGTGAFSEVVLAEEK (105)	ETLGTGAFSEVVLAEEK (106)	3.9	5
ESSIENEIAVLR (74)	ESSIENEIAVLR (55)	4	4
DLKPENLLYYSQDEESK (83)	DLKPENLLYYSQDEESK (79)	4	6
HENIVALEDIYESPNHLYLVMQLVSGGELFDR (110)	HENIVALEDIYESPNHLYLVMQLVSGGELFDR (129)	4.2	7
GKESSIENEIAVLR (113)	GKESSIENEIAVLR (103)	4.6	4
	GKESSIENEIAVLR Phospho (ST) (94)	4.6	6
LHLGSSLD (38)		5.1	2
DASTLIR (36)	DASTLIR (36)	6.2	2
IMISDFGLSK (43)	IMISDFGLSK (42)	6.2	2
IMISDFGLSK Oxidation (M) (40)	IMISDFGLSK Oxidation (M) (63)	6.2	2
GDVMSTACGTPGYVAPEVLAQKPYSK Phospho (ST) (42)		6.2	5
IFEFK (26)		6.4	2
GFYTEK (16)		6.4	2
LFEQILK (35)	LFEQILK (35)	6.4	2
QVLDAVYYLHR (65)	QVLDAVYYLHR (59)	7.3	2
HPWIAGDTALNK (38)		7.4	2
NIHESVSAQIR (68)	NIHESVSAQIR (40)	7.4	2
KLHLGSSLD (58)		7.4	2
KIFEFK (36)		9.4	2
	NLMEKDPNKR (66)	9.4	3
MGIVHR (43)		10.3	1
QAFNATAVVR (54)		10.4	1

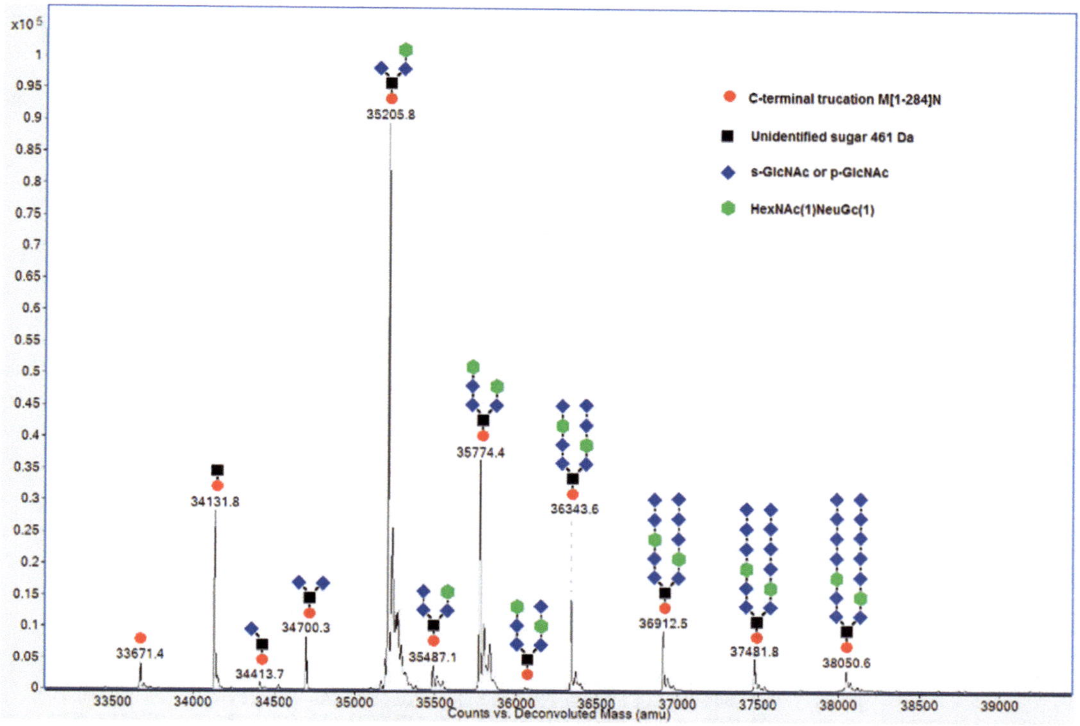

Fig. 5 Glycosylation envelope and derived glycan structure for an integral membrane protein

(Subheading 3.1) as shown in Fig. 5. Theoretically, this should allow detection of tryptic glycopeptides by LC-MSMS with asparagine, serine, or threonine specific corresponding mass shifts. In practice, such glycopeptides are not detected. The method described below utilizes the disappearance of *N*-linked glycopeptides and deamidation of asparagine in order to map their position.

1. Divide the target protein sample in two.
2. Deglycosylate one fraction using PNGase-F following the manufacturer's instructions.
3. Perform in solution tryptic digest and LC-MSMS on both fractions following the protocols described in Subheadings 3.3 and 3.4.
4. Ensure that the MS1 mass tolerance parameter in Mascot is set to >1.0 Da (*see* **Note 8**).
5. Map the *N*-glycosylation consensus motifs (N-X-S/T) within the target protein using a Web-based prediction tool.
6. Compare the MSMS results of untreated and deglycosylated protein. Deglycosylation specific peptides containing a consensus asparagine and with a mass discrepancy of +1 Da are consistent with *N*-glycosylation at that site. An example is shown in Fig. 6.

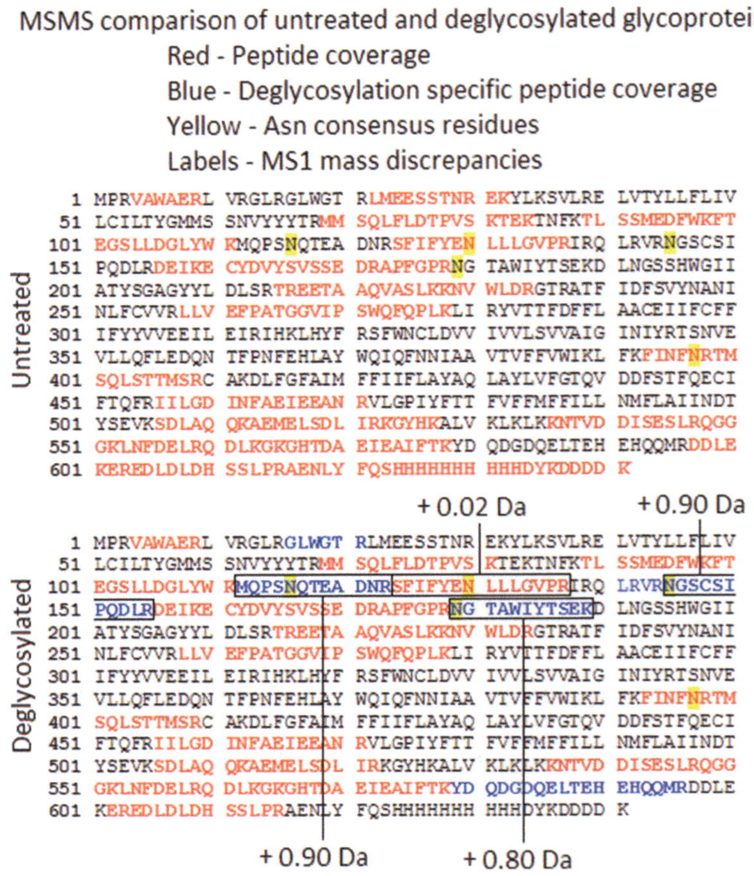

Fig. 6 MSMS comparison of untreated and de-glycosylated glycoprotein indicating sites at N116, N145 & N179. *red* – peptide coverage; *blue* – de-glycosylation specific peptide coverage; *yellow* – Asn consensus residues; labels – MS1 mass discrepancies)

3.7 Native Mass Spectrometry

Efficient desalting and desolvation, while maintaining the protein in its native state, along with sufficient material to allow detection of higher m/z species, are the factors critical for successful native analyses. Specialized equipment is not essential. Under native electrospray conditions, proteins acquire fewer charges and consequently generate a charge envelope at higher m/z. Given that older ESI instruments typically scan up to m/z 3500, this allows for detection of native monomeric proteins up to 50 kDa. Modern instruments offer greater sensitivity and consequently require less protein. Also they scan to m/z 20,000 or above allowing detection of megadalton complexes. The method described below was developed on an Agilent 6530 QTOF with a standard ESI source (*see* **Note 9**). Approximately 50 μg is required for proteins and complexes up to 50 kDa in a maximum volume of 75 μL of any buffer. Larger proteins and complexes will require more material. Samples should be analyzed as soon as possible following purification and stored on ice.

1. Change the instrument context tab from Acquisition to Tune.
2. Change the instrument mode to 1 GHz, high mass range 20,000 m/z and wait for 20 min for instrument to stabilize.
3. Take three Micro Bio-Spin 6 columns per sample and equilibrate in 50 mM ammonium acetate, pH 6.5 following the manufacturer's instructions.
4. Add up to 75 μL of sample and then buffer exchange following the manufacturer's instructions.
5. Collect the eluant and repeat this step a further two times, then hold on ice.
6. Calibrate in 1 GHz mode and apply calibration (*see* **Note 10**).
7. Return to Acquisition context.
8. Connect a 200 μL gastight syringe to the source via a fillport and 20 μm red PEEK tubing.
9. Open the source chamber and flush the nebulizer with 200 μL of acetonitrile to remove tune mix ions. Place absorbent tissue in front of the nebulizer tip to collect the acetonitrile spray and allow this to evaporate in a fume cupboard. Repeat.
10. Flush nebulizer with 200 μL of water, then close the source chamber (*see* **Note 11**).
11. Fill the syringe with ammonium acetate buffer and mount the syringe in a syringe pump set to 6 μL/min.
12. Load the acquisition method described in Table 8.
13. Make a "dummy injection" using the method parameters described below and begin data acquisition.

Table 8
Native MS acquisition parameters for Agilent 6530

Parameter	Value
LC injection	Blank
LC pump flow	0.0 mL/min
Ion mode	Positive
Nebulizer pressure	17 psi
Drying gas flow rate	5 L/min
Drying gas temperature	325 °C
Capillary voltage	3500 V
Fragmentor voltage	430 V
Skimmer voltage	65 V
Octopole RF voltage	750 V

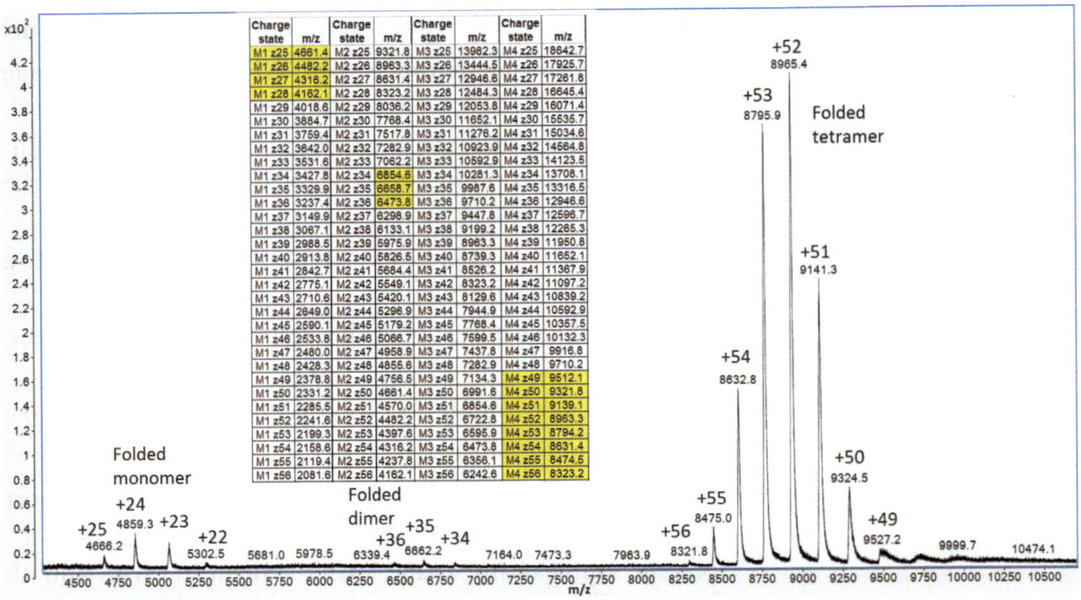

Fig. 7 Native m/z spectrum for tetrameric beta-galactosidase with charge table (inset)

14. When the total ion chromatogram (TIC) rises above background noise and the spray is stable, remove and empty the syringe, then quickly load with sample and replace in the syringe pump.
15. If high m/z species with good signal to noise are observed in real time, acquire spectra for 2–3 min. If no high m/z species are observed, continue acquisition until the syringe is empty.
16. Stop acquisition, then flush the syringe, fillport and tubing with water followed by ammonium acetate buffer.
17. Repeat **steps 6–9** for each new sample.
18. Using proprietary MS analysis software, sum spectra across the length of TIC above background to obtain the m/z spectrum. This spectrum is the native data.
19. Deconvolute the spectrum and obtain the observed (native) monomeric mass.
20. Compare this with the intact mass obtained under denaturing conditions. This will reveal non-covalent binding of ligands such as nickel or zinc.
21. Use the observed native mass as the input for an ion table (described in Subheading 2.6). Do not use the theoretical mass or the denatured observed mass.
22. Using the ion table, match each ion series in the m/z spectrum with a series of at least three consecutive ions in the ion table and assign a multimeric state to each distribution. An example of this is shown in Fig. 7.

23. Fit one or more Gaussian curves to the ion distribution for each multimeric state. In general, there are two or three ion distributions for the monomer, representing denatured, partially unfolded and folded conformations. Multimers generally possess a single ion distribution (*see* **Note 12**).

3.8 Integral Membrane Proteins

Integral membrane proteins present unique challenges in mass spectrometry because of the requirement for detergent to extract and solubilize and also because of the inherently low abundance of IMPs. As powerful ion suppressants, detergents are incompatible with MS and it becomes essential to separate them from the protein being analyzed. This can be done is a variety of ways, but the simplest method is to excise the protein from an SDS-PAGE gel and perform tryptic digest MSMS analysis as described in Subheadings 3.2 and 3.4. Membrane protein samples should not be boiled prior to SDS-PAGE. The peptide coverage for membrane proteins is lower than that for soluble proteins because membrane spanning regions are usually devoid of positively charged K and R residues necessary for tryptic cleavage. In addition, denaturation is generally incomplete due to strong hydrophobic interactions in the polypeptide chain, rendering these regions inaccessible to proteolysis.

While membrane proteins may be identified using MSMS analysis, confirmation of structure requires measurement of intact mass. Offline removal of detergents prior to mass spectrometry can be achieved by precipitation, but this is frequently ineffective for the small amounts and low concentrations which are characteristic for membrane protein purification. Resuspension of precipitated protein in an MS compatible buffer may also prove difficult. We have pioneered the use of LC-MS for in-line detergent removal and intact mass measurement of IMPs at μg quantities [10]. The technique described here gives good results for the majority of common membrane protein–detergent combinations.

1. Membrane proteins need to be solubilized using mild nonionic detergents with sugar-based head groups such as n-dodecyl β-D-maltoside (DDM) or octyl glucose neopentyl glycol (OGNG) and should be less than 0.1%.
2. Adjust the concentration of the protein to 1 mg/mL using 1% (v/v) formic acid. Depending upon the HPLC sample loop volume, further dilute the sample to achieve a final protein loading of 1 μg. Best results are obtained from concentrated protein samples where the detergent has been substantially diluted. Higher protein loading may be required for larger (>80 kDa) proteins or glycoproteins.
3. Program the LC-parameters according to Table 9 using an RP-18e Chromolith 25 × 2 mm column (*see* **Note 13**).

Table 9
LC parameters for membrane protein intact mass analysis

Time (min)	Buffer A (%)	Buffer B (%)	Flow (mL/min)	Max pressure (bar)
0.00	95	5	0.4	200
3.00	1	99	0.4	200
4.00	1	99	0.4	200
4.01	0	100	1.0	200
5.70	0	100	1.0	200
5.71	95	5	0.4	200
6.50	95	5	0.4	200

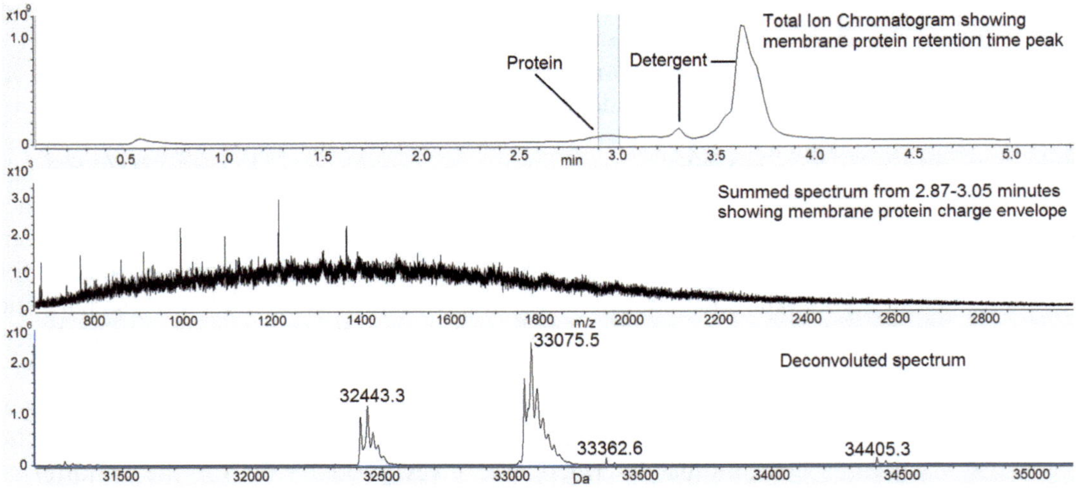

Fig. 8 Typical intact mass data for integral membrane protein SAP-B showing different stages of data analysis (contrast this with Figure 3)

4. Program the mass spectrometer using the acquisition parameters in Table 2 but changing the fragmentation voltage to 400.
5. Run blanks before and after each sample (*see* **step 8**).
6. Open the acquired data file using proprietary software, in this case Masshunter Qualitative Analysis version 7.0 (Agilent).
7. Use the "walk chromatogram" feature to identify which peaks in the TIC correspond to detergent and which contain a protein m/z peak distribution (Fig. 3). Expect a high signal to noise or detector saturation for detergents, which will be singly charged. The >1000 m/z region may need to be scaled separately to visualize protein signal.

8. If no chromatographic separation of protein and detergent is achieved, analyze the preceding blank LC-MS run. Frequently the blank is detergent free, and sufficient protein carries over in the LC to enable successful analysis (*see* **Note 14**).
9. Sum spectra across only the TIC region where protein is observed and apply maximum entropy deconvolution. An example of membrane protein analysis is shown in Fig. 8.
10. Perform data interpretation as described in Subheading 3.1.

4 Notes

1. The Chromolith column described in Subheading 2.7 has a polymer housing and must be used with a finger-tight fitting, not a stainless steel high-pressure connector. Flow rate ramping is used to prevent leaks here.
2. The method in Subheading 3.2 is optimized for high-throughput protein identification [11].
3. The optimum protein:typsin ratio is 20:1 (w/w).
4. In general, Mascot scores >100 and with two or more peptide identifications are statistically valid hits. Manually evaluate all other Mascot results. Experience is needed to do this reliably, but the following checklist is useful:

   Is the protein or homolog hit expected?

   Are there two or more peptide hits for the protein?

   Are the peptides the top scoring peptides (red)?

   Are the individual peptides scores not significant (<20), moderately significant (>20) or highly significant (>50)?

   Are the delta mass values (parent ion MS error for ion trap) <0.3 Da?

   Are the scores for short peptides moderately significant (>20) and long peptides highly significant (>50)?

   Do b and y ions predominate?

   Is there a run of consecutive b and y ions (combined) of seven or more?

   Are the major peaks in the spectrum represented in the ion table?

   Are the MSMS errors small and clustering near zero?

   Are the MSMS errors small and trending on a line (systematic calibration error)?

   Are the MSMS mass outliers close to +1 Da or −1 Da (C13 isotopes)?

   If the majority of ions are b++ or y++ is the peptide +2 or greater?

Do the b++ or y++ ions form a run of consecutive ions of seven or more?

Are the b and y ions absent where the b++ or y++ are present?

If there are gaps in the ion run, do they occur either side of a proline?

An example of high confidence Mascot peptide data is shown in Fig. 4.

5. Metal oxide affinity enrichment is peptide sequence dependent and cannot be predicted in advance. In our experience, TiO_2 and ZrO_2 give the best results. Where we have used both matrices in parallel we observed significant peptide overlap, but neither captured them all, hence our preference for home-made dual matrix spin columns.
6. The ideal particle size for MOA spin columns is 10 μm. Use of ZrO_2 with a sieve size less than 6 μm leads to compaction and column blocking.
7. Successful phosphorylation mapping is dependent upon generating sufficient MSMS coverage to span the all phosphorylation sites. Digestion using other enzymes may be required. In our hands chymotrypsin and pepsin also give good MSMS coverage without excessive overlap.
8. PNGase F action involves deamidation of Asn to Asp, resulting in a mass addition of 1 Da. It is essential therefore to ensure that the MS1 mass tolerance parameter in Mascot is set to >1.0 Da. This mass discrepancy can be used to validate MSMS glycopeptide matches [12].
9. The native method in Subheading 3.7 will work on any instrument with similar ion optics, i.e., Agilent ESI-TOF.
10. For native MS, the instrument requires tuning at high m/z for optimum sensitivity. Only calibrant ions 922–2722 Da are used for calibration and extrapolated to higher m/z values. Mass accuracy is better than 20 ppm and resolution 15,000 FWHM at 450 kDa. Incomplete desolvation, in-source fragmentation, and complex dissociation have not been observed by us using these methods.
11. The source used for native MS is a standard ESI source with steel capillary nebulizer. Care should be taken following the flushing protocols to ensure the needle does not block with precipitated protein or ammonium acetate crystals.
12. Data analysis remains the most challenging and time-consuming aspect of native MS. As no commercially available software exists for this purpose, users need to construct their own software tools or rely on shareware.

13. For membrane protein analysis, use of methanol as mobile phase B is critical to detergent–protein separation. We have never achieved successful LC-MS using acetonitrile.
14. Presence of membrane proteins in the post-run blank suggests that protein has precipitated on-column.

Acknowledgments

The SGC is a registered charity (number 1097737) that receives funds from AbbVie, Bayer Pharma AG, Boehringer Ingelheim, Canada Foundation for Innovation, Eshelman Institute for Innovation, Genome Canada, Innovative Medicines Initiative (EU/EFPIA) [ULTRA-DD grant no. 115766], Janssen, Merck & Co., Novartis Pharma AG, Ontario Ministry of Economic Development and Innovation, Pfizer, São Paulo Research Foundation-FAPESP, Takeda, and Wellcome Trust [092809/Z/10/Z].

References

1. Karas M, Bachmann D, Hillenkamp F (1985) Influence of the wavelength in high-irradiance ultraviolet-laser desorption mass-spectrometry of organic-molecules. Anal Chem 57:2935–2939
2. Dole M, Mack LL, Hines RL et al (1968) Molecular beams of macroions. J Chem Phys 49:2240–2249
3. Lewis JK, Wei J, Siuzdak G (2006) Matrix-Assisted Laser Desorption/Ionization Mass Spectrometry in Peptide and Protein Analysis. Encyclopedia of Analytical Chemistry. Published Online: 15 SEP 2006 DOI: 10.1002/9780470027318.a1621
4. Anderson NL, Anderson NG (1998) Proteome and proteomics: new technologies, new concepts, and new words. Electrophoresis 19:1853–1861
5. Cole RB (2011) Electrospray and MALDI mass spectrometry: fundamentals, instrumentation, practicalities, and biological applications. John Wiley & Sons, Hoboken, NJ
6. Garcia BA, Mollah S, Ueberheide BM et al (2007) Chemical derivatization of histones for facilitated analysis by mass spectrometry. Nat Protoc 2:933–938
7. Ong S-E, Blagoev B, Kratchmarova I et al (2002) Stable isotope labeling by amino acids in cell culture, SILAC, as a simple and accurate approach to expression proteomics. Mol Cell Proteomics 1:376–386
8. Thingholm TE, Jørgensen TJ, Jensen ON et al (2006) Highly selective enrichment of phosphorylated peptides using titanium dioxide. Nat Protoc 1:1929–1935
9. Rye PT, Frick LE, Ozbal CC et al (2011) Advances in label-free screening approaches for studying sirtuin-mediated deacetylation. J Biomol Screen 16:1217–1226
10. Berridge G, Chalk R, D'Avanzo N et al (2011) High-performance liquid chromatography separation and intact mass analysis of detergent-solubilized integral membrane proteins. Anal Biochem 410:272–280
11. Chalk R, Berridge G, Shrestha L et al (2014) High-throughput mass spectrometry applied to structural genomics. Chromatography 1: 159–175
12. Khoshnoodi J, Hill S, Tryggvason K et al (2007) Identification of N-linked glycosylation sites in human nephrin using mass spectrometry. J Mass Spectrom 42:370–379

Chapter 26

How to Determine Interdependencies of Glucose and Lactose Uptake Rates for Heterologous Protein Production with *E. coli*

David J. Wurm, Christoph Herwig, and Oliver Spadiut

Abstract

Induction by lactose is known to have a beneficial effect on the expression of soluble recombinant proteins in *E. coli* harboring the T7 expression system (e.g., *E. coli* BL21(DE3)). As lactose is a metabolizable inducer, it needs to be supplied continuously to prevent depletion and thus only partial induction. Overfeeding and accumulation of lactose or glucose on the other hand can lead to osmotic stress. Thus, it is of utmost importance to know the possible feeding ranges. Here, we show a fast method using a simple mechanistic model to characterize *E. coli* strains harboring the T7 expression system regarding their ability to take up lactose and glucose. This approach reduces experimental work and the gained data allows running a stable and robust bioprocess without accumulation of lactose or glucose.

Key words *Escherichia coli* BL21(DE3), pET expression system, Lactose induction, Specific lactose uptake rate, Mechanistic model, Recombinant protein production

1 Introduction

E. coli BL21(DE3) in combination with the well-known pET expression system is frequently used for recombinant protein production [1–3]. The pET system can be induced by lactose or its molecular mimic IPTG [4]. Since IPTG is a non-metabolizable inducer, it is commonly used as it only has to be added once and is known to promote strong induction. Nevertheless, IPTG puts a high metabolic burden on the cells and often leads to formation of inactive protein aggregates, the so-called inclusion bodies (IBs). Lactose, on the other hand, can serve as inducer and secondary carbon source and is described in literature to favor the production of soluble target protein [5–10]. However, as it is metabolized by the cells, a continuous supply of lactose is necessary [11].

When performing fed-batch cultivations, accumulation of lactose and glucose has to be avoided as it can lead to osmotic stress

Nicola A. Burgess-Brown (ed.), *Heterologous Gene Expression in E. coli: Methods and Protocols*, Methods in Molecular Biology, vol. 1586, DOI 10.1007/978-1-4939-6887-9_26, © Springer Science+Business Media LLC 2017

for the cells [12]. Therefore, it is of utmost importance to know the maximum specific uptake rate of lactose ($q_{s,lac}$) to prevent overfeeding and thus sugar accumulation. In previous studies we found that $q_{s,lac}$ is dependent on the specific uptake rate of glucose ($q_{s,glu}$) [13] (Fig. 1). This means that at a certain $q_{s,glu}$, $q_{s,lac}$ must not be exceeded and cultivations can only be performed on or underneath the curve shown in Fig. 1.

The correlation between $q_{s,glu}$ and $q_{s,lac}$ is dependent on the used strain and the expressed product and therefore has to be evaluated for every strain to guarantee stable process conditions (own unpublished data). We were able to mechanistically describe the correlation between $q_{s,glu}$ and $q_{s,lac}$ for a recombinant *E. coli* strain producing a single chain antibody fragment with the pET expression system (*see* Eq. 1). The parameters $q_{s,lac,noglu}$ and $q_{s,glu,crit}$ can be evaluated by one experiment each. To fit the other parameters $q_{s,lac,max}$, K_A and n, at least three supplementary fed-batch fermentations need to be carried out.

$$q_{s,lac}\,(g/g*h) = q_{s,lac,max}\cdot \max\left(\left(1-\frac{q_{s,glu}}{q_{s,glu,crit}}\right)^{n},0\right)\cdot\left(\frac{q_{s,glu}}{q_{s,glu}+K_A}+\frac{q_{s,lac,\text{ noglu}}}{q_{s,lac,max}}\right) \quad (1)$$

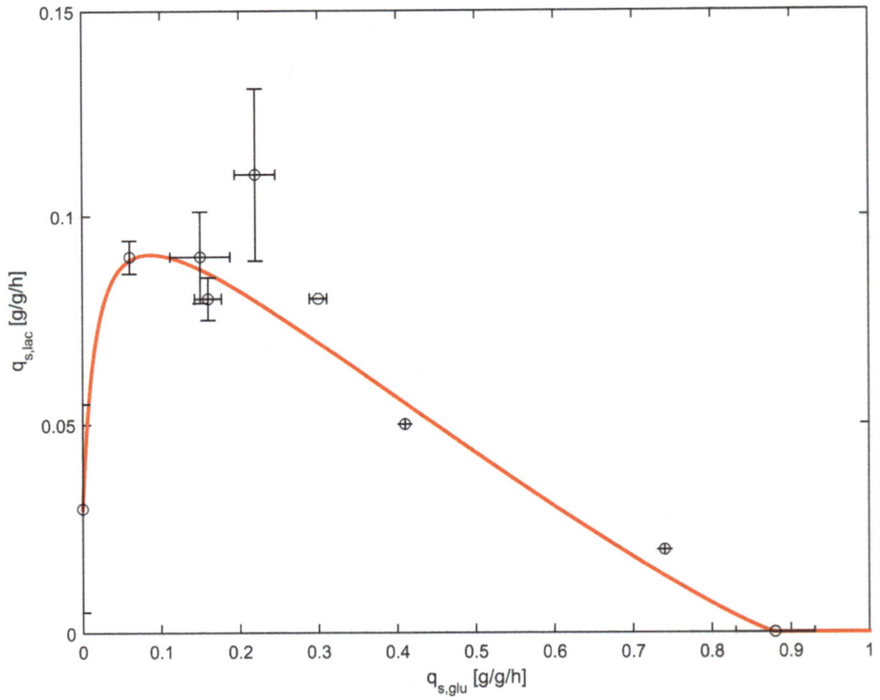

Fig. 1 The maximum specific lactose uptake rate ($q_{s,lac}$) as a function of the specific glucose uptake rate ($q_{s,glu}$) for a recombinant *E. coli* strain producing a single chain antibody fragment with the pET expression system; fitted curve according to the developed mechanistic model (Eq. 1) with parameters $q_{s,lac,max}$ = 0.088 g/g/h, $q_{s,glu,crit}$ = 0.88 g/g/h, $q_{s,lac,noglu}$ = 0.034 g/g/h, K_A = 0.019 g/g/h, n = 1.16 [13]

$q_{s,lac}$, Specific lactose uptake rate (g/g/h).

$q_{s,lac,max}$, Theoretical maximum specific lactose uptake rate (g/g/h).

$q_{s,glu}$, Specific glucose uptake rate (g/g/h).

$q_{s,glu,crit}$, Critical specific glucose uptake rate up to which lactose is consumed (g/g/h).

$q_{s,lac,noglu}$, Specific lactose uptake rate at $q_{s,glu} = 0$ (g/g/h).

K_A, Affinity constant for the specific lactose uptake rate (g/g/h).

n, Type of inhibition (noncompetitive, uncompetitive, competitive).

Here, we present a fast approach to characterize an *E. coli* strain regarding the correlation between $q_{s,glu}$ and $q_{s,lac}$ which allows reducing the experimental work to five cultivations. The gained data allow carrying out fed-batch cultivations without accumulating glucose or lactose, thus assuring a stable and robust bioprocess.

2 Materials

2.1 Stock Solutions

Prepare all media and solutions with analytical grade reagents and deionized water. Stock solutions can be stored at room temperature (*see* **Note 1**). Further information about the medium can be found elsewhere [14].

1. $MgSO_4$ stock (500×): 600.00 g/L $MgSO_4 \cdot 7H_2O$. Sterilize by autoclaving.
2. Fe(III)citrate stock (100×): 10.00 g/L Fe(III)citrate. Sterilize by autoclaving.
3. EDTA stock (100×): 0.84 g/L EDTA. Sterilize by autoclaving.
4. $Zn(CH_3COO)_2$ stock (200×): 2.60 g/L $Zn(CH_3COO)_2 \cdot 2H_2O$. Sterilize by filtration (0.2 μm).
5. Thiamine HCl stock (1000×): 4.50 g/L thiamine HCl. Sterilize by filtration (0.2 μm).
6. TE stock (200×): 0.50 g/L $CoCl_2 \cdot 6H_2O$, 3.00 g/L $MnCl_2 \cdot 4H_2O$, 0.24 g/L $CuCl_2 \cdot 2H_2O$, 0.60 g/L H_3BO_3 and 0.50 g/L $Na_2MoO_4 \cdot 2H_2O$. Sterilize by filtration (0.2 μm).
7. Antibiotics stock (1000×): depending on the antibiotics resistance of the used organism, Sterilize by filtration (0.2 μm).
8. Sterile antifoam (e.g., PPG). Sterilize by autoclaving.
9. Glucose solution for pre-culture medium (0.1 L): 8.80 g of glucose monohydrate. Sterilize by autoclaving.
10. Glucose solution for batch medium (0.1 L): 22.0 g of glucose monohydrate. Sterilize by autoclaving.

11. Glucose solution for feed medium (0.96 L): 275 g of glucose monohydrate. Sterilize by autoclaving.
12. Lactose solution for pulse medium (0.96 L): 210 g of lactose monohydrate. Sterilize by autoclaving.
13. Salt solution for pre-culture and batch medium (0.91 L): 13.30 g of KH_2PO_4, 4.00 g of $(NH_4)_2HPO_4$, and 1.70 g of citric acid. Dissolve salts, set pH to 7.2 with NaOH and sterilize by autoclaving.

2.2 Media

For preparation of media *see* **Notes 3–8**.

1. Pre-culture medium (1 L): 0.1 L of glucose solution for pre-culture, 0.91 L of salt solution for pre-culture, 2.0 mL of $MgSO_4$ stock (500×), 10 mL of Fe(III)citrate stock (100×), 10.0 mL of EDTA stock (100×), 5.0 mL of $Zn(CH_3COO)_2$ stock (200×), 5 mL of TE stock (200×), 1 mL of TE stock (200×) and 1 mL of antibiotics stock (1000×). After cooling the solutions to room temperature, combine them aseptically.
2. Batch medium (1 L): 0.1 L of glucose solution for batch medium, 0.91 L of salt solution for batch medium, 2.0 mL of $MgSO_4$ stock (500×), 10 mL of Fe(III)citrate stock (100×), 10.0 of mL EDTA stock (100×), 5.0 mL of $Zn(CH_3COO)_2$ stock (200×), 5 mL of TE stock (200×), 1 mL of TE stock (200×), 1 mL of antibiotics stock (1000×) and 0.1 mL of sterile Antifoam. After cooling the solutions to room temperature, combine them aseptically.
3. Glucose feed medium (1 L): 0.96 L of glucose solution for feed medium, 20.8 mL of $MgSO_4$ stock (500×), 2.5 mL of Fe(III)citrate stock (100×), 9.7 mL of EDTA stock (100×), 3.8 mL of $Zn(CH_3COO)_2$ stock (200×) and 5 mL of TE stock (200×). After cooling the solutions to room temperature, combine them aseptically (*see* **Note 9**).
4. Lactose pulse medium (1 L): 0.96 L of lactose solution for pulse medium, 20.8 mL of $MgSO_4$ stock (500×), 2.5 mL of Fe(III)citrate stock (100×), 9.7 mL of EDTA stock (100×), 3.8 mL of $Zn(CH_3COO)_2$ stock (200×) and 5 mL of TE stock (200×). After cooling the solutions to room temperature, combine them aseptically (*see* **Note 10**).

2.3 Equipment

For standard fed-batch experiments the following equipment is at least required:

1. Baffled shake flask (e.g., 2.5 L Ultra-yield flasks, UYF) and shaker with temperature control.
2. Bioreactor (e.g., 10 L working volume stainless steel Sartorius BIOSTAT Cplus bioreactor).

3. Stirring, gassing (air and oxygen), pH and temperature control in the bioreactor.
4. Probes for monitoring pH and pO_2 in the bioreactor.
5. Offgas analyzer for CO_2 and O_2 (e.g., DASGIP GA gas analyzer; Eppendorf).
6. Pumps and tubing for base and feed addition.
7. Balances (base balance and feed balance, connected to the process management system).
8. Process information management system (PIMS; e.g., Lucullus).
9. Spectrophotometer, centrifuge, dry oven and analytical balance for sample preparation and dry cell weight (DCW) determination.
10. High-performance liquid chromatography (HPLC) for exact determination of sugars (glucose, lactose, galactose) and metabolites (acetate) equipped with a column that can separate sugars (e.g. Supelcogel C-610 H ion exchange column (Sigma-Aldrich)) and a refractive index detector (e.g., Agilent Technologies).

3 Methods

3.1 Pre-culture

Start the pre-culture by inoculating 500 mL of pre-culture medium with 1.5 mL of frozen glycerol stock. The pre-culture is carried out in 2.5 L Ultra-yield flasks at 37 °C and 180 rpm for 20 h.

3.2 Batch Cultivation for Biomass Generation

1. After setting the pH of the salts solution in the bioreactor to 7.2 with NaOH, the bioreactor has to be autoclaved. As soon as the bioreactor is cooled down, add sterile stock solutions and sugar solution aseptically.
2. The temperature control is set to 35 °C, pH control to 7.2 (12.5% NH_4OH), gassing to 7.5 L/min, stirring to 1400 rpm, pressure to 0.5 bar and dissolved oxygen is maintained above 30% by increasing the ratio of pure oxygen to pressurized air in the gas inlet (*see* **Note 11**).
3. Aseptically transfer the pre-culture (10% (v/v) of total volume) to the bioreactor.
4. Temperature, pH, stirring, DO_2, pump setpoints, gassing setpoints, ratio of air in gassing, ratio of pure oxygen in gassing, signal of feed and base balance, CO_2 and O_2 in the offgas are monitored throughout the fermentation by a PIMS.
5. The end of the batch phase can be determined by a drop in the CO_2 signal (*see* **Note 13**).

3.3 Fed-Batch Cultivation for Biomass Generation

1. The fed-batch phase for biomass generation can be started after the CO_2 signal drops and as soon as the CO_2 signal stays constant. To start the fed-batch, the amount of biomass has to be known to be able to calculate the feed rate to obtain a constant $q_{s,glu}$.
2. Depending on the used strain and product, the correlation between OD_{600} and biomass dry cell weight (DCW) can differ and therefore should be evaluated for every used strain. If no correlation is available yet for the used strain, the biomass concentration can be roughly estimated by Eq. 2 (*see* **Note 12**).

$$DCW\left(\mathrm{g}/\mathrm{L}\right) = OD_{600} * 0.45 \quad (2)$$

DCW, Dry cell weight (g/L).

OD_{600}, Optical density at 600 nm.

3. Alternatively, if the biomass yield is known (this can be analyzed by the data after the first fermentation), the biomass concentration can be calculated by Eq. 3.

$$DCW\left(\mathrm{g}/\mathrm{L}\right) = \frac{c_{glu,0} * V_{\mathrm{Batch,start}} * Y_{X/S}}{V_{\mathrm{Batch},end}} \quad (3)$$

$c_{glu,0}$, Glucose concentration in reactor at start of Batch (g/L).

$V_{Batch,start}$, Volume of reactor at start of Batch (L).

$V_{Batch,end}$, Volume of reactor at end of Batch (L).

4. The duration of the fed-batch phase can be stirred by changing $q_{s,glu}$ and thus μ. The total amount of biomass during the fed-batch fermentation at a certain time point t can be calculated by Eq. 4 (*see* **Note 14**).

$$X\left(\mathrm{g}\right) = c_{x,0} * V_0 * e^{\left(q_s * Y_{x/s} * (t - t_0)\right)} \quad (4)$$

X, Total biomass at time point t (g).

$c_{x,0}$, Biomass concentration at start of fed-batch (g/L).

q_s, Specific uptake rate of glucose (g/g/h).

$Y_{x/s}$, Biomass yield (g/g).

t_0, Time point of fed-batch start (h)

t, Time (h).

5. The Feed rate F (g/h) of the glucose feed can be calculated by Eq. 5.

$$F\left(\mathrm{g}/\mathrm{h}\right) = \frac{q_s * X^* \rho_{\mathrm{Feed}}}{c_{\mathrm{s,Feed}}} \quad (5)$$

F, Feed rate (g/h).

ρ_{Feed}, Density feed (g/L).

$c_{s,Feed}$, Feed concentration (g/L).

6. After the desired amount of biomass is reached the feed can be stopped and the experiments to determine the correlation between $q_{s,glu}$ and $q_{s,lac}$ can be started.

3.4 Determination of Correlation Between $q_{s,lac}$ and $q_{s,glu}$

To determine the correlation between $q_{s,lac}$ and $q_{s,glu}$, five experiments have to be performed to be able to evaluate all parameters of Eq. 1. Pre-culture, batch cultivation for biomass generation, and fed-batch cultivation for biomass generation have to be carried out equally and only the last phase of the cultivation is different in each experiment.

Experiment 1: Determination of $q_{s,glu}$ at $q_{s,glu,crit}$ (green triangle, Fig. 2), which is the specific uptake rate of glucose when both glucose and lactose are in excess. Due to carbon catabolite repression, only glucose and no lactose will be metabolized as long as glucose is present in excess ($q_{s,lac} = 0$).

Experiment 2: Determination of $q_{s,lac}$ at $q_{s,lac,noglu}$ (blue square, Fig. 2), which is the specific uptake rate of lactose when lactose is in excess and no glucose is supplied ($q_{s,glu} = 0$).

Experiment 3–5: Determination of at least three $q_{s,lac}$ at certain $q_{s,glu}$ between $q_{s,glu} = 0$ and $q_{s,glu} = q_{s,glu,crit}$ (red diamonds in red ellipse, Fig. 2), when lactose is in excess and glucose is supplied in limiting amounts, to be able to fit the parameters $q_{s,lac,max}$, K_A and n for the mechanistic model describing the correlation between $q_{s,lac}$ and $q_{s,glu}$ (as shown in Eq. 1 and Fig. 2).

For all experiments, lactose concentration has to be kept in excess between 5 and 15 g/L at all times. This can be reached by at-line lactose measurements and consecutive lactose pulsing as soon as the lactose concentration falls below 5 g/L.

3.4.1 Experiment 1

To evaluate $q_{s,glu,crit}$ (green triangle, Fig. 2), a mixed pulse of glucose and lactose has to be supplied to reach a final concentration of 15 g/L of both sugars. So the uptake of glucose, $q_{s,glu,crit}$, can be evaluated (*see* **Note 17**).

3.4.2 Experiment 2

To evaluate $q_{s,lac}$ when no glucose is available ($q_{s,glu} = 0$, blue square, Fig. 2), a lactose pulse to reach a final lactose concentration of 15 g/L is applied and no glucose is added. As lactose gets metabolized very slowly when only lactose is available, a single pulse is sufficient.

3.4.3 Experiments 3–5

To evaluate $q_{s,lac}$ at $0 < q_{s,glu} < q_{s,glu,crit}$ (red diamonds, Fig. 2), apply a lactose pulse to 15 g/L final concentration and start a glucose feed supplying glucose at a constant $q_{s,glu}$ by calculating the feed rate according to Eqs. 4 and 5 (*see* **Note 15**). Depending on the supplied $q_{s,glu}$ the uptake of lactose will be higher than in Experiments

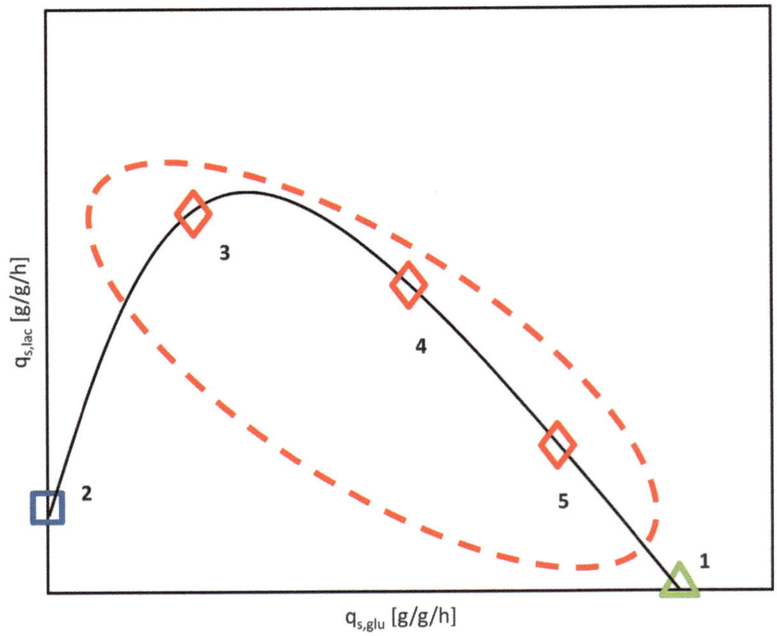

Fig. 2 Schematic correlation between $q_{s,lac}$ and $q_{s,glu}$; $q_{s,glu,crit}$ (*green triangle*, specific uptake rate of lactose when glucose and lactose are in excess), $q_{s,lac,noglu}$ (*blue square*, specific uptake rate of lactose when lactose is in excess and no glucose is supplied) and $q_{s,lac}$ (*red diamonds*, specific uptake rate of lactose when lactose is in excess and glucose is supplied in limiting amounts)

1 and 2, and thus lactose concentration in the media has to be measured frequently to avoid depletion of lactose.

3.5 Analysis of Biomass Concentration

1. Determination of DCW: Take 1 mL of culture broth and centrifuge it in pre-dried and pre-weighed Eppendorf tubes. Centrifuge (5000 × *g*, 4 °C, 10 min), wash pellet with 0.9% (w/v) NaCl solution, centrifuge again (5000 × *g*, 4 °C, 10 min) and determine DCW gravimetrically after drying at 105 °C for at least 72 h.

3.6 Analysis of Sugar Uptake and Metabolite Formation

1. Determination of sugar and metabolite concentration: Take fermentation supernatant after centrifugation of culture broth (14,000 × *g*, 4 °C, 15 min) and analyze samples by HPLC. Prepare standards for glucose, lactose, galactose (0–20 g/L), and acetate (0–2.5 g/L) to generate standard curves. In case the Supelcogel C-610 H ion exchange column is used (see above), inject 10 μL onto the ion exchange column and elute isocratically with 0.1% H_3PO_4 at a flow rate of 0.5 mL/h.

3.7 Correlation between $q_{s,glu}$ and $q_{s,lac}$

1. Calculate $q_{s,glu}$ and $q_{s,lac}$ as described in **Notes 16** and **17**.
2. Now you can plot $q_{s,lac}$ against $q_{s,glu}$. The curve can be interpolated using Eq. 1.

3. $q_{s,glu,crit}$ and $q_{s,lac,noglu}$ are evaluated by Experiments 1 and 2.
4. To fit the rest of the model parameters ($q_{s,lac,max}$, K_A and n) the error between all data points and the curve has to be minimized. This can be done with the program Matlab using the Nelder–Mead simplex method [14].
5. Plot $q_{s,lac}$ as a function of $q_{s,glu}$ and Eq. 1 to obtain the possible ranges for $q_{s,lac}$ at $q_{s,glu}$.

4 Notes

1. Do not autoclave stock solutions, which are described to be sterile filtered as they might not be heat stable.
2. Autoclaving of salt solution for batch medium can be performed inside the bioreactor.
3. Due to evaporation during autoclaving of the batch and pre-culture medium, the volume before autoclaving can be increased by about 5% to compensate this loss.
4. Autoclave sugars and salts separately. Otherwise amines and reducing sugars react via Maillard reaction which makes the solution turn brown.
5. Determine the exact sugar concentration after autoclaving in the batch medium by HPLC. For determination of exact sugar concentrations in the feed and pulse medium after autoclaving, high dilutions are necessary which can lead to a high error. You can develop a correlation between density of the feed and feed concentration by measuring the density of feeds with different sugar concentrations without autoclaving them. Plot the sugar concentration against the density of the feeds and make a linear correlation. Afterwards the exact concentration of the feed can be determined by measuring the density of the feed.
6. After storing the pre-culture medium for a few days, precipitation can be observed. To our knowledge this does not matter as these precipitates dissolve again during cultivation.
7. Glucose and lactose are mostly supplied as monohydrate. Do not forget to consider that when calculating the required amount of glucose (+10% w/v) and lactose (+5% w/v).
8. Wait until medium is cooled down before addition of stock solutions and antibiotics as most of them are heat sensitive.
9. The glucose concentration in the feed medium can be varied depending on $q_{s,glu}$. The limitations are minimum and maximum pump speed. Too low concentrations of glucose lead to high dilution in the bioreactor which should be avoided; too high concentrations of glucose can lead to very low feed rates and thus discontinuous pumping of the feed. Furthermore,

above 400 g/L, the solution gets very viscous and hard to pump. Therefore it has be tested whether the setup can pump the feed solution when using highly concentrated feeds prior to cultivation and the pump ranges have to be evaluated.

10. The maximum concentration of the lactose pulse medium is 200 g/L due to solubility reasons.
11. Overpressure in the fermenter is not a must, but helps the oxygen transfer, thus reducing the need for stirring, gassing and pure oxygen.
12. As the correlation between DCW and OD_{600} can change for different strains and products, it needs to be evaluated for the used *E. coli* strain. To establish the correlation between DCW and OD_{600} take culture broth, dilute it to stay within the linear range of the photometer and measure OD_{600}. Plot OD_{600} against DCW and make a linear regression to be able to estimate the DCW from OD_{600}.
13. If only base is used to control the pH during the cultivation, the pH will increase at the end of batch due to acetate consumption. The acetate has been produced during the batch phase. This can be compensated by using not only base control but also acid control. Otherwise the pH can also be set to 7.2 manually with HCl by addition via a syringe.
14. The duration of the fed-batch phase for biomass generation can be varied by changing the parameter $q_{s,glu}$. Upper limitation for $q_{s,glu}$ is the maximum specific uptake rate of the strain at 35 °C. We recommend running the batch phase on day 1, the fed-batch phase for biomass generation overnight (about 10–14 h) and conduct the cultivation phase on glucose and/or lactose for characterizing the strain, where high sampling frequencies are required, on the next day.
15. Conversion of lactose to biomass can be neglected. Higher uptake of lactose leads to stronger induction and thus a lower biomass yield. These two effects mostly cancel each other out.
16. The *E. coli* BL21(DE3) strain is a mutant that cannot metabolize galactose. Lactose is taken up into the cell by lactose permease and subsequently split into glucose and galactose by beta-galactosidase. Glucose is taken up and galactose is excreted to the medium. Therefore, the production rate of galactose is equivalent to the uptake rate of lactose and can also be used for determination of $q_{s,lac}$. We tested the influence of accumulating galactose on the specific growth rate and the specific sugar uptake rates up to a concentration of 20 g/L and did not see any impact of the galactose in the medium.
17. The volumetric uptake rate of glucose and lactose ($R_{s,glu}$, $R_{s,lac}$), the volumetric biomass formation rate (R_X), the volumetric

formation rate of galactose ($R_{P,gal}$), the specific uptake rate of glucose and lactose ($q_{s,glu}$, $q_{s,lac}$), the specific growth rate (μ), and the biomass yield ($\Upsilon_{X/S}$) can be calculated by Eqs. 6–12.

$$R_{S,glu}\left(g/h\right)=\frac{c_{glu,0}*V_0-c_{glu,1}*V_1+V_{\text{feed}}*c_{\text{feed}}}{t_1-t_0} \quad (6)$$

$$R_{S,lac}\left(g/h\right)=\frac{c_{lac,0}*V_0-c_{lac,1}*V_1}{t_1-t_0} \quad (7)$$

$$R_X\left(g/h\right)=\frac{c_{X,0}*V_0-c_{X,1}*V_1}{t_1-t_0} \quad (8)$$

$$R_{P,gal}\left(g/h\right)=\frac{c_{gal,1}*V_1-c_{gal,0}*V_0}{t_1-t_0} \quad (9)$$

$$q_S\left(g/g*h\right)=\frac{R_S}{\left(c_{X,0}*V_0+c_{X,1}*V_1\right)*0.5} \quad (10)$$

$$\mu\left(1/h\right)=\frac{R_X}{\left(c_{X,0}*V_0+c_{X,1}*V_1\right)*0.5} \quad (11)$$

$$\Upsilon_{X/S}\left(g/g\right)=\frac{R_X}{R_{S,\text{total}}} \quad (12)$$

References

1. Rosano GL, Ceccarelli EA (2014) Recombinant protein expression in *Escherichia coli*: advances and challenges. Front Microbiol 5:172
2. Studier FW, Moffatt BA (1986) Use of bacteriophage-T7 Rna-polymerase to direct selective high-level expression of cloned genes. J Mol Biol 189:113–130
3. Baneyx F (1999) Recombinant protein expression in *Escherichia coli*. Curr Opin Biotechnol 10:411–421
4. Neubauer P, Hofmann K (1994) Efficient use of lactose for the lac promotercontrolled overexpression of the main antigenic protein of the foot and mouth disease virus in *Escherichia coli* under fed-batch fermentation conditions. FEMS Microbiol Rev 14:99–102
5. Neubauer P, Hofmann K, Holst O et al (1992) Maximizing the expression of a recombinant gene in *Escherichia coli* by manipulation of induction time using lactose as inducer. Appl Microbiol Biotechnol 36:739–744
6. Pei XL, Wang QY, Li CL et al (2011) Efficient production of a thermophilic 2-deoxyribose-5-phosphate aldolase in glucose-limited fed-batch cultivations of *Escherichia coli* by continuous lactose induction strategy. Appl Biochem Biotechnol 165:416–425
7. Gombert AK, Kilikian BV (1998) Recombinant gene expression in *Escherichia coli* cultivation using lactose as inducer. J Biotechnol 60:47–54
8. Zou C, Duan X, Wu J (2014) Enhanced extracellular production of recombinant *Bacillus deramificans* pullulanase in *Escherichia coli* through induction mode optimization and a glycine feeding strategy. Bioresour Technol 172:174–179
9. Bashir H, Ahmed N, Khan MA, Zafar AU, Tahir S, Khan MI, Khan F, Husnain T (2016) Simple procedure applying lactose induction and one-step purification for high-yield production of rhCIFN. Biotechnol Appl Biochem 63(5):708–714. doi:10.1002/bab.1426
10. Fruchtl M, Sakon J, Beitle R (2015) Expression of a collagen-binding domain fusion protein: effect of amino acid supplementation, inducer type, and culture conditions. Biotechnol Prog 31:503–509
11. Striedner G, Cserjan-Puschmann M, Potschacher F et al (2003) Tuning the transcription rate of recombinant protein in strong *Escherichia coli* expression systems through repressor titration. Biotechnol Prog 19:1427–1432

12. Shiloach J, Fass R (2005) Growing *E. coli* to high cell density—a historical perspective on method development. Biotechnol Adv 23:345–357
13. Wurm DJ, Veiter L, Ulonska S, Eggenreich B, Herwig C, Spadiut O (2016) The *E. coli* pET expression system revisited-mechanistic correlation between glucose and lactose uptake. Appl Microbiol Biotechnol 100 (20):8721–8729. doi:10.1007/s00253-016-7620-7
14. DeLisa MP, Li JC, Rao G et al (1999) Monitoring GFP-operon fusion protein expression during high cell density cultivation of *Escherichia coli* using an on-line optical sensor. Biotechnol Bioeng 65:54–64
15. Lagarias JC, Reeds JA, Wright MH et al (1998) Convergence properties of the Nelder-Mead simplex method in low dimensions. SIAM J Optimiz 9:112–147

Chapter 27

Interfacing Biocompatible Reactions with Engineered *Escherichia coli*

Stephen Wallace and Emily P. Balskus

Abstract

Biocompatible chemistry represents a new way of merging chemical and biological synthesis by interfacing nonenzymatic reactions with metabolic pathways. This approach can enable the production of nonnatural molecules directly from renewable starting materials via microbial fermentation. When developing a new biocompatible reaction certain criteria must be satisfied, i.e., the reaction must be (1) functional in aqueous growth media at ambient temperature and pH, (2) nontoxic to the producing microorganism, and (3) have negligible effects on the targeted metabolic pathway. This chapter provides a detailed outline of two biocompatible reaction procedures (hydrogenation and cyclopropanation), and describes some of the chemical and microbiological experiments and considerations required during biocompatible reaction development.

Key words Biocompatible chemistry, Synthetic biology, Metabolic engineering, Hydrogenation, Cyclopropanation

1 Introduction

The fields of organic chemistry and metabolic engineering have traditionally represented two independent approaches to small molecule production [1–3]. However, as synthetic biology continues to impact the chemical sciences, the potential benefits of combining tools from both disciplines are beginning to be realized [4–6]. A recent approach that merges these two synthetic strategies is biocompatible chemistry: "*non-enzymatic reactions capable of structurally modifying small molecule metabolites as they are produced by living organisms*" (Fig. 1a and b) [7–11]. Interfacing these reactions with engineered metabolic pathways has recently been explored as a way to access nonnatural molecules directly from renewable starting materials [9, 11]. To the best of our knowledge, the use of nonenzymatic reagents in microbial fermentation dates back to Neuberg's steered fermentation of

Nicola A. Burgess-Brown (ed.), *Heterologous Gene Expression in E. coli: Methods and Protocols*, Methods in Molecular Biology, vol. 1586, DOI 10.1007/978-1-4939-6887-9_27,

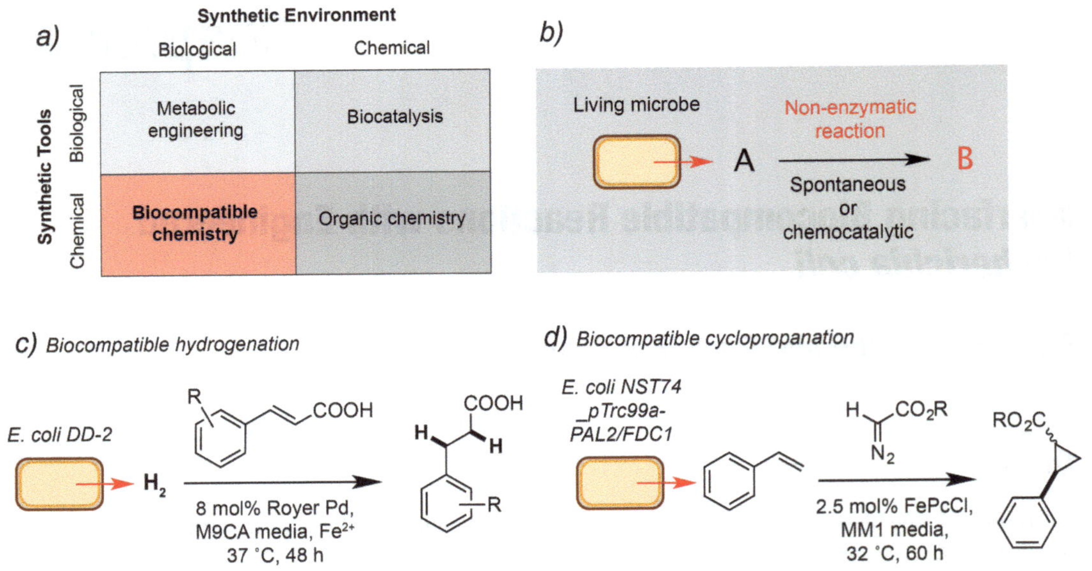

Fig. 1 Biocompatible chemistry merges organic chemistry and metabolic engineering (**a**), and the general reaction concept (**b**). The biocompatible hydrogenation and cyclopropanation reactions (**c**) and (**d**)

Saccharomyces cerevisiae during the First World War, in which sodium bisulfite was used to intercept the metabolite acetaldehyde, the upstream intermediate to ethanol, forming a stable adduct [12, 13]. This reaction resulted in an accumulation of NADH, which redirected metabolic flux to produce an alternative fermentation product, glycerol. Since this report nearly 100 years ago, the development of new nonenzymatic reactions capable of manipulating metabolic pathways has received little attention, despite significant developments in organic chemistry and biochemistry during this time. More recently, our laboratory and others have begun to target new microbial metabolites for biocompatible reaction development using modern synthetic methods [7–11]. In this chapter, we provide a detailed outline of two biocompatible reaction procedures: a Pd-catalyzed hydrogenation using hydrogen produced by engineered *E. coli*; and a Fe-catalyzed cyclopropanation using styrene produced by engineered *E. coli* [8, 9] (Fig. 1c and d).

2 Materials

Prepare all solutions using ultrapure water purified using a MilliQ (MQ) system. Unless stated otherwise, all catalysts and reagents are purchased from commercial suppliers and are of analytical grade.

2.1 Growth Media

1. Luria–Bertani (LB): Add 15 g of Bacto tryptone, 7.5 g of yeast extract, and 15 g of NaCl to 1 L of MQ water in a 2 L Pyrex bottle and autoclave at 121 °C for 20 min. Upon cooling the

solution to room temperature, make up to a final volume of 1.5 L under aseptic conditions using autoclaved MQ water. Store at room temperature.

2. M9-glucose: Add 9 g of Na_2HPO_4, 4.5 g of KH_2PO_4, 1.5 g of NH_4Cl, and 0.8 g of NaCl to 1 L of MQ water in a 2 L Pyrex bottle and autoclave at 121 °C for 20 min. Upon cooling the solution to room temperature, add 150 μL of a 10 mg/mL filter-sterilized aqueous solution of thiamine hydrochloride, 3 mL of a 1 M filter-sterilized aqueous solution of $MgSO_4$, 150 μL of a 1 M filter-sterilized aqueous solution of $CaCl_2$, and 38 mL of a 20% (w/v) of an autoclaved aqueous solution of D-glucose. Make up to a final volume of 1.5 L under aseptic conditions using autoclaved MQ water. Store at 4 °C.
3. M9CA-glucose: M9CA-glucose medium is prepared analogously to M9-glucose medium, except that 7.5 g of casamino acids should be added before autoclaving the solution.
4. MM1-glucose: Add 0.3 g of KH_2PO_4 and 1 g of K_2HPO_4 to 500 mL of MQ water and autoclave at 121 °C for 20 min. Upon cooling the solution to room temperature, add 100 mL of a 40 g/L autoclaved aqueous solution of $(NH_4)_2SO_4$ and 100 mL of a 250 g/L filter-sterilized aqueous solution of MOPS under aseptic conditions. Adjust the pH of the resulting solution to 7.4 using a 28% aqueous solution of NH_3 in water. Under aseptic conditions, add 100 mL of a 20% of an autoclaved aqueous solution of 20% (w/v) D-glucose, 10 mL of a 50 g/L filter-sterilized aqueous solution of $MgSO_4 \cdot 7H_2O$, and 10 mL of ATCC Trace Mineral Supplement (consisting of 0.5 g/L EDTA, 3 g/L $MgSO_4 \cdot 7H_2O$, 0.5 g/L $MnSO_4 \cdot 7H_2O$, 1 g/L NaCl, 0.1 g/L $FeSO_4 \cdot 7H_2O$, 0.1 g/L $Co(NO_3)_2 \cdot 6H_2O$, 0.01 g/L $AlK(SO_4)_2$, 0.01 g/L H_3BO_3, 0.01 g/L $Na_2MoO_4 \cdot 2H_2O$, 0.001 g/L Na_2SeO_3, 0.1 g/L $Na_2WO_4 \cdot 2H_2O$, and 0.02 g/L $NiCl_2 \cdot 6H_2O$). Adjust the total volume to 1 L using autoclaved MQ water. Filter-sterilize the final medium using a 0.2 μm filter under vacuum. Store the medium at room temperature.

2.2 Biocompatible Hydrogenation Equipment

1. Hugate tubes, septa, screw caps and aluminum crimp seals (ChemGlass Inc.).
2. Media bottles capped with polytetrafluoroethylene flat septa (Bellco Glass Inc.).
3. Royer Pd catalyst (3% Pd on polyethylenimine/SiO_2 (40–200 mesh), Strem Chemicals, GFS Chemicals, or synthesized from $PdCl_2$ as described in Subheading 3.2.2).
4. Antibiotic solutions: Ampicillin (1000×): 50 mg of ampicillin in 1 mL of MQ water. Spectinomycin (1000×): 25 mg of spectinomycin in 1 mL of MQ water. Chloramphenicol

(1000×): 12.5 mg of chloramphenicol in 1 mL of MQ water. Sterilize the solutions using a 0.2 μm filter before use.

5. 0.05 M ammonium iron(II) sulfate hexahydrate solution: 19.6 mg in 1 mL of MQ water. Sterilize the solution using a 0.2 μm filter before use.
6. Isopropyl β-D-1-thiogalactopyranoside (IPTG) solution (1000×): 119.2 mg in 1 mL of MQ water. Sterilize the solution using a 0.2 μm filter before use.
7. 10 mL of LB and 1 L of M9CA-glucose medium (*see* Subheading 2.1).

2.3 Biocompatible Cyclopropanation Equipment

1. Hugate tubes, septa, screw caps, and aluminum crimp seals (ChemGlass Inc.).
2. Sealable Erlenmeyer flasks and glass stoppers (Chemglass. Inc.)
3. Ampicillin solution (1000×): 50 mg of ampicillin in 1 mL MQ water. Sterilize the solution using a 0.2 μm filter before use.
4. Isopropyl β-D-1-thiogalactopyranoside (IPTG) solution (1000×): 48 mg IPTG in 1 mL MQ water. Sterilize the solution using a 0.2 μm filter before use.
5. Iron(III) phthalocyanine chloride (FePcCl).
6. Diazo component: either ethyl diazoacetate (EDA, ≥13 wt.% dichloromethane), benzyl diazoacetate (+ca. 10% dichloromethane as a stabilizer), *tert*-butyl diazoacetate.
7. 20 mL of LB and 1 L of MM1-glucose medium (*see* Subheading 2.1).
8. 10 mM 1,3,5-trimethoxybenzene (10×): 16.8 mg TMB in 10 mL hexanes.
9. 6 M NaOH(aq): 16 g NaOH(s) in 100 mL MQ water (dissolve in a 250 mL Pyrex bottle and autoclave at 121 °C for 20 min and store at room temperature).

2.4 Metabolite Analysis

For hydrogen quantification by GC, headspace samples are obtained using a gastight syringe. Peak areas are used to determine H_2 concentrations by comparing them to a standard curve generated by analyzing samples containing known concentrations of hydrogen gas. For styrene and cyclopropane quantification by GC, 1,3,5-trimethoxybenzene (TMB) is used as an internal standard. Standard curves are constructed over a range of six analyte concentrations (2–21 μmol for H_2, and 10 μM–5 mM for styrene/cyclopropane) providing a linear relationship between $pA_{metabolite}$, $pA_{TMB}{}^*$, and $c_{metabolite}$ of the form $y = mx + c$, where pA = peak area, c = concentration, $y = pA_{metabolite}/pA_{TMB}{}^*$). *Where appropriate.

3 Methods

3.1 Biocompatible Hydrogenation

3.1.1 Overview of Strain Engineering

The hydrogen-producing *E. coli* DD-2 strain was obtained from Prof. Pam Silver and Dr. Daniel Ducat (Harvard Medical School, Boston, MA, USA). A detailed account of the metabolic engineering that allows the production of hydrogen gas to be coupled to the breakdown of glucose has been described previously [14]. In brief, electrocompetent *E. coli* BL21(DE3) is transformed with three plasmids: (1) a pCDF-duet vector encoding for the maturation factors HydEF and HydG from *Chlamydomonas reinhardtii*, (2) a pACYC-duet vector encoding for a pyruvate ferredoxin oxidoreductase (PFOR) from *Desulfovibrio africanus*, and (3) a pET-duet vector encoding for an fusion protein consisting of a [Fe-Fe] hydrogenase and ferrodoxin from *Clostridium acetylbutylicum* connected at the hydrogenase *C*-terminus via a $(Gly_4Ser)_2$ amino acid linker (Fig. 2). All genes are codon-optimized for expression in *E. coli*. Insulation of this pathway from background hydrogenase activity is achieved by creating an *E. coli* BL21(DE3)Δ*hycE*Δhya*B*Δ*hybC* knockout via sequential P1 transduction from Keio collection mutants into BL21(DE3) followed by excision of the Kan^R resistance marker [15].

3.1.2 Small-Scale Biocompatible Hydrogenation (7 mL Volume)

1. Inoculate a LB:glycerol stock (stored at −80 °C) of *E. coli* BL21(DE3)Δ*hycE*Δhya*B*Δ*hybC* harboring the three plasmids outlined in Subheading 3.1.1 into 5 mL of M9CA-glucose containing 50 µg/mL ampicillin, 25 µg/mL spectinomycin and 12.5 µg/mL chloramphenicol.
2. Incubate aerobically for 12–15 h at 37 °C with shaking at 200 rpm.

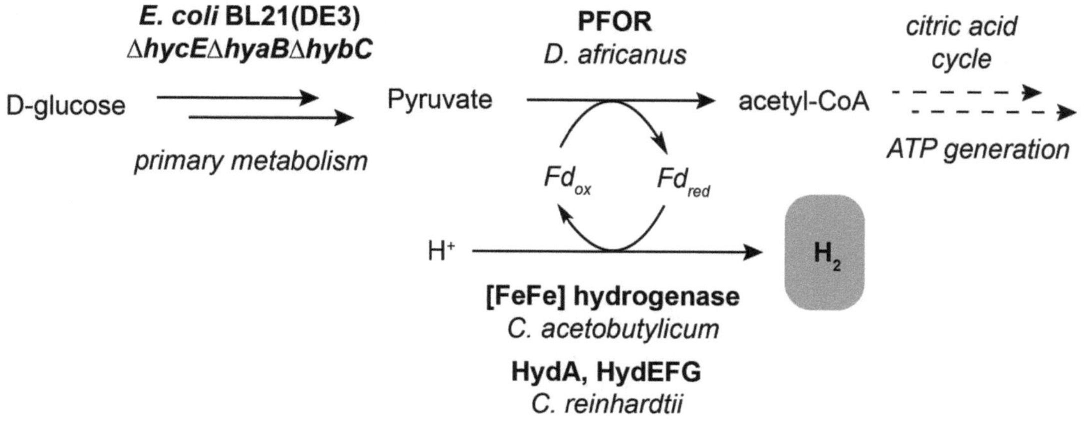

Fig. 2 The engineered hydrogen production pathway in *E. coli* BL21(DE3). *Fd* ferredoxin, *PFOR* pyruvate ferredoxin oxidoreductase

3. Transfer 70 μL of the saturated overnight culture into an autoclaved Hungate tube containing 7 mL of M9CA-glucose containing ampicillin, spectinomycin, and chloramphenicol (concentrations as in **step 1**).
4. Incubate under aerobic conditions at 37 °C with shaking at 190 rpm until OD_{600} = 0.5.
5. Under aseptic conditions, transfer the culture into a 16 mL Hungate tube containing 0.035 mmol alkene and 0.0028 mmol Royer Pd catalyst (12.5 mg, 0.08 equiv.) (*see* **Note 1**).
6. Tightly seal the Hungate tube using a butyl rubber septum and screw cap and vortex for 5 s.
7. Sparge the mixture with N_2(g) for 20 min using a 4 in. needle (inlet) and a smaller needle (outlet).
8. Add 50 μM $Fe(NH_4)_2(SO_4)_2{\bullet}6H_2O$ (7 μL of a 0.05 M filter-sterilized aqueous stock solution) and 500 μM IPTG (7 μL of a 0.5 M filter-sterilized aqueous stock solution) using a gas-tight syringe.
9. Place the tube horizontally on a rotating platform shaker, secure it tightly using adhesive tape and incubate at 37 °C with shaking at 190 rpm for 18 h.
10. Acidify the reaction mixture using 40 μL of concentrated aqueous hydrochloric acid and extract the product with 2 × 10 mL of ethyl acetate (*see* **Note 2**).
11. Combine the organic extracts, dry over sodium sulfate, filter and concentrate in vacuo.
12. Calculate the reaction conversion based on the ratio of relative signal integrations in the crude ^1H NMR spectrum (MeOH-d_4).

3.1.3 Large-Scale Biocompatible Hydrogenation (877 mL Volume)

1. Inoculate an LB:glycerol stock (stored at −80 °C) of *E. coli* BL21(DE3)Δ*hycE*,ΔhyaB,Δ*hybC* harboring the three plasmids outlined in Subheading 3.1.1 into 10 mL (2 × 5 mL volumes) of M9CA-glucose containing 50 μg/mL ampicillin, 25 μg/mL spectinomycin, and 12.5 μg/mL chloramphenicol.
2. Incubate aerobically for 12–15 h at 37 °C with shaking at 200 rpm.
3. Transfer 9 mL of the saturated overnight culture into an autoclaved 4 L Erlenmeyer flask containing 877 mL of M9CA-glucose and antibiotics (concentrations as in **step 1**).
4. Incubate under aerobic conditions at 37 °C with shaking at 190 rpm until OD_{600} = 0.5.
5. Under aseptic conditions, transfer the culture into a 1 L media bottle containing 8.77 mmol alkene, 3.11 g of Royer Pd catalyst (0.71 mmol, 0.08 equiv.), and 50 mM $Fe(NH_4)_2(SO_4)_2{\bullet}6H_2O$

(877 μL of a 0.05 M filter-sterilized aqueous stock solution) (*see* **Note 1**).

6. Cap the media bottle with two polytetrafluoroethylene flat septa, wrap with thread sealant Teflon tape, and seal tightly with an open-top screw cap.
7. Sparge the mixture for 20 min with N_2(g) using a 20 gauge, 12 in. needle (inlet) and a 22 gauge, 4 in. needle (outlet) to make the culture and headspace anaerobic.
8. Add 500 μM IPTG (877 μL of a 0.5 M filter-sterilized aqueous stock solution) using a gastight syringe.
9. Place the reaction vessel horizontally on a rotating platform shaker, secure it tightly using adhesive tape, and incubate at 37 °C with shaking at 190 rpm for 48 h.
10. Remove the serum bottle from the shaker, cool to room temperature and concentrate to a volume of ca. 300 mL in vacuo (*see* **Note 2**). Acidify the reaction mixture using 2 mL of concentrated aqueous hydrochloric acid and extract the product with 3 × 150 mL of ethyl acetate in a separating funnel.
11. Wash the combined organic extracts with 75 mL of brine, dry over sodium sulfate, filter the solution, and concentrate in vacuo.
12. Purify the product by flash chromatography on silica gel (typically 9:1 ethyl acetate–hexanes).

3.2 Royer Catalyst Reactivation and Synthesis

The activity of the hydrogenation catalyst, 3% Pd on polyethylenimine/SiO_2 (40–200 mesh) (Royer catalyst), varied depending on which commercial supplier we used, and on which batch from each supplier we used. We therefore developed a simple procedure for reactivating less active batches of commercial catalyst and also developed a new procedure for synthesizing the Royer catalyst, both of which are outlined below.

3.2.1 Reactivation of the Catalyst

1. Place the catalyst in an open glass container (e.g., a 20 mL borosilicate scintillation vial) and heat at 110 °C under vacuum for 24 h.
2. Cool to room temperature in a desiccator and store here until required.
3. Determine the Pd content (% w/w) by ICP-OES (typically 2.39–2.44% w/w).

3.2.2 Synthesis of the Royer Catalyst

1. Prepare a stock solution of $Pd(OAc)_2$ by dissolving 200 mg of $PdCl_2$ in 1 mL of concentrated aqueous hydrochloric acid by gently heating the solution with a heat gun.
2. Add 49 mL of 15% aqueous sodium acetate solution to give a final solution containing 2.4 mg of palladium per milliliter.

Add 12.5 mL of this solution to a 25 mL round-bottomed flask containing 970 mg of Royer® anion exchange resin PEI-100.

3. Sonicate the mixture for 30 min, during which time the supernatant will turn nearly colorless.
4. Decant the supernatant and wash the Pd-loaded resin with 2 × 5 mL of deionized water, resuspend the resin in 1 mL of deionized water and cool the mixture to 0 °C in an ice–water bath.
5. Add 1 mL of an ice-cooled solution of 1.32 M sodium borohydride (249 mg of $NaBH_4$(s) in 5 mL of deionized water). Swirl the mixture rapidly for 1 min and then allow the catalyst suspension to sit at 0 °C in an ice-water bath for 30 min.
6. Decant the supernatant and then wash the catalyst with 2 × 5 mL of deionized water. Repeat **steps 4–6**.
7. Incubate the catalyst with 10 mL of 30% (v/v) aqueous formic acid solution for 15 h.
8. Wash the catalyst with 10 mL of deionized water, 2 × 10 mL of methanol, then dry briefly under high vacuum before transferring the catalyst to a vacuum oven at 110 °C for 48 h.
9. Determine the Pd content (% w/w) of the resulting black solid by ICP-OES analysis (the mean palladium content of four independent catalyst syntheses in our lab is 2.82 ± 0.18% w/w).

3.3 Biocompatible Cyclopropanation

3.3.1 Overview of Strain Engineering

The styrene-producing *E. coli* strain is constructed by transforming electrocompetent *E. coli* NST74 with a modified pTrc99A plasmid containing a codon-optimized L-phenylalanine ammonia lyase (PAL2) from *Arabidopsis thaliana* and a ferulic acid decarboxylase (FDC1) from *Saccharomyces cerevisiae* (Fig. 3) [16]. The strain, *E. coli* NST74, is a metabolically evolved feedback-deregulated L-phenylalanine overproducer [*aroH*367, *tyrR*366, *tna*-2, *lacY*5, *aroF*394(fbr), *malT*384, *pheA*101(fbr), *pheO*352, *aroG*397(fbr)] and is commercially available (ATCC®, 31884™). The plasmid was obtained from Professor David R. Nielsen (Arizona State University, Tempe, AZ, USA).

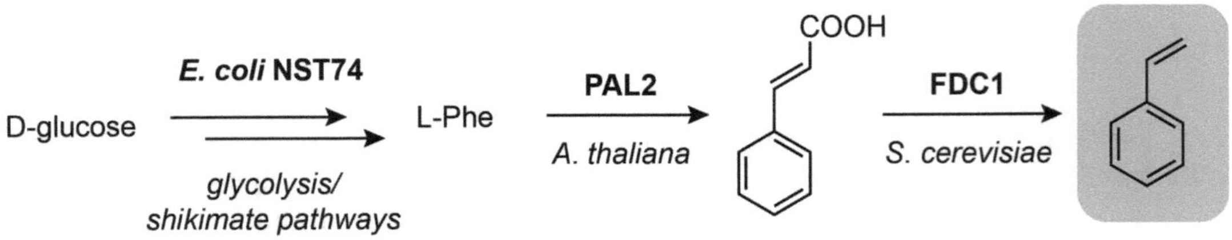

Fig. 3 The engineered styrene production pathway in *E. coli* NST74. *PAL2* phenylalanine ammonia lyase, *FDC1* ferulic acid decarboxylase

3.3.2 Small-Scale Biocompatible Cyclopropanation (100 mL Volume)

1. Inoculate an LB:glycerol stock (stored at −80 °C) of *E. coli* NST74 harboring the pTrc99a-*PAL2/FDC1* plasmid into 5 mL of LB containing 100 μg/mL ampicillin.
2. Incubate aerobically for 12–15 h at 32 °C, with shaking at 220 rpm.
3. Transfer 2 mL of the saturated overnight culture into an autoclaved 500 mL Erlenmeyer flask containing 100 mL of MM1-glucose and 100 μg/mL ampicillin.
4. Seal the flask with a glass stopper and incubate at 32 °C with shaking at 220 rpm until OD_{600} = 0.6–0.8 (*see* **Note 3**).
5. Add 0.2 mM IPTG (100 μL of a 200 mM filter-sterilized aqueous stock solution), 2.5 mg of iron(III) phthalocyanine chloride (4.1 μmol, 0.025 equiv.), and 1 mM of the desired diazo compound (0.1 mmol, 0.6 equiv.).
6. Seal the flask with a glass stopper and incubate at 32 °C with shaking at 220 rpm for 60 h (*see* **Notes 3** and **4**).
7. Add further portions of 1 mM diazo compound (0.1 mmol, 0.6 equiv.) at 12, 24, 36, and 48 h after **step 6** (3 equiv. in total).
8. To determine the concentration of styrene and cyclopropane over the course of the fermentation, transfer 0.8 mL of the culture into a 2 mL Eppendorf vial. Add 720 μL of hexanes and 1 mM 1,3,5-trimethoxybenzene (80 μL of a 10 mM stock solution in hexanes).
9. Vortex the biphasic sample for 20 min by attaching it to a flatbed shaker using adhesive tape.
10. Remove the cell material via centrifugation and analyze 1–2 μL of the organic phase by GC, as outlined in Subheading 2.4.

3.3.3 Large-Scale Biocompatible Cyclopropanation (800 mL Volume)

1. Inoculate a LB:glycerol stock (stored at −80 °C) of *E. coli* NST74 harboring the pTrc99a-*PAL2/FDC1* into 20 mL (4 × 5 mL volumes) of LB containing 100 μg/mL ampicillin.
2. Incubate aerobically for 12–15 h at 32 °C, with shaking at 220 rpm.
3. Transfer 16 mL of the saturated overnight cultures into an autoclaved 4 L Erlenmeyer flask containing 800 mL of MM1-glucose and 100 μg/mL ampicillin.
4. Seal the flask with a glass stopper and incubate at 32 °C with shaking at 220 rpm until OD_{600} = 0.6–0.8 (*see* **Note 3**).
5. Add 0.2 mM IPTG (800 μL of a 200 mM filter-sterilized aqueous stock solution), 20 mg of iron(III) phthalocyanine chloride (33 μmol, 0.025 equiv.), and 1 mM of the desired diazo compound (0.8 mmol, 0.6 equiv.).

6. Seal the flask with a glass stopper and incubate at 32 °C with shaking at 220 rpm for 60 h (*see* **Notes 3** and **4**).
7. Add further portions of 1 mM diazo compound (0.8 mmol, 0.6 equiv.) at 12, 24, 36, and 48 h after **step 6** (3 equiv. in total).
8. To determine the concentration of styrene and cyclopropane at any point, follow **steps 8–10** in Subheading 3.3.2.
9. After 60 h, split the culture into four 200 mL volumes and extract each with 3 × 100 mL of chloroform in a separating funnel.
10. Combine the organic extracts and concentrate to a volume of ca. 500 mL in vacuo.
11. Wash the combined organic extracts with 200 mL of brine, dry over sodium sulfate, filter the solution and concentrate in vacuo.
12. Purify the product by flash chromatography on silica gel (typically 0–50% ethyl acetate in hexanes) (*see* **Note 5**).

3.4 Biocompatible Reaction Development

3.4.1 Assaying the Effect of Growth Media and E. coli Cells on a Reaction

1. To five 25 mL Hungate Tubes containing the reaction components (typically 5 mM substrate), add 12.5 mL of either (1) H_2O, (2) 0.1 M K_2PO_4(aq) buffer (pH 7.4), (3) M9-glucose, (4) M9CA-glucose, or (5) LB (*see* **Note 6**).
2. Seal the tubes with butyl rubber septa and aluminum crimp seals, and vortex for 5 s.
3. Place the reaction tubes horizontally on a rotating platform shaker, secure them tightly with adhesive tape, and incubate at 37 °C with shaking at 190 rpm for the desired time.
4. Add 5 mL of ethyl acetate to each tube, vortex for 5 s and then allow the phases to separate for 5 min under gravity by standing the tubes vertically in a rack.
5. Transfer the organic phase into a borosilicate scintillation vial or a round-bottomed flask using a Pasteur pipette.
6. Repeat **steps 4** and **5** three times.
7. Concentrate the organic extract in vacuo (*see* **Note** 7).
8. Dissolve the crude residue in 1.5 mL of $CDCl_3$ containing 8 mM TMB and dry the sample using anhydrous sodium sulfate.
9. Filter the reaction into a NMR tube by passing it through a Pasteur pipette containing a small plug of cotton wool.
10. Quantify the starting material and product concentrations by ^1H NMR spectroscopy.
11. To assay the effect of *E. coli* cells on the reaction, transform electrocompetent *E. coli* BL21(DE3) cells with an unmodified

plasmid (e.g., pET29b(+)). Inoculate a LB:glycerol stock (−80 °C) of this transformant into 5 mL of LB containing the appropriate antibiotic (e.g., 50 μg/mL kanamycin for pET29b(+)) and incubate overnight at 37 °C with shaking at 200 rpm. Transfer 2 mL of the saturated overnight culture into an autoclaved 500 mL Erlenmeyer flask containing 200 mL of M9CA-glucose medium and antibiotic. Incubate at 32 °C with shaking at 220 rpm until OD_{600} = 0.5 (typically 3–4 h). Use 12.5 mL of this culture as the reaction media, and repeat **steps 1–10** accordingly.

3.4.2 Assaying the Effect of a Reaction and Its Components on *E. coli* Cells

1. At the end of the desired reaction (for example, after **step 9** of Subheading 3.1.2, **step 9** of Subheading 3.1.3, **step 6** of Subheading 3.3.2, **step 6** of Subheading 3.3.3, or **step 11** of Subheading 3.4.1), transfer 100 μL of the culture into an autoclaved 1.5 mL Eppendorf vial containing 900 μL of the appropriate growth media and vortex for 2 s (*see* **Note 8**).
2. Transfer 100 μL of this solution into a separate autoclaved 1.5 mL Eppendorf tube containing 900 μL of growth media.
3. Repeat **step 2** a further six times to afford samples of 10^1–10^8-fold dilutions.
4. Transfer 100 μL of each sample onto LB agar plates containing appropriate antibiotic(s) and spread evenly using a sterile cell spreader.
5. Incubate the plates at 37 °C overnight.
6. Use plates containing between 10–100 colonies to calculate the number of colony-forming units (CFU's) present in the original reaction culture (*see* **Note 9**).

4 Notes

1. The range of unsaturated substrates that can be reduced using the biocompatible hydrogenation procedure is outlined below in Fig. 4.
2. During the hydrogenation reaction there is a significant buildup of pressure in the reaction vessel. Care should therefore be taken when opening the tube/flask after the reaction is complete.
3. For the cyclopropanation, it is essential that the seed cultures are grown in Erlenmeyer flasks sealed with glass stoppers in order to avoid the loss of styrene via evaporation. A culture–headspace volume ratio of 1:5 should therefore be used to maintain an aerobic atmosphere. In experiments where regular sampling is not being done, the cultures should be opened and aerated for 1 min every 12 h. This precaution is not required

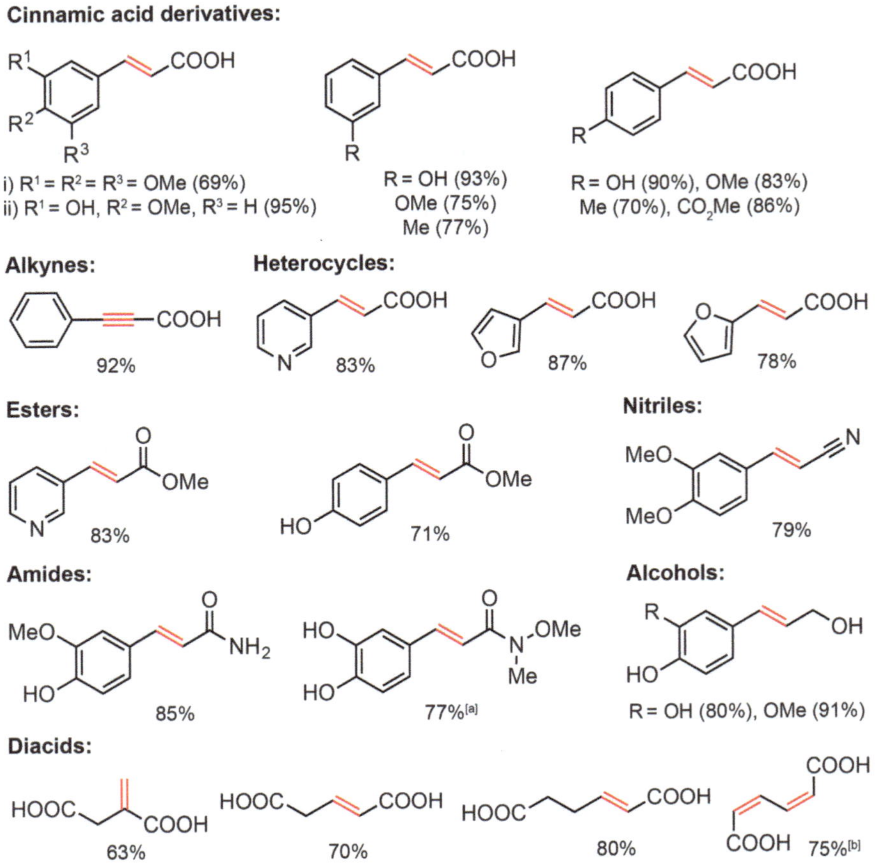

Fig. 4 Functional group tolerance of the biocompatible hydrogenation reaction. Positions of reduction are highlighted in *red*. Yields are shown in parenthesis and are determined by ^{1}H NMR spectroscopy as outlined in Subheading 3.1.2. Reaction conditions: 5 mM substrate, 8 mol% Royer catalyst, 90 mL culture volume. (**a**) Yield of isolated product for a large-scale reaction as outlined in Subheading 3.1.3, using 16 mol% Royer catalyst. (**b**) 2.5 mM substrate

when growing the overnight starter cultures (e.g., **steps 1–2** of Subheading 3.1.2, and **steps 1–2** of Subheading 3.1.3).

4. MM1 medium does not have the buffering capacity to maintain a pH of 7.4 over the course of the experiment. The pH of the culture should therefore be frequently monitored (ca. every 12 h) using accurate pH strips and readjusted to pH 7.4 if necessary using an autoclaved aqueous solution of 6 M NaOH(s) (typically 1–2 mL per 100 mL of culture).
5. All cyclopropanes are isolated as a mixture of *syn-* and *anti-*diastereomers. It is possible to partially separate diastereomers using a slower gradient of increasing ethyl acetate in the mobile phase during flash chromatography.
6. All aqueous media should also contain typically 0.2 mM IPTG (for screening purposes) and appropriate antibiotics that will need to be present in the final fermentation.

7. For small-scale reactions, be careful not to leave the sample under vacuum for too long after the solvent has been removed to avoid evaporating your sample. This will result in inaccurate quantifications. Keep the water bath of the rotary evaporator at room temperature to minimize this.
8. It is important to do this experiment under strictly aseptic conditions, ideally in a laminar flow biosafety cabinet.
9. Suitable control experiments to accompany this analysis are: (1) *E. coli* cultures grown in the absence of any reaction components, (2) cultures grown in the absence of one or more of the reaction components, and (3) dead cells produced by heating the diluted cell samples at 95 °C for 15 min prior to plating out the cells.

Acknowledgments

This work was generously supported by Harvard University, the Corning Foundation, the Searle Scholars Program, NIH grant DP2 GM105434, and the European Commission (Marie Curie IOF Fellowship to S.W.).

References

1. Hoffmann RW (2013) Natural product synthesis: changes over time. Angew Chem Int Ed Engl 52:123–130
2. Keasling JD (2013) Manufacturing molecules through metabolic engineering. Science 330:1355–1358
3. Keasling JD, Mendoza A, Baran PS (2012) Synthesis: a constructive debate. Nature 492:188–189
4. Suastegui M, Matthiesen JE et al (2016) Combining metabolic engineering and electrocatalysis: application to the production of polyamides from sugar. Angew Chem Int Ed Engl 55:2368–2373
5. Wallace S, Balskus EP (2014) Opportunities for merging chemical and biological synthesis. Curr Opin Biotechnol 30:1–8
6. Wallace S, Schultz EE, Balskus EP (2015) Using non-enzymatic chemistry to influence microbial metabolism. Curr Opin Chem Biol 25:71–79
7. Lee Y, Umeano A, Balskus EP (2013) Rescuing auxotrophic microorganisms with nonenzymatic chemistry. Angew Chem Int Ed Engl 53:11800–11803
8. Sirasani G, Tong L, Balskus EP (2014) A biocompatible alkene hydrogenation merges organic synthesis with microbial metabolism. Angew Chem Int Ed Engl 53:7785–7786
9. Wallace S, Balskus EP (2015) Interfacing microbial styrene production with a biocompatible cyclopropanation reaction. Angew Chem Int Ed Engl 54:7106–7109
10. Domaille DW, Hafenstine GR et al (2016) Catalytic upgrading in bacteria-compatible conditions via a biocompatible aldol condensation. ACS Sustain Chem Eng 4:671–675
11. Wallace S, Balskus EP (2016) Designer micelles accelerate flux through engineered metabolism in *E. coli* and support biocompatible chemistry. Angew Chem Int Ed Engl 55:6023–6027
12. Neuberg C, Hirsch J (1919) Uber den Verlauf der alkoholischen-Garung bei alkalischer Reaktion. II. Garung mit lebender Hefe in alkalischen Losungen. Biochem Z 96:175–202
13. Neuberg C, Hirsch J (1919) Die dritte Vergarungform des Zurkers. Biochem Z 100:304–322
14. Agapakis CM, Ducat DC et al (2010) Insulation of a synthetic hydrogen metabolism circuit in bacteria. J Biol Eng 4:3
15. Baba T, Ara T et al (2006) Construction of *Escherichia coli* K-12 in-frame, single-gene knockout mutants: the Keio collection. Mol Syst Biol 2:0008
16. McKenna R, Nielsen DR (2011) Styrene biosynthesis from glucose by engineered *E. coli*. Metab Eng 13:544–554

Index

A

B

C

Nicola A. Burgess-Brown (ed.), *Heterologous Gene Expression in E. coli: Methods and Protocols*, Methods in Molecular Biology, vol. 1586, DOI 10.1007/978-1-4939-6887-9, © Springer Science+Business Media LLC 2017

N

O

P

Q

R

S

T

U

V

MIX
Papier aus verantwortungsvollen Quellen
Paper from responsible sources
FSC® C105338
FSC
www.fsc.org

If you have any concerns about our products,
you can contact us on
ProductSafety@springernature.com

In case Publisher is established outside the EU,
the EU authorized representative is:
Springer Nature Customer Service Center GmbH
Europaplatz 3, 69115 Heidelberg, Germany

Printed by Libri Plureos GmbH
in Hamburg, Germany